Advances in Intelligent and Soft Computing

147

Editor-in-Chief

Prof. Janusz Kacprzyk
Systems Research Institute
Polish Academy of Sciences
ul. Newelska 6
01-447 Warsaw
Poland
E-mail: kacprzyk@ibspan.waw.pl

T0138116

For further volumes:
http://www.springer.com/series/4240

Bing-Yuan Cao and Xiang-Jun Xie (Eds.)

Fuzzy Engineering and Operations Research

 Springer

Editors

Bing-Yuan Cao
School of Mathematics and Information
Science
Guangzhou University
Key Laboratory of Mathematics and
Interdisciplinary Sciences of Guangdong
Higher Education Institutes
Guangzhou University
Guangdong
China

Xiang-Jun Xie
School of Science
Southwest Petroleum University
Chengdu
China, People's Republic

ISSN 1867-5662
ISBN 978-3-642-28591-2
DOI 10.1007/978-3-642-28592-9
Springer Heidelberg New York Dordrecht London

e-ISSN 1867-5670
e-ISBN 978-3-642-28592-9

Library of Congress Control Number: 2012932485

Printed on acid-free paper

Springer is part of Springer Science+Business Media (www.springer.com)

Preface

This book is the proceedings from submissions by the 5th International Conference on Fuzzy Information and Engineering (ICFIE2011) during Oct. 15–17, 2011 in Chengdu, China and by the 1st academic conference in establishment of Guangdong Province Operations Research Society (GDPORSC) on Oct. 20, 2011 in Guangzhou, China. The conference proceedings is published by *Advances in Intelligent and Soft Computing* (AISC, ISSN: 1867-5662), Springer.

This year, we have received more than 150 submissions. Each paper has undergone a rigorous review process. Only high-quality papers are included.

The 5th ICFIE2011, built on the success of previous conferences, and the GDPORSC, first held, are major Symposiums, respectively, for scientists, engineers practitioners and Operation Research (OR) researchers in China to present their updated results, developments and applications in all areas of fuzzy information and engineering and OR. It aims to strengthen relations between industry research laboratories and universities, and to create a primary symposium for world scientists in Fuzziology and OR fields as follows:

1) Fuzzy optimization, logic and information;
2) The mathematical theory of fuzzy systems;
3) Fuzzy engineering application and soft computing method;
4) OR and fuzziology;
5) Advances.

The book, containing 62 papers, is divided into five main parts:

In Part I, there are 18 papers on "fuzzy optimization, logic and information".

In Part II, we have 10 papers on "the mathematical theory of fuzzy systems".

In Part III, 25 papers on "fuzzy engineering application and soft computing method" appear.

In Part IV, we have 7 papers on "OR and fuzziology".

In Part V, we have 2 papers on "guess and review".

In addition to large numbers of submissions, we are blessed with the presence of five renowned keynote speakers and several distinguished panelists and we organized workshops in Chengdu. In Guangzhou, Prof Jing-zhong Zhang, an academician of the Chinese Academy of Sciences, Prof. Hao Wang, General and the former political commissar, in China National University of Defense Technology, Prof. Ya-xiang Yuan, an academician and President of Operation Research Society of China, Hao-biao Yang, Secretary-General of Guangdong Province Association for Science and Technology attended the GDPORSC and made important reports.

On behalf of the Organizing Committee, we appreciate Southwest Petroleum University, China, Fuzzy Information and Engineering Branch of China Operation Research Society for the former host and Guangzhou University, China, for the latter host; Fuzzy Information and Engineering Branch of International Institute of General Systems Studies, China (IIGSS-GB) and Iran Mazandran University for Co-sponsorships. We are showing gratitude to the members of the organizing committees, the steering Committees, and the program Committees for their hard work. We wish to express our heart-felt appreciation to the keynote and panel speakers, workshop organizers, session chairs, reviewers, and students. In particular, we are thankful to Yan Yang, Ci-yuan Xiao, Qian Chen, Fang Xu and Zhi-ping Zhu, who have contributed a lot to the development of this issue. We appreciated all the authors and participants for their great contributions that made this conference possible and all the hard work worthwhile. Meanwhile, we are thankful to School of Management from Sun Yat-sen University, Manager Xu Dong from Lang Wine Guangzhou office and Manager Yao-wu Li from Guang-er-feng Wine in Dongguan, China, and Present Qi-ling Peng from Yunfu Secondary School, Guangdong, for sponsoring the GDPORSC.

Finally, we thank the publisher, Springer, for publishing the ICFIE 2011 proceedings as Journal AISC (Notes: Our series of conference proceedings by Springer, like *Advances in Soft Computing* (ASC 40, AISC 78 and AISC 82, have been included into EI and all are indexed by ISTP), and thank the supports coming from Journal *Fuzzy Information and Engineering* by Springer.

Bing-yuan Cao
Xiang-jun Xie
P.R. China

Organization

Organizing Committee of ICFIE2011

Conference Chairs
Bing-yuan Cao (China)
Xiang-jun Xie (China)

Honorary Chairs
Lotfi A. Zadeh (USA)
Ying-ming Liu (China)
Zhi-min Du (China)

Steering Committee
J.C. Bezdek (USA)
Z. Bien (Korea)
D. Dubois (France)
Gui-rong Guo (China)
M.M. Gupta (Canada)
Xin-gui He (China)
Abraham Kandel (Hungary)

E. Mamdani (UK)
R.P. Nikhil (India)
M. Sugeno (Japan)
Hao Wang (China)
P.Z. Wang (USA)
H.J. Zimmermann (Germany)

Program Committee

Chair
Kai-qi Zou (China)

Co-chairs
Shu-Cherng Fang (USA)
Hadi Nasseri (Iran)

Yong Shi (China)
Ping Guo (China)

Members
K. Asai (Japan)
J.P. Barthelemy (France)
Tian-you Chai (China)
Guo-qing Chen (China)

Mian-yun Chen (China)
Shui-li Chen (China)
Ovanes Chorayan (Russia)
Sen-lin Cheng (China)

H.P. Deng (Australia)
M. Fedrizzi (Italy)
Jia-li Feng (China)
Yin-jun Feng (China)
Si-cong Guo (China)
Ming-hu Ha (China)
Li-yan Han (China)
Cheng-ming Hu (USA)
Bao-qing Hu (China)
Hiroshi Inoue (Japan)
Lin-min Jia (China)
Guy Jumarie (Canada)
J. Kacprzyk (Poland)
Jim Keller (USA)
E.E. Kerre (Belgium)
K.H. Kim (USA)
N.Kuroki (Japan)
D.Lakov (Bulgaria)
Tsu-Tian Lee (China Taiwan)
Hong-xing Li (China)
Jun Li (China)
Tai-fu Li (China)
Yu-cheng Li (China)
T.Y. Lin (SUSA)
Bao-ding Liu (China)
Zhi-qiang Liu (Hongkong)
Zeng-liang Liu (China)
Ming Ma (China)
Sheng-quan Ma (China)
D. Dutta Majumder (Calcutta)
Hong-hai Mi (China)
M. Mizumoto (Japan)
Zhi-wen Mo (China)
J. Motiwalla (Singapore)

M. Mukaidono (Japan)
J. Mustonen (Finland)
Shohachiro Nakanishi (Japan)
Jinping Ou (China)
Witold Pedrycz (USA)
H. Prade (France)
D.A. Ralescu (USA)
E.Sanchez (France)
Victor V. Senkevich (Russia)
Qiang Shen (UK)
Kai-quan Shi (China)
Zhen-ming Song (China)
Lan Su (China)
Enric Trillas (Spain)
Xi-zhao Wang (China)
Xue-ping Wang (China)
Zhen-yuan Wang (USA)
Xin-jiang Wei (China)
Ber-lin Wu (China Taiwan)
Yun-dong Wu (China)
Ci-yuan Xiao (China)
Ruo-ning Xu (China)
Yang Xu (China)
Ze-shui Xu(China)
R.R. Yager (USA)
T. Yamakawa (Japan)
Bing-ru Yang (China)
Liang-zhong Yi (China)
Xue-hai Yuan (China)
Qiang Zhang (China)
Cheng-yi Zhang(China)
Nai-rao Zhang (China)
Yun-jie Zhang (China)

Local Arrangements Chair
Ling Wang (China)

Co-chairs
Song-bai Zhu (China)
Yong-qing Peng (China)

Secretary
Yu-bin Zhong (China)
Yuan Cheng (China)
Yan Yang (China)

Members

Qian Chen (China)
Yi Qiu (China)
Yue Wu (China)

Wen-jun Xiong (China)
Xiao-ying Yang (China)
Yong-hua Zhou (China)

Publicity Chair
Ci-yuan Xiao (China)

Co-chair
Yu-bin Zhong (China)

Member
Qian Chen (China)

Organizing Committee of GDORC

Conference Chair
Bing-yuan Cao (China)

Honorary Chair
Jing-zhong Zhang (China)

Steering Committee

Zhi-feng Hao (China)
Guo-qing Wang (China)
Zhong-fei Li (China)
Shi-zhong Bai (China)

Dong-hui Li (China)
Hao Wang (China)
Ya-xiang Yuan (China)
Hadi Nasseri (Iran)

Publication Chair

Cao Bing-yuan

Co-chair
Xiang-jun Xie (China)

Member
Yan Yang (China)
Fang Xu (China)

Contents

Part II: The Mathematical Theory of Fuzzy Systems

Part III: Fuzzy Engineering Application and Soft Computing Method

Part IV: OR and Fuzziology

Part V: Guess and Review

Part I
Fuzzy Optimization, Logic
and Information

Initial Basic Solution for the Fuzzy Primal Simplex Algorithm Using a Two-Phase Method

S.H. Nasseri[1,*], Z. Alizadeh[1], and B. Khabiri[2]

[1] Department of Mathematics, University of Mazandaran, Babolsar, Iran
    nasseri@umz.ac.ir
[2] Department of Mathematics, Islamic Azad University, Jouybar Branch, Jouybar, Iran

Abstract. Fuzzy primal and dual simplex algorithms have been recently proposed to solve the linear programming with fuzzy variables (FVLP) problems. The fuzzy primal simplex method has been developed with the assumption that an initial basic feasible solution is at hand. In many cases, such a solution is not readily available, and some work may be needed to get the fuzzy primal method started. Furthermore, there exists a shortcoming in the fuzzy dual simplex algorithm when the dual feasibility or equivalently the primal optimality is not at hand and in this case we can not use the fuzzy dual simplex method for solving FVLP problem. In this paper, we propose a fuzzy two-phase method involving fuzzy artificial variables, to obtain an initial fuzzy basic feasible solution to a slightly modified set of constraints. Then the fuzzy primal simplex method is used to eliminate the fuzzy artificial variables and to solve the original problem.

Keywords: Fuzzy basic feasible solution, fuzzy primal simplex method, linear programming with fuzzy variables, ranking function, trapezoidal fuzzy number.

1   Introduction

Fuzzy programming approach [8, 19] is useful and efficient to treat a programming problem under uncertainty. While classical and stochastic programming approach may require a lot of cost to obtain the exact coefficient value or distribution, fuzzy programming approach does not (see [8]). From this fact, fuzzy programming approach will be very advantageous when the coefficients are not known exactly but vaguely specified by human expertise. The concept of fuzzy mathematical programming on general level was first proposed by Tanaka et al. [23]. The first formulation of fuzzy linear programming (FLP) is proposed by Zimmermann [26]. A review of the literature concerning fuzzy mathematical programming as well as comparison of fuzzy numbers can be seen in Klir and Yuan [10] and also Lai and Hwang [11]. Several authors considered various types of the FLP problems and proposed several approaches for solving them. Safi et al. [22] introduced some definitions

---

* Corresponding author.

B.-Y. Cao and X.-J. Xie (Eds.): Fuzzy Engineering and Operations Research, AISC 147, pp. 3–15.
springerlink.com   © Springer-Verlag Berlin Heidelberg 2012

in the geometry of two-dimensional fuzzy linear programming. After defining the optimal solution based on these definitions, they used the geometric approach for obtaining optimal solution(s) and showed that the algebraic solutions obtained by Zimmermann method and our geometric solutions are the same. Vijay et al. [24] introduced a generalized model for a two person zero sum matrix game with fuzzy goals and fuzzy payoffs via fuzzy relation approach and then showed that it was equivalent to two semi-infinite optimization problems. Mehra et al. [16] based on the specified grades of satisfaction, proposed two new concepts of (α, β)- acceptable optimal solution and (α, β)-acceptable optimal value of a fuzzy linear fractional programming problem with fuzzy coefficients, and developed a method to compute them. Inuiguchi [7] formulated fuzzy linear programming as a necessity measure optimization model. He showed that the problem can be reduced to a semi-infinite programming problem and solved by a combination of a bisection method and a relaxation procedure. He also proposed an algorithm in which the bisection method and the relaxation procedure converge simultaneously. Rommelfanger [21] used a new approach that represented a general interactive solution process for solving multicriteria linear programming systems with crisp, fuzzy or stochastic data. Jimnez and Bilbao [9] addressed the problem of solving multi-objective linear-programming problems,by assuming that the decision maker has fuzzy goals for each of the objective functions. They showed that,in the case that one of our goals is fully achieved, a fuzzy-efficient solution may not be Pareto-optimal and therefore proposed a general procedure to obtain a non-dominated solution,which is also fuzzy-efficient. A review of some convenient methods can be found in Mahdavi-Amiri et al. [14]. However, some authors have used the concept of comparison of fuzzy numbers to solve the fuzzy linear programming problems. In effect, most convenient methods are based on the concept of comparison of fuzzy numbers by using ranking functions [1, 2, 3, 4, 5, 6, 12, 13, 14, 17]. Of course, ranking functions have been proposed by researchers to suit their requirements of the problem under consideration and conceivably there are no generally accepted criteria for application of ranking functions. Nevertheless, usually in such methods authors define a crisp model which is equivalent to the FLP problem and then use optimal solution of the model as the optimal solution of the FLP problem. Recently linear programming with fuzzy variables (FVLP) problems have been attracted some interests. Some methods have been developed for solving these problems by introducing and solving certain auxiliary problems. Moreover, Mahdavi-Amiri et al. [14] have recently developed a new fuzzy primal and dual simplex algorithm for solving the FVLP problem. As they have mentioned the fuzzy primal simplex algorithm need an initial basic feasible solution (see [14]). In some cases, such a solution is not readily available and hence in this paper we propose a fuzzy two-phase method involving fuzzy artificial variables, to obtain an initial fuzzy basic feasible solution to a slightly modified set of constraints.

This paper is organized as follows: In Section 2, we give some necessary concepts of fuzzy set theory. A review of the fuzzy variable linear programming problem is given in Section 3. We present fuzzy primal simplex algorithm, using the fuzzy simplex tableau to solve the fuzzy variable linear programming problems in Section 4. we introduced a new approach to obtain an initial fuzzy basic feasible solution for solving FVLP problems and explain it by an illustrative example in Section 5. Finally, we conclude in Section 6.

2 Preliminaries

2.1 Fuzzy Arithmetic Operators and Ranking

In this section, we give some basic definitions on fuzzy numbers comparisons as well as the main concepts needed in the rest of the paper.

Definition 2.1. *Let $\mathbb{R}$ be the universal set. $\tilde{a}$ is called a fuzzy set in X if $\tilde{a}$ is a set of ordered pairs $\tilde{a} = \{(x, \mu_{\tilde{a}}(x)) | x \in \mathbb{R}\}$, where $\mu_{\tilde{a}}(.)$ is membership function of $\tilde{a}$ and assigns to each element $x \in \mathbb{R}$, a real number $\mu_{\tilde{a}}(x)$ in the interval $[0, 1]$.*

Definition 2.2. *The α-cut or α-level of a fuzzy set $\tilde{a}$ is defined as an ordinary set $[\tilde{a}]_\alpha$ for which the degree of its membership function exceeds the level α, that is, $[\tilde{a}]_\alpha = \{x \in \mathbb{R} \mid \mu_{\tilde{a}}(x) \geq \alpha\}$.*

Definition 2.3. *The support of a fuzzy set $\tilde{a}$ is a set of elements in $\mathbb{R}$ for which $\mu_{\tilde{a}}(x)$ is positive, that is, $supp\, \tilde{a} = \{x \in \mathbb{R} \mid \mu_{\tilde{a}}(x) > 0\}$.*

Definition 2.4. *A fuzzy set $\tilde{a}$ is called convex if for each $x, y \in \mathbb{R}$ and each $\lambda \in [0, 1]$, $\mu_{\tilde{a}}(\lambda x + (1 - \lambda)y) \geq min\, \{\mu_{\tilde{a}}(x), \mu_{\tilde{a}}(y)\}$.*

Definition 2.5. *A fuzzy number is a convex normalized fuzzy set of the real line $\mathbb{R}$; whose membership function is piecewise continuous. The set of fuzzy numbers on $\mathbb{R}$ is denoted by $F(\mathbb{R})$.*

Definition 2.6. *An LR type flat fuzzy number [18], is denoted as $\tilde{a}=(a^L, a^U, \alpha, \beta)_{LR}$, if*

$$
\mu_{\tilde{a}}(x) = \begin{cases} L\left(\frac{a^L - x}{\alpha}\right) & for \quad a^L - \alpha \leq x \leq a^L \\ 1 & for \quad a^L \leq x \leq a^U \\ R\left(\frac{x - a^U}{\beta}\right) & for \quad a^U \leq x \leq a^U + \beta \\ 0 & else \end{cases} \tag{1}
$$

where the symmetric non-increasing function $L : [0, \infty) \to [0, 1]$ is the left shape function, that $L(0) = 1$. Also, a right shape function $R(.)$ is similarly defined as $L(.)$. Trapezoidal fuzzy numbers (TRFN) are special cases of LR fuzzy numbers with $L(x) = R(x) = 1 - x$ and the following membership function:

$$
\mu_{\tilde{a}}(x) = \begin{cases} \frac{x-(a^L-\alpha)}{\alpha} & for \quad a^L - \alpha \leq x \leq a^L \\ 1 & for \quad a^L \leq x \leq a^U \\ \frac{(a^U+\beta)-x}{\beta} & for \quad a^U \leq x \leq a^U + \beta \\ 0 & else \end{cases} \tag{2}
$$

Now, we define arithmetic on trapezoidal fuzzy numbers. Let $\tilde{a} = (a^L, a^U, \alpha, \beta)$ and $\tilde{b} = (b^L, b^U, \gamma, \theta)$ be two trapezoidal fuzzy numbers. Define,

$$
x > 0, \ x \in \mathbb{R}; \quad x\tilde{a} = (xa^L, xa^U, x\alpha, x\beta)
$$
$$
x < 0, \ x \in \mathbb{R}; \quad x\tilde{a} = (xa^U, xa^L, -x\beta, -x\alpha)
$$

$$
\tilde{a} + \tilde{b} = (a^L + b^L, a^U + b^U, \alpha + \gamma, \beta + \theta).
$$

An effective approach for ordering the elements of $F(\mathbb{R})$ is to define a ranking function $R : F(\mathbb{R}) \rightarrow \mathbb{R}$ which maps each fuzzy number into the real line, where a natural order exists. We define orders on $F(\mathbb{R})$ by:

$$
\tilde{a} \succeq \tilde{b} \quad if \ and \ only \ if \quad R(\tilde{a}) \geq R(\tilde{b}) \tag{3}
$$
$$
\tilde{a} \succ \tilde{b} \quad if \ and \ only \ if \quad R(\tilde{a}) > R(\tilde{b}) \tag{4}
$$
$$
\tilde{a} \simeq \tilde{b} \quad if \ and \ only \ if \quad R(\tilde{a}) = R(\tilde{b}) \tag{5}
$$

where $\tilde{a}$ and $\tilde{b}$ are in $F(\mathbb{R})$. Also we write $\tilde{a} \preceq \tilde{b}$ if and only if $\tilde{b} \succeq \tilde{a}$.
We restrict our attention to linear ranking functions, that is, a ranking function R such that

$$
R(k\tilde{a} + \tilde{b}) = kR(\tilde{a}) + R(\tilde{b}) \tag{6}
$$

for any $\tilde{a}$ and $\tilde{b}$ belonging to $F(\mathbb{R})$ and any $k \in \mathbb{R}$.
We consider the linear ranking functions on $F(\mathbb{R})$ as:

$$
R(\tilde{a}) = c_L a^L + c_U a^U + c_\alpha \alpha + c_\beta \beta, \tag{7}
$$

where $\tilde{a} = (a^L, a^U, \alpha, \beta)$, and $c_L, c_U, c_\alpha, c_\beta$ are constants, at least one of which is nonzero. A special version of the above linear ranking function was first proposed by Yager [25] as follows:

$$
R(\tilde{a}) = \frac{1}{2} \int_0^1 (\inf \tilde{a}_\lambda + \sup \tilde{a}_\lambda) \, d\lambda \tag{8}
$$

which reduces to

$$
R(\tilde{a}) = \frac{a^L + a^U}{2} + \frac{1}{4}(\beta - \alpha). \tag{9}
$$

Then, for trapezoidal fuzzy numbers $\tilde{a} = (a^L, a^U, \alpha, \beta)$ and $\tilde{b} = (b^L, b^U, \gamma, \theta)$, we have

$$
\tilde{a} \succeq \tilde{b} \quad if \ and \ only \ if \quad a^L + a^U + \frac{1}{2}(\beta - \alpha) \geq b^L + b^U + \frac{1}{2}(\theta - \gamma). \tag{10}
$$

3   Linear Programming with Fuzzy Variables

Definition 3.1. *A linear programming with fuzzy variables (FVLP) is defined as follows* [2, 13]:

$$\max \tilde{z} \simeq c\tilde{x}$$
$$s.t. \ A\tilde{x} \simeq \tilde{b} \tag{1}$$
$$\tilde{x} \succeq \tilde{0}$$

where $c \in \mathbb{R}^n, \tilde{x} \in (F(\mathbb{R}))^n, A \in \mathbb{R}^{m \times n}, \tilde{b} \in (F(\mathbb{R}))^m$.
Note that an FVLP problem is a linear programming problem in fuzzy environment in which the decision making variables and Right-Hand-Sides are fuzzy numbers.

Definition 3.2. *We say that a vector* $\tilde{x} \in (F(\mathbb{R}))^n$ *is a feasible solution to* (1) *if and only if* $\tilde{x}$ *satisfies the constraints of the problem.*

Definition 3.3. *A feasible solution* $\tilde{x}_*$ *is an optimal solution for* (1), *if for all feasible solution* $\tilde{x}$ *for* (1), *we have* $c\tilde{x}_* \succeq c\tilde{x}$.

3.1   Fuzzy Basic Feasible Solution

Here, we briefly describe fuzzy basic feasible solution (FBFS) for the FVLP problem (1) as established by Mahdavi-Amiri and Nasseri [13]. Consider the FVLP problem (1). Let $A = [a_{ij}]_{m \times n}$. Assume rank(A)=m. Partition A as $[B \ \ N]$ where B, $m \times m$, is nonsingular. It is obvious that rank(B)=m. Let y_j be the solution to $By = a_j$. It is apparent that the basic solution

$$\tilde{x}_B \simeq (\tilde{x}_{B_1}, ..., \tilde{x}_{B_m})^T \simeq B^{-1}\tilde{b}, \ \tilde{x}_N \simeq \tilde{0} \tag{2}$$

is a solution of $A\tilde{x} = \tilde{b}$. In fact, $\tilde{x} = (\tilde{x}_B^T \ \ \tilde{x}_N^T)^T$. If $\tilde{x}_B \succeq \tilde{0}$, then the basic solution is feasible and the corresponding fuzzy objective value is: $\tilde{z} \simeq c_B\tilde{x}_B$, where $c_B = (c_{B_1}, ..., c_{B_m})$. Now, corresponding to every nonbasic variable $\tilde{x}_j, 1 \leq j \leq n, j \neq B_i, i = 1, ..., m$, define

$$z_j = c_B y_j = c_B B^{-1} a_j. \tag{3}$$

The following result concerns the non-degenerated problems, where every fuzzy basic variable corresponding to every basis B is positive .

Theorem 3.3. (Optimality conditons) *Assume the FVLP problem is non-degenerate. If a basic solution* $\tilde{x}_B = B^{-1}\tilde{b}$, $\tilde{x}_N \simeq \tilde{0}$ *is feasible to* (1) *and* $z_j \geq c_j$ *for all* $j, 1 \leq jn$, *then the fuzzy basic solution is an fuzzy optimal solution to* (1).

Proof. See Mahdavi-Amiri and Nasseri [13].
    Maleki et al. [15] proposed a method for solving FVLP problems by use of an auxiliary problem. They discussed some relations between the FVLP problem

and the auxiliary problem and used the results for solving the FVLP problem by an algorithm based on the solution of the auxiliary problem. Mahdavi-Amiri and Nasseri [13] developed the duality results for the FVLP problem. They showed that the auxiliary problem in Maleki et al. [15] is indeed the dual of the FVLP problem. Based on the results obtained, they presented a dual simplex algorithm for solving the FVLP problems directly using the primal simplex tableau. Moreover, they recently have discussed and developed the fuzzy primal simplex algorithm for solving the FVLP problems (see in Mahdavi-Amiri et al. [13]).

4 Fuzzy Simplex Method for the FVLP Problems

In this section, we are going to state a fuzzy primal simplex algorithm for FVLP problems which is established by Mahdavi-Amiri and Nasseri [13].

4.1 Simplex Method in the Tableau Format

Consider the FVLP problem (1), rewritten in the following form:

$$\max \tilde{z} \simeq c_B \tilde{x}_B + c_N \tilde{x}_N$$
$$\text{s.t. } B\tilde{x}_B + N\tilde{x}_N \simeq \tilde{b} \tag{1}$$
$$\tilde{x}_B, \tilde{x}_N \succeq \tilde{0}.$$

We can write $\tilde{x}_B \simeq B^{-1}\tilde{b} - B^{-1}N\tilde{x}_N$, $\tilde{z} \simeq c_B(B^{-1}\tilde{b} - B^{-1}N\tilde{x}_N) + c_N\tilde{x}_N$ and hence $\tilde{x}_B + B^{-1}N\tilde{x}_N \simeq B^{-1}\tilde{b}$, and $\tilde{z} + (c_B B^{-1}N - c_N)\tilde{x}_N \simeq c_B B^{-1}\tilde{b}$. Letting $\tilde{x}_N \simeq \tilde{0}$, we have $\tilde{x}_B = \tilde{y}_0 = B^{-1}\tilde{b}$, and $\tilde{z} \simeq c_B\tilde{y}_0$. Thus, we write the above FVLP problem in the following tableau format (Table 4.1).

Table 4.1. The FVLP simplex tableau

$basis \ \tilde{x}_B$		$\tilde{x}_N$	R.H.S.
$\tilde{z}$	0	$z_N - c_N = c_B B^{-1}N - c_N$	$\tilde{y}_{00} = c_B B^{-1}\tilde{b}$
$\tilde{x}_B$	I	$Y = B^{-1}N$	$\tilde{y}_0 = B^{-1}\tilde{b}$

Table 4.1 gives us all the information needed to proceed with the fuzzy primal simplex method. The cost row in the above tableau is:

$$y_{0j} = 0, j = B_i, 1 \le i \le m,$$

$$y_{0j} = c_B y_j - c_j = z_j - c_j,$$

$$1 \le j \le n, j \ne B_i, 1 \le i \le m.$$

According to the optimality conditions for these problems we are at the optimal solution if $y_{0j} \le 0$, for all $j \ne B_i, 1 \le i \le m$. On the other hand, if $y_{ok} > 0$, for some

$k \neq B_i, 1 \leq i \leq m$, then the problem is either unbounded or an exchange of a basic variable $\tilde{x}_{B_r}$ for some r, and a nonbasic variable $\tilde{x}_k$ can be made to decreasing the rank of the objective value (under nondegeneracy assumption). The following theorems state the conditions for unboundeness of the FVLP problem and the conditions permitting the update of the tableau to a new tableau having a nonincreasing (decreasing under nondegeneracy assumption) rank of the objective value. The proofs given by Mahdavi-Amiri and Nasseri [13] and then we omit them here.

Theorem 4.1. *If in an FVLP simplex tableau, there is a column k (not in basis) for which $z_k - c_k < 0$ and $y_{ik} \leq 0, i = 1, ..., m$, then the FVLP problem is unbounded.*

Theorem 4.2. *If in an FVLP simplex tableau, a k exists such that $z_k - c_k < 0$ and there exists a basic index B_i such that $y_{ik} > 0$, then a pivoting row can be found so that pivoting on y_{rk} will yield a fuzzy feasible tableau with a corresponding nonincreasing (decreasing under nondegeneracy assumption) objective value.*

4.2 Pivoting and Change of Basis

If $\tilde{x}_k$ enters the basis $\tilde{x}_{B_r}$ and leaves the basis, then pivoting on y_{rk} in the simplex tableau is stated as follows:

    1) Divide row r by y_{rk}.
    2) For $i = 0, 1, ..., m$ and $i \neq r$, update the i th row by adding to it $-y_{ik}$ times the new i th row.

Note: The pivoting results in the fuzzy simplex tableau corresponding to the new basis.

4.3 The Main Steps of FVLP Simplex Algorithm

Algorithm 1: The Fuzzy Simplex Method for the FVLP Problem.

Assumption: A basic feasible solution with basis and the corresponding simplex tableau is at hand.

1. The basic feasible solution is given by $\tilde{x}_B = \tilde{y}_0$ and $\tilde{x}_N = \tilde{0}$. The fuzzy objective value is: $\tilde{z} = \tilde{y}_{00} = c_B \tilde{y}_0$.
2. Let $y_{0k} = \max_j \{y_{0j}\}$, $j = 1, ..., n$, $j \neq B_i$, $i = 1, ..., m$. If $y_{0k} \leq 0$ then stop; the current solution is optimal.
3. If $y_k \leq 0$, then stop; the problem is unbounded. Otherwise determine the index B_r of the variable $\tilde{x}_{B_r}$ leaving the basis as follows:

$$\frac{y_{r0}}{y_{rk}} = \min_{1 \leq i \leq m} \left\{ \frac{y_{i0}}{y_{ik}} \,|\, y_{ik} > 0 \right\}$$

where $y_{i0} = R(\tilde{y}_{i0})$, $i = 1, ..., m$.
4. Pivot on y_{rk} and update the fuzzy simplex tableau. Go to Step 2.

5   The Fuzzy Two-Phase Method for Solving FVLP Problems

5.1   The Initial Fuzzy Basic Feasible Solution

As in the last section we mentioned the fuzzy primal simplex method starts with a fuzzy basic feasible solution and moves to an improved fuzzy basic feasible solution, until the optimal point is reached or else unboundedness of the objective function is verified. However, in order to initialize the fuzzy primal simplex method, a basis B with $y_0 \simeq B^{-1}\tilde{b}$ must be available. We shall show that the fuzzy primal simplex method can always be initiated with a very simple basis, namely, the identity. We describe two following cases.

Case 1: Suppose that the constraints are of the form $A\tilde{x} \preceq \tilde{b}$, $\tilde{x} \succeq \tilde{0}$, where A is an $m \times n$ crisp matrix and $\tilde{b}$ is an $m \times 1$ nonnegative fuzzy vector. By adding the fuzzy slack vector $\tilde{x}_s$, the constraints can be put in the following standard form: $A\tilde{x} + \tilde{x}_s \simeq \tilde{b}$, $\tilde{x} \succeq \tilde{0}$, $\tilde{x}_s \succeq \tilde{0}$. Note that the new $m \times (m + n)$ constraint matrix $(A|I)$ has rank m, and a fuzzy basic feasible solution of this system is at hand, by letting $\tilde{x}_s \simeq \tilde{b}$ be the fuzzy basic vector, and $\tilde{x} \simeq \tilde{0}$ be the fuzzy nonbasic vector. Hence, at this initial fuzzy basic feasible solution, $\tilde{x}_s \simeq \tilde{b}$, $\tilde{x} \simeq \tilde{0}$ and the fuzzy primal simplex method now can be applied.

Case 2: In many situations, finding a starting fuzzy basic feasible solution is not straightforward as the case just described. To illustrate, suppose that the constraints are of the form $A\tilde{x} \preceq \tilde{b}$, $\tilde{x} \succeq \tilde{0}$, such that at least one of the elements of $\tilde{b}$ is negative. In this case, after introducing the fuzzy slack vector $\tilde{x}_s$, we cannot let $\tilde{x} \simeq \tilde{0}$, because $\tilde{x}_s \simeq \tilde{b}$ violates the nonnegativity requirements.

Another situation occurs when the constraints are of the form $A\tilde{x} \succeq \tilde{b}$, $\tilde{x} \succeq \tilde{0}$, where for at least an $i, i = 1, ..., m$, we have $\tilde{b}_i \succ \tilde{0}$. After subtracting the fuzzy slack vector $\tilde{x}_s$, we get $A\tilde{x} - \tilde{x}_s \simeq \tilde{b}$, $\tilde{x} \succeq \tilde{0}$, $\tilde{x}_s \succeq \tilde{0}$. Again, there is no obvious way of picking a basis B from matrix $(A| - I)$ with $1 0 \, y_0 \simeq B^{-1}\tilde{b}$.

Again consider the linear programming with fuzzy variables (FVLP) problem as defined in (1). It is simple to see that any FVLP problem can be transformed into a problem of the following form:

$$\max \tilde{z} \simeq c\tilde{x}$$
$$s.t. \ A\tilde{x} \simeq \tilde{b} \tag{1}$$
$$\tilde{x} \succeq 0$$

where $\tilde{b} \succeq \tilde{0}$. (if $\tilde{b}_i \prec \tilde{0}$, the ith row can be multiplied by -1). This can be accomplished by introducing fuzzy slack variables and by simple manipulations of the constraints and variables. If A contains an identity matrix, then an immediate fuzzy basic feasible solution is at hand, by simply letting B = I , and since $\tilde{b} \succeq \tilde{0}$, then $\tilde{b} \simeq B^{-1}\tilde{b} \succeq \tilde{0}$. Otherwise, something else must be done.

5.2 Fuzzy Artificial Variables

After manipulating the constraints and introducing fuzzy slack variables, suppose that the constraints are put in the format $A\tilde{x} \preceq \tilde{b}$, $\tilde{x} \succeq \tilde{0}$, where A is an $m \times n$ matrix and $\tilde{b} \succeq \tilde{0}$ is an $m \times 1$ fuzzy vector. Furthermore, suppose that A has no identity submatrix (if A has an identity submatrix then we have an obvious to get a starting fuzzy basic feasible solution). In this case, we shall resort to the fuzzy artificial variables to get a starting fuzzy basic feasible solution, and then use the fuzzy primal simplex method itself and get rid of these fuzzy artificial variables.

To illustrate, suppose that we change the restrictions by adding a fuzzy artificial vector $\tilde{x}_a$ leading to the system $A\tilde{x} + \tilde{x}_a \simeq \tilde{b}$, $\tilde{x} \succeq \tilde{0}$, $\tilde{x}_a \succeq \tilde{0}$. Note that by construction, we created an identity matrix corresponding to the fuzzy artificial vector. This gives an immediate fuzzy basic feasible solution of the new system, namely, $\tilde{x}_a \simeq \tilde{b}$ and $\tilde{x} \simeq \tilde{0}$. Even though we now have a starting fuzzy basic feasible solution and the fuzzy primal simplex method can be applied, we have in effect changed the problem. In order to get back to our original problem, we must force these fuzzy artificial variables to zero, because $A\tilde{x} \simeq \tilde{b}$ if and only if a $A\tilde{x} + \tilde{x}_a \simeq \tilde{b}$ with $\tilde{x}_a \simeq \tilde{0}$. In other words, fuzzy artificial variables are only a tool to get the fuzzy primal simplex method started; however, we must guarantee that these fuzzy variables will eventually drop to zero, if at all possible.

Remark 5.1. We emphasize that a fuzzy slack variable is introduced to put the problem in equality form, and the fuzzy slack variable can very well be positive, which means that the inequality holds as a strict inequality. Fuzzy artificial variables, however, are not legitimate variables, and they are merely introduced to facilitate the initiation of the fuzzy simplex method. These fuzzy artificial variables, however, must eventually drop to zero in order to attain feasibility in the original linear programming with fuzzy variables problem.

5.3 The Fuzzy Two-Phase Method

There are various methods in the literature of linear programming that can be used to eliminate the fuzzy artificial variables. One of these methods is to minimize the sum of the artificial variables. Here we propose this idea for linear programming problem with fuzzy variables. In fact, we define a new fuzzy objective function which is the sum of the fuzzy artificial variables in minimization form, subject to the constraints are $A\tilde{x} + \tilde{x}_a \simeq \tilde{b}$, $\tilde{x} \succeq \tilde{0}$, $\tilde{x}_a \succeq \tilde{0}$. If the original linear programming with fuzzy variables problem has a fuzzy feasible solution, then the fuzzy optimal value of this problem is zero, where all the fuzzy artificial variables drop to zero. More importantly, as the fuzzy artificial variables drop to zero, they leave the basis, and legitimate variables enter instead. Eventually, all fuzzy artificial variables leave the basis (this is not always the case, because we may have a fuzzy artificial variable in the basis at level zero). The basis then consists of legitimate variables. In other words, we get a fuzzy basic feasible solution for the original system $A\tilde{x} \simeq \tilde{b}$, $\tilde{x} \succeq \tilde{0}$, and the fuzzy simplex method can be stated with the original fuzzy objective function $c\tilde{x}$. If, on the other hand, after solving this problem we have a positive fuzzy artificial variable, then the original problem has no feasible solution. This procedure is called the fuzzy two-phase method. In the first phase we reduce fuzzy artificial variables to value zero or conclude that the original problem has no feasible solutions. In the former case, the second phase minimizes the original fuzzy objective function starting with the fuzzy basic feasible solution obtained at the end of Phase I. The fuzzy two-phase method is outlined below.

Phase I: Solve the following linear FVLP problem starting with the fuzzy basic feasible solution $\tilde{x} \simeq \tilde{0}$ and $\tilde{x}_a \simeq \tilde{b}$:

$$\min \ \tilde{x}_0 \simeq 1\tilde{x}_a$$
$$s.t. \ A\tilde{x} + \tilde{x}_a \simeq \tilde{b} \tag{2}$$
$$\tilde{x} \succeq 0, \ \tilde{x}_a \succeq \tilde{0}$$

where $1 = (1, ..., 1)$ is $1 \times m$ constant vector and $(\tilde{x}_a)^T \simeq (\tilde{x}_{1a}, \tilde{x}_{2a}, ..., \tilde{x}_{ma})$. This problem is called a fuzzy simplex Phase I problem. If at optimality $\tilde{x}_a \not\simeq \tilde{0}$, then stop; the original FVLP problem has no feasible solutions. Otherwise, let the fuzzy basic and nonbasic legitimate variables be $\tilde{x}_B$ and $\tilde{x}_N$.

Phase II: Solve the following linear FVLP problem starting with the fuzzy basic feasible solution $\tilde{x}_B \simeq B^{-1}\tilde{b}$ and $\tilde{x}_N \simeq \tilde{0}$:

$$\max \ \tilde{z} \simeq c_B\tilde{x}_B + \tilde{c}_N\tilde{x}_N$$
$$s.t. \ B\tilde{x}_B + N\tilde{x}_N \simeq \tilde{b} \tag{3}$$
$$\tilde{x}_B \succeq 0, \ \tilde{x}_N \succeq \tilde{0}$$

The foregoing linear programming with fuzzy variables problem is of course equivalent to the original problem. Here, we illustrate our method to solve an FVLP problem.

Example 5.1. Consider the following FVLP:

$$\max \ \tilde{z} \approx -\tilde{x}_1 - \tilde{x}_2$$

$$s.t. \ \begin{cases} 2\tilde{x}_1 + 5\tilde{x}_2 \succeq (5, 8, 2, 5) \\ 3\tilde{x}_1 + 4\tilde{x}_2 \succeq (6, 10, 2, 6) \\ \tilde{x}_1, \tilde{x}_2 \succeq \tilde{0}. \end{cases}$$

By adding the shakes and artificial variables we obtain the Phase I problem as follows:

$$\min \ \tilde{x}_0 \approx \tilde{x}_{1a} + \tilde{x}_{2a}$$

$$s.t. \ \begin{cases} 2\tilde{x}_1 + 5\tilde{x}_2 - \tilde{x}_3 + \tilde{x}_{1a} \simeq (5, 8, 2, 5) \\ 3\tilde{x}_1 + 4\tilde{x}_2 - \tilde{x}_4 + \tilde{x}_{2a} \simeq (6, 10, 2, 6) \\ \tilde{x}_1, \tilde{x}_2, \tilde{x}_3, \tilde{x}_4, \tilde{x}_{1a}, \tilde{x}_{2a} \succeq \tilde{0}. \end{cases}$$

Then the first tableau of Phase I is as follows:

basis	$\tilde{x}_1$	$\tilde{x}_2$	$\tilde{x}_3$	$\tilde{x}_4$	$\tilde{x}_{1a}$	$\tilde{x}_{2a}$	R.H.S.
$\tilde{z}$	5	9	-1	-1	0	0	$(-18,-11,11,4)$
$\tilde{x}_{1a}$	2	5	-1	0	1	0	$(5,8,2,5)$
$\tilde{x}_{2a}$	3	4	0	-1	0	1	$(6,10,2,6)$

It is obvious that $\tilde{x}_2$ is an entering fuzzy variable and $\tilde{x}_{1a}$ is a leaving fuzzy variable. Then after pivoting the next tableau is given as:

basis	$\tilde{x}_1$	$\tilde{x}_2$	$\tilde{x}_3$	$\tilde{x}_4$	$\tilde{x}_{1a}$	$\tilde{x}_{2a}$	R.H.S.
$\tilde{z}$	$\frac{7}{5}$	0	$\frac{4}{5}$	-1	$\frac{-9}{5}$	0	$(\frac{-2}{5},6,6,\frac{38}{5})$
$\tilde{x}_2$	$\frac{2}{5}$	1	$\frac{-1}{5}$	0	$\frac{1}{5}$	0	$(1,\frac{8}{5},\frac{2}{5},1)$
$\tilde{x}_{2a}$	$\frac{7}{5}$	0	$\frac{4}{5}$	-1	$\frac{-4}{5}$	1	$(\frac{-2}{5},6,6,\frac{38}{5})$

So $\tilde{x}_1$ is an entering fuzzy variable and $\tilde{x}_{2a}$ is a leaving fuzzy variable. The last tableau is given in the below. Now we see all fuzzy artificial variables leaves the basis. So we can remove their columns.

basis	$\tilde{x}_1$	$\tilde{x}_2$	$\tilde{x}_3$	$\tilde{x}_4$	$\tilde{x}_{1a}$	$\tilde{x}_{2a}$	R.H.S.
$\tilde{z}$	0	0	0	0	-1	-1	$(0,0,0,0)$
$\tilde{x}_2$	1	0	-1	0	$\frac{3}{7}$	$\frac{-4}{7}$	$(\frac{-5}{7},\frac{12}{7},\frac{18}{7},\frac{19}{7})$
$\tilde{x}_1$	0	1	0	-1	$\frac{-4}{7}$	$\frac{5}{7}$	$(\frac{-2}{7},\frac{30}{7},\frac{30}{7},\frac{38}{7})$

Now consider the first tableau of Phase II which is constructed by the last tableau in the end of the Phase I as follows:

basis	$\tilde{x}_1$	$\tilde{x}_2$	$\tilde{x}_3$	$\tilde{x}_4$	R.H.S.
$\tilde{z}$	0	0	$\frac{6}{7}$	$\frac{10}{7}$	$(\frac{-300}{7},\frac{62}{7},\frac{428}{7},\frac{360}{7})$
$\tilde{x}_2$	1	0	-1	0	$(\frac{-5}{7},\frac{12}{7},\frac{18}{7},\frac{19}{7})$
$\tilde{x}_1$	0	1	0	-1	$(\frac{-2}{7},\frac{30}{7},\frac{30}{7},\frac{38}{7})$

Finally as we saw the current fuzzy simplex tableau is optimal based on the optimality condition.

6 Conclusion

Main contribution here is introducing the fuzzy two-phase method to find a starting fuzzy basic feasible solution for the fuzzy primal simplex algorithm. In particular, we have illustrated the mentioned method by solving a linear programming problem with fuzzy variables which have solved later by the fuzzy dual simplex algorithm in Mahdavi-Amiri and Nasseri [13].

Acknowledgement. The author thank to the Research Center of Algebraic Hyperstructure and Fuzzy Mathematics, Babolsar, Iran and also the first author tanks the National Elite Foundation, Tehran, Iran for their supports.

References

1. Ebrahimnejad, A., Nasseri, S.H.: Using complementary slackness property to solve linear programming with fuzzy parameters. Fuzzy Information and Engineering 3, 233–245
2. Ebrahimnejad, A., Nasseri, S.H., Hosseinzadeh Lotfi, F., Soltanifar, M.: A primal- dual method for linear programming problems with fuzzy variables. European Journal of Industrial Engineering 4(2), 189–209 (2010)
3. Ebrahimnejad, A., Nasseri, S.H., Hosseinzadeh Lotfi, F.: Bounded linear programs with trapezoidal fuzzy numbers. International Journal of Uncertainty, Fuzziness and Knowledge-Based Systems 18(3), 269–286 (2010)
4. Ebrahimnejad, A., Nasseri, S.H.: A dual simplex method for bounded linear programmes with fuzzy numbers. International Journal of Mathematics in Operational Research 2(6), 762–779 (2010)
5. Ebrahimnejad, A., Nasseri, S.H., Mansourzadeh, S.M.: Bounded primal simplex algorithm for bounded linear programming with fuzzy cost coefficients. International Journal of Operations Research and Information Systems 2(1), 96–120 (2011)
6. Fortemps, P., Roubens, M.: Ranking and defuzzification methods based on area compensation. Fuzzy Sets and Systems 82, 319–330 (1986)
7. Inuiguchi, M.: Necessity measure optimization in linear programming problems with fuzzy polytopes. Fuzzy Sets and Systems 158, 1882–1891 (2007)
8. Inuiguchi, M., Ramik, J.: Possibilistic linear programming: a brief review of fuzzy mathematical programming and a comparison with stochastic programming in portfolio selection problem. Fuzzy Sets and Systems 111(1), 3–8 (2000)
9. Jimenez, M., Bilbao, A.: Pareto-optimalsolutionsinfuzzymulti-objectivelinear programming. Fuzzy Sets and Systems 160, 2714–2721 (2009)
10. Klir, G.J., Yuan, B.: Fuzzy Sets and Fuzzy Logic: Theory and Applications. Prentice-Hall, PTR, New Jersey (1985)
11. Lai, Y.J., Hwang, C.L.: Fuzzy Mathematical Programming Methods and Applications. Springer, Berlin (1992)
12. Mahdavi-Amiri, N., Nasseri, S.H.: Duality in fuzzy number linear programming by use of a certain linear ranking function. Applied Mathematics and Computation 180, 206–216 (2006)
13. Mahdavi-Amiri, N., Nasseri, S.H.: Duality results and a dual simplex method for linear programming problems with trapezoidal fuzzy variables. Fuzzy Sets and Systems 158, 1961–1978 (2007)
14. Mahdavi-Amiri, N., Nasseri, S.H., Yazdani, A.: Fuzzy primal simplex algorithms for solving fuzzy linear programming problems. Iranian Journal of Operational Research 1(2), 68–74 (2009)
15. Maleki, H.R., Tata, M., Mashinchi, M.: Linear programming with fuzzy variables. Fuzzy Sets and Systems 109, 21–33 (2000)
16. Mehra, S.A., Chandra, C.R.: Bector: Acceptable optimality in linear fractional programming with fuzzy coefficient. Fuzzy Optimization and Decision Making 6, 5–16 (2007)

17. Nasseri, S.H., Ebrahimnejad, A.: A fuzzy dual simplex method for fuzzy number linear programming problem. Advances in Fuzzy Sets and Systems 5(2), 81–95 (2009)
18. Okada, S., Soper, T.: A shortest path problem on a network with fuzzy arc lengths. Fuzzy Sets and Systems 109, 129–140 (2000)
19. Rommelfanger, H.: Fuzzy linear programming and applications. European Journal of Oprational Research 92(3), 512–527 (1996)
20. Rommelfanger, H.: The advantages of fuzzy optimization models in practical use. Fuzzy Optimization and Decision Making 3(4), 295–309 (2004)
21. Rommelfanger, H.: A general concept for solving linear multicriteria programming problems with crisp, fuzzy or stochastic values. Fuzzy Sets and Systems 158, 1892–1904 (2007)
22. Safi, Z.M.R., Maleki, H.R., Zaeimazad, E.: A geometric approach for solving fuzzy linear programming problems. Fuzzy Optimization and Decision Making 6, 315–336 (2007)
23. Tanaka, H., Okuda, T., Asai, K.: On fuzzy mathematical programming. The Journal of Cybernetics 3, 37–46 (1974)
24. Vijay, V., Mehra, A., Chandra, S., Bector, C.R.: Fuzzy matrix games via a fuzzy relation approach. Fuzzy Optimization and Decision Making 6, 299–314 (2007)
25. Yager, R.R.: A procedure for ordering fuzzy subsets of the unit interval. Information Sciences 24, 143–161 (1981)
26. Zimmermann, H.J.: Fuzzy programming and linear programming with several objective functions. Fuzzy Sets and Systems 1, 45–55 (1978)

A New Approach to Duality in Fuzzy Linear Programming

S.H. Nasseri[1,*] and A. Ebrahimnejad[2]

[1] Department of Mathematical Sciences, University of Mazandaran, Babolsar, Iran
nasseri@umz.ac.ir
[2] Department of Mathematics, Qaemshahr Branch, Islamic Azad University, Qaemshahr,
Iran

Abstract. In a recently quoted paper [N. Mahdavi-Amiri and S.H. Nasseri, Duality results and a dual simplex method for linear programming problems with trapezoidal fuzzy variables, Fuzzy Sets and Systems 158 (2007)1961-1978], has been established the dual of linear programming problem with trapezoidal fuzzy variables as a fuzzy number linear programming, using certain general linear ranking functions. In this paper, we define a new dual problem for the linear programming problem with trapezoidal fuzzy variables as a linear programming problem with trapezoidal fuzzy variables and hence deduce the duality results such as weak duality, strong duality and complementary slackness theorems. In throughout of the paper, we apply the trapezoidal fuzzy numbers as a convenient fuzzy numbers which is more practical in comparison of other numbers (and also as a general form of triangular fuzzy numbers) and moreover use the certain linear ranking functions for the sake of illustrating the performance of our approach, but it is no restrict at all to use these linear ranking functions. In particular, we emphasize that the new duality definition and the associated duality results lead us to provide the primal and dual simplex algorithms for solving the linear programming problems with trapezoidal fuzzy numbers.

Keywords: Duality, linear programming with trapezoidal fuzzy numbers, linear ranking function, primal and dual simplex algorithms, trapezoidal fuzzy number.

1 Introduction

In contrast to classical linear programming, the concept of duality of fuzzy linear programming is not uniquely defined. From the historical perspective, Rodder and Zimmermann [18] have initially generalized max-min and min-max problems on fuzzy quantities to create two pairs of fuzzy dual linear programming problems. An economic interpretation of this duality in terms of market and industry is also included in [18]. Hamacher et al. [6] afterwards utilized sensitivity analysis in

---

* Corresponding author.

B.-Y. Cao and X.-J. Xie (Eds.): Fuzzy Engineering and Operations Research, AISC 147, pp. 17–29.
springerlink.com   © Springer-Verlag Berlin Heidelberg 2012

fuzzy linear programming. Bector and Chandra [1] found some inherent difficulties with the fuzzy dual formulations of [18]: in the special situation, Rodder and Zimmermann's approach doesn't lead to a standard primal-dual linear programming. Thus, they constructed a modified couple of fuzzy primal-dual model. Thereafter, Verdegay [20] stated the fuzzy dual problem with the help of parametric linear programming, and showed the under some suitable conditions, the fuzzy primal and dual problems have the same fuzzy solutions. Also, Bector et al. [1] introduced duality concepts for linear programming with fuzzy parameters and showed the equivalently of primal and dual results for a special problems in game theory. Their theorems are constructed with respect to a defuzzification function, which maps fuzzy numbers to real numbers. Moreover, Hashemi et al. [8] have introduced the weak duality theorem based on an alphabetic order function for fully fuzzified linear programming. Furthermore, Inuiguchi et al. [9] have proved some important dual theorems on linear programming using satisfying concepts. Next, Hashemi et al. [7] presented complementary slackness conditions for this fuzzy goal programming by using optimistic (lenient) and pessimistic (severe) operators on the objective function and the constraints of fuzzy problem to find optimal solutions keeping those viewpoints. Synchronously, Ramik [17] produced a similar approach for a general fuzzy linear programming with possibility and necessity measures. In addition, Zhong and Shi [22] gave duality concepts on fuzzy multi-criteria and multi-constraint linear programming applying a parametric approach. Mahdavi-Amiri et al. [16], based on a new type of fuzzy arithmetic for symmetric trapezoidal fuzzy numbers which is proposed by Ganesan and Veeramani in [5], proposed the dual of a linear programming problem with symmetric trapezoidal fuzzy numbers without converting them to crisp linear programming problems. A fuzzy dual simplex for fuzzy linear programming problems is proposed in [15]. Maleki et al. [13] proposed using an auxiliary to solve linear programming problems with fuzzy variables. Nasseri et al. [14] applied a fuzzy primal simplex for linear programming problems with fuzzy variables which was stated in [12], to solve the flexible linear programming problems. Mahdavi-Amiri and Nasseri in [11] applied a linear ranking function (specially Yager's ranking function) to order trapezoidal fuzzy numbers and established the dual problem of a linear programming problem with fuzzy variables as fuzzy number linear programming and hence deduced duality results. Based on their established results, Ebrahimnejad et al. [4] proposed a primal-dual method for solving linear programming problems with fuzzy variables. In this paper, we give a new approach to define the dual of linear programing with fuzzy variables and then investigate the duality results. We emphasize that for the sake of illustrating the performance of our approach, it has been here developed using the trapezoidal fuzzy numbers and the certain linear ranking functions, but it is no restrict at all to use these linear ranking functions.

This paper is organized as follows: In Section 2, we give some necessary concepts of fuzzy set theory and then define the fuzzy linear programming problems in Section 3. In particular, we present a new definition of the dual problem of the fuzzy linear programming. We then develop the duality theorems and results in the last part of this section. Finally, we conclude in Section 4.

2 Preliminaries

We give here, some basic definitions on fuzzy numbers comparison that will be used for illustrating our approach. In below some selected ranking functions of fuzzy numbers shall be shortly shown with regard to a better understanding of the approach that will be applied. In this way, and for methodological reasons, in spite that the our approach to be designed will work correctly for any kind of fuzzy numbers, we prefer in the following to focus on the usual case of the trapezoidal fuzzy numbers as well as the triangular fuzzy numbers. Consequently, in this section, we introduce some well-known fuzzy numbers ranking indexes as well as the main concepts/definitions needed in the rest of the paper.

Fuzzy set is defined as a subset $\tilde{a}$ of universal set $X \subseteq \mathbb{R}$ by its membership function $\mu_{\tilde{a}}(.)$, which assigns to each element $x \in \mathbb{R}$, a real number $\mu_{\tilde{a}}(x)$ in the interval $[0,1]$.

The α-cut or α-level of a fuzzy set $\tilde{a}$, which plays an essential role in fuzzy optimization, is defined as an ordinary set $[\tilde{a}]_\alpha$ for which the degree of its membership function exceeds the level α.

A fuzzy set $\tilde{a} = (a^L, a^U, \alpha, \beta)$ is called a generalized left right fuzzy numbers (GLRFN), if its membership function has the following form (taken from [19])

$$\mu_{\tilde{a}}(x) = \begin{cases} L\left(\frac{a^L - x}{\alpha}\right) & \text{for} \quad a^L - \alpha \le x \le a^L \\ 1 & \text{for} \quad a^L \le x \le a^U \\ R\left(\frac{x - a^U}{\beta}\right) & \text{for} \quad a^U \le x \le a^U + \beta \\ 0 & \text{else} \end{cases} \tag{1}$$

where L and R are non-increasing functions defined on $[0,1]$ and satisfying the following conditions:

$$L(x) = R(x) = 1 \quad \text{if} \quad x \le 0$$
$$L(x) = R(x) = 0 \quad \text{if} \quad x \ge 0$$

For $a^L = a^U$, we have the classical definition of Left Right Fuzzy Numbers (LRFN) of Dubois and Prade ([3]).

Trapezoidal Fuzzy Numbers (TRFN) are special cases of fuzzy numbers with the following membership function:

$$\mu_{\tilde{a}}(x) = \begin{cases} \frac{x - (a^L - \alpha)}{\alpha} & \text{for} \quad a^L - \alpha \le x \le a^L \\ 1 & \text{for} \quad a^L \le x \le a^U \\ \frac{(a^U + \beta) - x}{\beta} & \text{for} \quad a^U \le x \le a^U + \beta \\ 0 & \text{else} \end{cases} \tag{2}$$

We denote a TRFN $\tilde{a}$ as $\tilde{a} = (a^L, a^U, \alpha, \beta)$ where $(a^L - \alpha, a^U + \beta)$ is the support of $\tilde{a}$ and $[a^L, a^U]$ its core, and the set of all TRFN by $F(\mathbb{R})$.

Now, we define arithmetic on TRFN. Let $\tilde{a} = (a^L, a^U, \alpha, \beta)$ and $\tilde{b} = (b^L, b^U, \gamma, \theta)$ be two trapezoidal fuzzy numbers. Define,

$$x \geq 0, \; x \in \mathbb{R}; \quad x\,\tilde{a} = (xa^L, xa^U, x\alpha, x\beta)$$
$$x < 0, \; x \in \mathbb{R}; \quad x\,\tilde{a} = (xa^U, xa^L, -x\beta, -x\alpha)$$

$$\tilde{a} + \tilde{b} = (a^L + b^L, a^U + b^U, \alpha + \gamma, \beta + \theta).$$
$$\tilde{a} - \tilde{b} = (a^L - b^U, a^U - b^L, \alpha + \theta, \beta + \gamma).$$

An effective approach for ordering the elements of $F(\mathbb{R})$ is to define a ranking function $R : F(\mathbb{R}) \to \mathbb{R}$ which maps each trapezoidal fuzzy number into the real line, where a natural order exists. We define orders on $F(\mathbb{R})$ by the following roles which was taken from [11] and [13]:

$$\tilde{a} \succeq \tilde{b} \quad \text{if and only if} \quad R(\tilde{a}) \geq R(\tilde{b}) \tag{3}$$
$$\tilde{a} \succ \tilde{b} \quad \text{if and only if} \quad R(\tilde{a}) > R(\tilde{b}) \tag{4}$$
$$\tilde{a} \simeq \tilde{b} \quad \text{if and only if} \quad R(\tilde{a}) = R(\tilde{b}) \tag{5}$$

where $\tilde{a}$ and $\tilde{b}$ are in $F(\mathbb{R})$. Also we write $\tilde{a} \preceq \tilde{b}$ if and only if $\tilde{b} \succeq \tilde{a}$. We restrict our attention to linear ranking functions, that is, a ranking function R such that

$$R(k\tilde{a} + \tilde{b}) = kR(\tilde{a}) + R(\tilde{b}) \tag{6}$$

for any $\tilde{a}$ and $\tilde{b}$ belonging to $F(\mathbb{R})$ and any $k \in \mathbb{R}$.

Remark 2.1. For any TRFN $\tilde{a}$, the relation $\tilde{a} \succeq \tilde{0}$ holds, if there exist $\varepsilon \geq 0$ and $\alpha \geq 0$ such that $\tilde{a} \succeq (-\varepsilon, \varepsilon, \alpha, \alpha)$. We realize that $R(-\varepsilon, \varepsilon, \alpha, \alpha) = 0$ (we also consider $\tilde{a} \simeq \tilde{0}$ if and only if $R(\tilde{a}) = 0$). Thus, without loss of generality, throughout the paper we let $\tilde{0} = (0, 0, 0, 0)$ as the zero TRFN.

The following lemma is given from [11] and we omit the proofs here.

Lemma 2.1. *Let R be any linear ranking function. Then,*

(i) $\tilde{a} \succeq \tilde{b}$ if and only if $\tilde{a} - \tilde{b} \succeq \tilde{0}$ if and only if $-\tilde{b} \succeq -\tilde{a}$.
(ii) If $\tilde{a} \succeq \tilde{b}$ and $\tilde{c} \succeq \tilde{d}$, then $\tilde{a} + \tilde{c} \succeq \tilde{b} + \tilde{d}$.

We consider the linear ranking functions on $F(\mathbb{R})$ as:

$$R(\tilde{a}) = c_L a^L + c_U a^U + c_\alpha \alpha + c_\beta \beta, \tag{7}$$

where $\tilde{a} = (a^L, a^U, \alpha, \beta)$, and $c_L, c_U, c_\alpha, c_\beta$ are constants, at least one of which is nonzero.

In the following, we review some of important linear ranking functions and without any loss of generality but, as said, for the sake of illustrating, we will only calculate some well-known ranking functions for trapezoidal fuzzy numbers.

a) The second Yager's ranking function [21] is given as follows:

$$Y_2(\tilde{a}) = \tfrac{1}{2} \int_0^1 (\inf[\tilde{a}]_\alpha + \sup[\tilde{a}]_\alpha) \, d\alpha,$$

which for arbitrary TRFN $\tilde{a}$ reduce to

$$Y_2(\tilde{a}) = \frac{1}{2} \left[a^L + a^U + \frac{\beta - \alpha}{2} \right]. \tag{8}$$

b) The first Campos and Munoz's ranking function [2] is given as follows:

$$CM_1^\lambda(\tilde{a}) = \int_0^1 (\lambda \inf[\tilde{a}]_\alpha + (1 - \lambda) \sup[\tilde{a}]_\alpha) \, d\alpha,$$

which for arbitrary TRFN $\tilde{a}$ reduce to

$$CM_1^\lambda(\tilde{a}) = a^L + \lambda \left[(a^U - a^L) + \frac{\alpha + \beta}{2} \right] - \frac{\alpha}{2}. \tag{9}$$

c) The second Campos and Munoz's ranking function [2] is given as follows:

$$CM_2^\lambda(\tilde{a}) = \int_0^1 \alpha(\lambda \inf[\tilde{a}]_\alpha + (1 - \lambda) \sup[\tilde{a}]_\alpha) \, d\alpha,$$

which for arbitrary TRFN $\tilde{a}$ reduce to

$$CM_2^\lambda(\tilde{a}) = a^L + \lambda[(a^U - a^L) + \frac{\alpha + \beta}{3}] - \frac{\alpha}{3}. \tag{10}$$

3 Fuzzy Linear Programming and Duality Results

Here, we first define a linear programming with trapezoidal fuzzy variables and based on our methodology for ranking function define a new from of duality for FVLP problem.

Definition 3.1. *A linear programming with trapezoidal fuzzy variables (FVLP) is defined as [10, 12, 13]:*

$$\begin{aligned} \min \ &\tilde{z} \simeq c\tilde{x} \\ s.t. \ &A\tilde{x} \succeq \tilde{b} \\ &\tilde{x} \succeq \tilde{0} \end{aligned} \tag{1}$$

where $c \in \mathbb{R}^n, \tilde{x} \in (F(\mathbb{R}))^n, A \in \mathbb{R}^{m \times n}, \tilde{b} \in (F(\mathbb{R}))^m$.

Definition 3.2. *We say that a vector $\tilde{x} \in (F(\mathbb{R}))^n$ is a feasible solution to (1) if and only if $\tilde{x}$ satisfies the constraints of the problem.*

Definition 3.3. *A feasible solution $\tilde{x}_*$ is an optimal solution for (1), if for all feasible solution $\tilde{x}$ for (1), we have $c\tilde{x}_* \preceq c\tilde{x}$.*

Definition 3.4. *(Fuzzy Basic Feasible Solution) Consider the FVLP problem,*

$$\min \tilde{z} \simeq c\tilde{x}$$
$$s.t. \ A\tilde{x} = \tilde{b} \tag{2}$$
$$\tilde{x} \succeq \tilde{0}$$

where the parameters of the problem are as defined in (1). Let $A = [a_{ij}]_{m \times n}$ and assume rank(A)=m. Partition A as $[B \quad N]$ where B, $m \times m$, is nonsingular. It is obvious that rank(B)=m. Let y_j be the solution to $By = a_j$. It is apparent that the basic solution

$$\tilde{x}_B \simeq (\tilde{x}_{B_1}, ..., \tilde{x}_{B_m})^T \simeq B^{-1}\tilde{b}, \ \tilde{x}_N \simeq \tilde{0} \tag{3}$$

is a solution of $A\tilde{x} = \tilde{b}$. In fact, $\tilde{x} = (\tilde{x}_B^T \ \tilde{x}_N^T)^T$. If $\tilde{x}_B \succeq \tilde{0}$, then the basic solution is feasible and the corresponding fuzzy objective value is: $\tilde{z} \simeq c_B\tilde{x}_B$, where $c_B = (c_{B_1}, ..., c_{B_m})$. Now, corresponding to every nonbasic variable $\tilde{x}_j, 1 \leq j \leq n, j \neq B_i, i = 1, ..., m,$ define

$$z_j = c_B y_j = c_B B^{-1} a_j. \tag{4}$$

Mahdavi-Amiri and Nasseri [11] proved the optimality conditions for FVLP (1) as follows and we omit the proof here.

Theorem 3.1. *(Optimality conditions) If a basic solution $\tilde{x}_B = B^{-1}\tilde{b}$, $\tilde{x}_N \simeq \tilde{0}$ is feasible to (2) and $z_j - c_j$ for all $j, 1 \leq j \leq n$, then the fuzzy basic solution is an fuzzy optimal solution to (2).*

Also, Mahdavi-Amiri and Nasseri [11] defined the dual of FVLP problem (1) as follows:

$$\max \tilde{z} \simeq w\tilde{b}$$
$$s.t. \ wA \leq c \tag{5}$$
$$w \geq 0$$

where $w = (w_1, ..., w_m) \in \mathbb{R}^m$ is including the crisp variables corresponding to constraints of problem (1). We see that the dual of FVLP problem (1) is indeed a fuzzy number linear programming (FNLP) problem and then solving it by use of fuzzy simplex algorithm [12] gives a crisp solution for the dual problem. In this section, we apply a new approach to introduce the dual problem (1) that gives a fuzzy dual solution.

Definition 3.5. *We say that the real number a corresponds to the fuzzy number $\tilde{a}$, with respect to a given linear ranking function R, if $a = R(\tilde{a})$.*

Clearly, corresponding to a trapezoidal fuzzy number $\tilde{a}$ there is a real number a such that $a = R(\tilde{a})$. Now, we show that corresponding to a real number a there is a trapezoidal fuzzy number $\tilde{a}$, with respect to a given linear ranking function R (Specially, with respect to the mentioned linear ranking function in Section 2) such that $a = R(\tilde{a})$.

Lemma 3.1. *Corresponding to given a real number t, there is a trapezoidal fuzzy number $\tilde{t}$ such that $Y_2(\tilde{t}) = t$.*

Proof. To obtain a trapezoidal fuzzy number $\tilde{t}$, it is need to choose two arbitrary real number α and β such that $\alpha = \beta$ and then let $a^L = t - \frac{\alpha}{2}$ and $a^U = t + \frac{\alpha}{2}$ in (8). Thus, we have

$$Y_2(\tilde{t}) = Y_2\left((t - \frac{\alpha}{2}, t + \frac{\alpha}{2}, \alpha, \alpha)\right) = \frac{1}{2}\left[t - \frac{\alpha}{2} + t + \frac{\alpha}{2} + (\frac{\alpha - \alpha}{2})\right] = t$$

Hence, for every real number t there is a trapezoidal fuzzy number as $(t - \frac{\alpha}{2}, t + \frac{\alpha}{2}, \alpha, \alpha)$.

Remark 3.1. Depending upon to need, one can also use a small α and β in the above theorem.

Lemma 3.2. *Corresponding to given a real number t, there is a trapezoidal fuzzy number $\tilde{t}$ such that $CM_1^\lambda(\tilde{t}) = t$.*

Proof. Clearly, for given a real number λ $(0 \leq \lambda \leq 1)$, there are two real numbers α and β such that $\lambda = \frac{\alpha}{\alpha+\beta}$. Hence, if we let $a^L = t - \frac{\alpha}{2}$ and $a^U = t + \frac{\beta}{2}$ in (9), thus we have

$$CM_1^{\frac{\alpha}{\alpha+\beta}}(\tilde{t}) = t - \frac{\alpha}{2} + \frac{\alpha}{\alpha+\beta}\left[t + \frac{\beta}{2} - t + \frac{\alpha}{2} + \frac{\alpha+\beta}{2}\right] - \frac{\alpha}{2} = t - \frac{\alpha}{2} + \alpha - \frac{\alpha}{2} = t$$

Therefore, in this case we have a trapezoidal fuzzy number $\tilde{t}$ as $(t - \frac{\alpha}{2}, t + \frac{\beta}{2}, \alpha, \beta)$, where α and β obtain from definition of λ.

Lemma 3.3. *Corresponding to given a real number t, there is a trapezoidal fuzzy number $\tilde{t}$ such that $CM_2^\lambda(\tilde{t}) = t$.*

Proof. Similar to previous theorem, given a real number λ $(0 \leq \lambda \leq 1)$, there are two real numbers α and β, such that $\lambda = \frac{\alpha}{\alpha+\beta}$. Hence, if we let $a^L = t - \frac{2\alpha}{3}$ and $a^U = t + \frac{2\beta}{3}$. From (10) we have

$$CM_2^{\frac{\alpha}{\alpha+\beta}}(\tilde{t}) = t - \frac{2\alpha}{3} + \frac{\alpha}{\alpha+\beta}\left[t + \frac{2\beta}{3} - t + \frac{2\alpha}{3} + \frac{\alpha+\beta}{3}\right] - \frac{\alpha}{3} = t - \frac{2\alpha}{3} + \alpha - \frac{\alpha}{3} = t$$

Therefore, in this case we have a trapezoidal fuzzy number $\tilde{t}$ as $(t - \frac{2\alpha}{3}, t + \frac{2\beta}{3}, \alpha, \beta)$.

We are now in a position to naturally extend the usual definition of a linear programming problem to the problem with fuzzy variables.

Definition 3.6. *The dual of FVLP problem (1) which is shown by DFVLP and is defined to be*

$$\max \ \tilde{u} \simeq \tilde{w}b$$
$$s.t. \ \tilde{w}A \preceq \tilde{c} \tag{6}$$
$$\tilde{w} \succeq \tilde{0}$$

where $A \in \mathbb{R}^{m \times n}, \tilde{w} \in (F(\mathbb{R}))^m$ and $\tilde{c} = (\tilde{c}_1, ..., \tilde{c}_n) \in \mathbb{R}^n$ such that $\tilde{c}_i$ is the fuzzy number corresponding to real number c_i with respect to a linear ranking function according to Lemma 3.1 - 3.3. Also,$b = (b_1, ..., b_m) \in \mathbb{R}^m$ where b_i is the real number corresponding to fuzzy number $\tilde{b}_i$ according to Definition 3.5.

For an illustration of the above approach we consider the following example. Note that here we use only the linear ranking function mentioned in Section 2 as a pattern of linear ranking functions on $F(\mathbb{R})$.

Example 3.1. Consider the following FVLP problem:

$$\min \ \tilde{z} \simeq 6\tilde{x}_1 + 10\tilde{x}_2$$

$$s.t. \ \begin{cases} 2\tilde{x}_1 + 5\tilde{x}_2 \succeq (5, 8, 2, 5) \\ 3\tilde{x}_1 + 4\tilde{x}_2 \succeq (6, 10, 2, 6) \\ \tilde{x}_1, \tilde{x}_2 \succeq \tilde{0}. \end{cases}$$

We first apply the second ranking function of Yager (8) to the fuzzy the fuzzy right hand side vector $\tilde{b}$. Thus, we have $Y_2(5, 8, 2, 5) = \frac{29}{4}$ and $Y_2(6, 10, 2, 6) = 9$. Also, according to Lemma 3.1 we obtain the fuzzy numbers $(4,8,2,2)$ and $(\frac{17}{2}, \frac{23}{2}, 3, 3)$ corresponding to cost coefficients 6 and 10, respectively. Thus, by definition of the dual problem we have its dual problem as follows:

$$\max \ \tilde{u} \simeq \frac{29}{4}\tilde{w}_1 + 9\tilde{w}_2$$

$$s.t. \ \begin{cases} 2\tilde{w}_1 + 3\tilde{w}_2 \preceq (4, 8, 2, 2) \\ 5\tilde{w}_1 + 4\tilde{w}_2 \preceq (\frac{17}{2}, \frac{23}{2}, 3, 3) \\ \tilde{w}_1, \tilde{w}_2 \succeq \tilde{0}. \end{cases}$$

Now, we apply the first ranking function of Campos and Munoz (9) with $\lambda = \frac{1}{2}$ that gives $\alpha = \beta = 1$. Thus, we have $CM_1^{\frac{1}{2}}(5, 8, 2, 5) = \frac{29}{4}$ and $CM_1^{\frac{1}{2}}(6, 10, 2, 6) = 9$. Also, according to Lemma 3.2 we obtain the fuzzy numbers $(\frac{11}{2}, \frac{13}{2}, 1, 1)$ and $(\frac{19}{2}, \frac{21}{2}, 1, 1)$ corresponding to the crisp cost coefficients 6 and 10, respectively. So, with respect to this ranking function the dual problem is given as follows:

$$\max \tilde{u} \simeq \tfrac{29}{4}\tilde{w}_1 + 9\tilde{w}_2$$

$$\text{s.t.} \quad \begin{cases} 2\tilde{w}_1 + 3\tilde{w}_2 \preceq (\tfrac{11}{2}, \tfrac{13}{2}, 1, 1) \\ 5\tilde{w}_1 + 4\tilde{w}_2 \preceq (\tfrac{19}{2}, \tfrac{21}{2}, 1, 1) \\ \tilde{w}_1, \tilde{w}_2 \succeq \tilde{0}. \end{cases}$$

In a similar way, applying the second ranking function of Campos and Munoz (10) with $\lambda = \tfrac{1}{2}$ gives $\alpha = \beta = 1$. Thus, we have $CM_2^{\frac{1}{2}}(5, 8, 2, 5) = 3$ and $CM_2^{\frac{1}{2}}(6, 10, 2, 6) = \tfrac{26}{3}$. Also, by Lemma 3.2 the fuzzy numbers corresponding to the crisp cost coefficients 6 and 10 with respect to this ranking function are $(\tfrac{14}{3}, \tfrac{20}{3}, 1, 1)$ and $(\tfrac{28}{3}, \tfrac{32}{3}, 1, 1)$, respectively. Therefore, the dual problem is given as follows:

$$\max \tilde{u} \simeq 3\tilde{w}_1 + \tfrac{26}{3}\tilde{w}_2$$

$$\text{s.t.} \quad \begin{cases} 2\tilde{w}_1 + 3\tilde{w}_2 \preceq (\tfrac{14}{3}, \tfrac{20}{3}, 1, 1) \\ 5\tilde{w}_1 + 4\tilde{w}_2 \preceq (\tfrac{28}{3}, \tfrac{32}{3}, 1, 1) \\ \tilde{w}_1, \tilde{w}_2 \succeq \tilde{0}. \end{cases}$$

We shall discuss here the relationship between the FVLP (1) and its corresponding dual problem (6).

Lemma 3.4. *The dual of the problem* (6) *is the problem* (1).

Proof. Since the problem (6) is an FVLP problem, we may consider Definition 3.6 for its dual. We write the problem (6) as follows:

$$\begin{aligned} &\min (-b)^T \tilde{w}^T \\ &\text{s.t.} \ (-A)^T \tilde{w}^T \succeq -\tilde{c}^T \\ &\qquad \tilde{w}^T \succeq \tilde{0} \end{aligned} \tag{7}$$

Now, using the fuzzy vector $\tilde{x}$ as the vector of the fuzzy dual variables, the dual of (7) is:

$$\begin{aligned} &\max \tilde{x}^T(-c)^T \\ &\text{s.t.} \ \tilde{x}^T(-A)^T \preceq -\tilde{b}^T \\ &\qquad \tilde{x}^T \succeq \tilde{0} \end{aligned} \tag{8}$$

where c and $\tilde{b}$ are real number and fuzzy number corresponding to fuzzy number $\tilde{c}$ and real number b with respect to a linear ranking function, respectively. The problem (8) is the same as:

$$\begin{aligned} &\min c\tilde{x} \\ &\text{s.t.} \ A\tilde{x} \succeq \tilde{b} \\ &\qquad \tilde{x} \succeq \tilde{0} \end{aligned} \tag{9}$$

exactly the original FVLP problem (1).

Remark 3.2. It is important to note that multiplication of two fuzzy numbers $\tilde{a}$ and $\tilde{b}$ does not produce fuzzy number. In these situations, using linear ranking function, we define $\tilde{a}\tilde{b} = a\tilde{b} = \tilde{a}b$, where a and b are real numbers corresponding to fuzzy numbers $\tilde{a}$ and $\tilde{b}$ with respect to linear ranking function, respectively.

Theorem 3.2. *(Weak duality) If $\tilde{\bar{x}}$ and $\tilde{w}$ are fuzzy feasible solutions to (1) and (6), respectively, then $c\tilde{\bar{x}} \preceq \tilde{w}b$.*

Proof. Multiplying $A\tilde{\bar{x}} \succeq \tilde{b}$ on the left by $\tilde{w} \succeq \tilde{0}$ and $\tilde{w}A \preceq \tilde{c}$ on the right by $\tilde{\bar{x}} \succeq \tilde{0}$ give $c\tilde{\bar{x}} \preceq \tilde{w}A\tilde{\bar{x}} \preceq \tilde{w}b$ or $c\tilde{\bar{x}} \preceq \tilde{w}b$. Thus, by Remark 3.2, we have $c\tilde{\bar{x}} \preceq \tilde{w}b$.

Corollary 3.1. *If $\tilde{x}^*$ and $\tilde{w}^*$ are feasible solutions to FVLP (1) and DFVLP (6), respectively, and $c\tilde{x}^* \simeq \tilde{w}^*b$, then $\tilde{x}^*$ and $\tilde{w}^*$ are optimal solutions to their respective problems.*

Proof. It is sufficient to show that for each feasible solution $\tilde{\bar{x}}$ to (1) and each feasible solution $\tilde{w}$ to (6), we have $c\tilde{x}^* \preceq c\tilde{\bar{x}}$ and $\tilde{w}^*b \succeq \tilde{w}b$. Since $\tilde{\bar{x}}$ is a feasible solution to (1), so by Theorem 3.2, $c\tilde{\bar{x}} \succeq \tilde{w}^*b \simeq c\tilde{x}^*$ that leads to $c\tilde{\bar{x}} \succeq c\tilde{x}^*$. In a similar way, we get $\tilde{w}^*b \succeq \tilde{w}b$. This completes the proof.

The following corollary is immediate consequence of Theorem 3.2.

Corollary 3.2. *If any one of the FVLP (1) or DFVLP (6) is unbounded, then the other problem has no feasible solution.*

Theorem 3.3. *(Strong duality) If any one of the FVLP problem (1) or DFVLP problem (6) has an optimal solution, then both problems have optimal solutions and the two optimal objective fuzzy values are equal. (In fact, if $\tilde{x}^*$ is fuzzy optimal solution of the primal problem then the vector $\tilde{w}^* = \tilde{c}_B B^{-1}$, where B is the optimal basis, is an optimal solution of the dual problem.)*

Proof. Assume the FVLP (1) has a fuzzy optimal solution as $(\tilde{x}_B^*, \tilde{0})$, where B is optimal basis.Let $\tilde{u} \succeq \tilde{0}$ be the fuzzy slack variables for the constraints $A\tilde{x} \succeq \tilde{b}$. The new equivalent problem to the FVLP (1) is

$$\min \tilde{z} \simeq c\tilde{x} + 0\tilde{u}$$
$$\text{s.t.} \ \ A\tilde{x} - \tilde{u} \simeq \tilde{b} \tag{10}$$
$$\tilde{x} \succeq \tilde{0}, \ \tilde{u} \succeq \tilde{0},$$

where $\tilde{u} = (\tilde{u}_1, .., \tilde{u}_m)^T$.

Now, let $\tilde{z}_j = \tilde{c}_B B^{-1} a_j = \tilde{w}^* a_j$ be the fuzzy number corresponding to real number $z_j = c_B B^{-1} a_j$ with respect to a linear ranking function, where a_j is the jth column of the coefficient matrix $[A \ \ I]$ in (10). From Theorem 3.1, we have

$$\tilde{z}_j - \tilde{c}_j \preceq \tilde{0} \quad j = 1, ..., n, n+1, ..., m.$$

Therefore, this follows that $\tilde{w}^* a_j - \tilde{c}_j \preceq \tilde{0}$ for $j = 1, ..., n$, that is, $\tilde{w}A \preceq \tilde{c}$. In addition, for $i = 1, ..., m$, we have, $\tilde{z}_{n+i} - \tilde{c}_{n+i} = \tilde{c}_B B^{-1} e_i - \tilde{0} = \tilde{w}^* e_i = \tilde{w}_i^*$ and hence, $\tilde{w}^* \succeq \tilde{0}$. We have just shown that $\tilde{z}_j - \tilde{c}_j \preceq \tilde{0}$, for $j = 1, ..., n + m$ implies that $\tilde{w}^* A \preceq \tilde{c}$ and $\tilde{w}^* \succeq \tilde{0}$, where $\tilde{w}^* = \tilde{c}_B B^{-1}$. That is $\tilde{w}^*$ is a feasible solution to (6). Moreover, we will have,

$$c\tilde{x}^* \simeq \tilde{c}\tilde{x}^* \simeq \tilde{c}_B \tilde{b} \simeq \tilde{c}_B B^{-1} \tilde{b} \simeq \tilde{w}^* \tilde{b} \simeq \tilde{w}^* b,$$

and thus, by Corollary 3.1, establishing the optimality of $\tilde{x}^*$ and $\tilde{w}^*$ for the FVLP (1) and DFVLP (6), respectively.

Theorem 3.4. *(Fundamental theorem of duality) For any FVLP problem and its corresponding DFVLP problem, exactly one of the following statements is true.*

1. *Both have optimal solutions $\tilde{x}^*$ and $\tilde{w}^*$ with $c\tilde{x}^* \simeq \tilde{w}^* b$.*
2. *One problem is unbounded and the other is infeasible.*
3. *Both problems are infeasible.*

We now state and prove an important result of duality theory, generally named as complementary slackness.

Theorem 3.5. *(Complementary slackness Theorem) Suppose $\tilde{u}$ and $\tilde{v}$ be the slack variables to the FVLP (1) and DFVLP (6), respectively. Let $(\tilde{x}^*, \tilde{u}^*)$ and $(\tilde{w}^*, \tilde{v}^*)$ be any feasible solutions to FVLP problem and its corresponding dual problem. Then $\tilde{x}^*$ and $\tilde{w}^*$ are respectively optimal if and only if*

$$\tilde{v}^* \tilde{x}^* \simeq \tilde{0}, \quad \tilde{w}^* \tilde{u}^* \simeq \tilde{0}. \tag{11}$$

Proof. Since $(\tilde{x}^*, \tilde{u}^*)$ is fuzzy feasible solution to the standard form of FVLP (1), therefore $A\tilde{x}^* - \tilde{u}^* \simeq \tilde{b}$. Multiplying this equation on the left by $\tilde{w}^* \succeq \tilde{0}$, we get

$$\tilde{w}^* A\tilde{x}^* - \tilde{w}^* \tilde{u}^* \simeq \tilde{w}^* \tilde{b} \tag{12}$$

Also, since $(\tilde{w}^*, \tilde{v}^*)$ is a fuzzy feasible solution to the DFVLP (6), thus $\tilde{w}^* A + \tilde{v}^* \simeq \tilde{c}$. Multiplying this equation on the right by $\tilde{x}^* \succeq \tilde{0}$, we get

$$\tilde{w}^* A\tilde{x}^* + \tilde{v}^* \tilde{x}^* \simeq \tilde{c}\tilde{x}^* \tag{13}$$

Subtraction of (12) from (13) yields

$$\tilde{v}^* \tilde{x}^* + \tilde{w}^* \tilde{u}^* \simeq \tilde{w}^* \tilde{b} - \tilde{c}\tilde{x}^* \tag{14}$$

Now for optimal solutions $(\tilde{x}^*, \tilde{u}^*)$ and $(\tilde{w}^*, \tilde{v}^*)$, from Theorem 3.3 and Remark 3.2 we have, $\tilde{w}^* b \simeq \tilde{w}^* \tilde{b} \simeq \tilde{c}\tilde{x}^* \simeq c\tilde{x}^*$. So, from (14), we obtain $\tilde{v}^* \tilde{x}^* + \tilde{w}^* \tilde{u}^* \simeq \tilde{0}$, that is, $\tilde{v}^* \tilde{x}^* \simeq \tilde{0}$ and $\tilde{w}^* \tilde{u}^* \simeq \tilde{0}$.

Conversely, if $\tilde{v}^* \tilde{x}^* \simeq \tilde{0}$ and $\tilde{w}^* \tilde{u}^* \simeq \tilde{0}$, then $\tilde{v}^* \tilde{x}^* + \tilde{w}^* \tilde{u}^* \simeq \tilde{0}$. Thus, from (14) we get $\tilde{w}^* \tilde{b} - \tilde{c}\tilde{x}^* \simeq \tilde{0}$, or $\tilde{w}^* \tilde{b} \simeq \tilde{c}\tilde{x}^*$. From Remark 3.2, this gives $\tilde{w}^* b \simeq c\tilde{x}^*$. Therefore, from Corollary 3.1 $(\tilde{x}^*, \tilde{u}^*)$ and $(\tilde{w}^*, \tilde{v}^*)$ are optimal solutions to FVLP (1) and DFVLP (6), respectively.

4 Conclusion

In this paper, we established a new concept of duality for linear programming problems with trapezoidal fuzzy variables by use of several linear ranking functions as a linear programming problem with trapezoidal fuzzy variables and hence deduced some duality results. This approach is useful when decision maker needs to a fuzzy dual solution. We emphasize that only for the sake of illustrating the performance of our approach, it has been here to use of certain linear ranking functions, but it is no restrict at all to use these linear ranking functions.

Acknowledgement. The authors thank to the Research Center of Algebraic Hyperstructure and Fuzzy Mathematics, Babolsar, Iran and also National Elite Foundation, Tehran, Iran for their supports.

References

1. Bector, C.R., Chandra, S.: On duality in linear programming under fuzzy environment. Fuzzy Sets and Systems 125, 317–325 (2002)
2. Campos, L., Munoz, A.: A subjective approach for ranking fuzzy numbers. Fuzzy Sets and Systems 29, 145–153 (1989)
3. Dubois, D.D., Prade, H.: Possibility theory: an approach to computerized processing of uncertainty. Plenum Press, New York (1988)
4. Ebrahimnejad, A., Nasseri, S.H., Hosseinzadeh Lotfi, F.: A primal- dual method for linear programming problems with fuzzy variables. European Journal of Industrial Engineering 4(2), 189–209 (2010)
5. Ganesan, K., Veeramani, P.: Fuzzy linear programming with trapezoidal fuzzy numbers. Ann. Oper. Res. 143, 305–315 (2006)
6. Hamacher, H., Liberling, H., Zimmermann, H.J.: Sensitivity analysis in fuzzy linear programming. Fuzzy Sets and Systems 1, 269–281 (1978)
7. Hashemi, S.M., Ghatee, M., Hashemi, B.: Fuzzy goal programming: Complementary slackness conditions and computational schemes. Applied Mathematics and Computation 179, 506–522 (2006)
8. Hashemi, S.M., Modarres, M., Nasrabadi, E., Nasrabadi, M.M.: Fully fuzzified linear programming, solution and duality. Journal of Intelligent and Fuzzy Systems 17, 253–261 (2006)
9. Inuiguchi, M., Ramik, J., Tanio, T., Vlach, M.: Satisficing solutions and duality in interval and fuzzy linear programming. Fuzzy Sets and Systems 135, 151–177 (2003)
10. Mahdavi-Amiri, N., Nasseri, S.H.: Duality in fuzzy number linear programming by use of a certain linear ranking function. Applied Mathematics and Computation 180, 206–216 (2006)
11. Mahdavi-Amiri, N., Nasseri, S.H.: Duality results and a dual simplex method for linear programming problems with trapezoidal fuzzy variables. Fuzzy Sets and Systems 158, 1961–1978 (2007)
12. Mahdavi-Amiri, N., Nasseri, S.H., Yazdani, A.: Fuzzy primal simplex algorithms for solving fuzzy linear programming problems. Iranian Journal of Operational Research 2, 68–84 (2009)

13. Maleki, H.R., Tata, M., Mashinchi, M.: Linear programming with fuzzy variables. Fuzzy Sets and Systems 109, 21–33 (2000)
14. Nasseri, S.H., Ebrahimnejad, A.: A fuzzy primal simplex algorithm and its application for solving the flexible linear programming problems. European Journal of Industrial Engineering 4(3), 372–389 (2010)
15. Nasseri, S.H., Ebrahimnejad, A.: A fuzzy dual simplex method for a fuzzy number linear programming problem. Advances in Fuzzy Sets and Systems 5(2), 81–95 (2010)
16. Nasseri, S.H., Mahdavi-Amiri, N.: Some duality results on linear programming problems with symmetric fuzzy numbers. Fuzzy Information and Engineering 1, 59–66 (2009)
17. Ramik, J.: Duality in fuzzy linear programming with possibility and necessity relations. Fuzzy Sets and Systems 157, 1283–1302 (2006)
18. Rodder, W., Zimmermann, H.J.: Duality in fuzzy linear programming. In: Fiacoo, A.V., Kortanek, K.O. (eds.) External Methods and System Analysis, Berlin, New York, pp. 415–429 (1980)
19. Tran, L., Duckstein, L.: Comparison of fuzzy numbers using a fuzzy distance measure. Fuzzy Set Systems 130, 331–341 (2002)
20. Verdegay, J.L.: A dual approch to solve the fuzzy linear programming problems. Fuzzy Sets and Systems 14, 131–141 (1984)
21. Yager, R.R.: A procedure for ordering fuzzy subsets of the unit interval. Inform. 24, 143–161 (1981)
22. Zhong, Y., Shi, Y.: Duality in fuzzy multi-criteria and multi-constraint level linear programming: A parametric approach. Fuzzy Sets and Systems 132, 335–346 (2002)

Interval Number Model for Portfolio Selection with Liquidity Constraints

Man-yi Tan

Chengdu Neusoft University, Chengdu, 611844, China
tan_my@163.com

Abstract. Considering liquidity constraints in Markowitz portfolio selection model, this paper develops a fuzzy portfolio selection model with liquidity constraints and profit rate, risk rate and turnover ratio of securities are described by interval fuzzy number. Based on the results of interval programming, this problem is converted into a linear programming with parameter. Finally, a numerical example is given to illustrate the validity of the method.

Keywords: Portfolio selection, interval number, fuzzy linear programming, interval programming.

1 Introduction

Portfolio selection is how to configure the position of marketable securities and meet some extent tradeoff between risks and profits for investors. Securities market is an extremely complex system. Both profits and risk are all uncertain, so all investors should have to make their decisions in an uncertain environment. In 1952, Markowitz established a portfolio selection model with mean and variance, and this is a beginning of the modern portfolio selection theory [1].

Investors often have an objective intention for risks and profit level in securities market. The past profit rates and risks are only a reference of future profit and risks because the future market is always changing. For the expected profits and risks are fuzzy, portfolio selection model can be converted an interval programming when the profits and risks of securities are described by interval number. Many scholars have studied interval programming [2-5] and portfolio selection problems with interval numbers [6-13], and obtained many research results. Wang et al. extended Markowitz's model to interval programming model [6]. Using fuzzy constraints in Chen et al. (2007) [11], the Markowitz portfolio selection model would be transformed into a fuzzy linear programming model, and build an interval number fuzzy portfolio selection model where the expected profit rates and risk rates of securities are described by interval numbers. To deal with a portfolio selection problem with fuzzy expected return rates, Chen et al. (2009) [12] transformed a traditional optimization problem with variance constraints into a multiobjective parameter linear programming problem and proposed a fuzzy two-stage algorithm to solve it. Through introducing interval number to describe the future

B.-Y. Cao and X.-J. Xie (Eds.): Fuzzy Engineering and Operations Research, AISC 147, pp. 31–39.
springerlink.com   © Springer-Verlag Berlin Heidelberg 2012

profit, liquidity and β, Chen et al. (2010) [13] proposed a portfolio model with basing on interval number.

Based on Chen et al. (2007) [11], this paper introduces liquidity constraints in the traditional Markowitz model and develops a new fuzzy portfolio selection model with considering liquidity constraint. Profit rate, risk rate and turnover ratio are all uncertainty in this model, and described by interval numbers.

2 Establishment of Interval Fuzzy Portfolio Selection Model with Liquidity

The following is the Markowitz mean-variance portfolio selection model:

$$\max \quad f(x) = \sum_{i=1}^{n} E(r_i) x_i$$

$$s.t. \quad \sum_{i=1}^{n}\sum_{j=1}^{n} \sigma_{ij} x_i x_j \leq w, \tag{P_1}$$

$$\sum_{i=1}^{n} x_i = 1,$$

$$0 \leq x_i \leq u_i, i = 1, 2, \ldots, n,$$

$x = (x_1, \ldots, x_n)^T$ can be a portfolio, $x_i (i = 1, 2, \ldots, n)$ is the investment ratio of security i, $r_i (i = 1, 2, \ldots, n)$ is the profit rate of security i, $E(r_i)(i = 1, 2, \ldots, n)$ is the expected profit of security i in holding period. $\sum_{i=1}^{n}\sum_{j=1}^{n} \sigma_i x_i x_j$ is the variance of portfolio $x = (x_1, \ldots, x_n)^T$ (to measure risk of portfolio). $\sigma_{ij} (i, j = 1, 2, \ldots, n)$ is covariance of security i and security j, $\sigma_i (i = 1, 2, \ldots, n)$ is the standard variance of security i. w is the upper limit of risk for investors, and $u_i (i = 1, 2, \ldots, n)$ is the upper limit of investment for security i.

Profit and risk can be considered two major factors of concern to investors in investment decision theory in general, as Markowitz's portfolio selection model has suggested. However, the liquidity of securities also can not be ignored in real investment practice. Liquidity of security is the cashability of security, and there are many methods to measure it at present. The major methods that widely be used include trading shares, number of transactions, transaction amount, turnover and velocity of circulation. Turnover of security is ratio of the stock trading volume (or transaction amount) and float caps (or market capitalization), and it fully reflects the liquidity of security. To characterize the liquidity with turnover of security, l_i is the turnover of security i, so the turnover of the portfolio $(x_1, \ldots, x_n)$ is

$\sum\limits_{i=1}^{n} l_i x_i$. The investors usually propose an acceptable lower limit l_0 for turnover

of portfolio, so $\sum\limits_{i=1}^{n} l_i x_i \geq l_0$. This will ensure the liquidity of portfolio and make

the funds of investors be safety.

Considering the liquidity of portfolio, the traditional Markowitz portfolio selection model (P_1) can be transformed into the following model:

$$\max \ f(x) = \sum_{i=1}^{n} E(r_i) x_i$$

$$s.t. \ \sum_{i=1}^{n} \sum_{j=1}^{n} \sigma_{ij} x_i x_j \leq w,$$

$$\sum_{i=1}^{n} l_i x_i \geq l_0, \tag{P_2}$$

$$\sum_{i=1}^{n} x_i = 1,$$

$$0 \leq x_i \leq u_i, i = 1,2,...,n.$$

The above model includes quadratic constraint, and this make it difficult for solving. According the research of Elton and Gruber et al. [14-15], we suppose correlation coefficients of different securities are same.

Let $\rho_{ij} = \rho, i, j = 1,2,...,n; i \neq j (\rho \geq 0)$,

$$\sum_{i=1}^{n} \sum_{j=1}^{n} \sigma_i x_i x_j = \sum_{i=1}^{n} \sum_{j=1}^{n} \rho \sigma_i \sigma_j x_i x_j = \rho(\sum_{i=1}^{n} \sigma_i x_i)^2 + (1-\rho)\sum \sigma_i^2 x_i^2 .$$

The second part of the last item is non-system risk in the above equation. According to the empirical study of Sharpe [16], when the non-system risk of portfolio is far smaller than system risk and especially, the amount of security in portfolio is enough, we can simplify the variance constraints using by fuzzy constraints. The

following is specific operation: $\sum\limits_{i=1}^{n} \sum\limits_{j=1}^{n} \sigma_i x_i x_j \leq w$ can be converted into

$\rho(\sum\limits_{i=1}^{n} \sigma_i x_i)^2 \prec w$, further convert to $\sum\limits_{i=1}^{n} \sigma_i x_i \prec \sqrt{\dfrac{w}{\rho}} = M$, that is,

$\sum\limits_{i=1}^{n} \sigma_i x_i \prec M$. Here $\prec$ is fuzzy less-than. The membership function of fuzzy inequality is

$$\mu(x) = \begin{cases} 1, & 0 \le \sum_{i=1}^{n} \sigma_i x_i \le M - d, \\[2em] \dfrac{M - \sum_{i=1}^{n} \sigma_i x_i}{d}, & M - d \le \sum_{i=1}^{n} \sigma_i x_i \le M, \\[2em] 0, & otherwise. \end{cases}$$

Let d be a tolerance degree of investors.

According to the membership function of fuzzy inequality $\prec$ and related conclusions, risk constraints of portfolio can be expressed as $\sum_{i=1}^{n} \sigma_i x_i \le M + d(1-\alpha)$, α is the membership of investors, and gradually changes from 0 to 1. so the evolution of model (P$_2$) is

$$\max \quad f(x) = \sum_{i=1}^{n} E(r_i) x_i$$

$$s.t. \quad \sum_{i=1}^{n} \sigma_i x_i \le M + d(1-\alpha),$$

$$\sum_{i=1}^{n} l_i x_i \ge l_0, \tag{P_3}$$

$$\sum_{i=1}^{n} x_i = 1,$$

$$0 \le x_i \le u_i, i = 1,2,...,n.$$

Due to the uncertainty for the future profit, liquidity and risk of securities, their changes are fuzzy, and can be regarded as fuzzy phenomena to solve. In this paper, fuzziness is described by interval numbers. Let

$$r_i = [\underline{r_i}, \overline{r_i}], \sigma_i = [\underline{\sigma_i}, \overline{\sigma_i}], l_i = [\underline{l_i}, \overline{l_i}]$$

to make the parameters of problem (P_3) fuzzifications. We can develop interval fuzzy portfolio selection model with liquidity of securities:

$$\max \ f(x) = \sum_{i=1}^{n} [\underline{r_i}, \overline{r_i}] x_i$$

$$s.t. \ \sum_{i=1}^{n} [\underline{\sigma_i}, \overline{\sigma_i}] x_i \leq [\underline{M} + d(1-\alpha), \overline{M} + d(1-\alpha)],$$

$$\sum_{i=1}^{n} [\underline{l_i}, \overline{l_i}] x_i \geq [\underline{l_0}, \overline{l_0}], \qquad (2) \qquad (LIPM) \quad (1)$$

$$\sum_{i=1}^{n} x_i = 1,$$

$$0 \leq x_i \leq u_i, i = 1, 2, ..., n,$$

where $[\underline{M}, \overline{M}]$ and $[\underline{l_0}, \overline{l_0}]$ are constant. $\underline{M}$ is the pessimistic affordable risk level of investors, and $\overline{M}$ is the optimistic affordable risk level. $\underline{l_0}$ is pessimistic affordable liquidity level, and $\overline{l_0}$ is optimistic affordable liquidity level.

The interval numbers of objective function in $(LIPM)$ represents portfolio uncertain profits, the left of (1) is risks of portfolio described by interval numbers of standard variance, the right represents affordable risk interval. The left of (2) represents uncertainty of liquidity, the right is affordable liquidity interval. So $(LIPM)$ is an interval programming problem to maximize uncertain profit of investors under the conditions of uncertain risk constraints and uncertain liquidity constraints of securities. The uncertainty is described by interval numbers among them. The above problem could not exist classic optimal solution due to introducing interval order relation in constraints. $(LIPM)$ is an optimal problem with interval coefficient.

Without liquidity constraints, $(LIPM)$ can be degenerated into an interval number fuzzy portfolio selection model, it is considered by Chen et al. (2007).

3　Solution of Interval Fuzzy Portfolio Selection Model with Liquidity

Mark up interval number $A = [\underline{a}, \overline{a}]$ with $A = [m(A), w(A)]$. The midpoint of A, $m(A) = (\underline{a} + \overline{a})/2$, is called as position coefficient of A and reflects size of A. The half wide of A, $w(A) = (\underline{a} - \overline{a})/2$, is called as flexible coefficient, reflects uncertain degree of A. let $B = [\underline{b_i}, \overline{b_i}]$.

Definition 3.1 [8]. $\lambda(A \le B) = \dfrac{m(B) + m(A)}{w(B) + w(A)}$ *is called as satisfaction degree*

for $A \le B$.

Lemma 3.1 [8]. *At the level (* λ_0 *) of satisfaction degree,* $\displaystyle\sum_{i=1}^{n}[\underline{a_i}, \overline{a_i}] \ge [\underline{b_i}, \overline{b_i}]$

can be transformed into the following crisp constraints

$$\sum_{i=1}^{n}[(1+\lambda_0)\underline{a_i} + (1-\lambda_0)\overline{a_i}] \ge (1-\lambda_0)\underline{b_i} + (1+\lambda_0)\overline{b_i}.$$

From Lemma 3.1, given λ_0-level of satisfaction degree, risk constraints (1) and liquidity constraints (2) in $(LIPM)$ should be converted to crisp constraints (3) and (4):

$$\sum_{i=1}^{n}[(1+\lambda_0)\underline{\sigma_i} + (1-\lambda_0)\overline{\sigma_i}]x_i \le (1-\lambda_0)[\underline{M} + d(1-\alpha)] + (1+\lambda_0)[\overline{M} + d(1-\alpha)],$$

$$\tag{3}$$

$$\sum_{i=1}^{n}[(1+\lambda_0)\underline{l_i} + (1-\lambda_0)\overline{l_i}]x_i \ge (1-\lambda_0)\underline{l_0} + (1+\lambda_0)\overline{l_0}. \tag{4}$$

Definition 3.2. [5] $\displaystyle\sum_{i=1}^{n}[\underline{r_i} + \theta(\overline{r_i} - \underline{r_i})]x_i$ *is called as* θ*-level solution of linear*

programming in target range of objective function $f(x) = \displaystyle\sum_{i=1}^{n}[\underline{r_i}, \overline{r_i}]x_i$.

In this way, given optimal level θ of target range and satisfaction degree λ_0 of interval inequality constraints, $(LIPM)$ is equivalent to the following linear programming with parameters:

$$\max f(x) = \sum_{i=1}^{n}[\underline{r_i} + \theta(\overline{r_i} - \underline{r_i})]x_i$$

$$s.t. \ \sum_{i=1}^{n}[(1+\lambda_0)\underline{\sigma_i} + (1-\lambda_0)\overline{\sigma_i}]x_i \le (1-\lambda_0)[\underline{M} + d(1-\alpha)] + (1+\lambda_0)[\overline{M} + d(1-\alpha)],$$

$$\sum_{i=1}^{n}[(1+\lambda_0)\underline{l_i} + (1-\lambda_0)\overline{l_i}]x_i \ge (1-\lambda_0)\underline{l_0} + (1+\lambda_0)\overline{l_0}, \qquad (LIPM_0)$$

$$\sum_{i=1}^{n}x_i = 1,$$

$$0 \le x_i \le u_i, i = 1, 2, ..., n.$$

When $\theta = 0.5$, the objective function of $(LIPM_0)$

is $f(x) = \sum_{i=1}^{n} (\underline{r_i} + \overline{r_i})x_i / 2$. This is midpoint of interval numbers, or position coefficient of interval numbers. That is, midpoint of interval numbers is measured to the size of target, it make fuzzy target be crisp.

4 Numerical Examples

The following gives a numerical example for illustrating the interval number fuzzy portfolio selection model with liquidity constraints and the method to solve the linear programming. The data is mainly taken from the literature 11 and 13, see also Table 1, Table 2 and Table 3.

Table 1. Securities profit rate interval

security	1	2	3	4	5
$[\underline{r_i},\overline{r_i}]$	[0.0050,0.0060]	[0.0060,0.0080]	[0.0120,0.0140]	[0.0050,0.0065]	[0.0050,0.0085]
stock	6	7	8	9	
$[\underline{r_i},\overline{r_i}]$	[0.0050,0.0084]	[0.0060,0.0080]	[0.0040,0.0060]	[0.0200,0.0300]	

Table 2. Securities risk rate interval

security	1	2	3	4	5
$[\underline{\sigma_i},\overline{\sigma_i}]$	[0.030, 0.040]	[0.050, 0.080]	[0.080, 0.096]	[0.045, 0.070]	[0.035, 0.054]
stock	6	7	8	9	
$[\underline{\sigma_i},\overline{\sigma_i}]$	[0.04,0.065]	[0.04,0.054]	[0.03,0.050]	[0.08,0.094]	

Table 3. Securities liquidity interval

security	1	2	3	4	5
$[\underline{l_i},\overline{l_i}]$	[0.22, 0.34]	[0.33, 0.54]	[0.32, 0.44]	[0.25, 0.37]	[0.14, 0.25]
stock	6	7	8	9	
$[\underline{l_i},\overline{l_i}]$	[0.31, 0.57]	[0.16, 0.34]	[0.23, 0.46]	[0.12, 0.26]	

Let $M = [\underline{M},\overline{M}] = [0.046,0.056]$, d=0.004, α =0.9, $u_i = 0.4$, θ =0.7, λ_0 =0.7. We can solve the linear programming problem $(LIPM 0)$ with Matlab 7.5. When

$[l_0, \bar{l}_0] = [0.20, 0.35]$, the optimal portfolio is $x = (0.0000\ 0.4000\ 0.0659\ 0.0000$
$0.0000\ 0.4000\ 0.0000\ 0.0171\ 0.1170)$, and the optimal value is 0.0100. when
$[l_0, \bar{l}_0] = [0.15, 0.25]$, the optimal portfolio is $x = (0.3017\ \ 0.0000\ 0.0000\ 0.0000$
$0.0000\ 0.3112\ 0.0000\ 0.0000\ 0.3871)$, and the optimal value is 0.0145. If there is
no liquidity constraints, the optimal portfolio is $x = (0.3019\ 0.0000\ 0.0000$
$0.0000\ 0.2981\ 0.0000\ 0.0000\ 0.0000\ 0.4000)$, and the optimal value is 0.0147.

5 Conclusion

These factors in securities investment all are uncertain. It is very consistent with
real condition to describe the uncertainty by interval numbers. The author has
converted the fuzzy portfolio selection problem with liquidity constraints into a
linear programming problem with parameters. The numerical examples have
shown the method for solving this problem is valid. By the results, we know that it
has an obvious influence on investment whether or not to consider liquidity constraints, and different levels of liquidity have also significant implications. These
conclusions have important guidance for investment decisions in practice

References

1. Markowitz, H.: Portfolio selection. Journal of Finance 7, 77–91 (1952)
2. Ishibuchi, H., Tanaka, H.: Multiobjective programming in optimization of the interval
   objective function. European Journal of Operational Research 48, 219–225 (1990)
3. Chanas, S., Kuchta, D.: Multiobjective programming in optimization of the interval
   objective functions – a generalized approach. European Journal of Operational Research 94, 594–598 (1996)
4. Tong, S.: Interval number and fuzzy number linear programming. Fuzzy Sets and Systems 66, 301–306 (1994)
5. Da, Q.L., Liu, X.W.: Interval number linear programming and its satisfactory solution.
   Systems Engineering Theory & Practice 4, 3–7 (1999)
6. Wang, S.Y., Zeng, J.H., Lai, K.K.: Portfolio selection models with transaction costs:
   crisp case and interval number case. In: Proceedings of the 5th International Conference on Optimization Techniques and Applications, Hong Kong, pp. 943–950 (2001)
7. Lu, Y.J., Tang, X.W., Zhou, Z.F.: Interval number linear programming method for the
   portfolio investment. Journal of Systems Engineering 19, 33–37 (2004)
8. Ida, M.: Portfolio selection problem with interval coefficients. Applied Mathematics
   Letters 16, 709–713 (2003)
9. Chen, H.Y., Zhao, Y.M.: Research on interval number portfolio investment model.
   College Mathematics 23, 21–25 (2007)
10. Yue, W., He, X.S.: The application of interval programming in portfolio selection.
    Value Engineering 9, 63–66 (2007)
11. Chen, G.H., Chen, S., Wang, S.Y.: Interval number fuzzy portfolio selection model.
    Systems Engineering 8, 34–37 (2007)

12. Chen, G.H., Chen, S., Fang, Y., Wang, S.Y.: Model for portfolio selection with fuzzy return rates. Systems Engineering Theory and Practice 7, 8–15 (2009)
13. Chen, G.H., Liao, X.L.: A model for portfolio selection based on interval programming. Journal of Liaoning Technical University(Natural Science) 10, 835–838 (2010)
14. Elton, E.J., Gruber, M.J.: Estimating the dependence structure of share prices. Journal of Finance 28, 1203–1232 (1973)
15. Elton, E.J., Gruber, M.J., Urich, T.J.: Are betas best. Journal of Finance 5, 1375–1384 (1978)
16. Sharpe, W.F.: A simplified model for portfolio analysis. Management Science 9, 277–293 (1963)

Application of Fuzzy Clustering Method in the Crime Data Analysis

Kai-qi Zou[1], Xiu-min Zhou[2], and Feng-xin Liu[1]

[1] College of Information Engineering, Dalian University, Dalian 116622, China
[2] Department of Information Engineering, Hu Lun Bei Er Vocational Technical College, Hu Lun Bei Er 021000, China
zoukq@vip.sina.com, zhouxiumin300@126.com, lfxlzh@163.com

Abstract. As is known to all, crime is a worldwide phenomenon, in specific thorough study to the crime problems which can help the government and related departments get the actual condition of an objective evaluation about crime, correctly determine the special policies and measures, so as to achieve the purpose of reducing crime levels. At the same time, it can promote the justice department according to the actual situation of crime and adjust the law enforcement, on one hand the actual crime situation strengthen control of inflation;on the other hand strengthen investigation measures constantly breakthroughs hidden case, reducing hidden case, reducing crime dark hidden case, to reduce crime number and maintain social stability purposes. So as to realize the law enforcement departments change from the passive prevention into active prevention.

Keywords: Crime data analysis, fuzzy neural network,clustering analysis.

1 Preface

The nature of the harmonious society is the social various organizations, social stratum can harmonious operation and develop coordinating in each other. However reality shows that the operation and development in society, various contradictions and conflicts often affect social harmony degree, including this conflict and contradiction is the most intense form of crime. Therefore, a national crime height reflects a nation's harmonious degree, the study of crime rates for social stability, for the development of the country has a very important significance.

In recent years, the security situation in China's social is still more severe [1]. Because of it has not completely out of the financial crisis which leads leading some group employment difficulty, and then increased inequality, relative poverty population increase, plus a variety of various social conflicts masses event happens, dimension stability not relieve pressure. China's violent crimes, crimes against

B.-Y. Cao and X.-J. Xie (Eds.): Fuzzy Engineering and Operations Research, AISC 147, pp. 41–50.
springerlink.com

property crime and economic crimes will still maintain high-risk situation. In the economic crisis conditions, not only the unemployment rise, but the crime rate.

In human society, crime is very sensitive and complex social phenomenon. For many years the researchers studied from the angle of criminology, the formation mechanism and social crime control method, but the theoretical study always with real situation exists some bias,so it can not reflect social crime real situations, investigate its reason is the number of criminal dark existence. So-called dark number refers to the actual crime has happened, but because of all sorts of reasons, or have been compromised but not exposed or have not by the public security and judicial organs shall investigate and deal with crime quantity. Therefore mining criminal dark number problem is facing very serious problems in our country.

Since the 1970s, crime statistics as criminal theory research provided a good tool of the crime, but there was greater distance compare with mechanism and internal rules cognitive needs. In today's society, in the national economic and social development process, steadily crime will be affected by many factors. Investigate its reasons, main performance for the following two points: first social stability can't get enough security, people security needs cannot completely meet; second is there are some problems in the economic development process, although all the economic indexes obviously growth, but inflation pressure, insufficient employment pressure, house prices higher various economic problems cause many social problems, thus which make crimes increases unceasingly, and social stability and economic development to a negative influence, it not to be ignored. Therefore, the study of the structure and law of criminal question becomes the social stability and speeds up the economic development of the moment.

Foreign scholars adopt spatial data mining, knowledge vector analysis and neural network methods to establish various models to study the crime. Arun Kulkarni etc [2] introduce fuzzy neural network model, and use a new algorithm applied in previous data focus on knowledge discovery; Memon, Q.A etc [3] who created the automatic crime analysis system that based on artificial neural network, but the system is still based on application of the statistical data; Tony H. Grubesic etc [4] put into use fuzzy clustering method to the mining crime hotspots; Loial, V. etc [5] with fuzzy semantic mapping method to study crime problems.

In 2004, Liu Cheng [6] put forward efense space theory, mainly from the crime prevention of environmental design angle, only got discussed from the geographical environment to the perspectives of crime which was not comprehensive. In 2005, Zuoqiang Chen etc [7] put forward xplore space analysis method for potential crime spot; they analyzed just from the concept and did not angle of in-depth research. In 2008, Ping He [8] proposed crime spatial analysis theory; he systematically expounded the concept and criminal space research significance. In 2010, Taiwan scholar Tun-Sheng etc [9] established such people MWP intelligent decision support model framework and through the analysis of previous crime data to predict criminal trends. Some scholars use the grey system theory to study crime and crime forecast, through the stage data analysis gives grey degrees, therefore, it avoids incomplete of information, and it makes some achievements.

2 Fuzzy Clustering Neural Networks

2.1 Fuzzy Clustering Analysis

The fuzzy clustering analysis based on the characteristics between objective things, similarity and relatives' degree and by establishing the fuzzy similarity relation to classify objective things of a kind of mathematical method. It is an important branch of the supervision and pattern recognition, the development of the theory of fuzzy clustering in production practice promoted its application, and in turn the needs of practical application promotes the theory of fuzzy clustering to enrich and perfect. Along with the development of the theory, fuzzy clustering has in many fields---pattern recognition, data mining, computer vision and fuzzy control etc has been widely used, this is also a recent rapidly developing a hot spot of research[10].

In fuzzy clustering, a sample X is divided into c fuzzy subsets $X_1, X_2, \cdots, X_c$ of the membership function, and samples from $\{0,1\}$ expanded to $[0,1]$ interval, and it meets the following conditions:

$$\begin{cases} \mu_{ik} \in [0,1], \\ \sum_{i=1}^{c} \mu_{ik} = 1, & \forall k; \\ 0 < \sum_{k=1}^{n} \mu_{ik} < n, & \forall i. \end{cases} \quad (1)$$

By (1) known $\overset{c}{\underset{i=1}{U}} \operatorname{supp}(X_i) = X$, here's supp means take support set of fuzzy sets[11].

In the real world, whether it belongs to a kind of a thing, its boundaries tend not to be clear, and in a certain extent has great fuzziness, fuzzy set theory is mathematical method which officially depicts and solves this type of fuzzy problem. With the fuzzy set theory formation, development and deepening, Ruspini[12] was first put forward the concept of fuzzy division, fuzzy clustering method is rapidly becoming the mainstream of cluster analysis.

Using of this concept people put forward many clustering methods, the typical: based on similarity relation and fuzzy relation method (including polymerization and anti-secession law)[13], based on fuzzy equivalence relation relay closure methods[14], and based on data sets and the convex decomposition, dynamic planning and is difficult to identify relations method.

2.2 Fuzzy Clustering Neural Network

In recent years, an important direction is the fuzzy set theory combined with neural network and fuzzy neural network is the establishment of artificial neural network research, including an important branch of clustering analysis is used for the research of fuzzy neural network[15]. Although at present the fuzzy clustering neural network research has achieved many results, fuzzy neural network clustering features has greatly improved, but the fuzzy clustering neural network research is still in its infancy and it has many problems. Such as how to avoid network into the local minimum value, how to avoid the fuzzy rules combination explosion, how to avoid the effect of initial condition and input mode to network successively and how to establish a suitable fuzzy clustering of optimizing the structure of the neural network model and so on[16].

Because of the traditional competitive learning algorithm which can deal with ordinary against crime data sets, against crime space spatial clustering sample characteristic data is often take interval-valued situation, we first establish crime interval-valued fuzzy neural network and its competition network structure as shown in Figure 1:

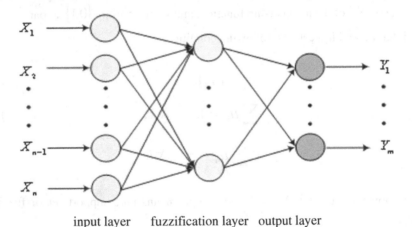

input layer   fuzzification layer   output layer

Fig. 1. Fuzzy competitive neural network structure

The network consists of three layers, which the first layer for input layer,the input data for the pretreatment of the training sample of after a crime, input variables for X_i, among $(i = 1, \cdots, n)$. The second is fuzzification layer, X_i by several fuzzy variables respectively. The third is competition layer, crime sample data after the training and winning neurons competition output for 1, the rest output for 0.

The algorithm for winning neuron competition is: assume the input mode of input layer is $X = (X_1, X_2, \cdots, X_n)$, in $X_i = [\underline{x_i}, \overline{x_i}] \in I[0,1], (i = 1,2,\cdots,n)$, all of the neuron competition layer corresponding right vector $w_j = (w_{j1}, w_{j2}, \cdots, w_{jn}), (j = 1,2,\cdots,m)$; the most similar with X right vector convicted of competitive win neurons. Finally the learning rule according to Winner-Take-All interpreting value, Adjust rules:

$$w_{j^*}(t+1) = w_{j^*}(t) + \alpha(f(w_{j^*_1}, X), f(w_{j^*_2}, X), \cdots, f(w_{j^*_n}, X)), \qquad (2)$$

$$f(w_{j^*_i}, X) = \text{sgn}(\underline{x_i} - w_{j^*_i})d(w_{j^*_i}, X). \qquad (3)$$

In the formula (2), type $\alpha \in [0,1]$ means the learning efficiency, through full training in this method, competition after the weight vectors of wins neurons become an input mode clustering center, which neurons wins will output 1 in competition layer, what kind is this mode output.

3 Application of Fuzzy Clustering Neural Network Method in the Crime Data Analysis

To analyze the samples dimension, sample size, grid size and the relationship between the density threshold and puts forward an effective grid and the density of the parameters set methods and applies in the crime analysis.

Based on data which provided by the public security bureau for sample build the sample space S , using the method of half the sample space grid divided into a grid unit, through analyzing the density of mesh, determine the rallying point of sample, through the rallying point of samples to determine the approximate location of clustering center, namely clustering center position.

Set $A = \{A_1, A_2, \cdots, A_n\}$ is a n bounded domain collection, so $S = A_1 \times A_2 \times \cdots \times A_n$ is a sample space dimension of n. Set samples is $V = \{v_1, v_2, \cdots, v_m\}$, in $v_i = \{v_{i1}, v_{i2}, \cdots, v_{in}\}$, $v_{ij} \in A_j$, through the input parameter $\xi_1, \xi_2, \cdots, \xi_n$, it can make each of the sample space s one-dimensional respectively divided into $\xi_1, \xi_2, \cdots, \xi_n$ interval, thus the whole space will divide into the limited class at the intersection of a rectangular element, each of these can be described as a rectangular element $U_i = \{u_{i1}, u_{i2}, \cdots, u_{im}\}$ $(i = 1, 2, \cdots, n)$, in $u_{ij} = [l_{ij}, h_{ij})$ is a left closed right and open interval, we call U_i for grid units.

Grid unit U_i density $D(U_i)$ defined as: $D(U_i) =$ the number of grid cell samples / Total sample size; grid unit U_i was dense unit only when $D(U_i) > \tau$, one density threshold τ for input parameters.

At any of the grid unit in the space $U_i = \{u_{i1}, u_{i2}, \cdots, u_{id}\}$, if U_i for intensive unit, so p called a rallying point in the U_i geometric center. The definition by rallying point, in the original sample points of dense area distribution, rallying point distribution is dense too; Contrast at the original sample point's sparse area, the distribution of rallying point is sparse too. Therefore rallying point is better to reflect the distribution of clustering sample, so that we can put concentrated extract from sample clustering center into a from rallying point concentrated extract clustering center.

3.1   Algorithm Described

Input parameters are the meshing $\xi_1, \xi_2, \cdots, \xi_n$, we will make the corresponding interval to each -axes intelligently, and the sample space S is divided into the grid shape. Setting density threshold is τ, and to get through a rough division meticulous make grid division, until the sample density of each of the grid $< \tau$. In order to make the grid differentiate continuing reduction in volume, when half each grid should rotate. In this process, we remember the t times for the grid division is $grid\{(I_1^{t1}, \cdots, I_N^m), t\}$, t_i on the i dimension said grouped frequency.

Calculation every $D(U_i)$, whether $D(U_i)$ is more than the given density threshold, if more than a given density threshold, the grid unit is intensive unit. At this time, that p for a rallying point of the geometric center U_i. Repetitive execution, until all the grid units were judged. Because the rallying point can reflect the distribution of clustering sample, according to the definition of rallying point, in the original sample points dense area, distribution of the distribution of rallying point is dense too; the original sample point sparse area, the distribution of rallying point is sparse too. So it can extract from sample concentration clustering center is transformed from the rallying point concentrated extract clustering center. Sample clustering center has the characteristics of data points, so dense with the maximum density rallying point p must be sample clustering center, from concentration of the rallying point, we can delete rallying point the same class of rallying point, so the whole of this category from rallying point concentrated deleted. But not all rallying point connected with some of the rallying point in the same class. According to the distribution of each sample is commonly by clustering center outwards density reduced gradually, by comparing two nearby rallying point to determine whether the density change belonging to the same class. From the current clustering center to outside, if rallying point density begins to increase, says we have been searching to this border. Thus we can get the clustering center centralized point is sample clustering center; the number of clustering center is the class number of samples.

Set a clustering prototype for $P = \{p_1, p_2, \cdots, p_c\}$, fuzzy clustering neural network of weight vectors $m_j \in R^s, j = 1, 2, \cdots, c$. The network's input data for classification of data $x_k \in R^s$, and the number of neurons in the network output for prototype initialization that is got from prototype number.

3.2 Fuzzy Clustering Neural Network Structure

Proper selection fuzzy clustering neural network model can make greatly improve clustering speed. This article chooses three-layer feed forward neural network model network, the first layer is the number of neurons equal to the dimension s of data, the second and third layer, the number of neurons is equal to the number of data c. Fuzzy clustering is shown in Figure 2.

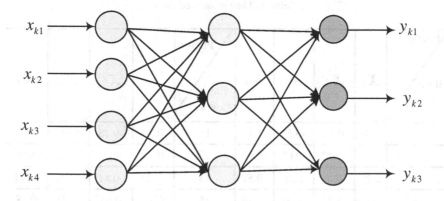

Fig. 2. Fuzzy clustering

The first layer and the second input and output data and weights are interval inside $[0,1]$. The first layer and the second neuron right connectivity for $\{\omega_{ij}, 1 \le i \le c, 1 \le j \le s\}$, which represents all kinds of clustering center. The second floor and the third between neurons connected right for constant. The first is input layer whose neurons are the sample data. The neurons of the second layer which realize all kinds of input sample and matching arithmetic. The neurons of the third layer for fuzzy output neurons, its output said the sample belongs to all kinds of input levels.

The paper is based on the analysis of the sample data, and uses what based on clustering center half grid extraction algorithm, which can predict the classification number and sample clustering center, and avoid classification number and sample clustering center of clustering effect.

3.3   The Simulation Experiment

All data normalized, data see Table 1.After normalized, we can determine inputs after meshing parameters for 0.2, the sample space grid rough classification. Input density threshold $\tau = 0.2$, Using the method of half grid to results which obtained from the rough division and then make meticulous division. Then we make the extraction of clustering center. The extraction of clustering center is as follows table 1:

1) According to the definition of rallying point extracted from clustering sample, the experimental samples taken the number of rallying point for 4. Namely, we must carry on the clustering of clustering prototype sample space number for 4.

2) With the maximum density from starting the rallying point, gradually determine sample clustering center. Get clustering center as shown in Table 2.

Table 1. The normalized data

Factor / Elements	x_1	x_2	x_3	x_4	Factor / Elements	x_1	x_2	x_3	x_4
1	-0.42	0.27	-0.55	-1.11	11	0.50	-0.18	0.40	-0.41
2	-0.83	-1.12	-0.67	-0.98	12	0.25	-0.92	0.59	-0.98
3	2.25	-0.58	2.51	-0.87	13	-0.25	-0.81	-0.04	-0.71
4	-0.83	-1.14	-0.65	-1.39	14	0.17	0.65	-0.21	0.28
5	0.00	2.87	-0.61	1.40	15	-0.58	-0.82	-0.47	-0.67
6	3.67	1.17	1.64	0.20	16	-0.58	0.36	-0.68	0.89
7	-0.50	-0.44	-0.47	-0.72	17	1.67	-1.06	2.88	-1.47
8	1.08	-0.65	1.4	-0.90	18	0.75	0.19	0.36	0.48
9	-0.58	-0.33	-0.53	0.01	19	0.17	0.61	-0.22	0.70
10	0.08	0.09	-0.09	0.30	20	-0.58	0.13	-0.63	0.37

Table 2. Clustering center

0.120	3.221	0.034	1.650
0.070	1.720	0.045	0.472

Using clustering center of income and rallying point number to initialize and design network structure. In the neural network, the output layer uses 4 neurons, initial adopted for 0.2, after training. Clustering results such as shown in Table 3.

Table 3. Clustering result

Class	No.
1	1,2,4,5,7,9,10,13,14,15,16,19,20
2	6,8
3	3,17
4	11,12,18

4 Results Analysis

In order to better a crime to civil classification and dark number and to provide investors high-precision crime dark several classification analysis basis and results, this paper adopts currently popular half grid extraction based on clustering center of fuzzy clustering method of neural network structure optimization clustering analysis, cluster effect is obvious. According to the analysis result, for the first kind of data in the crime, police detection dark number is not high; for the second, third and fourth such data in catch scale, police raids ability, the technical level have certain advantages, strong ability and have certain development potential, it can sum up experience and is worth reference and exchange experience.

5 Conclusion

This paper based on data which provided by the public security bureau for the sample space, the research work on around the crime space systems. Using fuzzy clustering neural network analysis method is realized crime data classification problem, prediction potential case occurrence of trend and development situation of the properties, and do analysis to the developing situation predict, have been supervised in advance ,which make crime cloth police crime prevention and control technology is more perfect. Based on the classification of the sample space analysis and simulation experiment results, it can be concluded that the above method in crime data analysis is of guiding. To sum up, this research for establishing harmonious society and maintain social stability and national development have very important significance.

Acknowledgements. Thanks to the support by National Natural Science Foundation of China (No. 60873042).

References

1. Yu, L.: For the first time in nearly a decade violent crime growth in China. China News Service (2010)
2. Arun, K., Sara, M.: Intelligent data analysis: developing new methodologies through pattern discovery and recovery. Fuzzy Neural Network Models for Knowledge Discovery 15, 103–119 (2009)

3. Memon, Q.A., Mehboob, S.: Crime investigation and analysis using neural nets. In: 7th International Multi Topic Conference, INMIC, pp. 346–350 (2003)
4. Grubesic, T.H.: On the application of fuzzy clustering for crime hot spot detection. Journal of Quantitative Criminology 22, 77–105 (2006)
5. Loia, V., Mattiucci, M., Senatore, S., Veniero, M.: Computer crime investigation by means of fuzzy semantic maps. Web Intelligence and Intelligent Agent Technologies, 183–186 (2009)
6. Li, C.: Defense space theory and crime prevention of environmental design beforehand. Science and Technology Journal of Huazhong University 21, 88–92 (2004)
7. Chen, Z. Q., Lin, S.Y., Zhang, H.Y.: Discuss with spatial analysis method downtown potential crime places. In: The Taiwan Geography Annual Meeting and Academic Seminar of Information Institute Proceedings (2005)
8. He, P.: Crime spatial analysis theory and prevention and control technology research. Modern Education Press, Beijing (2008)
9. Li, S.T., Kuo, S.C., Tsai, F.C.: An intelligent decision-support model using FSOM andrule extraction for crime pvention. Expert Systems with Applications 37, 7108–7119 (2010)
10. Du, S.P., Chen, F.H., Yan, Q.: Based on neural network of clustering analysis. Acoustic and Electronic Engineering 1, 189–194 (2000)
11. Kwang, H.L., Lee, K.M.: Fuzzy hyper-graph and fuzzy partition. IEEE Trans. SMC 1, 196–201 (1995)
12. Ruspini, E.H.: A new approach to clustering. Inf. Cont. 15, 22–32 (1969)
13. Tamura, S., Higuchi, S., Tanaka, K.: Pattern classification based on fuzzyrelations. IEEE SMC 1, 217–242 (1971)
14. Le, Z.: Fuzzy relation compositions and pattern recognition. Inf Sci. 89, 107–130 (1996)
15. Zou, K.Q.: The creative calculation based on neural network models. Computer Application and Research 9, 12–15 (2004)
16. Wu, Y.S., Long, W.: Modeling research of production logistics system based on fuzzy graph theory. Journal of Applied Sciences 4, 417–419 (2005)

Properties and Algorithms for Fuzzy Geometric Programming

Bing-yuan Cao

School of Mathematics and Information Science, Guangzhou University.
Key Laboratory of Mathematics and Interdisciplinary Sciences of Guangdong Higher
Education Institutes, Guangzhou University, 510006, China
caobingy@163.com

Abstract. Fuzzy reverse posynomial geometric programming, on the basis of previous work, is studied in the properties and algorithms with two algorithms advanced in this paper: a direct algorithm and a dual algorithm. Meanwhile, its optimal solution is proposed, and an improved imagination is proposed to the direct algorithm.

Keywords: Reversed posynomial, fuzzy geometric programming, algorithm, guess.

1 Introduction

In 1987, the author advanced a fuzzy posynomial geometric programming (FPGP) first [1]; In 1989, again he devoloped a fuzzy dual geometric progratnming (GP) [2]   Consider: [3]

$$(\tilde{P}) \qquad \widetilde{\min}\ g_0(x)$$
$$\text{s.t.}\ \ g_i(x) \lesssim 1 \quad (1 \leqslant i \leqslant p),$$
$$x > 0,$$

where $x = (x_1, x_2, \cdots, x_m)^{\mathrm{T}}$ is an $m-$dimensional variable vector, $g_i(x)$ are polynomials of x, i.e., $g_i(x) = \sum\limits_{k=1}^{J_i} v_{ik}(x) = \sum\limits_{k=1}^{J_i} c_{ik} \prod\limits_{l=1}^{m} x_l^{\gamma_{ikl}} (0 \leqslant i \leqslant p)$, and c_{ik} and $\gamma_{ikl}(1 \leqslant k \leqslant J_i, 0 \leqslant i \leqslant p, 1 \leqslant l \leqslant m)$ are fuzzy coefficients and fuzzy exponents, respectively, all of them being real numbers, with $c_{ik} > 0$, we call $(\tilde{P})$ an FPGP.

And we call

$$(\tilde{D}) \qquad \max\ \tilde{d}(w) = \prod_{i=0}^{p} \prod_{k=1}^{J_i} \left(\frac{\tilde{c}_{ik}}{w_{ik}}\right)^{w_{ik}} \prod_{i=1}^{p} \left(w_{ik}\right)^{w_{ik}}$$
$$\text{s.t.}\ \ w_{00} = 1$$
$$\Gamma^{\mathrm{T}} w = 0$$
$$w \geqslant 0$$

B.-Y. Cao and X.-J. Xie (Eds.): Fuzzy Engineering and Operations Research, AISC 147, pp. 51–61.
springerlink.com

a dual programming of $(\tilde{P})$, where $w = (w_{01}, \cdots, w_{0J_0}, \cdots, w_{p1}, \cdots, w_{pJ_p})^{\mathrm{T}}$ is a J−dimensional variable vector, $w_{i0} = w_{i1} + w_{i2} + \cdots + w_{iJ_i} (i = 0, 1, \cdots, p)$ and

$$
\Gamma = \begin{pmatrix}
\gamma_{011} & \cdots & \gamma_{01l} & \cdots & \gamma_{01m} \\
\cdots & & \cdots & & \cdots \\
\gamma_{0J_01} & \cdots & \gamma_{0J_0l} & \cdots & \gamma_{0J_0m} \\
\cdots & & \cdots & & \cdots \\
\gamma_{p11} & \cdots & \gamma_{p1l} & \cdots & \gamma_{p1m} \\
\cdots & & \cdots & & \cdots \\
\gamma_{pJ_p1} & \cdots & \gamma_{pJ_pl} & \cdots & \gamma_{pJ_pm}
\end{pmatrix}
$$

is a fuzzy exponent matrix. We stipulate $(w_{ik})^{w_{ik}}|_{w_{ik}=0} = 1$.

In Section 1, an FPGP model and dual one is advanced In Section 2, some main results of FPGP is introduced in these years. In Section 3, a solution is given to the fuzzy reversed posynomial geometric programming (FRPGP) problem and its Lagrange one. In Section 4, a direct algorithm is developed with dual one. In Section 5, research direction and guess is given.

2 Prevlous Main Results

In the past years, we have studied the FPGP before obtaining the following results:

1) Change FPGP $(\tilde{P})$ into a determination GP with parameter variates by means of fuzzy valued sets:

$$
\max_{x \in R^+}\{\tilde{Y}^*(x, \beta) = \min_{0 \le i \le p} \tilde{A}_i^*(x, \beta)\}, \tag{1}
$$

where $\tilde{Y}^* = \tilde{A}_0^* \bigcap \tilde{F} = \tilde{A}_0^*(x, \beta)$ is an objective function defined on R concerning fuzzy optimal solution set of $\tilde{A}_0^*$; $\tilde{A}_i^*(x, \beta)$ is a restraint function on R about fuzzy feasible solution sets of $\tilde{A}_i^*(1 \le i \le p)$.

The writer comes up with the problem below:

i)[3] It doesn't increase a difficult degree when changing $(\tilde{P})$ into (1).

ii)[1][2] When $\tilde{A}_i^*(0 \le i \le p)$ continues and strictly non-decreases, there exist the following states.

a) Any fuzzy local minimum solution (FLMs) in $(\tilde{P})$ is its fuzzy global minimum solution (FGMS).

b) A strict FLMS of $(\tilde{P})$ is its strict FGMS.

c) If $(\tilde{P})$ is a strong fuzzy convex programming, its FLMS is unique FGMS.

iii)[1][2][3] Give a prime algorithm and a dual one to $(\tilde{P})$ and we adjust some problems in algorithm in [4].

iv) Obtaining a fuzzy strong dual result in the FPGP.

2) The study of geometirc programming model with fuzzy coefficient.

When coefficients in $(\tilde{P})$ are interval form fuzzy number [6], $(\tilde{P})$ is equivalent to

$$\min \bigcup_{\alpha \in [0,1]} \alpha \Big[\sum_{k=1}^{J_0} c_{ik\alpha}^- \prod_{l=1}^{m} x_l^{\gamma_{0kl}}, \ \sum_{k=1}^{J_0} c_{ik\alpha}^- c_{ik\alpha}^+ \prod_{l=1}^{m} x_l^{\gamma_{0kl}} \Big]$$

$$\text{s.t. } \bigcup_{\alpha \in [0,1]} \alpha \Big[\sum_{k=1}^{J_i} c_{ik\alpha}^- \prod_{l=1}^{m} x_l^{\gamma_{ikl}}, \ \sum_{k=1}^{J_i} c_{ik\alpha}^+ \prod_{l=1}^{m} x_l^{\gamma_{ikl}} \Big],$$

$$\subseteq \bigcup_{\alpha \in [0,1]} \alpha [1_\alpha^-, 1_\alpha^+] \ (1 \leqslant i \leqslant p),$$

$$x > 0.$$

3) GP with fuzzy variation [13].

i) By means of a disposal of T-fuzzy data in paper [12] like the disposal of fuzzy linear programming in [13], we build a GP model with T-fuzzy data.

ii) GP with trapezoidal fuzzy variation.

4) Having proved feasible multi-objective fuzzy geometric programming (MOFGP) model advanced in [15] by two operators, we study the MOFGP of soft constraint and fluctuating objective or that with fuzzy coefficients. [16]

5) In application, the author succeeds in putting the theory into optimal selection of waste-water-disposal scheme in electric plant [19] and the option of economical supply radius in transformer substation. The example shows that, the waste-water-disposal model built by this method can get rid of waste water BOD 98% through 4 steps, which 15 inferior to nation standard (97.1 %) and annual total expens is smallest with only RMB 406.85 (thousand). More and more experts in this field pay attention to paper [19] and [20].

3 FRPGP and Its Lagrange Problem

1) FRPGP and Dual Form.

Definition 1. Let us call the GP defined in paper [1] a prime posynomial GP. While we call an inequality constraint problem with opposite direction

$$(\tilde{P}) \qquad \widetilde{\min}\ \tilde{g}_0(x)$$
$$\text{s.t. } \tilde{g}_i(x) \precsim 1 (1 \leqslant i \leqslant p') \tag{2}$$
$$\tilde{g}_i(x) \succsim 1 (p'+1 \leqslant i \leqslant p)$$
$$x > 0$$

an FRPGP. Here, $x = (x_1, x_2, \cdots, x_m)^{\mathrm{T}}$ is an $m-$dimensional variable vector, "$\precsim$" and "$\succsim$" correspond to " approximately $\leqslant$ " and " approximately $\geqslant$", respectively, and all $\tilde{g}_i(x) = \sum_{k=1}^{J_i} \tilde{v}_{ik}(x)(0 \leqslant i \leqslant p)$ are fuzzy posynomials of x, whereas

$$\tilde{v}_{ik}(x) = \begin{cases} \tilde{c}_{ik} \prod_{l=1}^{m} x_l^{\gamma_{ikl}}, & (1 \leqslant k \leqslant J_i; 0 \leqslant i \leqslant p'), \\ \tilde{c}_{ik} \prod_{l=1}^{m} x_l^{-\gamma_{ikl}}, & (1 \leqslant k \leqslant J_i; p'+1 \leqslant i \leqslant p), \end{cases}$$

where its coefficients $\tilde{c}_{ik} \geqslant 0$ may be freely fixed in a closed value interval $[c_{ik}^-, c_{ik}^+](0 \leq i \leqslant p, 1 \leqslant k \leqslant J_i, 1 \leq l \leq m)$, respectively. Here $c_{ik}^- < c_{ik}^+$, $c_{ik}^- \geq 0$ and c_{ik}^+ belonging to left and right endpoints in the intervals respectively are all real numbers. $-\gamma_{ikl}$ are exponents corresponding to each item x_l in the reversed direction inequality $\tilde{g}_i(x) \gtrsim 1$.

Theorem 1. *Let $\tilde{A}_i(0 \leq i \leq p)$ be a continual and strictly monotonous fuzzy valued function, a dual programming of $(\tilde{P})$ is*

$$(\tilde{D}) \quad \widetilde{\max}\ \tilde{d}(w) = \left(\frac{\tilde{a}_{0k}}{w_{00}}\right)^{w_{00}} \prod_{i=0}^{p'} \prod_{k=1}^{J_i} \left(\frac{\tilde{c}_{ik}}{\tilde{a}_{ik}w_{ik}}\right)^{w_{ik}} \prod_{i=p'+1}^{p} \prod_{k=1}^{J_i}$$

$$\left(\frac{\tilde{c}_{ik}\tilde{a}_{ik}}{w_{ik}}\right)^{-w_{ik}} \prod_{i=1}^{p'} w_{i0}^{w_{i0}} \prod_{i=p'+1}^{p} w_{i0}^{-w_{i0}}$$

$$s.t. \quad w_{00} = \sum_{k=1}^{J_0} w_{0k} = 1,$$

$$\Gamma^{*T} w = 0,$$

$$w \geqslant 0,$$

where Γ^* still represents an exponent matrix, i.e.,

$$\Gamma^* = \begin{pmatrix} \gamma_{011} \cdots & \gamma_{01l} \cdots & \gamma_{01m} \\ \cdots & \cdots & \cdots \\ \gamma_{0J_01} \cdots & \gamma_{0J_0l} \cdots & \gamma_{0J_0m} \\ \cdots & \cdots & \cdots \\ \gamma_{p'J_{p'}1}^{-1} \cdots & \gamma_{p'J_{p'}l}^{-1} \cdots & \gamma_{p'J_{p'}m}^{-1} \\ -\gamma_{p'+1J_{p'+1}1}^{-1} \cdots & -\gamma_{p'+1J_{p'+1}l}^{-1} \cdots & -\gamma_{p'+1J_{p'+1}m}^{-1} \\ \cdots & \cdots & \cdots \\ -\gamma_{pJ_p1} \cdots & -\gamma_{pJ_pl} \cdots & -\gamma_{pJ_pm} \end{pmatrix}.$$

Here $w = (w_{00}, w_{01}, \ldots, w_{0J_0}, w_{p1}, \ldots, w_{pJ_p})^T$ is a J'–dimensional fuzzy variable vector $(J' = 1 + J_0 + \cdots + J_p)$, and $w_{i0} = w_{i1} + w_{i2} + \cdots + w_{iJ_i}$; $-w_{ik}$ and $-w_{i0}$ denote a reversed direction inequality $\tilde{g}_i(x) \gtrsim 1$ corresponding to factors $\left(\frac{\tilde{c}_{ik}}{w_{ik}}\right)^{-w_{ik}}$ and $w_{i0}^{-w_{i0}}$ in the upper-right-corner exponent.

It is easy to prove the theorem in the similar way in Theorem 4.1 in [3].

2) Extension of FRPGP.

i) Developing a GP solution with (·,c) type fuzzy parametric fuzzy one [9]. If an objective expectation value b_0 is presented by decision makers, then (2) can be turned into the following form:

objection
$$\tilde{g}_0 = \tilde{b}_0 G_{00} - \sum_{k=1}^{J_0} \tilde{c}_{0k} G_{0k}(x) \gtrsim 0,$$

constraint
$$\begin{cases} \tilde{g}_i = G_{00} - \sum_{k=1}^{J_i} \tilde{c}_{ik} G_{ik}(x) \gtrsim 0 & (1 \leqslant i \leqslant p'), \\ -\tilde{g}_i = -G_{00} + \sum_{k=1}^{J_i} \tilde{c}_{ik} G_{ik}(x) \gtrsim 0 & (p'+1 \leqslant i \leqslant p), \\ x > 0, \end{cases}$$
$$G_{00} = 1, G_{ik}(x) = \prod_{l=1}^{m} x_l^{\gamma_{ikl}} \qquad (0 \leqslant i \leqslant p).$$

The fuzzy functions are denoted by

$$\tilde{g}_i(x) = \tilde{c}G = (\alpha^{\mathrm{T}} G_i, c^{\mathrm{T}} G_i);$$

where $G_i = (\prod_{l=1}^{m} x_l^{\gamma_{i1l}}, \prod_{l=1}^{m} x_l^{\gamma_{i2l}}, \dots, \prod_{l=1}^{m} x_l^{\gamma_{iJ_il}})^{\mathrm{T}}$ and its membership function is

$$\mu_{\tilde{g}_i}(g_i) = \begin{cases} 1 - \frac{|g_i - \alpha^{\mathrm{T}} G_i|}{c^{\mathrm{T}}|G_i|}, & G_i \neq 0, \\ 1, & G_i = 0, g_i = 0, \\ 0, & G_i = 0, g_i \neq 0, \end{cases}$$

where $|G_i| = (|G_{i1}|, \dots, |G_{iJ_i}|)^T$ and $\mu_{\tilde{g}_i}(g_i) = 0$ for $c^T|G_i| \leqslant |g_i - \alpha^T G_i|$.

$$(2) \Longleftrightarrow \max h$$
$$\text{s.t. } (\alpha_{ik}^* - hc_{ik}^*)^{\mathrm{T}} G_i(x) \geqslant 0 \; h \in [0,1],$$
$$x > 0 \; (1 \leqslant k \leqslant J_i, 0 \leqslant i \leqslant p).$$

ii) Developing a GP solution with flat fuzzy parametric fuzzy one. Turning the problem into a defined perameter reversed posynomial GP with an effective algorithm by use of Paper [9] $PD(\tilde{F}, \tilde{g}) = \dfrac{\overline{F} - \sum\limits_{k=1}^{J_0} c_{0k}^{-} \prod\limits_{l=1}^{m} x_l^{\gamma_{0kl}}}{\sigma_{\overline{F}}^{+} + \sum\limits_{k=1}^{J_0} \sigma_{c_{0k}}^{-} \prod\limits_{l=1}^{m} x_l^{\gamma_{0kl}}}$. Let $PD(\cdot) \geq \theta$. Then

$$(2) \Longleftrightarrow \min \frac{\sum\limits_{k=1}^{J_0} c_{0k}^{-} \prod\limits_{l=1}^{m} x_l^{\gamma_{0kl}} - \overline{F}}{\sum\limits_{k=1}^{J_0} \sigma_{c_{0k}}^{-} \prod\limits_{l=1}^{m} x_l^{\gamma_{0kl}} + \sigma_{\overline{F}}^{+}}$$
$$\text{s.t. } \sum_{k=1}^{J_i} c_{ik}^{+} \prod_{l=1}^{m} x_l^{\gamma_{ikl}} \leqslant 1^{+},$$
$$\sum_{k=1}^{J_i} (c_{ik}^{+} + \sigma_{c_{ik}}^{+}) \prod_{l=1}^{m} x_l^{\gamma_{ikl}} \leqslant 1^{+} + \sigma_1^{+},$$
$$\sum_{k=1}^{J_i} (c_{ik}^{-} - \sigma_{c_{ik}}^{-}) \prod_{l=1}^{m} x_l^{\gamma_{ikl}} \geqslant 1^{-} - \sigma_1^{-},$$
$$\sum_{k=1}^{J_i} c_{ik}^{-} \prod_{l=1}^{m} x_l^{\gamma_{ikl}} \geqslant 1^{-} \; (1 \leqslant i \leqslant p),$$
$$x_1, x_2, \dots, x_m > 0.$$

3) Lagrange Problem in FRPGP.

Definition 2. Let us call the GP defined in paper [1] a prime posynomial GP. While we call an inequality constraint problem with opposite direction $\tilde{M}_p = \{\inf \tilde{g}_0(x) | \text{s.t. } \tilde{g}_i(x) \lesssim 1 \ (1 \leqslant i \leqslant p'); \tilde{g}_i(x) \gtrsim 1 \ (p'+1 \leqslant i \leqslant p), x > 0\}$.

Definition 3. Suppose $\tilde{A}_i (0 \leqslant i \leqslant p)$ to be a CSM function; we call $\tilde{g}_i(x^*) = 1$ the fuzzy equality, if its membership is $\tilde{A}_i(\tilde{g}_i(x) - 1) = 1$. We call the problem a fuzzy Lagrange problem for (2), i.e., finding a fuzzy feasible solution x^* to (2) and $\mu^* = (\mu_1^*, \mu_2^*, \ldots, \mu_p^*)^\mathrm{T} \geqslant 0$ satisfying $\mu_i^*(\tilde{g}_i(x^*) - 1) = 0 (1 \leqslant i \leqslant p)$, such that a fuzzy

Lagrange function $\tilde{L}(x, \mu) = \tilde{g}_0(x) + \sum_{i=1}^{p'} \mu_i(\tilde{g}_i(x) - 1) + \sum_{i=p'+1}^{p} \mu_i(1 - \tilde{g}_i(x))$

satisfies $\nabla_x \tilde{L}(x^*, \mu^*) = 0$.

Theorem 2. *Let x^* be a fuzzy feasible solution to (2). If $\mathcal{I} = \{i | \tilde{g}_i(x^*) = 1, 0 \leqslant i \leqslant p\}$ denotes a subscript set of fuzzy effective constraints and if $\tilde{A}_i(0 \leqslant i \leqslant p)$ is a CSM function, then there exists μ^* such that (x^*, μ^*) becomes a fuzzy solution to the Lagrange problem $\Longleftrightarrow$ to all vectors there exists $x > 0$ satisfying*

$$\sum_{l=1}^{m} \tilde{\Gamma}_{il}(\log x_l - \log x_l^*) \lesssim 0 \ (i \in \mathcal{I}); \tag{3}$$

then

$$\tilde{g}_0(x^*) \lesssim \tilde{g}_0(x). \tag{4}$$

Here

$$\tilde{\Gamma}_{il} = \sum_{k=1}^{J_i} \gamma_{ikl} \tilde{v}_{ik}(x^*) \ (i \in \mathcal{I}, 1 \leqslant l \leqslant m). \tag{5}$$

Proof. Let $\tilde{A}_i(0 \leqslant i \leqslant p)$ [3] be a CSM function. Then $(2) \Longleftrightarrow$

$$\begin{aligned}
& \min \ \tilde{A}_0(\tilde{g}_0(x)) \\
& \text{s.t. } \tilde{A}_i(\tilde{g}_i(x) - 1) \geqslant \alpha \ (1 \leqslant i \leqslant p'), \\
& \qquad \tilde{A}_i(1 - \tilde{g}_i(x)) \geqslant \alpha \ (p'+1 \leqslant i \leqslant p), \\
& \qquad x > 0, \alpha \in [0, 1],
\end{aligned} \tag{6}$$

while $\mathcal{I} \Longleftrightarrow \mathcal{I}' = \{i | \tilde{A}_i(\tilde{g}_i(x^*) - 1) = 0 \ (1 \leqslant i \leqslant p)\}$,

$$(3) \Longleftrightarrow \tilde{A}_i[\Gamma_{il}(\log x_l - \log x_l^*)] \geqslant \alpha (i \in \mathcal{I}'), \tag{7}$$

$$(4) \Longleftrightarrow \tilde{A}_0[\tilde{g}_0(x^*) - \tilde{g}_0(x)] \geqslant \alpha. \tag{8}$$

From the condition of the theorem, we know that x^* is a fuzzy feasible solution to (2), which is equivalent to $\bar{x}^*$ being a parameter feasible solution to (6) [3]. Therefore, for subscript set $\mathcal{I}'$ and any $\alpha, \beta \in [0, 1]$ there exists μ^* enabling $(\bar{x}^*, \mu^*)$ to be a Lagrange problem solution with parameters $\alpha, \beta \iff$ all vectors $x > 0$ satisfying (7) can satisfy (8) from Theorem 1.4.5 in [17].

The theorem holds due to the arbitrariness of α, β on $[0, 1]$.

Proposition 1. *Let $\tilde{A}_i(0 \leqslant i \leqslant p)$ be the CSM function. On the assumption of a constraint complete lattice, a local fuzzy optimum solution to FRPGP (2) must be its part of a fuzzy solution to the Lagrange problem.*

It is easy to prove the proposition in the similar way of Theorem 2. Its converse proposition do not hold (example omitted).

4 Algorithm

There exist two algorithms in FRPGP, a balanced solution one and a direct one. For the former we use fuzzy harmonic mean inequaltity and property of fuzzy balanced solution, change (2) into fuzzy GP ($\tilde{P}(\alpha)$) with weight vector $\alpha(> 0)$, and then find it, which we will discuss in another paper. Here we only study the latter.

Lemma 1. [18] *If constraint $\tilde{A}$ is a strictly convex fuzzy set on R^n, i.e., $\forall \lambda \in (0, 1), x^1, x^2 \in R^n, x^1 \neq x^2, \tilde{A}[\lambda x^1 + (1 - \lambda)x^2] > \tilde{A}(x^1) \wedge \tilde{A}(x^2)$, then $\tilde{M} \triangleq \sup_{x \in A_0} |\tilde{B}(\tilde{g}_0(x))|$ is called continuouation.*

Lemma 2. *If $\tilde{M}$ is differentiable to arbitrate $\lambda \in (0, 1)$ concerning $x \in D \subset R^n$, then it is continuous on D.*

It can be proved that there exists a unique fixed point

$$\alpha = \tilde{M}(\bar{\alpha}) = \sup_{x \in A_\alpha} |\tilde{B}(\tilde{g}_0(x))|$$

if $\tilde{M}$ is a Continous function on $[0, 1]$.

I. Direct Algorithm

Let $g_i(x)$ be differentiable, then we can get a direct algorithm step for (2):

1^0 Let k = 1, and determine α, and h by $l - hd = \alpha$.

2^0 Calculat $\tilde{M}^{(k)}(\bar{\alpha}) = \sup_{x \in A_{\alpha_k}} |\tilde{B}(\tilde{g}_0(x))|$.

3^0 Calculate $\varepsilon_k = \alpha_k - \tilde{M}^{(k)}$, If $\varepsilon_k > \varepsilon$, we go to Step 2^0, otherwise to 5^0.

4^0 Select $r_k \in [0, 1]$ properly. Let $\alpha_{k+1} = \alpha_k - r_k \varepsilon_k$ and $k = k + 1$. Go to 2^0.

5^0 When $\bar{\alpha} = \alpha_k$, find x such that $\sup_{x \in A_{\alpha_k}} |\tilde{B}(\tilde{g}_0(x))| = \tilde{M}^{(k)}(\bar{\alpha})$. It is properly to take $\alpha_k \in [0.9, 1]$ when $\tilde{g}_0(x)$ strictly increases monotonously, otherwise to take an $\alpha_k \in [0.75, 0.9]$ If $b(> 0)$ is very larger, larger, smaller, very small, it is properly to take 0.02, 0.2, 2 and 20. To r_k selection, when $\varepsilon_l \gg \varepsilon_2$, we may choose $r_k = 0.5$. If $\varepsilon_1 \searrow \varepsilon_2$ changes a little, we can take $r_k \in [0.618, 1]$, if $\varepsilon_1 \ll \varepsilon_2$, take $r_k \in [0.382, 0.4]$ properly.

II. Dual Algorithm

We shall exhibit steps for fuzzy dual algorithms to the FRPGP $(\tilde{P})$ in the case of a DD $D > 0$:

1^0 Change $(\tilde{P})$ into two determined reverse GP [11] as follows:

$$
\begin{aligned}
\min\ &g_0(\bar{x}') \\
\text{s.t.}\ &g_i(\bar{x}') \leqslant 1\,, (1 \leqslant i \leqslant p') \\
&g_i(\bar{x}') \geqslant 1\,, (p'+1 \leqslant i \leqslant p), \beta \in [0,1] \\
&x > 0
\end{aligned}
$$

and

$$
\begin{aligned}
\min\ &g_0(\bar{x}') \\
\text{s.t.}\ &g_i(\bar{x}') \leqslant 1 + d_i,\ (1 \leqslant i \leqslant p'), \\
&g_i(\bar{x}') \geqslant 1 + d_i, (p'+1 \leqslant i \leqslant p), \beta \in [0,1], \\
&x > 0.
\end{aligned}
$$

2^0 Write out their dual form

$$
\max\ d(\bar{w}) = \left(\frac{1-\alpha}{\bar{w}_{00}}\right)^{\bar{w}_{00}} \prod_{i=0}^{p'} \prod_{k=1}^{J_i} \left(\frac{\tilde{c}_{ik}^{-1}(\beta)/(1+d_i)}{\bar{w}_{ik}}\right)^{\bar{w}_{ik}}
$$
$$
\prod_{i=p'+1}^{p} \prod_{k=1}^{J_i} \left(\frac{\tilde{c}_{ik}^{-1}(\beta)/(1+d_i)}{\bar{w}_{ik}}\right)^{-\bar{w}_{ik}} \prod_{i=1}^{p'} \bar{w}_{i0}^{\bar{w}_{i0}} \prod_{i=p'+1}^{p} \bar{w}_{i0}^{-\bar{w}_{i0}} \tag{9}
$$

$$
\begin{aligned}
\text{s.t.}\ &\bar{w}_{00} = 1, \\
&\Gamma^{\mathrm{T}} \bar{w} = 0, \alpha, \beta \in [0,1], \\
&\bar{w} \geqslant 0,
\end{aligned}
$$

where d_i is selected as 0 or $d_i(> 0)$.

3^0 Find (9) for $d_i = 0$ or for $d_i = d_i$, and we can find two optimal values g_0^* and g_0^{*1}. As $D > 0$, we can find a feasible solution to (9) by means of optimizing methods such as a brief gradient, a gradient projection, feasible direction or variable elimination. Especially for $D = 1$, we apply search methods (or methods of 0.618, twice or three times in interpolation) into obtaining its optimal solution.

4^0 Find an ordinary GP

$$
\begin{aligned}
\max\ &\alpha = \min(1 - \alpha) \\
\text{s.t.}\ &g_0(\bar{x}') \leqslant \tilde{B}_0^{-1}(\alpha), \\
&g_i(\bar{x}') \leqslant \tilde{B}_i^{-1}(\alpha)(1 \leqslant i \leqslant p'), \\
&g_i(\bar{x}') \geqslant \tilde{B}_i^{-1}(\alpha)(p'+1 \leqslant i \leqslant p), \alpha, \beta \in [0,1], \\
&x > 0,
\end{aligned}
$$

where $z_0 = g_0^{*1}$ (we had better suppose $g_0^{*1} < g_0^*$), $d_0 = g_0^{*1} - g_0^*$ and its dual programming denotes (9). And (9) optimal parameter solutions are obtained by using one method of Step 3 to (9); accordingly the value of α, β is determined by $d(\bar{w}) = 1 - \alpha$.

5^0 According to formula

$$\bar{w}_{ik} = \begin{cases} v_{00}(\bar{x})/(1-\alpha), & (i=0, k=0, \alpha \neq 0), \\ \bar{w}_{i0}v_{ik}(\bar{x}), & (1 \leqslant k \leqslant J_i, 0 \leqslant i \leqslant p), \end{cases}$$

we find $\bar{x}$, i.e., it is the optimal solution to (9) with respect to parameters α, β, respectively. In application, we stop when we come to this step.

6^0 $\forall \alpha, \beta \in [0,1]$, a fuzzy optimal solution x to $(\tilde{P})$ is compounded by $\bar{x}$.

5   Guess and Further Research

The research just begins in FRPFG, and it is more difficult and complicated, needing a lot of work, in comparison with the FPG.

1   First, let us advance a guess.

Generally, the deformed prime programming in FRPGP is not a fuzzy convex programming of z [11], hence many good properties fail to remain in the FPGP, it is difficult for us to study the FRPGP (Example omitted), but the writer makes some progress in research for it (the research concerned in another paper).

In Proposition 1, if x^* is a part of the solution to Lagrange problem, then when (2) is fuzzy convex, or $p = M$ in (2), x^* is a global fuzzy optimal solution to the programming. But in (2), when $p \neq M$, x^* is not necessarily a global fuzzy optimal solution.

The FRPGP (2) is not the necessary a fuzzy convex programming, but it can be proved that, under a certain condition, a local optimal solution (or satisfactory) of the FRPGP is its golbal optimal (or satisfactory) one.

2   Advance a few problems for thinking.

i) Iterative algorithm is used in FPGP for solution, depending on the choice of membership function. If the choice is improper, the result concludes a difference dramatically. Is it proper to adopt the membership function like the following since the FGP is composed of exponent polynomial?

$$\tilde{B}_i(x) = \begin{cases} 1, & g_i(x) \leqslant 1 \\ e^{-\frac{1}{d_i}\left(g_i(x)-1\right)}, & 1 < g_i(x) \leqslant 1 + d_i \end{cases}$$

for $1 \leqslant i \leqslant p'$, and

$$\tilde{B}_i(x) = \begin{cases} 0, & g_i(x) \leqslant 1 \\ 1 - e^{-\frac{1}{d_i}\left(g_i(x)-1\right)}, & 1 < g_i(x) \leqslant 1 + d_i \end{cases}$$

for $p' + 1 \leqslant i \leqslant p$, where $d_i \geqslant 0$ is a maximum flexible index of $i-$th function $g_i(x)$.

ii)[7]   Building a GP model with L-R coefficient. Let the coefficient $(\tilde{P})$ be an L-R number and $\tilde{c}_{ik} = (c_{ik}, \underline{c}_{ik}, \overline{c}_{ik})_{LR}$, as $L = R$, then $(\tilde{P})$ is turned into a reversed posynomial mmultiobjective GP

$$\min \ g_0(x), \max \ \underline{g}_0(x), \min \ \overline{g}_0(x)$$
$$\text{s.t.} \ \ g_i(x) \leqslant 1, \ \ \underline{g}_i(x) \geqslant 1, \ \ \overline{g}_i(x) \leqslant \overline{1},$$
$$x > 0 \ (1 \leqslant i \leqslant p),$$

where $\tilde{g}_i^*(x) = (\sum\limits_{k=1}^{J_i} c_{ik} \prod\limits_{l=1}^{m} x_l^{\gamma_{ikl}}, \sum\limits_{k=1}^{J_i} \underline{c}_{ik} \prod\limits_{l=1}^{m} x_l^{\gamma_{ikl}}, \sum\limits_{k=1}^{J_i} \overline{c}_{ik} \prod\limits_{l=1}^{m} x_l^{\gamma_{ikl}})_{LR}$ $(0 \leqslant i \leqslant p)$ and $g_i^*(x)$ is $(g_i^*(x), \underline{g}_i^*(x), \overline{g}_i^*(x))$, $\tilde{c}_{ik}^*$ is $(c_{ik}, \underline{c}_{ik}, \overline{c}_{ik})$ and pointing out that a fuzzy GP can be turned into a multiobjective linear programming solution by means of fuzzy geometric inequality.

When $\tilde{g}_i(x) = \sum\limits_{k=1}^{J_i} \tilde{v}_{ik}(x)(0 \leqslant i \leqslant p)$ are fuzzy posynomials of x, where its coefficients $\tilde{c}_{ik} \geqslant 0$ and exponents $\tilde{\gamma}_{ikl}$ may be freely fixed in a closed value interval $[c_{ik}^-, c_{ik}^+], [\gamma_{ikl}^-, \gamma_{ikl}^+](0 \leq i \leqslant p, 1 \leqslant k \leqslant J_i, 1 \leq l \leq m)$, respectively, where $c_{ik}^- < c_{ik}^+$ and $\gamma_{ikl}^- < \gamma_{ikl}^+$; c_{ik}^-, c_{ik}^+ and $\gamma_{ikl}^-, \gamma_{ikl}^+$ belonging to left and right endpoints in the intervals respectively are all real numbers. How do we solve?

6 Conclusion

We have obtained many properties and some algorithms to FPGP. Can they be extended and transplanted into the FRPGP? The writer thinks most of them can do so and some can be done also after changing in the future research, wishing more scholars who are interested in it do more hard researching work on it.

Acknowledgement. Thanks to the support by National Natural Science Foundation of China (No. 70771030).

References

1. Cao, B.Y.: Solution and theory of questions for a kind of fuzzy positive geometric program. In: Proc. of 2nd IFSA Congress, Tokyo, vol. 1, pp. 205–208 (1987)
2. Cao, B.Y.: Study of fuzzy positive geometric programming dual form. In: Proc. of 3rd IFSA Congress, Seattle, pp. 775–778 (1989)
3. Cao, B.Y.: Fuzzy geometric programming (I). Fuzzy Sets and Systems 53, 135–154 (1993)
4. Asai, K.: An introduction to the theory of fuzzy systems. Peking Norm. University Press, Peking (1982)
5. Cao, B.Y.: Extended fuzzy geometric programming. J. of Mathematics(USA) 2, 285–293 (1993)
6. Cao, B.Y.: Fuzzy strong dual results for fuzzy posynomial geometric programming. In: Proc. of 3rd IFSA Congress, Seoul, vol. 1, pp. 588–591 (1995)

7. Cao, B.Y.: Posynomial geometric programming with L-R fuzzy coefficients. Fuzzy Sets and Systems 67, 267–276 (1994)
8. Cao, B.Y.: Further study of posynomial geometric programming with fuzzy coefficients. Mathematics Applicata 5(4), 119–120 (1992)
9. Cao, B.Y.: The study of geometric programming with $(\cdot, c)$-fuzzy parameters. J. of Changsha, Univ. of Electric Power (Natural Sci. Ed.) 1, 15–21 (1995)
10. Cao, B.Y.: Study for a kind of regression forecasting model with fuzzy da-tums. J. of Mathematical Statistics and Applied Probability 4(2), 182–189 (1989)
11. Cao, B.Y.: Fuzzy geometric programming. Kluwer Academic Publishers (2002)
12. Cao, B.Y.: Study on non-distinct self-regression forecast model. Chinese Sci. Bull. 35(13), 1057–1062 (1990); (also to see Kexue Tong-bao, 34(17), 1291–1294 (1989))
13. Cao, B.Y.: New model with T-fuzzy variations in linear programming. Fuzzy Sets and Systems 78, 289–292 (1996)
14. Cao, B.Y.: Research for a geometric programming model with T-fuzzy variable. J. of Fuzzy Mathematics 5(3), 625–632 (1997)
15. Verma, R.K.: Fuzzy geometric programming with several objective function. Fuzzy Sets and Systems 35, 115–120 (1990)
16. Cao, B.Y.: Types of non-distinct multiobjective geometric programming. Hunan Annals of Mathematics 15(1), 99–106 (1995)
17. Wu, F., Yuan, Y.Y.: Geometric programming. Math. in Practice and Theory, 1–2 (1982)
18. Tanaka, H., et al.: On fuzzy mathomatical programming. J. Cybern. 3 (4), 37–46 (1973)
19. Cao, B.Y.: Fuzzy geometric programming optimum seeking of scheme for waste-water disposal in power plant. In: Proc. of FUZZ-IEEE/IFES 1995, pp. 793–798 (1995)
20. Yu, Y.Y., Cao, B.Y., et al.: The application of geometric and fuzzy geometric programming in option of economic supply radius of transformer substation. In: Proc. of the ICIK 1995, pp. 245–249 (1995)
21. Wilde, D.J., Beightler, C.S.: Foundations of optimization, pp. 76–109. Prentice Hall Co. Inc., Englewood Cliffs (1967)
22. Cao, B.Y.: Optimal models and methods with fuzzy quantity. Springer, Heidelberg (2010)

A Kind of Fuzzy Relation Programming with Fuzzy Coefficients

Xiao-wen Zhou and Bing-yuan Cao[*]

School of Mathematics and Information Science, Guangzhou University,
Key Laboratory of Mathematics and Interdisciplinary Sciences of Guangdong
Higher Education Institutes, Guangzhou University,
Guangzhou, Guangdong 510006, China
Zhouxiaowen14@126.com, caobingy@163.com

Abstract. A new fuzzy relation programming with fuzzy coefficients using max-product composition is considered. Since the object is a fuzzy relation function which the coefficients are also fuzzy numbers, we convert the problem into a fuzzy relation geometric programming by using a linear ranking function. Then we study the fuzzy relation geometric programming and capture the optimal solution. For illustration purpose, an numerical example is provided.

Keywords: Fuzzy relation programming, max-product composition, fuzzy relation geometric programming, triangular fuzzy number, ranking function.

1 Introduction

The notion of fuzzy relation equations based upon the max-min composition was first investigated by Sanchez [13]. Then fuzzy relation equations with max-product composition was proposed by Cao [1] in 1987. In 1998, Bourke and Fisher [17] extended the study of an inverse solution of a system of fuzzy relation equations with max-product composition. Their results showed that the complete sets can be characterized by one maximum solution and a number of minimal solutions. A fuzzy relation geometric programming was proposed by Yang and Cao [5]. They discussed optimal solutions with two kinds of object functions cased on different fuzzy operators.

Motivated by the work [6], we are interested in studying the optimization problem with a fuzzy relation objective function subject to a system of fuzzy relation equations with the max-product composition which the all coefficients are fuzzy numbers. In order to solve this optimization problem, we consider a ranking function [15] to order triangular fuzzy numbers. Then the optimization problem is converted into a classical fuzzy relation geometric programming. By solving the fuzzy relation geometric programming, we can obtain the optimal solution to the original problem which the all coefficients are triangular fuzzy numbers. To illustrate the proposed method, numerical example is solved.

---

[*] Corresponding author.

B.-Y. Cao and X.-J. Xie (Eds.): Fuzzy Engineering and Operations Research, AISC 147, pp. 63–73.
springerlink.com   © Springer-Verlag Berlin Heidelberg 2012

2 Preliminaries

In this section, we review some necessary backgrounds and notions of fuzzy sets theory (taken from [14, 15]).

Definition 2.1 [14]. *The characteristic function* μ_A *of a crisp set* $A \subseteq X$ *assigns a value either 0 or 1 to each member in X. This function can be generalized to a function* $\mu_{\tilde{A}}$ *such that the value assigned to the element of the universal set X fall within a specified range i.e.* $\mu_{\tilde{A}} : X \to [0,1]$. *The assigned value indicate the membership grade of the element in the set* A.

The function $\mu_{\tilde{A}}$ *is called the membership function and the set*
$\tilde{A} = \{(x, \mu_{\tilde{A}}(x)) \mid x \in X\}$ *defined by* $\mu_{\tilde{A}}(x)$ *for each* $x \in X$ *is called a fuzzy set.*

We denote the set of all triangular fuzzy numbers by $F(R)$.

Definition 2.2 [14]. *A fuzzy number* $\tilde{A} = (\alpha, a, \beta)$ *is said to be a triangular fuzzy number if its membership function is given*

$$\mu_{\tilde{A}}(x) = \begin{cases} \dfrac{x-\alpha}{a-\alpha}, \alpha \le x \le a; \\ \dfrac{x-\beta}{\beta-a}, a \le x \le \beta; \\ 0, \qquad \text{otherwise.} \end{cases}$$

Definition 2.3 [14]. *Let* $\tilde{A} = (\alpha, a, \beta)$ *and* $\tilde{B} = (\gamma, b, \theta)$ *be two triangular fuzzy numbers. Then*

(i) $\tilde{A} + \tilde{B} = (\alpha + \gamma, a + b, \beta + \theta)$

(ii) $\tilde{A} - \tilde{B} = (\alpha + \theta, a - b, \beta + \gamma)$,

(iii) $-\tilde{A} = (-\beta, -a, -\alpha)$,

(iv) $k \ne 0, k \in R, k\tilde{A} = \begin{cases} (k\alpha, ka, k\beta), k > 0; \\ (k\beta, ka, k\alpha), k < 0. \end{cases}$

Theorem 2.1 [3]. *If* $\tilde{A}, \tilde{B}$ *are fuzzy numbers, then* $\tilde{A} \vee \tilde{B}, \tilde{A} \wedge \tilde{B}$ *are also fuzzy numbers.*

Definition 2.4 [15]. *A ranking function is a function* $R : F(R) \to R$ *is a set of fuzzy numbers defined on set of real numbers, which maps each fuzzy number into* *the real line, where a natural order exists. Let* $\tilde{A} = (\alpha, a, \beta)$ *be a triangular fuzzy number. Then* $\mathsf{R}(\tilde{A}) = \dfrac{\alpha + 2a + \beta}{4}$.

Theorem 2.5 [12]. (1) $\tilde{A} \geq \tilde{B} \Leftrightarrow \alpha + 2a + \beta \geq \gamma + 2b + \theta,$

$$(2)\ \tilde{A} \leq \tilde{B} \Leftrightarrow \alpha + 2a + \beta \leq \gamma + 2b + \theta.$$

Corollary 2.1. $\mathsf{R}(\tilde{A}) \geq \mathsf{R}(\tilde{B}) \Leftrightarrow \tilde{A} \vee \tilde{B} = \tilde{A}$.

3 Fuzzy Number Nonlinear Programming Problems

Theorem 3.1[6]. *The following equation*

$$A \circ x = b \tag{1}$$

is called a fuzzy relation equation, where $x = (x_1, x_2, \cdots, x_n)^T$, $0 \leq x_j \leq 1$ ($j = 1, 2, \cdots, n$) *is an n-dimensional vector,* $A = (a_{ij})_{m \times n}, 0 \leq a_{ij} \leq 1$ *is a* $m \times n$ *fuzzy matrix,* $b = (b_1, b_2, \cdots, b_m)^T, 0 \leq b_i \leq 1, (i = 1, 2, \cdots, m)$ *is an m-dimensional vector, and* " $\circ$ " *is the max-product composition, which means* $\overset{n}{\underset{j=1}{\vee}}(a_{ij} \cdot x_j) = b_i (1 \leq i \leq m)$.

Definition 3.1 [6]. *We call the following model*

$$\max\ f(x) = (c_1 \cdot x_1^{r_1}) \vee \cdots \vee (c_n \cdot x_n^{r_n})$$
$$s.t.\ \ A \circ x = b \tag{2}$$
$$0 \leq x_j \leq 1, j = 1, 2, \cdots, n$$

fuzzy relation geometric programming, where
$c = (c_1, c_2, \cdots, c_n), (c_j \geq 0, j = 1, 2, \cdots, n)$ *and*
$r = (r_1, r_2, \cdots, r_n), (r_j \in R, j = 1, 2, \cdots, n)$ *are n-dimensional vectors.*

Now we consider the following problem

$$\max\ \tilde{f}(x) = (\tilde{c}_1 \cdot x_1^{r_1}) \vee \cdots \vee (\tilde{c}_n \cdot x_n^{r_n})$$
$$s.t.\ \ \tilde{A} \circ x = \tilde{b}, \tag{3}$$
$$0 \leq x_j \leq 1, j = 1, 2, \cdots, n,$$

where $\tilde{A} = (\tilde{a}_{ij})_{m \times n}, \tilde{c} = (\tilde{c}_1, \tilde{c}_2, \cdots, \tilde{c}_n), \tilde{b} = (\tilde{b}_1, \tilde{b}_2, \cdots, \tilde{b}_m)^T$ and $\tilde{a}_{ij} = (\alpha_{ij}, a_{ij},$
$\beta_{ij}), \tilde{c}_j = (\varphi_j, c_j, \tau_j), \tilde{b}_i = (\gamma_i, b_i, \theta_i) \in F$ (R) for $i = 1, \cdots, m$, $j = 1, \cdots, n$.

Note 1: Since the constraints of (3) are fuzzy relation equations, then
$\varphi_i < c_i < \tau_i$, $0 \leq \alpha_{ij} < a_{ij} < \beta_{ij} \leq 1, 0 \leq \gamma_i < b_i < \theta_i \leq 1$, for $i = 1, \cdots, m$,
$j = 1, \cdots, n$.

Definition 3.3. *Any set of x_j which satisfies the set of constraints (3) is called a feasible solution for (3). Let Q be the set of all feasible solutions of (3). We shall say that $x^0 \in Q$ is an optimal feasible solution for (3) if $\tilde{f}(x^0) \geq \tilde{f}(x)$ for all $x \in Q$.*

Lemma 3.1. *Problem (3) and the following problem are equivalent:*

$$\max \overline{f}(x) = (\overline{c}_1 \cdot x_1^{r_1}) \vee \cdots \vee (\overline{c}_n \cdot x_n^{r_n})$$
$$s.t. \quad \overline{A} \circ x = \overline{b}, \tag{4}$$
$$0 \leq x_j \leq 1, j = 1, 2, \cdots, n,$$

where $\overline{c} = (\overline{c}_j)_{1 \times n} = (\mathrm{R}\,(\tilde{c}_j))_{1 \times n}, \overline{b} = (\mathrm{R}\,(\tilde{b}_i))_{m \times 1}, \overline{A} = (\overline{a}_{ij})_{m \times n} = (\mathrm{R}\,(\tilde{a}_{ij}))_{m \times n},$
and $\mathrm{R}\,(\tilde{c}_j) = \frac{1}{4}(\varphi_j + 2c_j + \tau_j), \mathrm{R}\,(\tilde{b}_i) = \frac{1}{4}(\gamma_i + 2b_i + \theta_i)$, $\mathrm{R}\,(\tilde{a}_{ij}) = \frac{1}{4}(\alpha_{ij} + 2a_{ij} + \beta_{ij})$.

Proof. Let Q_1 and Q_2 be the set of all feasible solution of Problems (3) and (4) respectively.

Then $x \in Q_1$ if and only if

$$\bigvee_{j=1}^{n} (\tilde{a}_{ij} \cdot x_j) = \tilde{b}_i (1 \leq i \leq m),$$

if and only if

$$\mathrm{R}\left[\bigvee_{j=1}^{n} (\tilde{a}_{ij} \cdot x_j)\right] = \mathrm{R}\,(\tilde{b}_i)(1 \leq i \leq m),$$

if and only if

$$\bigvee_{j=1}^{n}(\mathsf{R}\,(\tilde{a}_{ij}\cdot x_{j})) = \mathsf{R}\,(\tilde{b}_{i})(1\le i\le m),$$

if and only if

$$\bigvee_{j=1}^{n}[\frac{1}{4}(\alpha_{ij}\cdot x_{j}+2a_{ij}\cdot x_{j}+\beta_{ij}\cdot x_{j})] = \frac{1}{4}(\gamma_{i}+2b_{i}+\theta_{i})(1\le i\le m),$$

if and only if

$$\bigvee_{j=1}^{n}[\frac{1}{4}(\alpha_{ij}+2a_{ij}+\beta_{ij})\cdot x_{j}] = \frac{1}{4}(\gamma_{i}+2b_{i}+\theta_{i})(1\le i\le m),$$

if and only if

$$\bigvee_{j=1}^{n}(\overline{a}_{ij}\cdot x_{j}) = \overline{b}_{i}(1\le i\le m),$$

if and only if

$$\overline{A}\circ x = \overline{b}.$$

Hence $x\in Q_{2}$.

Now suppose that x^{0} is an optimal feasible solution for Problem (3), then for all $x\in Q_{1}$, we have $\tilde{f}\left(x^{0}\right)\ge \tilde{f}\left(x\right)$
if and only if

$$\bigvee_{j=1}^{n}(\tilde{c}_{j}\cdot x_{j}^{0^{r_{j}}}) \ge \bigvee_{j=1}^{n}(\tilde{c}_{j}\cdot x_{j}^{r_{j}})(j=1,\cdots,n),$$

if and only if

$$\mathsf{R}\left[\bigvee_{j=1}^{n}(\tilde{c}_{j}\cdot x_{j}^{0^{r_{j}}})\right] \ge \mathsf{R}\left[\bigvee_{j=1}^{n}(\tilde{c}_{j}\cdot x_{j}^{r_{j}})\right](j=1,\cdots,n),$$

if and only if

$$\bigvee_{j=1}^{n}[\mathsf{R}(\tilde{c}_{j}\cdot x_{j}^{0^{r_{j}}})] \ge \bigvee_{j=1}^{n}[\mathsf{R}(\tilde{c}_{j}\cdot x_{j}^{r_{j}})](j=1,\cdots,n),$$

if and only if

$$\bigvee_{j=1}^{n}[\frac{1}{4}(\varphi_{j}\cdot x_{j}^{0^{r_{j}}}+2c_{j}\cdot x_{j}^{0^{r_{j}}}+\tau_{j}\cdot x_{j}^{0^{r_{j}}})] \ge \bigvee_{j=1}^{n}[\frac{1}{4}(\varphi_{j}\cdot x_{j}^{r_{j}}+2c_{j}\cdot x_{j}^{r_{j}}+\tau_{j}\cdot x_{j}^{r_{j}})]$$

$$(j=1,\cdots,n),$$

if and only if

$$\bigvee_{j=1}^{n}[\frac{1}{4}(\varphi_j+2c_j+\tau_j)\cdot x_j^{0^{r_j}}]\geq \bigvee_{j=1}^{n}[\frac{1}{4}(\varphi_j+2c_j+\tau_j)\cdot x_j^{r_j}](j=1,\cdots,n),$$

if and only if

$$\bigvee_{j=1}^{n}(\overline{c}_j\cdot x_j^{0^{r_j}})\geq \bigvee_{j=1}^{n}(\overline{c}_j\cdot x_j^{r_j})(j=1,\cdots,n).$$

Hence $\overline{f}(x^0)\geq \overline{f}(x)$.

So we can conclude that x^0 is an optimal feasible solution for Problem (4).

Lemma 3.1 shows that we can reduce Problem (3) to a classical fuzzy relation geometric programming.

4   Solutions to Optimization Model (3)

Suppose $X(\overline{A},\overline{b})=\{(x_1,\cdots,x_n)^T\in R^n\mid \overline{A}\circ x=\overline{b},0\leq x_j\leq 1, j=1,2,\cdots,n\}$

is a solution set of   (4) . We define $x^1\leq x^2 \Leftrightarrow x_j^1\leq x_j^2 (j=1,2,\cdots,n)$,

$\forall x^1,x^2\in X(\overline{A},\overline{b})$. Such a definition of " $\leq$ " is a partial order relation $X(\overline{A},\overline{b})$.

Definition 4.1. *If* $\exists \hat{x}\in X(\overline{A},\overline{b})$, *such that* $x\leq \hat{x}, \forall x\in X(\overline{A},\overline{b})$, *then* $\hat{x}$ *is called the greatest solution to problem (4). If* $\exists \breve{x}\in X(\overline{A},\overline{b})$, *such that* $\breve{x}\geq x, \forall x\in$

$X(\overline{A},\overline{b})$, *then* $\breve{x}$ *is called the smallest solution to problem (4). And if* $\exists \breve{x}\in X(\overline{A},\overline{b})$, *when* $x\leq \breve{x}$, *then* $x=\breve{x}$, $\breve{x}$ *is called a minimal solution to (1)*

$$\breve{x}_j = \bigwedge_{j=1}^{n}\overline{a}_{ij}\alpha^{-1}\overline{b}_i, \tag{5}$$

where
$$\overline{a}_{ij}\alpha^{-1}\overline{b}_i = \begin{cases} \dfrac{\overline{b}_i}{\overline{a}_{ij}}, if\ \overline{a}_{ij}>\overline{b}_i; \\ 1\ \ ,if\ \overline{a}_{ij}\leq \overline{b}_i. \end{cases}(i=1,\cdots,m, j=1,\cdots,n),$$

If $\hat{x}=(\hat{x}_1,\cdots,\hat{x}_n)^T$ is a solution to Problem (4), we can easily prove that $\hat{x}$ must be the greatest solution.

We assume that the solution set of Problem (4) contains a finite number of minimal solutions. If we denote all minimal solution to (4) by $\breve{X}(\overline{A},\overline{b})$, then the solution set of problem (4) can be denoted as follows:

$$X(\overline{A},\overline{b}) = \bigcup_{\breve{x} \in \breve{X}(\overline{A},\overline{b})} \{x \mid \breve{x} \le x \le \hat{x}, x \in R^m\}. \tag{6}$$

The optimal value of the object function $\overline{f}(x)$ of Problem (4) is related to the sign of the exponent r_j of x_j and the coefficients $\overline{c}_j (1 \le j \le n)$. Now we discuss optimization (4) by considering the following three cases.

Lemma 4.1. *If* $\overline{c}_j \cdot r_j \ge 0$ $(1 \le j \le n)$, *then the greatest solution* $\hat{x}$ *to Problem (4) is optimal solution to Problem (4).*

Proof. Since $\overline{c}_j \cdot r_j \ge 0$ $(1 \le j \le n)$, then

$$\frac{d(\overline{c}_j \cdot x_j^{r_j})}{dx_j} = \overline{c}_j \cdot r_j \cdot x_j^{r_j-1} \ge 0 \ (1 \le j \le n)$$

for each x_j with $0 \le x_j \le 1$. Then $\overline{c}_j \cdot x_j^{r_j}$ is a monotone increasing function of x_j.

Moreover, $\forall x \in X(\overline{A},\overline{b})$, when $x \le \hat{x}$, then $\overline{c}_j \cdot x_j^{r_j} \le \overline{c}_j \cdot \hat{x}_j^{r_j}$, so that $\overline{f}(x) \le \overline{f}(\hat{x})$.

Hence $\hat{x}$ is an optimal solution to Problem (4).

Lemma 4.2. *If* $\overline{c}_j \cdot r_j < 0$ $(1 \le j \le n)$, *then a minimal solution* $\breve{x}$ *to Problem (4) is an optimal solution to Problem (4).*

Proof. Since $\overline{c}_j \cdot r_j < 0$ $(1 \le j \le n)$, then

$$\frac{d(\overline{c}_j \cdot x_j^{r_j})}{dx_j} = \overline{c}_j \cdot r_j \cdot x_j^{r_j-1} < 0 \ (1 \le j \le n)$$

for each x_j with $0 \le x_j \le 1$. Then $\overline{c}_j \cdot x_j^{r_j}$ is a monotone decreasing function of x_j.

Moreover, $\forall x \in X(\overline{A},\overline{b})$, according to Formula (6), there exists $\breve{x} \in \breve{X}(\overline{A},\overline{b})$, such that $x \ge \breve{x}$, that is $x_i \ge \breve{x}_i$. Hence $\overline{c}_j \cdot x_j^{r_j} \le \overline{c}_j \cdot \breve{x}_j^{r_j}$, that is the optimal solution to Problem (4) must exist in $\breve{X}(\overline{A},\overline{b})$. Let

$$\overline{f}(\breve{x}^*) = \max\{f(\breve{x}) \mid \breve{x} \in \breve{X}(\overline{A},\overline{b})\},$$

where $\breve{x}^* \in \breve{X}(\overline{A},\overline{b})$. Then $\forall x \in X(\overline{A},\overline{b}), \overline{f}(x) \le \overline{f}(\breve{x}^*)$. Therefore, $\breve{x}^*$ is an optimal solution to Problem (4).

Note 2.1: In a more general case $\breve{x}^*$ may not be unique.

As for general situation, $\overline{c}_j \cdot r_j \ (1 \le j \le n)$ is either a positive value or a negative one. Let

$$R_1 = \{i \mid \overline{c}_j \cdot r_j \ge 0, 1 \le j \le n\},$$
$$R_2 = \{i \mid \overline{c}_j \cdot r_j < 0, 1 \le j \le n\}.$$

Then $R_1 \cap R_2 = \varnothing, R_1 \cup R_2 = I$, where $I = \{1, 2, \cdots, n\}$.

Let $\overline{f}_1(x) = \underset{j \in R_1}{\vee} \overline{c}_j \cdot x_j^{r_j}, \overline{f}_2(x) = \underset{j \in R_2}{\vee} \overline{c}_j \cdot x_j^{r_j}$. Then $\overline{f}(x) = \overline{f}_1(x) \vee \overline{f}_2(x)$.

Therefore, based on the above, we have the following two optimization problems:

$$\max \overline{f}_1(x) = \underset{j \in R_1}{\vee} \overline{c}_j \cdot x_j^{r_j}$$

$$s.t. \ \overline{A} \circ x = \overline{b} \tag{7}$$

$$0 \le x_j \le 1, j = 1, 2, \cdots, n$$

and

$$\max \overline{f}_2(x) = \underset{j \in R_2}{\vee} \overline{c}_j \cdot x_j^{r_j}$$

$$s.t. \ \overline{A} \circ x = \overline{b}, \tag{8}$$

$$0 \le x_j \le 1, j = 1, 2, \cdots, n.$$

By Lemma 4.1, $\hat{x}$ is an optimal solution to optimization problem (7). By Lemma 4.2, $\exists \breve{x}^* \in \breve{X}(\overline{A}, \overline{b})$ and $\breve{x}^*$ is an optimal solution to optimization Problem (8).

Let

$$x_j^* = \begin{cases} \hat{x}_j, j \in R_1; \\ \breve{x}_j^*, j \in R_2. \end{cases} \tag{9}$$

We have the following theorem.

Theorem 4.1. *If some* $\overline{c}_j \cdot r_j \ (1 \le j \le n)$ *are positive value, while others are negative, then* x^* *is an optimal solution to Problem (4).*

Proof. $\forall x \in X(\overline{A}, \overline{b})$, according to Formula (6), $\exists \breve{x} \in \breve{X}(\overline{A}, \overline{b})$, such that $\breve{x} \le x \le \hat{x}$. By Lemma 4.1 and 4.2, we have

$$\overline{f}(x) = \overline{f}_1(x) \vee \overline{f}_2(x) \le \overline{f}_1(\hat{x}) \vee \overline{f}_2(\breve{x}) \le \overline{f}_1(\hat{x}) \vee \overline{f}_2(\breve{x}^*) = \overline{f}(x^*).$$

So x^* is an optimal solution to optimization Problem (4).

According to Lemma 3.1, x^* is an optimal solution to optimization Problem (3).

The methods for solving fuzzy relation equation with max-product composition can be seen in Cao [1] and Hu [2].

5 Numerical Example

Example 5.1. We consider the following fuzzy relation programming problem:

$$\max \ [(-1,1,2)\cdot x_1^{-2}]\vee[(-6,1,2)\cdot x_2^{\frac{1}{2}}]\vee[(-8,-4,0)\cdot x_3^{-\frac{1}{3}}]\vee[(2,4,8)\cdot x_4^{3}]$$
$$s.t. \ \tilde{A}\circ x=\tilde{b},$$
$$0\le x_j \le 1, j=1,2,\cdots,4,$$

$$(10)$$

where

$$\tilde{A}=\begin{bmatrix} (0.5,0.7,0.8) & (0.05,0.15,0.2) & (0.7,0.8,0.9) & (0.7,0.8,0.9) \\ (0.2,0.4,0.6) & (0.1,0.3,0.4) & (0.01,0.09,0.1) & (0.1,0.5,0.6) \\ (0.1,0.3,0.4) & (0.05,0.1,0.25) & (0.3,0.4,0.5) & (0.15,0.25,0.4) \end{bmatrix},$$

$$\tilde{b}=[(0.2,0.6,0.8), \ (0.05,0.35,0.4), \ (0.1,0.3,0.4)]^{T}.$$

From Lemma 3.1 we can have the following equivalent problem

$$\max \ (0.75\cdot x_1^{-2})\vee(-0.5\cdot x_2^{\frac{1}{2}})\vee(-4\cdot x_3^{-\frac{1}{3}})\vee(4.5\cdot x_4^{3})$$
$$s.t. \ \overline{A}\circ x=\overline{b},$$
$$0\le x_j \le 1, j=1,2,\cdots,4,$$

$$(11)$$

where $\overline{A}=\begin{bmatrix} 0.675 & 0.1375 & 0.8 & 0.8 \\ 0.4 & 0.275 & 0.0725 & 0.425 \\ 0.275 & 0.125 & 0.4 & 0.2625 \end{bmatrix},$

$\overline{b}=[0.55, \ 0.2875, \ 0.275]^{T}$. It is easy to see that $R_1=\{2,4\}, R_2=\{1,3\}$.

Therefore, we have the following two problems:

$$\max \ (-4\cdot x_3^{-\frac{1}{3}})\vee(4.5\cdot x_4^{3})$$
$$s.t. \ \overline{A}\circ x=\overline{b}$$
$$0\le x_j \le 1, j=1,2,\cdots,4$$

$$(12)$$

and

$$\max \quad (0.75 \cdot x_1^{-2}) \vee (-0.5 \cdot x_2^{\frac{1}{2}})$$

$$s.t. \quad \overline{A} \circ x = \overline{b}, \tag{13}$$

$$0 \le x_j \le 1, j = 1, 2, \cdots, 4.$$

From the Formula (6), we have

$$\hat{x} = (\frac{23}{32}, 1, \frac{11}{16}, \frac{23}{34})^T.$$

Since $\overline{A} \circ \hat{x} = \overline{b}$, then $\hat{x}$ is the greatest solution to $\overline{A} \circ x = \overline{b}$. By Lemma 4.1, we know $\hat{x}$ is the optimal solution to Problem (12).

Then we use the method from Cao [1] to solve the fuzzy relation equation $\overline{A} \circ x = \overline{b}$, we obtain two minimal solutions:

$$\breve{x}_1 = (\frac{23}{32}, 0, \frac{11}{16}, 0)^T, \breve{x}_2 = (0, 0, \frac{11}{16}, \frac{23}{34})^T.$$

From Lemma 4.2, we know $\breve{x}_1$ is the optimal solution to Problem (13). Therefore, $\breve{x}^* = \breve{x}_1$. By Formula (9), we have the optimal solution to Problem (11)

$$x^* = (\frac{23}{32}, 0, \frac{11}{16}, \frac{23}{34})^T$$

and the optimal value is $\overline{f}(x^*) = 1.3930$.

From Lemma 3.1, we know that x^* is also the optimal solution to Problem (10), and the optimal value is $\tilde{f}(x^*) = (0.6192, 1.2384, 2.4768)$.

6   Conclusion

In this paper, we proposed a new fuzzy relation programming with fuzzy coefficients. By using ranking function, it is converted into a fuzzy relation geometric programming. We solved the fuzzy relation geometric programming, and obtained the optimal solution to the new fuzzy relation programming. To illustrate the proposed method numerical example is solved.

Acknowledgements. This research is supported by the National Natural Science Foundation of China (No.70771030).

References

1. Cao, B.Y.: The theory and practice of fuzzy relative equation in max-product. BUSEFAL 35, 124–131 (1988)
2. Cao, B.Y.: Applied fuzzy mathematics and systems. Science Press, Beijing (2005)
3. Hu, B.Q.: The fuzzy mathematical theory basis. Wuhan University Press, Wuhan (2004)
4. Maleki, H.R., Tata, M., Mashinchi, M.: Linear programming with fuzzy variables. Fuzzy Sets and Systems 109, 21–33 (2000)
5. Yang, J.H., Cao, B.Y.: Geometric programming with fuzzy relation equation constraints. Fuzzy Systems and Mathematics 20, 110–115 (2006)
6. Zhou, X.G., Zhu, Z.X.: Fuzzy relation geometric programming with max-product composition. Journal of Hunan University of Science Engineering 27, 134–138 (2006)
7. Kumar, A., Kaur, J., Singh, P.: A new method for solving fully fuzzy linear programming problems. Applied Mathematical Modeling 35, 817–823 (2011)
8. Mahdavi-Amiri, N., Nasseri, S.H.: Duality in fuzzy number linear programming by use of a certain linear ranking function. Applied Mathematics and Computation 180, 206–216 (2006)
9. Hu, R.J., Cao, B.Y.: A method for solving fuzzy posynomial geometric programming. Journal of Guangzhou University (Natural Science Edition) 19, 23–25 (2010)
10. Jiranut, L., Fang, S.C.: Optimization of fuzzy relation equations with max-product composition. Fuzzy Sets and Systems 118, 509–517 (2001)
11. Yang, J.H., Cao, B.Y.: Monomial geometric programming with fuzzy relation equation constraints. Fuzzy Optim. Decis. Making 6, 337–349 (2007)
12. Yang, J.H., Cao, B.Y.: Fuzzy relation linear programming with fuzzy objective coefficient. Mathematics in Practice and Theory 38, 105–112 (2008)
13. Sanchez, E.: Resolution of composite fuzzy relation equations. Inform. and Control 30, 38–48 (1976)
14. Kaufmann, A., Gupta, M.M.: Introduction to fuzzy arithmetic theory and application. Van Nostrand Reinhold, New York (1985)
15. Liou, T.S., Wang, M.J.: Ranking fuzzy numbers with integral value. Fuzzy Set. Syst. 50, 247–255 (1992)
16. Zimmermann, H.J.: Fuzzy set theory and its applications. Kluwer Academic Publishers, Boston (1991)
17. Bourke, M.M., Fisher, D.G.: Solution algorithms for fuzzy relational equations with max-product composition. Fuzzy Sets and Systems 94, 61–69 (1998)
18. Maleki, H.R., Tata, M., Mashinchi, M.: Linear programming with fuzzy variables. Fuzzy Sets and Systems 109, 21–33 (2000)

Composite Product of Grade Approximation Operators in the Same Direction

Xian-yong Zhang[1], Zhi-wen Mo[1], Chang Shu[1,2], and Li-hong Deng[1]

[1]College of Mathematics and Software Science, Sichuan Normal University,
   Chengdu 610068, China
[2]Information Management Institute, Chengdu University of Technology,
   Chengdu 610059, China
xianyongzh@sina.com.cn

Abstract. Related to the absolute quantitative information, grade approximation operators are important approximation operators in approximate space. This paper aims to explore composite product of grade approximation operators in the same direction in graded rough set model. First, the composite product of grade approximation operators in the same direction is defined. Next, direct algorithm is proposed and analyzed. Finally, the essence of the composite product are investigated and obtained, and furthermore, the basic properties are obtained. Besides, the extension in parameters of the composite product is explored too. Composite product of grade approximation operators in the same direction is illustrated by an example.

Keywords: Artificial intelligence, rough set theory, graded rough set, approximation operator, composite product, direct algorithm.

1 Introduction

Classical rough set model is first proposed by Pawlak in 1982 [1]. It is a data analysis theory, and a new mathematical tool for dealing with vague and incomplete information. However, classical rough set model has a limitation, which is the relationships between equivalence classes and sets are strict, and quantitative information about the degree of overlap of equivalence classes and sets is not taken into consideration. In fact, there is some degree inclusion relation between sets, and quantity of overlap of sets is very important information in applications. It is therefore not surprising that similar efforts have been attempted in rough set model to incorporate such information, such as the references [2-7]. Among all these improving models, graded rough set model [2] and variable precision rough set model [3] are two important models related to quantization.

In [2], Yao and Lin have explored the relationships between rough sets and modal logics, and proposed graded rough sets by graded modal logics, and proposed probabilistic rough sets by defining probabilistic modal logics. Probabilistic rough set model is an important model with probability method. As to graded

B.-Y. Cao and X.-J. Xie (Eds.): Fuzzy Engineering and Operations Research, AISC 147, pp. 75–84.
springerlink.com   © Springer-Verlag Berlin Heidelberg 2012

rough set model, it has expanded classical rough set model from grade. Grade is an important quantitative index, and related to the absolute quantization. There are some papers about graded rough set model, such as references [8-11]. [8-9] have studied the model and its properties, and [10] has studied graded rough set based on rough membership function, and [11] has studied logical AND operation of grade approximation operators in graded rough set model.

Similarly, precision is an important quantitative index, and related to the relative quantization. Variable precision rough set model has expanded classical rough set model from precision, and it has been a research focus, such as the references [12-19]. The combination research of graded rough set model and variable precision rough set model is an innovative field, and references [20-22] have already obtained some useful initial results.

It is just in this background that this paper is to explore composite product of grade approximation operators in the same direction in graded rough set model.

2 Preliminaries

(U, R) is an approximate space. Grade parameter k is 0 or natural number, and grade k R upper and lower approximations of A are defined by the following formulas:

$$\overline{R}_k A = \cup\{[x]_R : |[x]_R \cap A| > k\},$$

$$\underline{R}_k A = \cup\{[x]_R : |[x]_R| - |[x]_R \cap A| \le k\}.$$

$\overline{R}_k A$ is the union of equivalence classes, whose number of elements in A is more than k ones, while $\underline{R}_k A$ is the union of equivalence classes, whose number of elements out of A is at most k ones. If $\overline{R}_k A \ne \underline{R}_k A$, then A is called a rough set by grade k, otherwise A is called a definable set by grade k. $\overline{R}_k$ and $\underline{R}_k$ are called grade upper approximation operator and grade lower approximation operator respectively.

Graded rough set model has improved classical rough set model from grade.

$$\text{If } k = 0, \text{ then } \overline{R}_k A = \overline{R}A, \underline{R}_k A = \underline{R}A.$$

This means that graded rough set model has completely expanded classical rough set model.

3   Composite Product of Grade Approximation Operators in the Same Direction

3.1   Definition

Definition 1. *Composite product of* $\overline{R}_k$ *and* $\overline{R}_k$ *is noted as* $\overline{R}_k \circ \overline{R}_k$, *and composite product of* $\underline{R}_k$ *and* $\underline{R}_k$ *is noted as* $\underline{R}_k \circ \underline{R}_k$. *They are defined as follows:*

$$\forall A \subseteq U, \ (\overline{R}_k \circ \overline{R}_k)A = \overline{R}_k(\overline{R}_k A),$$

$$(\underline{R}_k \circ \underline{R}_k)A = \underline{R}_k(\underline{R}_k A).$$

Composite product of $\overline{R}_k$ and $\overline{R}_k$, and composite product of $\underline{R}_k$ and $\underline{R}_k$, both are called composite product of grade approximation operators in the same direction. They have the logical composite product operation of grade approximation operators in the same direction, and they are just the results of this logical operation. Their definitions are natural, and similar to the definitions of composite product of maps or functions. In nature, they are corresponding to two-level power action of grade approximation operators.

Composite product of $\overline{R}_k$ and $\overline{R}_k$ is the union of equivalence classes, whose number of elements in $\overline{R}_k A$ is more than k ones, while composite product of $\underline{R}_k$ and $\underline{R}_k$ is the union of equivalence classes, whose number of elements out of $\underline{R}_k A$ is at most k ones.

3.2   Direct Algorithm

In order to calculate composite product of grade approximation operators in the same direction, direct algorithm will be proposed naturally and analyzed deeply based on the definition.

Direct Algorithm:

　　(1) First, calculate $\overline{R}_k A$ and $\underline{R}_k A$;

　　(2) Next, calculate $(\overline{R}_k \circ \overline{R}_k)A$ and $(\underline{R}_k \circ \underline{R}_k)A$ by Definition 1.

The core task of the direct algorithm is to research whether each equivalence class belongs to a specific set, and the main calculation is comparison. For each equivalence class, it needs two input data: $|[x]_R|$ and $|[x]_R \cap A|$. Suppose there are n equivalence classes, then there are $2n$ input data. Now comparison is chosen as the basic operation to analyze the direct algorithm.

　　In direct algorithm, each equivalence class needs to show whether it belongs to $\overline{R}_k A$ and $\underline{R}_k A$ first, and it needs comparing twice, and one auxiliary variable:

$|[x]_R|-k$. After obtaining $\overline{R}_k A$ and $\underline{R}_k A$, it needs to show whether each equivalence class belongs to $\overline{R}_k(\overline{R}_k A)$ and $\underline{R}_k(\underline{R}_k A)$. In this process, $|[x]_R \cap \overline{R}_k A|$ should compare once with k, and for $|[x]_R \cap \overline{R}_k A|$ is equal to 0 or $|[x]_R|$, it needs no auxiliary variables; At the same time, $|[x]_R \cap \underline{R}_k A|$ should compare once with $|[x]_R|-k$, and for $|[x]_R \cap \underline{R}_k A|$ is equal to 0 or $|[x]_R|$, it also needs one auxiliary variable: $|[x]_R|-k$ but this auxiliary variable is same to the previous one.

Therefore, time complexity and space complexity of direct algorithm are as follows:

$$T(n) = 4n \text{ and } S(n) = n.$$

Obviously, the result is the same in the best case, or in the worst case, or in the average case.

3.3 An Example

(U, R) is an approximate space, and U/R is composed of 20 equivalence classes. Suppose $A \subseteq U$. The related statistical result of equivalence classes is shown in Table 1.

Table 1. Statistical result of equivalence classes

$[x]_m$	$|[x]_m|$	$|[x]_m \cap A|$	$|[x]_m|-|[x]_m \cap A|$
$[x]_1$	1	0	1
$[x]_2$	1	1	0
$[x]_3$	2	0	2
$[x]_4$	2	1	1
$[x]_5$	2	2	0
$[x]_6$	3	0	3
$[x]_7$	3	1	2
$[x]_8$	3	2	1
$[x]_9$	3	3	0
$[x]_{10}$	4	0	4
$[x]_{11}$	4	1	3
$[x]_{12}$	4	2	2
$[x]_{13}$	4	3	1
$[x]_{14}$	4	4	0
$[x]_{15}$	5	0	5
$[x]_{16}$	5	1	4
$[x]_{17}$	5	2	3
$[x]_{18}$	5	3	2
$[x]_{19}$	5	4	1
$[x]_{20}$	5	5	0

Now, composite product of $\overline{R}_k$ and $\overline{R}_k$, and composite product of $\underline{R}_k$ and $\underline{R}_k$, will be calculated in the case $k = 2$.

Direct algorithm:

(1) $\overline{R}_2 A$ is just composed of the following 6 equivalence classes:

$$[x]_9, [x]_{13}, [x]_{14}, [x]_{18}, [x]_{19}, [x]_{20},$$

and $\underline{R}_2 A$ is just composed of the following 14 equivalence classes:

$$[x]_1, [x]_2, [x]_3, [x]_4, [x]_5, [x]_7, [x]_8, [x]_9, [x]_{12}, [x]_{13}, [x]_{14}, [x]_{18},$$
$$[x]_{19}, [x]_{20};$$

(2) There are just the 6 equivalence classes which satisfy $|[x]_R \cap \overline{R}_2 A| > 2$:

$$[x]_9, [x]_{13}, [x]_{14}, [x]_{18}, [x]_{19}, [x]_{20},$$

and so $(\overline{R}_2 \circ \overline{R}_2) A$ is just the union of the 6 equivalence classes;

Similarly, it is the 14 equivalence classes that satisfy $|[x]_R \cap \underline{R}_2 A| \geq |[x]_R| - 2$:

$$[x]_1, [x]_2, [x]_3, [x]_4, [x]_5, [x]_7, [x]_8, [x]_9, [x]_{12}, [x]_{13}, [x]_{14},$$
$$[x]_{18}, [x]_{19}, [x]_{20},$$

and so $(\underline{R}_2 \circ \underline{R}_2) A$ is just the union of the 14 equivalence classes.

In this example, there are only 20 equivalence classes. Time complexity and space complexity of direct algorithm are as follows:

$$T(20) = 80 \text{ and } S(20) = 20.$$

It is very important to see that

$$(\overline{R}_2 \circ \overline{R}_2) A = \overline{R}_2 A \text{ and } (\underline{R}_2 \circ \underline{R}_2) A = \underline{R}_2 A.$$

Actually, this result is not occasional but predetermined. The corresponding study will be made in the next section.

3.4 Essence Analysis

Theorem 1. $(\overline{R}_k \circ \overline{R}_k) A = \overline{R}_k A$,

$$(\underline{R}_k \circ \underline{R}_k) A = \underline{R}_k A.$$

Proof. (1) If $[x]_R \subseteq (\overline{R}_k \circ \overline{R}_k)A$, then $|[x]_R \cap \overline{R}_k A| > k$ from the definition.
If $[x]_R \not\subseteq \overline{R}_k A$, then $|[x]_R \cap \overline{R}_k A| = 0$ from classification property and it contradicts with the above formula. Therefore,

$$[x]_R \subseteq \overline{R}_k A \text{ and } (\overline{R}_k \circ \overline{R}_k)A \subseteq \overline{R}_k A.$$

On the contrary, if $[x]_R \subseteq \overline{R}_k A$ then $|[x]_R \cap A| > k$ from the definition. Hence $|[x]_R \cap \overline{R}_k A| = |[x]_R| \ge |[x]_R \cap A| > k$, and $[x]_R \subseteq \overline{R}_k(\overline{R}_k A)$ from the definition. Therefore,

$$\overline{R}_k A \subseteq (\overline{R}_k \circ \overline{R}_k)A.$$

So $(\overline{R}_k \circ \overline{R}_k)A = \overline{R}_k A$.

(2) If $[x]_R \subseteq (\underline{R}_k \circ \underline{R}_k)A$, then $|[x]_R \cap \underline{R}_k A| \ge |[x]_R| - k$ from the definition. If $[x]_R \not\subseteq \underline{R}_k A$, then

$$|[x]_R \cap \underline{R}_k A| = 0 \text{ and } 0 \le |[x]_R \cap A| < |[x]_R| - k.$$

However, $0 \ge |[x]_R| - k$ and $0 \le |[x]_R \cap A| < |[x]_R| - k$ form a contradiction. Therefore,

$$[x]_R \subseteq \underline{R}_k A \text{ and } (\underline{R}_k \circ \underline{R}_k)A \subseteq \underline{R}_k A.$$

On the contrary, if $[x]_R \subseteq \underline{R}_k A$ then $|[x]_R \cap A| \ge |[x]_R| - k$ from the definition. Hence $|[x]_R \cap \underline{R}_k A| = |[x]_R| \ge |[x]_R| - k$, and $[x]_R \subseteq \underline{R}_k(\underline{R}_k A)$ from the definition. Therefore,

$$\underline{R}_k A \subseteq (\underline{R}_k \circ \underline{R}_k)A.$$

So $(\underline{R}_k \circ \underline{R}_k)A = \underline{R}_k A$.

Corollary 1. $\overline{R}_k(\overline{R}_k A) = \overline{R}_k A$,

$$\underline{R}_k(\underline{R}_k A) = \underline{R}_k A.$$

Theorem 1 has presented the essence of composite product of grade approximation operators in the same direction. In fact, the result of composite product of $\overline{R}_k$ and $\overline{R}_k$ is just grade upper approximation $\overline{R}_k A$, while the result of composite product of $\underline{R}_k$ and $\underline{R}_k$ is just grade lower approximation $\underline{R}_k A$.

Based on the theoretical result of the essence analysis, the previous direct algorithm may be improved. Correspondingly, the time and space complexities can be improve to

$$T(n) = 2n \text{ and } S(n) = n.$$

In classical rough set model, approximation operators $\overline{R}$ and $\underline{R}$ have the following idempotent properties:

$$\overline{R}(\overline{R}A) = \underline{R}(\overline{R}A) = \overline{R}A, \quad \overline{R}(\underline{R}A) = \underline{R}(\underline{R}A) = \underline{R}A.$$

At the same time, Corollary 1 has also shown that grade approximation operators $\overline{R}_k$ and $\underline{R}_k$ have the idempotent properties:

$$\overline{R}_k(\overline{R}_k A) = \overline{R}_k A \text{ and } (\underline{R}_k \underline{R}_k)A = \underline{R}_k A,$$

when the grade approximation operators are in the same direction.

3.5 Basic Properties

Proposition 1.

(1) $(\overline{R}_k \circ \overline{R}_k)\phi = \phi$,

$\quad (\underline{R}_k \circ \underline{R}_k)\phi = \cup\{[x]_R : |[x]_R| \leq k\}$,

$\quad (\overline{R}_k \circ \overline{R}_k)U = \cup\{[x]_R : |[x]_R| > k\}$,

$\quad (\underline{R}_k \circ \underline{R}_k)U = U$;

(2) $A \subseteq B$, then

$\quad (\overline{R}_k \circ \overline{R}_k)A \subseteq (\overline{R}_k \circ \overline{R}_k)B$,

$\quad (\underline{R}_k \circ \underline{R}_k)A \subseteq (\underline{R}_k \circ \underline{R}_k)B$;

(3) $(\overline{R}_k \circ \overline{R}_k)(A \cup B) \supseteq (\overline{R}_k \circ \overline{R}_k)A \cup (\overline{R}_k \circ \overline{R}_k)B$,

$\quad (\underline{R}_k \circ \underline{R}_k)(A \cup B) \supseteq (\underline{R}_k \circ \underline{R}_k)A \cup (\underline{R}_k \circ \underline{R}_k)B$;

(4) $(\overline{R}_k \circ \overline{R}_k)(A \cap B) \subseteq (\overline{R}_k \circ \overline{R}_k)A \cap (\overline{R}_k \circ \overline{R}_k)B$,

$\quad (\underline{R}_k \circ \underline{R}_k)(A \cap B) \subseteq (\underline{R}_k \circ \underline{R}_k)A \cap (\underline{R}_k \circ \underline{R}_k)B$;

(5) $(\overline{R}_k \circ \overline{R}_k)(\sim A) = \sim (\underline{R}_k \circ \underline{R}_k)A$,

$\quad (\underline{R}_k \circ \underline{R}_k)(\sim A) = \sim (\overline{R}_k \circ \overline{R}_k)A$;

(6) $k \geq l \Leftrightarrow$

$\quad (\overline{R}_k \circ \overline{R}_k)A \subseteq (\overline{R}_l \circ \overline{R}_l)A$,

$\quad (\underline{R}_k \circ \underline{R}_k)A \supseteq (\underline{R}_l \circ \underline{R}_l)A$.

Proposition 2.

(1) $(\overline{R}_k \circ \overline{R}_k)[(\overline{R}_k \circ \overline{R}_k)A] = (\overline{R}_k \circ \overline{R}_k)A \subseteq (\underline{R}_k \circ \underline{R}_k)[(\overline{R}_k \circ \overline{R}_k)A]$;

(2) $(\underline{R}_k \circ \underline{R}_k)[(\overline{R}_k \circ \overline{R}_k)A] = (\overline{R}_k \circ \overline{R}_k)A \cup (\cup\{[x]_R : |[x]_R| \leq k\})$;

(3) $(\overline{R}_k \circ \overline{R}_k)[(\underline{R}_k \circ \underline{R}_k)A] \subseteq (\underline{R}_k \circ \underline{R}_k)A = (\underline{R}_k \circ \underline{R}_k)[(\underline{R}_k \circ \underline{R}_k)A]$;

(4) $(\underline{R}_k \circ \underline{R}_k)A = (\overline{R}_k \circ \overline{R}_k)[(\underline{R}_k \circ \underline{R}_k)A] \cup (\cup\{[x]_R : |[x]_R| \leq k\})$.

Proposition 1 and Proposition 2 have obtained the basic properties of composite product of grade approximation operators in the same direction. Proposition 1 is with respect to the basic set operation, while proposition 2 is with respect to two-level power action of operation. Obviously, it is $|[x]_R| \leq k$ that hinders the idempotent properties.

3.6 Extension in Parameters

The above study has explored composite product of grade approximation operators in the same direction, but there are only one parameter k in it. It may introduce two parameters, and of course, the new study with two parameters is also the one "in the same direction" with respect to grade approximation operators.

Let k and l be two grade parameters. And then,

$$(\overline{R}_l \circ \overline{R}_k)A = \overline{R}_l(\overline{R}_k A) \text{ and } (\underline{R}_l \circ \underline{R}_k)A = \underline{R}_l(\underline{R}_k A)$$

will be studied.

Theorem 2. $(\overline{R}_l \circ \overline{R}_k)A = \cup\{[x]_R : |[x]_R| > l\} \cap \overline{R}_k A$,

$\quad\quad (\underline{R}_l \circ \underline{R}_k)A = \cup\{[x]_R : |[x]_R| \leq l\} \cup \underline{R}_k A$.

Theorem 3. $(\overline{R}_l \circ \overline{R}_k)A = \cup\{[x]_R : |[x]_R| > l, |[x]_R \cap A| > k\}$,

$\quad\quad (\underline{R}_l \circ \underline{R}_k)A = \cup\{[x]_R : |[x]_R| \leq l, \text{ or } |[x]_R \cap A| \geq |[x]_R| - k\}$.

Theorem 4. $(\overline{R}_l \circ \overline{R}_k)A \subseteq \overline{R}_k A$,

$\quad\quad (\underline{R}_l \circ \underline{R}_k)A \supseteq \underline{R}_k A$.

Theorem 5. $(\overline{R}_l \circ \overline{R}_k)[(\overline{R}_l \circ \overline{R}_k)A] = (\overline{R}_l \circ \overline{R}_k)A$,

$\quad\quad (\underline{R}_l \circ \underline{R}_k)[(\underline{R}_l \circ \underline{R}_k)A] = (\underline{R}_l \circ \underline{R}_k)A$.

Theorem 6. *When* $l \leq k$,

$\quad\quad (\overline{R}_l \circ \overline{R}_k)A = \overline{R}_k A$,

$\quad\quad (\underline{R}_l \circ \underline{R}_k)A = \underline{R}_k A$.

Theorem 2 and Theorem 3 have studied the essence of composite product of grade approximation operators in the same direction based on two parameters, and

Theorem 5 has obtained the idempotent properties. Theorem 4 and Theorem 6 have also shown that composite product of grade approximation operators in the same direction based on two parameters has partially expanded graded rough set model.

4 Conclusion

This paper has mainly studied composite product model of grade approximation operators in the same direction, and many theoretical properties are obtained. In nature, composite product model of grade approximation operators in the same direction is a kind of two-level combinations of grade approximation operators. Furthermore, other two-level combinations and multi-level combinations of grade approximation operators, as well as the algebraic system of grade approximation operators, are worth researching deeply.

Acknowledgements. Thanks to the support by National Natural Science Foundation of China (No. 10671030), National Natural Science Foundation for Young Scholars of China (No. 60803028), Science & Technology Pillar Program of Sichuan Province (09ZC1838), Young Scientific Research Fund of Sichuan Provincial Education Department in China (10ZB004), Scientific Research Fund of Sichuan Normal University (08KYL06).

References

1. Pawlak, Z.: Rough sets. International Journal of Computer and Information Sciences 11, 341–356 (1982)
2. Yao, Y.Y., Lin, T.Y.: Generalization of rough sets using modal logics. Intelligent Automation and Soft Computing: an International Journal 2(2), 103–120 (1996)
3. Ziarko, W.: Variable precision rough set model. Journal of Computer and System Sciences 46(1), 39–59 (1993)
4. Katzberg, J.D., Ziarko, W.: Variable precision extension of rough sets. Fundamenta Informaticae 27(8), 2–3 (1996)
5. Cattaneo, G.: Generalized rough sets: Preclusivity fuzzy-intuitionistic (BZ) lattices. Studia Logica 58, 47–77 (1997)
6. Yao, Y.Y., Wong, S.K.M.: A decision theoretic framework for approximating concepts. International Journal of Man-machine Studies 37, 793–809 (1992)
7. Yao, Y.Y.: Relational interpretations of neighborhood operators and rough set approximation. Information Sciences 111, 239–259 (1998)
8. Yao, Y.Y., Lin, T.Y.: Graded rough set approximations based on nested neighborhood systems. In: Zimmermann, H.J. (ed.) Proceedings of 5th European Congress on Intelligent Techniques and Soft Computing, vol. 1, pp. 196–200. Verlag Mainz, Aachen (1997)
9. Zhang, X.Y., Xie, S.C., Mo, Z.W.: Graded rough sets. Journal of Sichuan Normal University (Natural Science) 333(1), 12–16 (2010)

10. Xu, W.H., Liu, S.H., Wang, Q.R., Zhang, W.X.: The first type of graded rough set based on rough membership function. In: 2010 Seventh International Conference on Fuzzy Systems and Knowledge Discovery (FSKD), Yantai, China, pp. 1922–1926 (2010)

11. Zhang, X.Y., Xiong, F., Mo, Z.W., Shu, L.: Algorithms and algorithm analysis of logical AND operation of grade approximation operators. Advanced Materials Research 204-210, 1701–1704 (2011)

12. Xie, F., Lin, Y., Ren, W.W.: Optimizing model for land use/land cover retrieval from remote sensing imagery based on variable precision rough sets. Ecological Modelling 222(2), 232–240 (2011)

13. Pan, X., Zhang, S.Q., Zhang, H.Q., et al.: A variable precision rough set approach to the remote sensing land use/cover classification. Computers and Geosciences 36(12), 1466–1473 (2010)

14. Xie, G., Yue, W.Y., Wang, S.Y., et al.: Dynamic risk management in petroleum project investment based on a variable precision rough set model. Technological Forecasting and Social Change 77(6), 891–901 (2010)

15. Inuiguchi, M., Yoshioka, Y., Kusunoki, Y.: Variable-precision dominance-based rough set approach and attribute reduction. International Journal of Approximate Reasoning 50(8), 1199–1214 (2009)

16. Wang, J.Y., Zhou, J.: Research of reduct features in the variable precision rough set model. Neurocomputing 72(10-12), 2643–2648 (2009)

17. Ningler, M., Stockmanns, G., Schneider, G., et al.: Adapted variable precision rough set approach for EEG analysis. Artificial Intelligence in Medicine 47, 239–261 (2009)

18. Li, W.H., Chen, S.B., Wang, B.: A variable precision rough set based modeling method for pulsed GTAW. The International Journal of Advanced Manufacturing Technology 36, 1072–1079 (2008)

19. Xie, G., Zhang, J.L., Lai, K.K., Yu, L.: Variable precision rough set for group decision-making: an application. International Journal of Approximate Reasoning 49(2), 331–343 (2008)

20. Zhang, X.Y., Mo, Z.W., Xiong, F.: Approximation of intersection of grade and precision. In: Cao, B.Y., Zhang, C.Y., Li, T.F. (eds.) Fuzzy Information and Engineering. AISC, vol. 54, pp. 526–530. Springer, Heidelberg (2008)

21. Zhang, X.Y., Mo, Z.W., Xiong, F.: Properties of approximation operators of logical AND operation of precision and grade. In: Chen, W., Li, S.Z., Wang, Y.L. (eds.) 2009 IEEE International Conference on Intelligent Computing and Intelligent Systems (ICIS 2009), vol. 1, pp. 33–37. IEEE (2009)

22. Zhang, X.Y., Mo, Z.W., Xiong, F.: Algorithms and algorithm analysis of logical OR operation of variable precision lower approximation operator and grade upper approximation operator. In: Ruan, D., Li, T.R., Xu, Y., Chen, G.Q., Kerre, E.E. (eds.) Computational Intelligence: Foundations and Applications. World Scientific Proceedings Series on Computer Engineering and Information Science, vol. 4, pp. 672–677. World Scientific, Singapore (2010)

A New Linguistic Aggregation Operator for Group Decision Making

Li Zou[1,4], Xin Liu[2], Zheng Pei[3], and Degen Huang[4]

[1] School of Computer and Information Technology, Liaoning Normal University, Dalian 116029, China
zoulicn@163.com
[2] School of Mathematics, Liaoning Normal University, Dalian 116029, China
[3] School of Mathematics and Computer Engineering, Xihua University, Chengdu, Sichuan 610039, China
[4] School of Computer Science and Technology, Dalian University of Technology, Dalian 116024, China

Abstract. Different linguistic aggregation methods have been proposed and applied in the linguistic decision making problems. Generally, weights for experts or criteria are considered in linguistic aggregation processes. In linguistic decision analysis, it can be noticed that some of initial linguistic values used by experts have priority over others linguistic values in evaluation processes. In this paper, we formalize the priority over initial linguistic values as weights for linguistic values, and propose a new linguistic aggregation operator including weights for linguistic values and weights for experts. We investigate the properties of the new linguistic aggregation operator. We particularly see that the new linguistic aggregation operator is extensions of the 2-tuple arithmetic mean, the 2-tuple weighted aggregation operator and the 2-tuple ordered weighted averaging operator.

Keywords: Computing with words, linguistic group decision making, 2-tuple fuzzy linguistic representation model.

1 Introduction

In many cases, information of decision making problems cannot be assessed precisely in a quantitative form but may be in a qualitative one, people use natural language instead of numerical values to express their evaluations of decision making problems [7], in this case, we need the linguistic approach to solve group decision making problems with linguistic assessment [5, 10, 13]. Recently, many linguistic approaches have been proposed and applied to solve problems with linguistic assessment, *e.g.*, environmental assessment [4], personnel management [6, 14], software developing [18], material selection [2], web information processing [12], sensor evaluation and fuzzy risk analysis [9, 15, 21, 22, 23]. In [8], Herrera, *et al* make a review of the developments of Computing with Words in decision making and explore different linguistic computational models that have been applied to the decision making field.

B.-Y. Cao and X.-J. Xie (Eds.): Fuzzy Engineering and Operations Research, AISC 147, pp. 85–95.
springerlink.com © Springer-Verlag Berlin Heidelberg 2012

In linguistic decision making analysis, the problems are associated with [7]: (1) The choice of the linguistic value set with its semantic; (2) The choice of the aggregation operator of linguistic information; (3) The choice of the best alternatives.

Ordering and weights for linguistic information are two important aspects in all linguistic aggregation operators. In this paper, we formalize the priority over initial linguistic values as weights for linguistic values, by considering weights for linguistic values as well as weights for experts in linguistic aggregation processes, we propose a new linguistic aggregation operator called the linguistic weighted ordered weighted aggregation operator. Formally, some interesting properties of the new linguistic aggregation operator are discussed, these properties show that the linguistic aggregation operator is extensions of the 2-tuple arithmetic mean, the 2-tuple weighted aggregation operator and the 2-tuple ordered weighted averaging operator. Moreover, we provide an an optimization model to obtain weights of the new linguistic aggregation operator, and compare the new linguistic aggregation operator with the 2-tuple weighted aggregation operator and the 2-tuple ordered weighted averaging operator in an illustrative example.

This paper is structured as follows: In Section 2, we briefly review the 2-tuple linguistic representation model and some other numerical aggregation operators. In Section 3, we analyze experts' perceptions for linguistic values and the habits of language use, and formalize them as weights for linguistic values. Then, we propose the new linguistic aggregation operator including weights for linguistic values and experts, we investigate the properties of the new linguistic aggregation operator, in which, we prove that the new linguistic aggregation operator is extensions of the 2-tuple arithmetic mean, the 2-tuple weighted aggregation operator and the 2-tuple ordered weighted averaging operator. Moreover, we provide an optimization model to obtain weights of the new linguistic aggregation operator. We draw some conclusions in Section 4.

2   Preliminaries

In this section, we briefly review the 2-tuple linguistic representation model and some aggregation operators, we refer to [1, 5, 11, 16, 19] for more details.

2.1   The 2-Tuple Linguistic Representation Model

The 2-tuple linguistic representation model be introduced by Herrera [5]. Let $S = \{s_0, \cdots, s_g\}$ be the initial finite linguistic value set. Formally, the 2-tuple linguistic representation model is formed by (s_i, α), in which, $s_i \in S(i \in \{0, 1, \cdots, g\})$ and $\alpha \in [-0.5, 0.5)$, $i.e.$, linguistic information is encoded in the space $S \times [-0.5, 0.5)$. Based on the representation (s_i, α), we can easily obtain the following symbolic translation of linguistic values from $\beta \in [0, g]$ to $S \times [-0.5, 0.5)$:

$$\Delta : [0, g] \to S \times [-0.5, 0.5),$$
$$\beta \longmapsto (s_i, \alpha), \tag{1}$$

in which, $i = round(\beta)$ ($round(\cdot)$ is the usual round operation), $\alpha = \beta - i \in [-0.5, 0.5)$. Intuitively, $\Delta(\beta) = (s_i, \alpha)$ expresses that s_i is the closest linguistic value to β, and α is the value of the symbolic translation. Additionally, there is a Δ^{-1} function such that from a 2-tuple it returns its equivalent numerical value $\beta \in [0, g]$.

$$\Delta^{-1} : S \times [-0.5, 0.5) \longrightarrow [0, g],$$
$$\Delta^{-1}(s_i, \alpha) = i + \alpha = \beta. \tag{2}$$

In fact, this model defines a set of transformation functions between linguistic values and 2-tuples linguistic representations as well as numeric values and 2-tuples linguistic representations. This makes us easily to process linguistic information by numeric value, *e.g.*, we have the following linguistic aggregation operators: Let a set of the 2-tuples linguistic representations be $x = \{(s_1, \alpha_1), \cdots, (s_n, \alpha_n)\}$ and $W = \{w_1, \cdots, w_n\}$ be an associated weights such that $w_i \in [0, 1]$ and $\sum_{i=1}^{n} w_i = 1$.

1. The 2-tuple arithmetic mean operator $\bar{x}^e$:

$$\bar{x}^e = \Delta(\sum_{i=1}^{n} \frac{1}{n} \times \Delta^{-1}(s_i, \alpha_i)) = \Delta(\frac{1}{n} \times (\sum_{i=1}^{n} \beta_i)).$$

2. The 2-tuple weighted aggregation operator $F^{\bar{w}}$:

$$F^{\bar{w}} = \Delta(\sum_{i=1}^{n} w_i \times \Delta^{-1}(s_i, \alpha_i)) = \Delta(\sum_{i=1}^{n} w_i \times \beta_i).$$

3. The 2-tuple ordered weighted aggregation operator F^e:

$$F^e((s_1, \alpha_1), (s_2, \alpha_2), \cdots, (s_n, \alpha_n)) = \Delta(\sum_{j=1}^{n} w_j \times \beta_j^*),$$

in which, β_j^* is the jth largest of $\{\beta_i = \Delta^{-1}(s_i, \alpha_i) | i = 1, 2, \cdots, n\}$.

2.2 The WA, OWA and WOWA Operator

Definition 1. *Let* $P = \{p_1, p_2, \cdots, p_n\}$ *be a weight vector of dimension n such that* $p_i \in [0, 1]$ *and* $\Sigma_i p_i = 1$. *A mapping* $f_{wm} : \mathbb{R}^n \longrightarrow \mathbb{R}$ *is called a weighted averaging (WA) operator of dimension n if*

$$f_{wm}(a_1, a_2, \cdots, a_n) = \Sigma_i p_i a_i.$$

Definition 2. *[19] Let w be a weigh vector of dimension n, $w = \{w_1, w_2, \cdots, w_n\}$ such that* $w_i \in [0, 1]$ *and* $\Sigma_i w_i = 1$. *A mapping* $f_{owa} : \mathbb{R}^n \longrightarrow \mathbb{R}$ *is called an ordered weighted averaging (OWA) operator of dimension n if*

$$f_{owa}(a_1, \cdots, a_n) = \Sigma_i w_i a_{\sigma(i)},$$

where $\{\sigma(1), \cdots, \sigma(n)\}$ is a permutation of $\{1, \cdots, n\}$ such that $a_{\sigma(i-1)} \geq$ $a_{\sigma(i)}$ for all $i = 2, \cdots, n$, i.e., $a_{\sigma(i)}$ is the ith largest element in the collection $\{a_1, \cdots, a_n\}$.

The weighted ordered weighted averaging operator (the *WOWA* operator) is used to aggregate numerical information. In the *WOWA* operator, there are two kinds of weights for information, one is used to explain weights for information sources, the other to information itself. Formally, the *WOWA* operator can be defined as follows:

Definition 3. *[16] Let P and W be weigh vectors of dimension n such that*

1. $p_i \in [0, 1]$ and $\sum_{i=1}^{n} p_i = 1$,
2. $w_i \in [0, 1]$ and $\sum_{i=1}^{n} w_i = 1$.

In this case, a mapping $f_{wowa} : R^n \longrightarrow R$ is a WOWA operator of dimension n if

$$f_{wowa}(a_1, \cdots, a_n) = \sum_{i=1}^{n} w_i a_{\sigma(i)}.$$

Where $\{\sigma(1), \cdots, \sigma(n)\}$ is a permutation of $\{1, \cdots, n\}$ such that $a_{\sigma(i-1)} \geq a_{\sigma(i)}$ for any $i = 2, \cdots, n$, i.e., $a_{\sigma(i)}$ is the ith largest element in the collection $\{a_1, \cdots, a_n\}$, the weight w_i is defined as

$$w_i = w^*(\sum_{j \leq i} p_{\sigma(j)}) - w^*(\sum_{j < i} p_{\sigma(j)})$$

with w^* a monotone increasing function that interpolates the points $(\frac{i}{n}, \sum_{j \leq i} w_j)$ together with the point $(0, 0)$.

3 A New Linguistic Aggregation Operator and Its Properties

Inspired by the 2-tuple linguistic representation model and the *WOWA* operator, in this section, we firstly analyze weights for linguistic values in group decision making. Then we propose a new linguistic aggregation operator including weights for linguistic values and experts, and investigate the properties of the new linguistic aggregation operator.

3.1 Weights for Linguistic Values in Group Decision Making

In linguistic group decision making, the initial finite linguistic value set is provided for experts to evaluate decision making problems, every expert selects linguistic values to evaluate alternatives according to his or her perceptions, attitudes, motivations and personalities. Due to different perception for linguistic values and the habit of language use, experts appreciate some of initial linguistic values than others in evaluation processes. Such appreciation can be understood as weights for linguistic values. Formally, we adopt the following three steps to obtain weights for linguistic values, let the initial finite linguistic value set be $S = \{s_0, \cdots, s_g\}$.

1. In linguistic group decision making, evaluations can be rewritten as the following linguistic evaluation matrix

$$R_{n \times m} = (r_{ij})_{n \times m} = \begin{pmatrix} r_{11} & r_{12} & \cdots & r_{1m} \\ r_{21} & r_{22} & \cdots & r_{2m} \\ \vdots & \vdots & \vdots & \vdots \\ r_{n1} & r_{n2} & \cdots & r_{nm} \end{pmatrix}_{n \times m},$$

in which, n is the number of alternatives, m is the number of experts, ith row vector of $R_{n \times m}$ expresses the evaluation of alternative i given by all experts, jth column vector expresses the evaluation of all alternatives provided by expert j, $r_{ij} \in S$ is the evaluation of ith alternative provided by jth expert.

2. In $R_{n \times m}$, for every $i \in \{1, 2, \cdots, n\}$ and s_k, we denote the set

$$L^i_{s_k} = \{r_{ij} | r_{ij} = s_k, j = 1, 2, \cdots, m\},$$

intuitively, $L^i_{s_k}$ means how many experts use linguistic value s_k to evaluate alternative i, i.e., $n^i_{s_k} = |L^i_{s_k}|$ (the cardinality of the set) can be used to explain experts'perception for s_k and the habit of language use when they use $S = \{s_0, \cdots, s_g\}$ to evaluate alternative i.

3. For every alternative i and s_k in $R_{n \times m}$, weight for linguistic value s_k is calculated by

$$w^i_{s_k} = \frac{n^i_{s_k}}{m}.$$

Obviously, if $s_k \in S$ doesn't appears in $R_{n \times m}$, then for every alternative i, $w^i_{s_k} = 0$, hence, it is easily to obtain $\sum_{k=1}^{|S|} w^i_{s_k} = 1 (i \in \{1, 2, \cdots, n\})$. Formally, the more $w^i_{s_k}$ is, the more experts appreciate linguistic value s_k to evaluate alternative i in decision making processes.

Example 1. [5] A distribution company needs to upgrade its computing system, so it hires a consulting company to survey the different possibilities existing on the market, to decide which is the best option for its needs. The options (alternatives) are shown in Table 1. The consulting company has a group of four consultancy departments (shown in Table 2). In each of departments, there is one expert provides evaluation for each alternative (shown in Table 3), these evaluations are assessed in the initial finite linguistic value set $S = \{s_0 = none(N), s_1 = very\ low(VL), s_2 = low(L), s_3 = medium(M), s_4 = high(H), s_5 = very\ high(VH), s_6 = perfect(P)\}$. According to Table 3, we can obtain the linguistic evaluation matrix is

$$R_{4\times 4} = \begin{pmatrix} r_{11} & r_{12} & r_{13} & r_{14} \\ r_{21} & r_{22} & r_{23} & r_{24} \\ r_{31} & r_{32} & r_{33} & r_{34} \\ r_{41} & r_{42} & r_{43} & r_{44} \end{pmatrix}_{4\times 4} = \begin{pmatrix} s_1 & s_3 & s_4 & s_4 \\ s_3 & s_4 & s_1 & s_4 \\ s_3 & s_1 & s_3 & s_2 \\ s_2 & s_4 & s_3 & s_2 \end{pmatrix}_{4\times 4},$$

for alternative x_3 and linguistic value s_3, $L_{s_3}^3 = \{r_{31}, r_{33}\}$, $n_{s_3}^3 = |L_{s_3}^3| = 2$ and $w_{s_3}^3 = \frac{2}{4} = \frac{1}{2}$. Similarly, $L_{s_1}^3 = \{r_{32}\}$, $n_{s_1}^3 = |L_{s_1}^3| = 1$ and $w_{s_1}^3 = \frac{1}{4}$. In practice, we can say that experts appreciate s_3 than s_1 in evaluation alternative x_3 due to $w_{s_3}^3 > w_{s_1}^3$.

Table 1. The four alternatives.

x_1	x_2	x_3	x_4
UNIX	WINDOWS−NT	AS /400	VMS

Table 2. The four consultancy departments.

d_1	d_2	d_3	d_4
Cost analysis	System analysis	Risk analysis	Technology analysis

Table 3. Evaluations provided by four experts.

Alternatives	Experts			
	e_1	e_2	e_3	e_4
x_1	s_1	s_3	s_4	s_4
x_2	s_3	s_4	s_1	s_4
x_3	s_3	s_1	s_3	s_2
x_4	s_2	s_4	s_3	s_2

Weights for experts are generally considered in group decision making, *i.e.*, different experts have a different importance in decision-making processes. From the information processing point of view, because experts provide linguistic values to evaluate decision making problems, hence, weights for experts can be understood by weights for information sources. On the other hand, linguistic values are used to evaluate decision making problem, weights for linguistic values can be understood by weights for information itself, they are concentrated on explaining experts' perceptions for linguistic values and the habits of language use.

3.2 The Linguistic Weighted Ordered Weighted Aggregation Operator

As we have mentioned in subsection 3.1, there are two kinds of weights information in linguistic decision making processes, *i.e.*, weights for linguistic values and weights for experts. Hence, when weights for linguistic values as well as experts are considered in linguistic aggregation processes, we have the linguistic weighted ordered weighted aggregation operator as following.

Definition 4. *Let the initial finite linguistic value set be* $S = \{s_0, \cdots, s_g\}$ $(s_{l_1}, s_{l_2}, \cdots, s_{l_m})$ *be linguistic values to be aggregated,* $(w_{s_{k_1}}, w_{s_{k_2}}, \cdots, w_{s_{k_v}})$ *and* $(p_{e_1}, p_{e_2}, \cdots, p_{e_m})$ *be weights for linguistic values and experts, respectively. The linguistic weighted ordered weighted aggregation operator (the LWOWA operator) of dimension* m *is defined as*

$$f_{lwowa}(s_{l_1}, s_{l_2}, \cdots, s_{l_m}) = (\omega_1 \odot s_{\sigma(l_1)}) \oplus (\omega_2 \odot s_{\sigma(l_2)}) \oplus \cdots \oplus (\omega_m \odot s_{\sigma(l_m)})$$
$$= s_\tau = (s_k, \alpha), \tag{1}$$

in which, $\{\sigma(l_1), \sigma(l_2), \cdots, \sigma(l_m)\}$ *is a permutation of* $\{1, 2, \cdots, m\}$ *such that* $s_{\sigma(l_{i-1})} \geq s_{\sigma(l_i)}$ *for all* $i = 2, \cdots, m$, *i.e.,* $s_{\sigma(l_i)}$ *is the* ith *largest linguistic value in the collection* $\{s_{l_1}, s_{l_2}, \cdots, s_{l_m}\}$, $k \in \{0, 1, \cdots, g\}$, $\alpha \in [-0.5, 0.5)$ *and*

$$k + \alpha = \tau = \sum_{i=1}^{m} \omega_i \times \sigma(l_i),$$

the weight ω_i *is decided by*

$$\omega_i = w^*\left(\sum_{j \leq i} p_{e_{\sigma(l_j)}}\right) - w^*\left(\sum_{j < i} p_{e_{\sigma(l_j)}}\right), \tag{2}$$

w^* *is a monotone increasing function that interpolates the points* $(\frac{b}{v}, \sum_{c \leq b} w_{s_{k_c}})$ *together with the point* $(0, 0)$, $b \in \{1, 2, \cdots, v\}$ *and* $c \in \{1, 2, \cdots, b\}$.

In Definition 4, $v \leq m$ because there may be several experts use the same linguistic value to evaluate decision making problems, in this case, $(s_{l_1}, s_{l_2}, \cdots, s_{l_m})$ is reduced as $(s_{l_1}, s_{l_2}, \cdots, s_{l_v})$ such that for any $b, c \in \{1, 2, \cdots, v\}$, $s_{l_b} \neq s_{l_c}$. In practice, if the set of linguistic values to be aggregated is the 2-tuple fuzzy linguistic representations, *i.e.,* $S_1 = \{(s_{l_1}, \alpha_1), (s_{l_2}, \alpha_2), \cdots, (s_{l_m}, \alpha_m)\}$, then (1) can be rewritten as

$$f_{lwowa}((s_{l_1}, \alpha_1), (s_{l_2}, \alpha_2), \cdots, (s_{l_m}, \alpha_m)) = \Delta\left(\sum_{j=1}^{m} \omega_j \times \beta_j^*\right), \tag{3}$$

in which, β_j^* is the jth largest of $\{\beta_i = \Delta^{-1}(s_{l_i}, \alpha_i) | i = 1, 2, \cdots, m\}$.

In the *LWOWA* operator, weights information is depended on weights for linguistic values and experts. In decision making process, weights for experts are depended on the importance degrees (the reliability degrees or the relevance degrees) of alternatives to experts. Weights for linguistic values are depended on all experts' perceptions for linguistic values and the habits of language use. Naturally, different weights for linguistic values are corresponding to different aggregation results. On the other hand, the monotone increasing function w^* is important in the *LWOWA* operator, formally, function w^* is decided by weights for linguistic values, *i.e.,* it is decided by experts' perceptions for linguistic values and the habits of language use, hence for the same weights for experts, different function w^* are corresponding to different aggregations. Selection of w^* is an optimization problem [17, 20], *i.e.,* the monotone increasing interpolated functions w^* is obtained by solving the following mathematical programming problem

$$\text{Maximize} : -\sum_{i=1}^{m} \omega_i \ln \omega_i$$

$$\text{subject to} : \sum_{i=1}^{m} \omega_i \frac{m-i}{m-1} = \alpha, \qquad (4)$$

$$\sum_{i=1}^{m} \omega_i = 1,$$

$$\omega_i \geq 0.$$

In which, ω_i is decided by (2), $-\sum_{i=1}^{m} \omega_i \ln \omega_i$ is the dispersion of f_{lwowa}, α is the desired attitudinal character.

Example 2. Continues Example 1. In Table 3, assume that weights for four experts are $(p_{e_1}, p_{e_2}, p_{e_3}, p_{e_4}) = (0.2, 0.3, 0.15, 0.35)$. Considering alternatives x_1 and x_2, we can obtain weights for linguistic values are $W_{x_1} = (w_{s_1}, w_{s_3}, w_{s_4}) = (0.25, 0.25, 0.5)$ and $W_{x_2} = (w_{s_3}, w_{s_4}, w_{s_1}) = (0.25, 0.5, 0.25)$, respectively.

In this example, functions $w^*_{x_1}$ and $w^*_{x_2}$ are selected as strictly piecewise linear functions which interpolate the points $(\frac{1}{3}, 0.25), (\frac{2}{3}, 0.5), (1, 1)$ and $(\frac{1}{3}, 0.25), (\frac{2}{3}, 0.75), (1, 1)$ together with the point $(0, 0)$, respectively, *i.e.*,

$$w^*_{x_1}(x) = \begin{cases} 0.75 \times x, & 0 \leq x \leq \frac{1}{3}, \\ 0.75 \times x, & \frac{1}{3} < x \leq \frac{2}{3}, \\ 1.5 \times x - 0.5, & \frac{2}{3} < x \leq 1. \end{cases}$$

$$w^*_{x_2}(x) = \begin{cases} 0.75 \times x, & 0 \leq x \leq \frac{1}{3}, \\ 1.5 \times x - 0.25, & \frac{1}{3} < x \leq \frac{2}{3}, \\ 0.75 \times x + 0.25, & \frac{2}{3} < x \leq 1. \end{cases}$$

According to $(p_{e_1}, p_{e_2}, p_{e_3}, p_{e_4}, p_{e_5}) = (0.2, 0.3, 0.15, 0.35)$, for alternative x_1, we have

$$\omega_1 = w^*(0.15) - w^*(0) = 0.1125,$$
$$\omega_2 = w^*(0.15 + 0.35) - w^*(0.15) = 0.2625,$$
$$\omega_3 = w^*(0.15 + 0.35 + 0.3) - w^*(0.15 + 0.35) = 0.325,$$
$$\omega_4 = w^*(1) - w^*(0.15 + 0.35 + 0.3) = 0.3.$$

The final evaluation of alternative x_1 based on the LWOWA operator is $f_{lwowa}(s_1, s_3, s_4, s_4) = s_{0.1125 \times 4 + 0.2625 \times 4 + 0.325 \times 3 + 0.3 \times 1} = s_{2.775} = (s_3, -0.225) = (M, -0.225)$. Similarly, for alternative x_2, we have

$$\omega_1 = w^*(0.15) - w^*(0) = 0.1125,$$
$$\omega_2 = w^*(0.15 + 0.35) - w^*(0.15) = 0.3875,$$
$$\omega_3 = w^*(0.15 + 0.35 + 0.3) - w^*(0.15 + 0.35) = 0.35,$$
$$\omega_4 = w^*(1) - w^*(0.15 + 0.35 + 0.3) = 0.15,$$

The final evaluation of alternative x_2 based on the LWOWA operator is $f_{lwowa}(s_3,$ $s_4, s_1, s_4) = s_{0.1125 \times 4 + 0.3875 \times 4 + 0.35 \times 3 + 0.15 \times 1} = s_{3.2} = (s_3, 0.2) = (M, 0.2)$.

In the following we analyze properties of the *LWOWA* operator when weight vectors $(w_{s_{k(1)}}, w_{s_{k(2)}}, \cdots, w_{s_{k(v)}})$ and $(p_{e_1}, p_{e_2}, \cdots, p_{e_m})$ are specialized.

Proposition 1. *The LWOWA operator f_{lwowa} satisfies the following propositioner-ties:*

1. *It is an aggregation operator which remains between the minimum and the maximum.*
2. *It satisfies idempotency.*
3. *It is commutative if and only if $p_{e_1} = \frac{1}{m}$ for all $i = 1, 2, \cdots, m$.*
4. *It is monotone in relation to the input values a_i.*
5. *It leads to dictatorship of the ith value when $\omega_i = 1$ and $\omega_j = 0$ for all $j = 1, \cdots, m$ but $j \neq i$.*

Proposition 2. *f_{lwowa} is the 2-tuple arithmetic mean operator $\bar{x}^e$ if it satisfies (1) for every $w_{s_{k(i)}}$, $w_{s_{k(i)}} = \frac{1}{v}$; (2) for every p_{e_j}, $p_{e_j} = \frac{1}{m}$; (3) w^* is a strictly piece-wise linear interpolation.*

Proof. Because w^* is a strictly piecewise linear interpolation, if $w_{s_{k(i)}} = \frac{1}{v}$, then $w^*(x) = x$. Hence, for every $p_{e_j} = \frac{1}{m}$, we have $\omega_k = \frac{1}{m}$, i.e., $f_{lwowa}((s_1, \alpha_1),$ $\cdots, (s_n, \alpha_n)) = \Delta(\sum_{j=1}^{n} \omega_j \cdot \beta_j^*) = \Delta(\sum_{j=1}^{n} \frac{1}{m} \times \beta_j^*) = \bar{x}^e((s_1, \alpha_1), \cdots, (s_n, \alpha_n))$.

4 Conclusion

From the practical point of view, group decision making is associated with multi-information sources fusion. In this paper, we propose the *LWOWA* operator to solve linguistic group decision making problems, which includes weights for linguistic values and weights for experts, the properties of the *LWOWA* operator shown that it is extensions of the 2-tuple arithmetic mean, the 2-tuple weighted aggregation operator and the 2-tuple ordered weighted averaging operator. In practice, the weights of the *LWOWA* operator can be obtained by solving a mathematical programming problem, this means that we can adjust influences of weights for experts by selecting the monotone increasing function of the *LWOWA* operator, hence, the *LWOWA* operator is an alternative linguistic aggregation operator in linguistic decision making problems.

Acknowledgement. This work is partly supported by national nature science foundation of China (Grant No.61105059, 61175055,61173100), Liaoning Excellent Talents in University(LJQ2011116), the research fund of Sichuan key laboratory of intelligent network information processing (SGXZD1002-10) and the key laboratory of the radio signals intelligent processing (Xihua university) (XZD0818-09).

References

1. Cabrerizo, F.J., Alonso, S., Herrera-Viedma, E.: A consensus model for group decision making problems with unbalanced fuzzy linguistic information. International Journal of Information Technology & Decision Making 8(1), 109–131 (2009)
2. Chen, S.M.: A new method for tool steel materials selection under fuzzy environment. Fuzzy Sets and Systems 92, 265–274 (1997)
3. Delgado, M., Verdegay, J.L.: On aggregation operations of linguistic labels. International Journal of Intelligent Systems 8, 351–370 (1993)
4. Geldermann, J., Spengler, T., Rentz, O.: Fuzzy outranking for environmental assessment. Case study: iron and steel making industry. Fuzzy Sets and Systems 115, 45–65 (2000)
5. Martinez, L., Ruan, D., Herrera, F.: Computing withWords in Decision support Systems: An overview on Models and Applications. International Journal of Computational Intelligence Systems 3(4), 382–395 (2010)
6. Herrera, F., Lopez, E., Mendana, C., Rodriguez, M.A.: A linguistic decision model for personnel management solved with a linguistic biojective genetic algorithm. Fuzzy Sets and Systems 118, 47–64 (2001)
7. Herrera, F., Herrera-Viedma, E., Martinez, L.: A fusion approach for managing multi-granularity linguistic term sets in decision making. Fuzzy Sets and Systems 114, 43–58 (2000)
8. Herrera, F., Alonso, S., Chiclana, F., Herrera-Viedma, E.: Computing With Words in Decision Making: Foundations, Trends and Prospects. Fuzzy Optimization and Decision Making 8(4), 337–364 (2009)
9. Huang, C.F., Ruan, D.: Fuzzy risks and an updating algorithm with new observations. Risk Analysis 28(3), 681–694 (2008)
10. Meng, D., Pei, Z.: The linguistic computational models based on index of linguistic label. The Journal of Fuzzy Mathematics 18(1), 9–20 (2010)
11. Merigó, J.M., Casanovas, M.: The uncertain generalized owa operator and its application to financial decision making. International Journal of Information Technology & Decision Making 10(2), 211–230 (2011)
12. Pei, Z., Ruan, D., Xu, Y., Liu, J.: Handling linguistic web information based on a multi-agent system. International Journal of Intelligent Systems 22, 435–453 (2007)
13. Pei, Z., Ruan, D., Liu, J., Xu, Y.: Linguistic Values based Intelligent Information Processing: Theory. In: Methods, and Application. Atlantis Computational Intelligence Systems, vol. 1. Atlantis press, World Scientific (2009)
14. Pei, Z., Xu, Y., Ruan, D., Qin, K.: Extracting complex linguistic data summaries from personnel database via simple linguistic aggregations. Information Sciences 179, 2325–2332 (2009)
15. Pei, Z.: Fuzzy risk analysis based on linguistic information fusion. ICIC Express Letters 3(3), 325–330 (2009)
16. Torra, V.: The weighted OWA operator. International Journal of Intellgent Systems 12, 153–166 (1997)
17. Wang, Y., Luo, Y., Liu, X.: Two new methods for determing OWA operator weights. Computer & Industrial Engineering 52, 203–209 (2007)
18. Wang, J., Lin, Y.I.: A fuzzy gruop decision making approach to select configuration items for software development. Fuzzy Sets and Systems 134, 343–363 (2003)
19. Yager, R.R.: On ordered weighted averaging aggregation operators in multicriteria decision making. IEEE Transactions on Systems Mean and Cybernetics 18, 183–190 (1998)

20. Yager, R.R.: On the dispersion measure of OWA operators. Information Sciences 179, 3908–3919 (2009)
21. Zou, L., Ruan, D., Pei, Z., Xu, Y.: A linguistic truth-valued reasoning approach in decision making with incomparable information. Intelligent and Fuzzy Systems 19(4-5), 335–343 (2008)
22. Zou, L., Liu, X., Wu, Z., Xu, Y.: A uniform approach of linguistic truth values in sensor evaluation. Fuzzy Optimization and Decision Making 7(4), 387–397 (2008)
23. Zou, L., Pei, Z., Liu, X., Xu, Y.: Semantic of Linguistic Truth-Valued Intuitionistic Fuzzy Proposition Calculus. International Journal of Innovative Computing. Information and Control 5(12), 4745–4752 (2009)

Fuzzy Multi-objective Programming Problem with Fuzzy Structured Element Solution

Yun-zhi Liu and Si-zong Guo

Liaoning Technical University, Institute of Mathematics and Systems Science,
Fuxin 123000, China
lyz19850521@126.com

Abstract. A class of fuzzy multiple objective programming problem with all fuzzy coefficients is discussed based on the fuzzy structured element method in this paper. By introducing the structured element weighted characteristic number, we define an order relation of fuzzy numbers. Then, we use of this characteristic number, the solving problems of the class of fuzzy multi-objective programming model can be translated into the other solving problems of a class of clear multi-objective programming model. We calculate the quasi-optimal feasible solution of the clear model by the method based on a class of linear weighted function. Finally, a numerical example is given to illustrate how to solve such fuzzy multi-objective programming problem.

Keywords: Structured element, structured element weighted characteristic number, order relation, linear weighted function, quasi-optimal feasible solution.

1 Introduction

There are a lot of fuzziness in many fields of science. Fuzzy programming is considered as a powerful tool to solve the optimization problem of the decision-making with fuzzy parameters. Therefore, many scholars have done a lot of research of the fuzzy programming. Bellman, Zadeh [1] first proposed the concept of decision-making under fuzzy environment. In the case of fuzzy parameters, Tong Shaocheng [2], etc. obtained the optimal solution of a class of linear programming model. Cadenas, Verdegay [3] considered a linear multi-objective programming problem with the fuzzy target coefficient, and given a solution by functions sorting. Maleki, Tata, Mashinchi [4] considered a linear programming with fuzzy variables, and given a solution of the linear programming by applying the concept of fuzzy number comparison. Professor Bao-ding Liu, etc. proposed stochastic programming and fuzzy programming unified theory [5], and laid the foundation for the optimization theory in generally uncertain environment. Meanwhile, we not only made a series of advances of the fuzzy programming problem in theory but also apply these theories to many fields.

In this paper, a class of fuzzy multiple objective programming with all fuzzy coefficients is discussed. And the quasi-optimal feasible solution of the above model is given based on the fuzzy structured element method.

B.-Y. Cao and X.-J. Xie (Eds.): Fuzzy Engineering and Operations Research, AISC 147, pp. 97–107.
springerlink.com   © Springer-Verlag Berlin Heidelberg 2012

2 Fuzzy Structured Element Method and Structured Element Weighted Ranking of Fuzzy Numbers

2.1 Fuzzy Structured Element Method

As prior knowledge, we first do the fuzzy structured element method to a brief. And the detailed content of the fuzzy structured element method can be found in [6].

Let E be a fuzzy set in the real number domain R, $E(x)$ be the membership function of E. We call that E is a fuzzy structured element in R, if

(i) $E(0) = 1$, $E(1+0) = E(-1-0) = 0$.

(ii) $E(x)$ is a monotone increasing right continuous function on [-1,0], and monotonic decreasing left continuous on (0,1].

(iii) $E(x) = 0$ ($-\infty < x < -1$ and $1 < x < +\infty$).

From the definition of fuzzy structured element may seen that E is a special fuzzy number.

E is called a regular fuzzy structured element, if

(i) $\forall x \in (-1,1), E(x) > 0$.

(ii) The membership function $E(x)$ is a strictly monotone increasing and continuous on [-1, 0], strictly monotone decreasing and continuous on (0, 1].

If $E(-x) = E(x)$, then E is called a symmetrical fuzzy structured element.

Let E be a fuzzy structured element, and $E(x)$ be its membership function, $f(x)$ be a bounded monotone function on [-1,1]. Then $f(E)$ is a bounded closed fuzzy number, and the membership function of $f(E)$ is $E(f^{-1}(x))$, where $f^{-1}(x)$ is the rotation symmetric function of $f(x)$ (if $f(x)$ is a continuous strictly monotone, $f^{-1}(x)$ is the inverse function of $f(x)$). Otherwise, given a regular fuzzy structured element E and arbitrary bounded fuzzy number $\tilde{A}$, there exist a bounded monotone function $f(x)$ in [-1,1], making $\tilde{A} = f(E)$. We call fuzzy number $\tilde{A}$ is generated by the structured element E.

2.2 Structured Element Weighted Ranking of Fuzzy Numbers

All bounded closed fuzzy numbers are denoted as $\tilde{N}_C(R)$.

Definition 2.1[8]. *Let $\tilde{A}$ be generated by the structured element E. Then $\tilde{A} = f(E)$, where E is a given fuzzy structured element, $f(x)$ is a bounded monotone function on [-1,1], then*

$$\rho(\tilde{A}) = \int_{-1}^{1} E(x)f(x)dx, \tag{1}$$

$\rho(\tilde{A})$ *is called structured element weighted characteristic number of $\tilde{A}$, and is simply called characteristic number of $\tilde{A}$.*

Definition 2.2. *Let $\tilde{A}_1, \tilde{A}_2 \in \tilde{N}_c(R)$ be generated by the structured element E. Then $\tilde{A}_i = f_i(E)(i = 1,2)$, where E is a given regular fuzzy structured element, $E(x)$ is its membership function, and f_1, f_2 are monotone functions with same monotonic formal on [-1,1], then order relation of two fuzzy numbers is defined as*

$$\tilde{A}_1 \le \tilde{A}_2 \Leftrightarrow \rho(\tilde{A}_1) \le \rho(\tilde{A}_2), \tag{2}$$

"$\le$" is called structured element weighted order of fuzzy numbers.

Easy to know, "$\le$" is the total order on $\tilde{N}_C(R)$.

According to the literature [6], let $\tilde{A}, \tilde{A}_1, \tilde{A}_2 \in \tilde{N}_c(R)$ be generated by the structured E, then $\tilde{A} = f(E)$, $\tilde{A}_1 = f_1(E)$, $\tilde{A}_2 = f_2(E)$. So we have $k\tilde{A} = kf(E)$, $\tilde{A}_1 + \tilde{A}_2 = (f_1 + f_2)(E)$. By Definition 2.1, we are easy to get Property 2.1.

Property 2.1. *ρ satisfy the following properties, that*

(i) *If $k \in R, \tilde{A} \in \tilde{N}_C(R)$, then $\rho(k\tilde{A}) = k\rho(\tilde{A})$.*

(ii) *If $\tilde{A}_1, \tilde{A}_2 \in \tilde{N}_C(R)$, then $\rho(\tilde{A}_1 + \tilde{A}_2) = \rho(\tilde{A}_1) + \rho(\tilde{A}_2)$.*

3 Fuzzy Multi-objective Programming Model and Solution

Definition 3.1. *If $x = (x_1,...,x_n) \in R^n$, $\tilde{c} = (\tilde{c}_1,...,\tilde{c}_m)$ and $\tilde{c}_i \in \tilde{N}_C(R)$, $i = 1,...,m$, then*

$$H(x;\tilde{c}) = H(x_1,...,x_n;\tilde{c}_1,...,\tilde{c}_m) = \sum_{i=1}^{m}\tilde{c}_ih_i(x) = \sum_{i=1}^{m}\tilde{c}_ih_i(x_1,...,x_n).$$

We call this function is a fuzzy linear combination function with respect x, *where* $h_i(x) = h_i(x_1, \ldots, x_n)$ *is any function with respect* x. *And all the fuzzy linear combination functions are denoted as* $FLC(R)$.

Consider the following fuzzy multi-objective programming model

$$\max \tilde{Z}_k = F_k\left(x_1, x_2, \cdots, x_n; \tilde{c}_1^{(k)}, \tilde{c}_2^{(k)}, \cdots, \tilde{c}_m^{(k)}\right), k = 1, 2, \cdots, n_1$$

$$s.t \begin{cases} G_i\left(x_1, x_2, \cdots, x_n; \tilde{d}_1^{(i)}, \tilde{d}_2^{(i)}, \cdots, \tilde{d}_l^{(i)}\right) \leq \tilde{D}_i, & i = 1, 2, \cdots, n_2, \\ x_i \geq 0, i = 1, 2, \cdots, n, \end{cases} \tag{3}$$

where $\tilde{Z}_k$ $(k = 1, \ldots, n_1)$ is the k-objective function, $F_k, G_i \in FLC(R)$, that

$$F_k\left(x_1, x_2, \cdots, x_n; \tilde{c}_1^{(k)}, \tilde{c}_2^{(k)}, \cdots, \tilde{c}_m^{(k)}\right) = \sum_{j=1}^{m} \tilde{c}_j^{(k)} f_j\left(x_1, x_2, \cdots, x_n\right),$$

$$G_i\left(x_1, x_2, \cdots, x_n; \tilde{d}_1^{(i)}, \tilde{d}_2^{(i)}, \cdots, \tilde{d}_l^{(i)}\right) = \sum_{t=1}^{l} \tilde{d}_t^{(i)} g_t\left(x_1, x_2, \cdots, x_n\right).$$

The coefficients of objective functions are $\tilde{c}_j^{(k)} \in \tilde{N}_C(R)$, the coefficients of constraints are $\tilde{d}_t^{(i)} \in \tilde{N}_C(R)$, and $j = 1, 2, \ldots, m$, $t = 1, 2, \cdots, l$. We note

$$x = (x_1, x_2, \ldots, x_n), \ \tilde{c}^{(k)} = \left(\tilde{c}_1^{(k)}, \tilde{c}_2^{(k)}, \cdots, \tilde{c}_m^{(k)}\right), \ \tilde{d}^{(i)} = \left(\tilde{d}_1^{(i)}, \tilde{d}_2^{(i)}, \cdots, \tilde{d}_l^{(i)}\right).$$

Model (3) also can be written as

$$\max \tilde{Z}_k = F_k\left(x; \tilde{c}^{(k)}\right), k = 1, 2, \cdots, n_1$$

$$s.t \begin{cases} G_i\left(x; \tilde{d}^{(i)}\right) \leq \tilde{D}_i, & i = 1, 2, \cdots, n_2, \\ x \geq 0 \end{cases}$$

and

$$F_k\left(x; \tilde{c}^{(k)}\right) = \sum_{j=1}^{m} \tilde{c}_j^{(k)} f_j(x),$$

$$G_i\left(x; \tilde{d}^{(i)}\right) = \sum_{t=1}^{l} \tilde{d}_t^{(i)} g_t(x).$$

In particular, when both $f_j(x)$ and $g_t(x)$ are liner functions with respect x, Model (3) will degenerate into a class of fuzzy multi-objective linear programming model.

We call the solution vector x that satisfies the model constraints is a feasible solution of Model (3). And all feasible solution sets of Model (3) are denoted as X. If $\bar{x} = (\bar{x}_1, ..., \bar{x}_p) \in X$, and there is no other feasible solution $x' = (x'_1, ..., x'_p) \neq \bar{x}$

$\in X$, there $F_k(\tilde{c}^{(k)}, x') \geq F_k(\tilde{c}^{(k)}, \bar{x})(k = 1, 2, ..., n_1)$, $\bar{x} = (\bar{x}_1, ..., \bar{x}_p) \in X$

is called an efficient solution of Model (3). And all efficient solution sets are denoted as $\bar{X}$.

Definition 3.2. *If the mapping $\psi : X \to R$, where X is the feasible solution sets of Model (3), then ψ is called a quasi-optimal function of all the objective functions. $x^* = (x_1^*, ..., x_p^*)$ is the quasi-optimal feasible solution of Model (3), if and only if $x^* \in \bar{X}$, $\psi(x^*) = \max\limits_{x \in X} \psi(x)$, where $\psi(x) = \psi(\tilde{Z}_1(x), ..., \tilde{Z}_{n_1}(x))$.
Here we will discuss how to find the quasi-optimal feasible solution of Model (3).*

Let $\rho(\tilde{A})$ be the characteristic number of $\tilde{A}$. Consider the following model

$$\max \rho(\tilde{Z}_k) = F_k\left(x_1, x_2, \cdots, x_n; \rho(\tilde{c}_1^{(k)}), \rho(\tilde{c}_2^{(k)}), \cdots, \rho(\tilde{c}_m^{(k)})\right)$$

$$k = 1, 2, \cdots, n_1 \tag{4}$$

$$s.t \begin{cases} G_i\left(x_1, x_2, \cdots, x_n; \rho(\tilde{d}_1^{(i)}), \rho(\tilde{d}_2^{(i)}), \cdots, \rho(\tilde{d}_l^{(i)})\right) \leq \rho(\tilde{D}_i), \ i = 1, 2, \cdots, n_2, \\ x_i \geq 0, i = 1, 2, \cdots, n, \end{cases}$$

simply denoted as

$$\max \rho(\tilde{Z}_k) = F_k\left(x; \rho(\tilde{c}^{(k)})\right), k = 1, 2, \cdots, n_1$$

$$s.t \begin{cases} G_i\left(x; \rho(\tilde{d}^{(i)})\right) \leq \rho(\tilde{D}_i), \ i = 1, 2, \cdots, n_2, \\ x \geq 0 \end{cases}$$

and

$$F_k\left(x; \rho(\tilde{c}^{(k)})\right) = \sum_{j=1}^{m} \rho(\tilde{c}_j^{(k)}) f_j(x),$$

$$G_i\left(x;\rho\left(\tilde{d}^{(i)}\right)\right)=\sum_{t=1}^{l}\rho\left(\tilde{d}_t^{(i)}\right)g_t(x).$$

Theorem 3.1. *If* x^* *is the quasi-optimal feasible solution of Model (4), then* x^* *must be the quasi-optimal feasible solution of Model (3).*

Proof. Let X_1 be the feasible solution sets of Model (3), and $\overline{X}_1$ be the efficient solution sets of Model (3). Let X_2 be the feasible solution sets of Model (4), and $\overline{X}_2$ be the efficient solution sets of Model (4). Then by the Property 2.1, there

$$\forall x\in X_2 \Leftrightarrow G_i\left(x;\rho\left(\tilde{d}^{(i)}\right)\right)=\sum_{t=1}^{l}\rho\left(\tilde{d}_t^{(i)}\right)g_t(x)\le\rho(\tilde{D}_i)$$

$$\Leftrightarrow \rho\left(G_i(x;\tilde{d}^{(i)})\right)=\rho\left(\sum_{t=1}^{l}\tilde{d}_t^{(i)}g_t(x)\right)\le\rho(\tilde{D}_i)\ ,$$

$$\Leftrightarrow G_i(x;\tilde{d}^{(i)})=\sum_{t=1}^{l}\tilde{d}_t^{(i)}g_t(x)\le\tilde{D}_i \Leftrightarrow x\in X_1$$

that $X_1=X_2$.

$\forall \overline{x}\in \overline{X}_2 \Leftrightarrow$ there is no other feasible solution $x'=\left(x_1',...,x_p'\right)\ne \overline{x}\in X_2$, $k=1,2,...,n_1$, there

$$F_k\left(x';\rho(\tilde{c}^{(k)})\right)=\sum_{j=1}^{m}\rho\left(\tilde{c}_j^{(k)}\right)f_j(x')\ge F_k\left(\overline{x};\rho(\tilde{c}^{(k)})\right)=\sum_{j=1}^{m}\rho\left(\tilde{c}_j^{(k)}\right)f_j(\overline{x})$$

$\Leftrightarrow$ there is no other feasible solution $x'=\left(x_1',...,x_p'\right)\ne \overline{x}\in X_2$, $k=1,2,...,n_1$, there

$$\rho\left(F_k(x';\tilde{c}^{(k)})\right)=\rho\left(\sum_{j=1}^{m}\tilde{c}_j^{(k)}f_j(x')\right)\ge\rho\left(F_k(\overline{x};\tilde{c}^{(k)})\right)=\rho\left(\sum_{j=1}^{m}\tilde{c}_j^{(k)}f_j(\overline{x})\right)$$

$\Leftrightarrow$ there is no other feasible solution $x'\ne \overline{x}\in X_1$, $k=1,2,...,n_1$, there

$$F_k(x';\tilde{c}^{(k)})=\sum_{j=1}^{m}\tilde{c}_j^{(k)}f_j(x')\ge F_k(\overline{x};\tilde{c}^{(k)})=\sum_{j=1}^{m}\tilde{c}_j^{(k)}f_j(\overline{x})$$

$\Leftrightarrow \overline{x}\in \overline{X}_1$

that $\overline{X}_1 = \overline{X}_2$.

If $x^* \in \overline{X}_2$ is the quasi-optimal feasible solution of Model (4), then there exists a quasi-optimal function ψ_2, that

$$\psi_2(x) = \psi_2\left(\rho\left(\tilde{Z}_1(x)\right), \rho\left(\tilde{Z}_2(x)\right), \ldots, \rho\left(\tilde{Z}_{n_1}(x)\right)\right),$$

make $\psi_2(x^*) = \max\limits_{x \in X} \psi_2(x)$, so that

$$\psi_1(x) = \psi_1\left(\tilde{Z}_1(x), \tilde{Z}_2(x), \ldots, \tilde{Z}_{n_1}(x)\right) = \psi_2(x) = \psi_2\left(\rho(\tilde{Z}_1(x)), \rho(\tilde{Z}_2(x)), \ldots, \rho(\tilde{Z}_{n_1}(x))\right),$$

then $\psi_1(x^*) = \max\limits_{x \in X} \psi_1(x)$, that ψ_1 is a quasi-optimal function of Model(3). So $x^* \in \overline{X}_2 = \overline{X}_1$ is the quasi-optimal feasible solution of Model (3).

By Theorem 3.1, we can see that the solving problems of Model (3) can be translated into the other solving problems of Model (4). In order words, the solving problems of a class of fuzzy multi-objective programming model can be translated into the other solving problems of a class of clear multi-objective pro-gramming model.

Then we will discuss how to find the quasi-optimal feasible solution of Model(4). The optimal solution of each single objective programming problem in Model(4) is denoted as x_k^*, and the optimal objective function value is denoted as $\rho\left(\tilde{Z}_k\right)^*$, $k = 1, \ldots, n_1$. Obviously, if x_k^* is the same one, then x_k^* is the quasi-optimal feasible solution, but the chances of this happening is very small. So we give the weight value λ_k ($\sum\limits_{k=1}^{n_1} \lambda_k = 1$) of each single objective base on the impor-tance of the single-objective goals in the overall. In general, the weight values are known. In exceptional circumstances, we can calculate the weight values based on some given conditions. In the case of each weight value is known, we construct a linear weighted function as

$$h(x) = \sum_{k=1}^{n_1} \lambda_k \rho(\tilde{Z}_k).$$

Then we put the linear weighted function as a new objective function, and construct a new single objective programming model

$$\max h(x) = \sum_{k=1}^{n_1} \lambda_k \rho(\tilde{Z}_k)$$

$$s.t \begin{cases} G_i\left(x; \rho(\tilde{d}^{(i)})\right) \le \rho(\tilde{D}_i), & i = 1, 2, \cdots, n_2, \\ x \ge 0. \end{cases} \tag{5}$$

Obviously, this model is a clear programming model.

Theorem 3.2. *If x^* is the only optimal solution of Model (5), then x^* must be the quasi-optimal feasible solution of Model (4).*

Proof. By the Definition 3.2, we can know that $h(x) = \sum_{k=1}^{n_1} \lambda_k \rho(\tilde{Z}_k)$ is a quasi-optimal function. Since x^* is the only optimal solution of Model (5), then

$$h(x^*) = \max_{x \in X} h(x).$$

If it is assumed that $x^* \notin \overline{X}$, then there is a $x' \ne x^* \in X$, $k = 1, 2, ..., n_1$, there

$$h(x') \ge h(x^*).$$

This is inconsistent with $h(x^*) = \max_{x \in X} h(x)$, if $h(x') = h(x^*)$, which is also inconsistent with x^* being the only optimal solution. Therefore, the null hypothesis does not hold, then $x^* \in \overline{X}$. By the Definition 3.2, x^* must be the quasi-optimal feasible solution of Model (4).

According to Theorem 3.1 and Theorem 3.2, if x^* is the only optimal solution of Model (5), then x^* must be the quasi-optimal feasible solution of Model (3).

4   Numerical Example

Fuzzy multi-objective programming model

$$\max \tilde{Z}_1 = -\tilde{1}x_1^2 + \tilde{2}x_2$$

$$\max \tilde{Z}_2 = 2.\tilde{5}x_1 + 0.\tilde{5}x_1x_2$$

$$s.t \begin{cases} \tilde{1}x_1^2 + \tilde{2}x_2 \le \tilde{6}, \\ \tilde{4}x_1 + \tilde{3}x_1x_2 \ge 1\tilde{2}, \\ \tilde{3}x_1 + \tilde{2}x_2 \ge \tilde{8}, \\ x_1, x_2 \ge 0, \end{cases}$$

or

$$\max \tilde{Z}_1 = -\tilde{1}x_1^2 + \tilde{2}x_2$$

$$\max \tilde{Z}_2 = 2.\tilde{5}x_1 + 0.\tilde{5}x_1x_2$$

$$s.t \begin{cases} \tilde{1}x_1^2 + \tilde{2}x_2 \le \tilde{6}, \\ -\tilde{4}x_1 - \tilde{3}x_1x_2 \le -\tilde{12}, \\ -\tilde{3}x_1 - \tilde{2}x_2 \le -\tilde{8}, \\ x_1, x_2 \ge 0. \end{cases}$$

It is assumed that each coefficients is linear generated by the triangular fuzzy structured element E. And each coefficient is expressed as

$$\tilde{c}_1^{(1)} = -\tilde{1} = (-1+0.5x)_{x=E}, \tilde{c}_2^{(1)} = \tilde{2} = (2+x)_{x=E};$$

$$\tilde{c}_1^{(2)} = 2.\tilde{5} = (2.5+0.5x)_{x=E}, \tilde{c}_2^{(2)} = 0.\tilde{5} = (0.5+x)_{x=E};$$

$$\tilde{d}_1^{(1)} = \tilde{1} = (1+0.5x)_{x=E}, \tilde{d}_2^{(1)} = \tilde{2} = (2+x)_{x=E};$$

$$\tilde{d}_1^{(2)} = -\tilde{4} = (-4+x)_{x=E}, \tilde{d}_2^{(2)} = -\tilde{3} = (-3+0.5x)_{x=E};$$

$$\tilde{d}_1^{(3)} = -\tilde{3} = (-3+x)_{x=E}, \tilde{d}_2^{(3)} = -\tilde{2} = (-2+x)_{x=E};$$

$$\tilde{D}^{(1)} = \tilde{6} = (6+x)_{x=E}, \tilde{D}^{(2)} = -\tilde{12} = (-12+x)_{x=E}, \tilde{D}^{(3)} = -\tilde{8} = (-8+x)_{x=E}$$

$E(x)$ is the membership function of the triangular fuzzy structured element E, there

$$E(x) = \begin{cases} 1+x, x \in [-1,0], \\ 1-x, x \in [0,1], \\ 0, others. \end{cases}$$

And $\lambda_1 = 0.3$ is the weight value of the objective function $\tilde{Z}_1$, $\lambda_2 = 0.7$ is the weight value of the objective function $\tilde{Z}_2$.

By Definition 2.1, we get the characteristic number of each fuzzy coefficient, there

$$\rho(\tilde{c}_1^{(1)}) = \int_{-1}^{1} E(x)(-1+0.5x)dx = \int_{-1}^{0}(1+x)(-1+0.5x)dx + \int_{0}^{1}(1-x)(-1+0.5x)dx = -1 \cdot$$

Similarly

$$\rho\left(\tilde{c}_2^{(1)}\right)=2,\ \rho\left(\tilde{c}_1^{(2)}\right)=2.5,\ \rho\left(\tilde{c}_2^{(2)}\right)=0.5;$$

$$\rho\left(\tilde{d}_1^{(1)}\right)=1,\ \rho\left(\tilde{d}_2^{(1)}\right)=2;$$

$$\rho\left(\tilde{d}_1^{(2)}\right)=-4,\ \rho\left(\tilde{d}_2^{(2)}\right)=-3;$$

$$\rho\left(\tilde{d}_1^{(3)}\right)=-3,\ \rho\left(\tilde{d}_2^{(3)}\right)=-2;$$

$$\rho\left(\tilde{D}^{(1)}\right)=6,\ \rho\left(\tilde{D}^{(2)}\right)=-12,\ \rho\left(\tilde{D}^{(3)}\right)=-8.$$

Thus, this fuzzy multi-objective programming model can be transformed into the following clear multi-objective model, as following

$$\max \rho\left(\tilde{Z}_1\right)=-x_1^2+2x_2$$

$$\max \rho\left(\tilde{Z}_2\right)=2.5x_1+0.5x_1x_2$$

$$s.t \begin{cases} x_1^2+2x_2 \le 6, \\ -4x_1-3x_1x_2 \le -12, \\ -3x_1-2x_2 \le -8, \\ x_1,x_2 \ge 0. \end{cases}$$

We construct the linear weighted function as

$$h(x)=\lambda_1\rho\left(\tilde{Z}_1\right)+\lambda_2\rho\left(\tilde{Z}_2\right)=-0.3x_1^2+1.75x_1+0.35x_1x_2+0.6x_2.$$

So we further create a single objective programming model as Model (5). There

$$\max h(x)=-0.3x_1^2+1.75x_1+0.35x_1x_2+0.6x_2$$

$$s.t \begin{cases} x_1^2+2x_2 \le 6, \\ -4x_1-3x_1x_2 \le -12, \\ -3x_1-2x_2 \le -8, \\ x_1,x_2 \ge 0. \end{cases}$$

By solving the above clear single objective programming model, we get it the optimal solution x_h^*, then we also get the quasi-optimal feasible solution $x^* = x_h^*$ of the original fuzzy multi-objective programming model.

5   Conclusion

Through the above examples, this solution presented in this paper is very simple for a class of fuzzy multi-objective programming problem with all fuzzy coefficients. So a very effective and simple solution of such fuzzy multi-objective programming problem is provided for various engineering fields.

Acknowledgements. Thanks to the support by Doctoral Program of the Ministry of Education's Specialized Research Fund of China (No.20102121110002).

References

1. Bellman, R.E., Zadeh, L.A.: Decision making in a fuzzy environment. Management Sci. 17, 141–164 (1970)
2. Tong, S.C.: Interval number and fuzzy number linear programming. Fuzzy Sets and Systems 66, 301–306 (1994)
3. Carlsson, C., Fuller, R.: Fuzzy multiple criteria decision making. Fuzzy Sets and Systems 78, 139–153 (1996)
4. Cadenas, J.M., Verdegay, J.L.: Using ranking functions in multiobjective fuzzy linear programming. Fuzzy Sets and Systems 111, 47–53 (2000)
5. Liu, B.D., Zhao, R.Q.: Stochastic Programming and Fuzzy Programming. Tsinghua University Press, Beijing (1998)
6. Guo, S.Z.: Principle of Fuzzy Mathematical Analysis Based on Structured Element. Northeastern University Press (2004)
7. Guo, S.Z.: Comparison and sequencing of fuzzy numbers based on the method of structured element. Systems Engineering Theory and Practice 3, 106–111 (2009)
8. Zhao, H.K., Guo, S.Z.: Bi-level Linear Programming with All-coefficient-fuzzy. Fuzzy Systems and Mathematics 3, 98–106 (2010)

Soft Sensor Modeling Based on Fuzzy System Optimization

Yan Sun[1], Li-Biao Zhang[2] and Ming Ma[1,*]

[1] College of Information Technology , Beihua University, Jilin 132013, China
  ma9063@163.com
[2] College of Computer Science and Technology, Jilin University, Changchun 130012, China

Abstract. In order to implement real-time control or optimization for variables key to the process, we need to build soft sensor, and the key step of it is soft sensor modeling. In this paper, the soft sensor modeling process based on Takagi-Sugeno (T-S) model and Differential Evolution (DE) were discussed. The proposed algorithm could evolve both the structure of T-S model and parameters, and effectively solves the problem of soft sensor modeling. The numerical experiments indicate the effectiveness of the algorithm.

Keywords: Soft sensor, T-S fuzzy model, differential evolution.

1 Introduction

In industrial processes, some important variables are difficult to be detected ,many hardware sensors usually involve significant time lags and high investment and maintenance costs, instead of hardware sensors, soft sensors [1], provide a convenient solution to solve this problem. In recent years, many algorithms have been proposed to build the soft sensor, including using artificial neural networks,and support vector machine etc[2-5].

Data-driven fuzzy modeling has become an active research area in recent years, and it has been applied to many fields, such as pattern recognition, data mining, classification, prediction, and process control, and etc [6-9]. In this paper, an algorithm based on Differential Evolution (DE) is proposed to optimize Takagi-Sugeno (T-S) model, the proposed algorithm could evolve both the topology and parameters, and to a problem it can obtain the near-optimal structure of T-S model. The experimental analysis and calculation shows that the algorithm is good in solving fuzzy system optimization, and effectively solves the problem of soft sensor.

The rest of this paper is organized as follows: The T-S model is introduced in Section 2. The proposed algorithm is described in Section 3. The simulation and experimental results are presented in Section 4. Finally, concluding remarks are given in Section 5.

---

* Corresponding author.

B.-Y. Cao and X.-J. Xie (Eds.): Fuzzy Engineering and Operations Research, AISC 147, pp. 109–114.
springerlink.com
© Springer-Verlag Berlin Heidelberg 2012

2   T-S Model

Takagi-Sugeno (T-S) model is a fuzzy system proposed by Takagi and Sugeno in 1985[10]. As a method of data-driven modeling, it has been successfully used in a wide variety of applications. In the model the ith fuzzy rule have the form

$$\mathbf{R_i} : x_1 \ is \ A_{i1}, ..., x_n \ is \ A_{in} \ then \ y_i = c_{i0} + c_{i1}x_{i1} + ... + c_{in}x_n, \qquad (1)$$

where n is the number of input variables. $i = 1 \ldots r$,and r is the number of if-then rules. A_{ij} is the antecedent fuzzy set of the ith rule. y_i is the consequence of the ith if-then rule $.c_{ij}(i = 1 \ldots r; j = 1 \ldots n)$ is real number. Then by using center of gravity method for defuzzification, we can represent the T-S system as:

$$\mathbf{y} = \frac{\sum\limits_{i=1}^{r} y_i \prod\limits_{j=1}^{n} \mu_{Aij}(x_i)}{\sum\limits_{i=1}^{r} \prod\limits_{j=1}^{n} \mu_{Aij}(x_i)}. \qquad (2)$$

3   The Proposed Algorithm

3.1   Differential Evolution

Differential Evolution (DE) is an optimization algorithm proposed by Storn and Price in 1997[11]. It has shown superior performance in both widely used benchmark functions and real-world applications [12-15]. It combines simple arithmetic operators with the classical events of crossover, mutation and selection to evolve from a randomly generated starting population to a final solution. DE executes its mutation by adding a weighted difference vector between two individuals to a third individual, and then the mutated individuals will do discrete crossover and produce offspring, in the final step, each offspring in the child population is evaluated for fitness on a parent, with only the stronger of the two surviving into the next generation.

Individuals in DE are represented by D-dimensional vectors $x_i, \forall i \in (1, \cdots, NP)$, where D is the number of optimization parameters and NP is the population size. The evolutionary operations of classical DE can be summarized as follow:

1) Mutation

$$v_i = x_{r1} + F \times (x_{r2} - x_{r3}), i = 1, \cdots, np, \qquad (3)$$

where r_1, r_2, and r_3 are three mutually distinct randomly drawn indices from $(1, \cdots, np)$, and also distinct from i, and $0 < F <= 2$.

2) Crossover

$$u_{ji} = \begin{cases} v_{ji}, \ if \ randb \leq CR \ or \ j = randr, \\ x_{ji}, \ if \ randb > CR \ or \ j \neq randr, \end{cases} \qquad (4)$$

where $i = 1, \cdots, NP, j = 1, \cdots, D, and\ CR \in (0,1)$ is the crossover rate, and randr is a random integer in [1, D].

This ensures at least some crossover, one component of u_i is selected at random to be from v_i.

3) Selection

If the objective value $f(u_i)$ is lower than $f(x_i)$, then u_i replaces x_i in the next generation. Otherwise, we keep x_i.

3.2 Description of the Algorithm

In the proposed algorithm, we use n dimensional real-valued vector to represent a solution, as shown in follows:

$$X = (x_1, x_2, \cdots, x_n),\tag{5}$$

where n is the number of all the real-valued parameters, it include the parameters of membership functions and consequence parameters $c_{ij} (i = 1, \cdots, r; j = 1, \cdots, n)$.

The following functions have been used for evaluation of DE.

$$F(x) = \frac{1}{\sum_K (O - T)^2},\tag{6}$$

where K is the number of sample, T is the teacher signal, and O is the output.

The proposed algorithm is formed of two phases. In the first phase, through mutation, crossover and fitness-based selection, it evolves from an initial randomly generated population to a solution. However DE can't be immediately applied to update parameters of membership functions, compared with the consequence parameter, the optimization of membership functions only adjusts a little. This paper has made some improvement on mutation and crossover, DE doesn't execute mutation, and the crossover is shown as follows:

$$u_{ji} = \begin{cases} r_{ji}, & if\ randb \leq CR\ or\ j = randr, \\ x_{ji}, & if\ randb > CR\ or\ j \neq randr, \end{cases}\tag{7}$$

where r is a randomly drawn index from $(1, \cdots, np)$, and distinct from i. $i = 1, \cdots, NP, j = 1, \cdots, D, and\ CR \in (0,1)$ is the crossover rate, and randr is a random integer in [1, D].

If the fixed precision is achieved, then go to the second phase. In the second phase, we used the effective sample to cut redundant fuzzy rules. In the T-S model an antecedent of the ith fuzzy rule has the form:

$$x_1\ is\ A_{i1}\ x_2\ is\ A_{i2}, ..., x_n\ is\ A_{in}.\tag{8}$$

The possibility that the ith rule will fire is given by the product of all the membership functions associated with the ith rule.

$$u_i(x_k) = \prod_{j=1}^{n} A_{ij}(x_j(k)). \tag{9}$$

A sample (x_k) is defined as the effective sample if $u_i(x_k) > a$, where a is a threshold.

Calculate the value for each sample according to equation (9), all the effective samples of ith fuzzy rule can be obtained. If a fuzzy rule has no effective sample or number of effective samples are small enough, that is to say the fuzzy rule is redundant, then cut the fuzzy rule.

Initialize the population again, and loop the above process until it achieves the termination condition, we will obtain a suitable T-S model.

3.3 The Execution of the Algorithm

The algorithm:

1. Randomly generate an initial population, and evaluate the fitness of each individual.

2. Mutation, crossover, selection.

3. Evaluate the fitness of each individual.

4. If the termination condition is achieved then stop, otherwise go to 5.

5. If the fixed precision is achieved then go to 6 otherwise go to 2.

6. Picked the best individual to optimize fuzzy rules according to the effective sample.

7. Changed the structure of all individuals according to the best individual, go to 1.

4 Numerical Simulations

We used the proposed algorithm to build the soft sensor modeling. Reaction temperature, propylene flow rate, air flow rate and ammonia gas flow rate are selected as the secondary variables to predict the yield coefficient of acrylonitrile.The 80 samples were used as training data, and the 40 samples as test data to validate the model's performance.

In our algorithm we used 3 fuzzy sets for each input, and the Gaussian membership function is used for each fuzzy subset. We randomly set the population size to 60, probability of crossover CR to 0.8, let F=0.6, and stopping condition: 3000 generations. Before the execution of the algorithm the T-S model has 81 fuzzy rules. The algorithm which immediately used DE to optimize T-S model is called algorithm 1, the proposed algorithm is called algorithm 2, and the comparative results between two algorithms are summarized in Table 1.

From this table we can observe that the proposed algorithm achieves the better result in the test data, with a relatively small number of fuzzy rules. The results have proved that the proposed algorithm is applicable and efficient.

Table 1. Comparative results

Algorithm	Number of Fuzzy rules	MSE(Training date)	MSE(Test date)
algorithm 1	81	0.29	1.64
algorithm 2	32	0.33	0.76

5 Conclusion

Based on DE and T-S model, the proposed algorithm can obtain suitable fuzzy rules and optimize parameters of model.It can solve the soft sensor modeling problem. A real-world simulation was presented to show the advantage of the algorithm.

References

1. Martin, G.: Consider soft sensors. Chemical Engineering Progress 66(7), 66–70 (1997)
2. Bhartiya, S., Whiteley, J.R.: Development of Inferential measurements Using Neural Networks. ISA Transactions 40(4), 307–323 (2001)
3. Chen, W., Li, J.M.: Adaptive Output-feedback Regulation for Nonlinear Delayed Systems Using Neural Network. International Journal of Automation and Computing 5(1), 103–108 (2008)
4. Yan, W.W., Shao, H.H., Wan, X.F.: Soft sensing modeling based on support vector machine and Bayesian model selection. Computers and Chemical Engineering 28(8), 1489–1498 (2004)
5. Zhang, Y., Su, H.Y., Liu, R.L., Chu, J.: Fuzzy Support Vector Regression Model of 4-CBA Concentration for Industrial PTA Oxidation Process. Chinese J. Chem. Eng. 13(5), 642–648 (2005)
6. Setnes, M., Roubos, H.: GA-fuzzy modeling and classification: complexity and performance. IEEE Trans. Fuzzy Systems 8(5), 509–522 (2000)
7. Mastorocostas, P.A., Theocharis, J.B., Petridis, V.S.: A constrained orthogonal least-squares method for generating TSK fuzzy models: application to short-term load forecasting. Fuzzy Sets and Systems 118(2), 215–233 (2001)
8. Xing, Z.Y., Jia, L.M., Yong, Z.: A Case study of data-driven interpretable fuzzy modeling. Acta Automatica Sinica 31(6), 815–824 (2005)
9. T-Sekouras, G., Sarimveis, H., Kavakli, E., Bafas, G.: A hierarchical fuzzy clustering approach to fuzzy modeling. Fuzzy Sets and Systems 150, 245–266 (2005)
10. Takagi, T., Sugeno, M.: Fuzzy identification of system s and its app lication to modeling and control. IEEE Trans. on Systems, Man and Cybernetics 15(1), 116–132 (1985)
11. Storn, R., Price, K.: Differential Evolution - A Simple and Efficient Heuristic Strategy for Global Optimization over Continuous Spaces. Journal of Global Optimization 11, 341–359 (1997)
12. Vesterstrom, J., Thomsen, R.: A Comparative Study of Differential Evolution. Particle Swarm Optimization, and Evolutionary Algorithms on Numerical Benchmark Problems. Evolutionary Computation 2, 1980–1987 (2004)

13. Storn, R.: System Design by Constraint Adaptation and Differential Evolution. IEEE Transactions on Evolutionary Computation 2, 82–102 (1999)
14. Yang, Z.Y., Tang, K., Yao, X.: Self-adaptive differential evolution with neighborhood search. In: Proc. of 2008 IEEE Congress on Evolutionary Computation, pp. 1110–1116 (2008)
15. Das, S., Abraham, A., Konar, A.: Automatic Clustering Using an Improved Differential Evolution Algorithm. IEEE Transactions on Systems, Man and Cybernetics, Part A: Systems and Humans 38(1), 218–237 (2008)

Based on the Qualitative Data EOWA Operator Multi-attribute Decision-Making Method and Its Application

Guang-Can Xiao[1,*] Yue-Ya Shi[2], Xiao Xiao[3], and Jun-Ran Zhang[4,**]

[1] School of Science, Southwest University of Science and Technology,
    Mianyang 621010, P.R. China
[2] School of Air Traffic Management, Civil Aviation Flight University of China,
    Guanghan 618300, P.R. China
[3] School of Electrical and Electronic Engineering,University of Adelaide, SA 5005,
    Australia
[4] School of Electrical Engineer and Information, Sichuan University,
    Chengdu 610065, P.R. China
    zhangjunran@126.com

Abstract. The article introduces the concept of qualitative data judgments and the scale of qualitative data assessment at first, and then it gives the algorithms related to qualitative data variables. Based on the qualitative data gathered in the extended ordered weighted averaging $EOWA$ operator, the research explores mathematical operation of $EOWA$ operator and the characteristics of qualitative data. Furthermore, multiple attribute decision making steps of $EOWA$ operator have been summarized.Finally, a practical example is given to illustrate the method is reliable and effective.

Keywords: Qualitative data, qualitative data assessment scale, $EOWA$ operator, multi-attribute decision making.

1 Introduction

According to the behavior of decision-makers, decision-making could be divided into two categories: when there is no fundamental interest conflict between policy makers, it is called decision-making; when there is a fundamental conflict of interest, it is called the game, on the other hand. In decision-making process, due to the variations of knowledge structure, cognitive level and personal experience of the policy makers, and many other factors, combining with the ambiguity of the thing itself and uncertainty, different preferences could been given by different decision-makers to the same question. Even the same decision-maker would give different preferences

---

* Guang-Can Xiao and Yue-Ya Shi are Co-first Authors, they contributed equally to this paper.
** Corresponding author. Tel:86-28-85422844.

B.-Y. Cao and X.-J. Xie (Eds.): Fuzzy Engineering and Operations Research, AISC 147, pp. 115–121.
springerlink.com   © Springer-Verlag Berlin Heidelberg 2012

to the same question at different times. The selection and presentation of the aggregate in group preference has become the key point of current research when people try to collect preferences from different decision makers to build a group preference. Therefore, how to express the preferences of decision makers properly, how to gather information from decision makers more effectively would make a direct impact on the efficiency and quality of decision-making. In decision analysis, judgment after two schemes comparison which offered by decision-makers is a common ordered qualitative data, that is, the form of preference information. However, due to people's thinking is fuzzy and uncertainty, and complexity of decision making, sometimes the pairwise comparison of things could not determine the exact value to represent during the decision-making, and then a strong form of the fuzzy language was given, these are, the ordered qualitative data (such as absolute poor, very poor, poor, fair, good, very good, satisfactory, excellent, welcome, etc). was used the natural language to evaluate of things or people in normal life, it is much more convenient. Therefore, the decision-making methods which use natural language phrases to judge the information given in a certain form (the qualitative data) have caused wide concern in recent years. Currently, from the view of the existing research results in this field, most of them focus on the integration method of qualitative data, and researches on the characteristics of qualitative data which was used to determine qualitative data are scant, especially in the problems in consistency judgment which similar to AHP, the research is nearly empty. In this paper, the problems in ordering qualitative data, using $EOWA$ operator, processing in multiple properties decision of qualitative data, were discussed in detail.

The organization of this paper: In section 2, it describes the $EOWA$ operator theory and its methods, including the computation and related properties; In section 3, it shows the decision making methods and procedures of building qualitative data; In section 4, an example is given: using $EOWA$ operator with multi-attribute decision making of qualitative data to analyze the decision-making process on logistics equipments purchasing; A short summary was presented at the end of the paper in section 5.

2   Description of Qualitative Data

Qualitative data analysis is an important content of statistical analysis. Qualitative data is a natural language when people evaluate someone or something. In the real world, due to the complexity of objective things, uncertainty and ambiguity of human thoughts, when people are constrained by a number of objective and subjective factors, they tend to assess the information and give an incomplete conclusion, such as qualitative data (or language information), these conclusions were of high ambiguity and uncertainty. Therefore, the research on incomplete information for the qualitative data of multiple attribute decision making problems has great theoretical and practical value. Consequently, the relevant concepts in judgment of qualitative data are given.

2.1 Qualitative Data Assessment Scale

Decision problem is from set A, a limited scheme set $A = \{A_1, A_2, \cdots, A_n\}$ $(A_i(i = 1, 2, \cdots, n)$ each one of $A = \{A_1, A_2, \cdots, A_n\}$ is called A scheme), Select the best scheme or sort them, $A_i(i = 1, 2, \cdots, n)$ stands for the i_{th} scheme in the list of A. Scheme optimization is based on the decision information provided by decision-makers (or called the linguistic judgment). When qualitative policy makers make a decision, they generally require appropriate measure of the qualitative data assessment scale. Therefore, we can set in orderly qualitative data assessment scale $S = \{s_\alpha : \alpha = -l, \cdots, l\}$, the term is generally odd number $(2l + 1, l \in Z)$.

Qualitative data assessment scale desirable this form:

Three scales:

$S = \{s_{-1}, s_0, s_1\}$. $s_{-1} = $ "medium $-$ poor", $s_0 = $ "medium", $s_1 = $ "medium $-$ good".

Five scales:

$S = \{s_{-2}, s_{-1}, s_0, s_1, s_2\}$. $s_{-2} = $ "fairly $-$ poor", $s_{-1} = $ "medium $-$ poor", $s_0 = $ "medium", $s_1 = $ "medium $-$ good", $s_2 = $ "fairly $-$ good".

11 scales:

$S = \{s_{-5}, s_{-4}, s_{-3}, s_{-2}, s_{-1}, s_0, s_1, , s_2, s_3, s_4, s_5\}$. $s_{-5} = $ "absolutely $-$ poor", $s_{-4} = $ "very $-$ poor", $s_{-3} = $ "poor", $s_{-2} = $ "fairly $-$ poor", $s_{-1} = $ "medium $-$ poor", $s_0 = $ "medium", $s_1 = $ "medium $-$ good", $s_2 = $ "fairly $-$ good", $s_3 = $ "good", $s_4 = $ "very $-$ good", $s_5 = $ "absolutely $-$ good".

Definition 1. Let $S = \{s_\alpha : \alpha = -l, \cdots, l\}$ is a qualitative data evaluation (or qualitative data symbols) sets, and satisfy the following properties:

(1) orderliness: $\alpha < \beta \Longrightarrow s_\alpha < s_\beta$; This symbol " $<$ " means s_α is less favourable than s_β;

(2) Negative operator: $neg(s_\alpha) = s_{-\alpha}$;

(3) Maximizing operator: $\alpha < \beta \Longrightarrow max\{s_\alpha, s_\beta\} = s_\beta$;

(4) Minimization operator: $\alpha < \beta \Longrightarrow min\{s_\alpha, s_\beta\} = s_\alpha$.

In order to avoid missing decision-making information and convenient calculation, we add a scale $S^* = \{s_\alpha : \alpha \in [-q, q], q > l\}$ in the original scale $S = \{s_\alpha : \alpha = -l, \cdots, l\}$. q is a sufficiently large natural number, and if $\alpha \in \{-l, \cdots, l\}$, as a result, s_α is called original terminology; if $\alpha \bar{\in} \{-l, \cdots, l\}$, $(\alpha \in [-q, q])$, s_α is called extend term. The extended scale terms still meet the requirements of [1-4].

2.2 Qualitative Data Assessment Scale Operation and Law

Definition 2. Set mapping $\oplus : S^* \to S^*$ as a result $s_\alpha, s_\beta \in S^*, \lambda, \lambda_1, \lambda_2 \in [0,1]$, thus

(1)$s_\alpha \oplus s_\beta = s_{\alpha+\beta}$;

(2)$s_\alpha \oplus s_\beta = s_{\alpha \oplus \beta}$;

(3)$\lambda s_\alpha = s_{\lambda\alpha}$;

(4)$\lambda(s_\alpha \oplus s_\beta) = \lambda s_\alpha \oplus \lambda s_\beta$;

(5)$(\lambda_1 + \lambda_2)s_\alpha = \lambda_1 s_\alpha + \lambda_2 s_\alpha$.

Obviously, this law has linear computation,

$\oplus_i^n s_\alpha = s_{\Sigma_i^n \alpha_i}$; $\oplus_i^n \lambda_i s_\alpha = s_{\Sigma_i^n \lambda_i \alpha_i}$,

where $\lambda_i \in [0,1](i = 1, 2, \cdots, n)$.

2.3 EOW A Function and Nature

Definition 3. Set mapping $EOWA$:

$(S^*)^n \to S^*$, if $EOWA_\varpi(s_{\alpha_1}, s_{\alpha_2}, \cdots, s_{\alpha_n}) = s_{\Sigma_i^n \omega_j \beta_j} = s_{\beta^*}$, Among them $\beta^* = \Sigma_j^n \omega_j \beta_j$, $\omega = (\varpi_1, \varpi_2, \cdots, \varpi_n)$ is associated with $EOWA$ weighted vector, $\varpi_j \in [0,1], (j = 1, 2, \cdots, n), \Sigma_j^n \varpi_j = 1$ and s_β is a group of orderly qualitative data element the first j, So that $EOWA$ function is the orderly development weighted average ($EOWA$) operator.

$EOWA$ Operator has the following properties:

Theorem 1. *(replacement invariants):*
$EOWA_\varpi(s_{\alpha_1}, s_{\alpha_2}, \cdots, s_{\alpha_n})$ $=$ $EOWA_\varpi(t_{\alpha_1}, t_{\alpha_2}, \cdots, t_{\alpha_n})$. *Including* $(t_{\alpha_1}, t_{\alpha_2}, \cdots, t_{\alpha_n})$ *is the qualitative variable data group* $(s_{\alpha_1}, s_{\alpha_2}, \cdots, s_{\alpha_n})$ *any replacement.*

Proof: Because $EOWA_\varpi(s_{\alpha_1}, s_{\alpha_2}, \cdots, s_{\alpha_n}) = \omega_1 s_{\beta_1} \oplus \omega_2 s_{\beta_2} \oplus \cdots \oplus \omega_n s_{\beta_n}$
$EOWA_\varpi(t_{\alpha_1}, t_{\alpha_2}, \cdots, t_{\alpha_n}) = \omega_1 t_{\beta_1} \oplus \omega_2 t_{\beta_2} \oplus \cdots \oplus \omega_n t_{\beta_n}$ and $(t_{\alpha_1}, t_{\alpha_2}, \cdots, t_{\alpha_n})$ is the qualitative variable data group $(s_{\alpha_1}, s_{\alpha_2}, \cdots, s_{\alpha_n})$ any replacement, thus $t_{\alpha_j} = s_{\beta_j}, (j = i, 2, \cdots, n)$. So
$EOWA_\varpi(s_{\alpha_1}, s_{\alpha_2}, \cdots, s_{\alpha_n}) = EOWA_\varpi(t_{\alpha_1}, t_{\alpha_2}, \cdots, t_{\alpha_n})$.

Theorem 2. *(homogeneity):*
If $\forall j(j = i, 2, \cdots, n)$, *have* $s_{\alpha_j} = s_\alpha$, *thus*
$EOWA_\varpi(s_{\alpha_1}, s_{\alpha_2}, \cdots, s_{\alpha_n}) = s_\alpha$.

Proof: Because $s_{\alpha_j} = s_\alpha, (j = i, 2, \cdots, n)$,then
$EOWA_\varpi(s_{\alpha_1}, s_{\alpha_2}, \cdots, s_{\alpha_n}) = \oplus_j^n(\omega_j s_j) = \oplus_j^n(\omega_j s_\alpha) = s_\alpha$.

Theorem 3. *(monotonicity):*
If $\forall j(j = i, 2, \cdots, n)$, *have* $s_{\alpha_j} \le t_{\alpha_j}$, *thus*
$EOWA_\varpi(s_{\alpha_1}, s_{\alpha_2}, \cdots, s_{\alpha_n}) \le EOWA_\varpi(t_{\alpha_1}, t_{\alpha_2}, \cdots, t_{\alpha_n})$.

Proof: Because $EOWA_\varpi(s_{\alpha_1}, s_{\alpha_2}, \cdots, s_{\alpha_n}) = \oplus_j^n(\omega_j t_{\beta_j})$,
$EOWA_\varpi(t_{\alpha_1}, t_{\alpha_2}, \cdots, t_{\alpha_n}) = \oplus_j^n(\omega_j t_{\beta_j})$, and
$\forall j(j = i, 2, \cdots, n)$, when $s_{\alpha_j} \le t_{\alpha_j}$, we obtain $s_{\beta_j} \le t_{\beta_j}$, thus
$EOWA_\varpi(s_{\alpha_1}, s_{\alpha_2}, \cdots, s_{\alpha_n}) \le EOWA_\varpi(t_{\alpha_1}, t_{\alpha_2}, \cdots, t_{\alpha_n})$.

Theorem 4. *(Intermediate-value property):*
$$(min)_i^n(s_{\alpha_i}) \leq EOWA_{\varpi}(s_{\alpha_1}, s_{\alpha_2}, \cdots, s_{\alpha_n}) \leq (max)_i^n(s_{\alpha_i}).$$

Proof: Now let $s_\beta = (max)_j^n(s_{\alpha_j}), s_\alpha = (min)_i^n(s_{\alpha_i})$,thus
$$EOWA_{\varpi}(s_{\alpha_1}, s_{\alpha_2}, \cdots, s_{\alpha_n}) = \oplus_j^n(\omega_j \beta_j) \leq \oplus_j^n(\omega_j s_\beta) = s_\beta = (max)_j^n(s_{\alpha_j}),$$
$$EOWA_{\varpi}(s_{\alpha_1}, s_{\alpha_2}, \cdots, s_{\alpha_n}) = \oplus_j^n(\omega_j \beta_j) \geq \oplus_j^n(\omega_j s_\alpha) = s_\beta = (min)_j^n(s_{\alpha_j}),$$
so, $(min)_i^n(s_{\alpha_i}) \leq EOWA_{\varpi}(s_{\alpha_1}, s_{\alpha_2}, \cdots, s_{\alpha_n}) \leq (max)_i^n(s_{\alpha_i})$.

Remark:
(1) If $\forall j (j = 1, 2, \cdots, n)$, have $\omega_j = \frac{1}{n}$, thus
$$EOWA_{\varpi}(s_{\alpha_1}, s_{\alpha_2}, \cdots, s_{\alpha_n}) = s_{\alpha^*}, \alpha^* = \frac{\sum_j^n(\alpha_j)}{n};$$
(2) If $\omega_1 = 1, \omega_j = 0, (j = 2, 3, \cdots, n)$,thus
$$EOWA_{\varpi}(s_{\alpha_1}, s_{\alpha_2}, \cdots, s_{\alpha_n}) = (max)_j^n(s_{\alpha_j});$$
(3) If $\omega_n = 1, \omega_j = 0, (j = 1, 2, \cdots, n - 1)$, thus
$$EOWA_{\varpi}(s_{\alpha_1}, s_{\alpha_2}, \cdots, s_{\alpha_n}) = (min)_j^n(s_{\alpha_j}).$$

3   Qualitative Data Decision-Making Method Steps

Based on the qualitative data $EOWA$ multiple attribute decision making method of the operator, specific procedure is as follows:

Step 1: For multiple attribute decision making problems, policymakers give out plans $A_i(i = 1, 2, \cdots, n)$ and then evaluate $r_{ij}(i = 1, 2, \cdots, n, j = 1, 2, \cdots, m)$ under qualitative data attribute $B_j(j = 1, 2, \cdots, m)$ and get an appraisal matrix $R = (r_{ij})_{n \times m}, r_{ij} \in S^*$;

Step 2: Use $EOWA$ operator to gather the i_{th} row accessed information from the matrix $R = (r_{ij})_{n \times m}, r_{ij} \in S^*$, and we can get the comprehensive attribute value $Z_i(\varpi)(i = 1, 2, \cdots, n)$ of the decision-making plan u_i. Where $Z_i(\varpi) = EOWA_{\varpi}(r_{i1}, r_{i2}, \cdots, r_{im})$;

Step 3: For all decision-making plans using $Z_i(\varpi)(i = 1, 2, \cdots, n)$, sort the results and get the optimal solution.

4   Case Study

Suppose a logistics company will purchase a forklift, there are four options: A_1, A_2, A_3, A_4. These options are assessed based on three main factors: economic, functional and operational, which are denoted by B_1, B_2, B_3, the weight of these three factors is $\omega_1, \omega_2, \omega_3$, respectively. Company invites experts to evaluate the importance of different options and each factor. Experts use evaluation scale of linguistic variables (11 scales) $S = s_{-5}, \cdots, s_5$ to express their evaluations. Relevant data are presented as follows (Table 1).

Table 1. Expert evaluation opinions

	A_1	A_2	A_3	A_4	r_j
B_1	S_2	S_4	S_2	S_4	1.0
B_2	S_2	S_1	S_1	S_0	0.7
B_3	S_4	S_1	S_2	S_4	0.8

We use the following formula to determine the weights of each factor: $\omega_j = \frac{r_j}{\sum_k^m (r_k)}$ $(j = 1, 2, \cdots, m)$.

Then $\omega = (\omega_1, \omega_2, \omega_3) = (0.4, 0.28, 0.32)$.

The results are as follows (table 2)

Table 2. Expert evaluation opinions and weighting

	A_1	A_2	A_3	A_4	ω_j
B_1	S_2	S_4	S_2	S_4	0.4
B_2	S_2	S_1	S_1	S_0	0.28
B_3	S_4	S_1	S_2	S_4	0.32

$$Z_i(\varpi) = EOWA_\varpi(r_{i1}, r_{i2}, \cdots, r_{im}), (i = 1, 2, \cdots, n).$$
$$Z_1(\varpi) = EOWA_\varpi(r_{11}, r_{12}, \cdots, r_{1m}) = \oplus_j^m (\omega_j r_{1j}) = s_{2.64}.$$
$$Z_2(\varpi) = EOWA_\varpi(r_{21}, r_{12}, \cdots, r_{2m}) = \oplus_j^m (\omega_j r_{2j}) = s_{2.20}.$$
$$Z_3(\varpi) = EOWA_\varpi(r_{31}, r_{32}, \cdots, r_{3m}) = \oplus_j^m (\omega_j r_{3j}) = s_{1.52}.$$
$$Z_4(\varpi) = EOWA_\varpi(r_{41}, r_{42}, \cdots, r_{4m}) = \oplus_j^m (\omega_j r_{4j}) = s_{2.88}.$$

Sort the above result, it can be obtained that $A_4 \succ A_1 \succ A_2 \succ A_3$. As a result, the optimal solution is A_4.

5 Conclusion

This paper gave the scale in qualitative data evaluation and $EOWA$ operator of uncertainty qualitative data, and also discussed the characteristics of $EOWA$ operator and multiple attribute decision-making methods of qualitative data operator. Then procedures in multiple attribute decision-making methods of qualitative data operator were concluded. Finally, a practical example was given to illustrate the effectiveness and reliability of this methods which consistent with objective reality.

Acknowledgement. This work were supported by National Natural Science Foundation of China (No.81000605, 81110108007), Young Teachers Science Foundations of Sichuan University (No.2009SCU11178),Science Foundations of Southwest Science and Technology University (No.11zgc201, 09zx7102) and General Project of Civil Aviation Flight University of China (No.J2010-30).

References

1. Xu, Z.S.: Uncertain Multiple Attribute Decision Making: Method and Applications, pp. 169–176. Tsinghua University Press (2005)
2. Agresti, A.: Analysis of Ordinal Categorical Data. John Wiley Sons (1984)
3. Zhang, R.T., Liang, X.J.: Qualitative material statistical analysis. Princeton university press, China Statistical Publishing House(China Statistics Press) (2008)
4. Hwang, C.L., Yoon, K.: Multiple Attributes Decision Making Methods and Applications, pp. 12–23. Springer, Heidelberg (1981)
5. Keeney, R.L., Raiffa, H.: Decision with Multiple Objectives. Wiley, New York (1976)

An Improved Algorithm for NSM in Weighted Fuzzy Reasoning

Jun Shen and Jun-hong Miao

College of Mathematics and Statistics, Hainan Normal University, Haikou, 571158, China
yyq50@163.com

Abstract. In the fuzzy expert systems, the performance of fuzzy reasoning methods is an important factor related to the capability of the system.This paper indicates the limitation of the existing weighted fuzzy reasoning method in [1] and proposes an improved reasoning method.

Keywords: Fuzzy expert system, weighted fuzzy production rule, fuzzy reasoning, similarity measure.

1 Introduction

Fuzzy production rules (FPRs) are widely used in expert systems to represent fuzzy imprecise ambiguous and vague concepts. Many fuzzy reasoning methods have been developed which draw conclusions based on the theory of approximate reasoning. A number of them are based on Zadeh's well-known compositional rule of inference (CRI) method [2]. Despite its success in various rule-based system applications, CRI has been criticized to be too complex and its underlying semantic to be unclear by Turksen and Zhong [3,4]. They propose a similarity-based fuzzy reasoning method called approximate analogical reasoning schema (AARs) to compute the fuzzy value of the deduced consequent and offer a way to interpret its linguistic meaning better than the approach of the CRI-based method. Later, many researchers do many works in fuzzy reasoning based similarity measure. Especially, Yeung and Tsang [1,5-7] proposed DS(Distance Similarity) and NSM (New Similarity Measure) and made lots of valuable results in this area. Ha Ming-hu found the limitations of Yeung and Tsang's DS and NSM and proposed them.

This paper indicates the limitation of the NSM on the other hand and proposes an improved algorithm, also gives an example to show the rationality of the improved algorithm.

2 Yeung and Tsang's NSM Algorithm

2.1 Weighted Fuzzy Production Rule

Weighted fuzzy production rule has three types: simple WFPR, conjunctive WFPR, disjunctive WFPR. Here we only discuss conjunctive WFPR, other types may be got as the same method.

B.-Y. Cao and X.-J. Xie (Eds.): Fuzzy Engineering and Operations Research, AISC 147, pp. 123–128.
springerlink.com   © Springer-Verlag Berlin Heidelberg 2012

Conjunctive WFPR :

$R : IF\ V_1\ is\ A_1\ AND\ V_2\ is\ A_2 ... AND\ V_n\ is\ A_n\ THEN\ U\ is\ B$

$CF_R, LW_1, LW_2, ..., LW_n, \lambda_{A_1}, \lambda_{A_2}, ..., \lambda_{A_n}, GW(R)$

Fact 1: $V_1\ is\ A_1', CF_{F_1}$

Fact 2: $V_2\ is\ A_2', CF_{F_2}$

$\vdots$

Fact n: $V_n\ is\ A_n', CF_{F_n}$

Conclusion: $U\ is\ B', CF_{B'}$

$V_1, V_2, ..., V_n$ and U are variables, $A_1, A_2, ..., A_n$ and B are the fuzzy values of the variables $V_1, V_2, ..., V_n$ and U respectively. $CF_R, CF_{F_1}, CF_{F_2}, ... CF_{F_n}$ are the certainty factors of the rule, the observed facts, and the deduced result respectively. LW_i is the local weight of the proposition" $V_i\ is\ A_i$ " and each $LW_i \in (0,1]$, and λ_{A_i} is the threshold value for the similarity measure between A_i and the observed fact A_i'. $GW(R)$ denotes the global weights assigned to R.

When $n = 1$, this conjunctive $WFPR$ is reduced to a simple $WFPR$ as the following:

$R : If\ V\ is\ A\quad THEN\quad U\ is\ B, CF_R, \lambda_A, L_W, G_{W(R)}$

Fact: $V\ is\ A', CF_F$

Conclusion: $U\ is\ B', CF_{B'}$.

2.2 Yeung and Tsang`s NSM (New Similarity Measure) Algorithm

Yeung and Tsang defined a new similarity measure and proposed a better way to interpret the linguistic meaning of the fuzzy value, the certainty factor and the consequent.

The definition of NSM:

$$SM_{A_i}(A_i', A_i) = \begin{cases} \dfrac{1}{2} * \left(1 + \dfrac{M(A_i') - M(A_i)}{Max[M(A_i'), M(A_i)]}\right) & , \max[\min(A_i', A_i)] = 1, \\ \dfrac{1}{2} * \dfrac{Min(A_i', A_i)}{Max(A_i', A_i)} & , \quad others. \end{cases}$$

Yeung and Tsang`s NSM (New Similarity Measure) algorithm:

Step1: Compute the overall similarity measure:

$$SM_w = \sum_{i=1}^{n} \left[SM_{A_i}(A_i', A_i) * \frac{LW_i}{\sum_{j=1}^{n} LW_j} \right].$$

Step2: obtain linguistic meaning of B' by matching SM_{max} against the values in the given table (table 1).

Table 1. Similarity interval and its meaning

Similarity interval	linguistic
(0.9, 1.0)	extremely
(0.7,0.9)	very very
(0.5,0.7)	very
[0.5,0.5]	$A' = A$ then $B' = B$
(0.4,0.5)	quite
(0.3,0.4)	more or less
(0.2,0.3)	rather
(0.1,0.2)	basically not
(0.0,0.1)	not

Step3: Compute B'

$$B' = \begin{cases} Min\left\{1, \dfrac{B}{2*SM_w}\right\}, if\ 0 < SM_w < \dfrac{1}{2}, \\ B*2*(1-SM_w), if\ \dfrac{1}{2} \le SM_w < 1. \end{cases}$$

Step4: Let $LW_{max} = Max\{LW_1, LW_2, \cdots, LW_n\}$,

$$CF_{max} = CF\ of\ max[LW_1, LW_2, \cdots, LW_n],$$

$$CF_{B'} = CF_R * \underset{1 \le i \le n}{Min}\left[CF_{F_i} + (CF_{max} - CF_{F_i}) * \frac{LW_{max} - LW_i}{LW_{max}} \right].$$

One example is presented to demonstrate the limitation of NSM with the following given fuzzy sets:

$$A = "tall" = (0.2, 0.4, 0.6, 0.8, 0.9, 1, 1),$$

$$B = "strong" = (0.3, 0.4, 0.5, 0.6, 0.7, 0.9, 1.0).$$

Example 1: $R: IF$ x *is tall*,*THEN* x *is strong*

(i)   Fact 1: x is A_1, $A_1 = (0.1, 0.3, 0.5, 0.7, 0.8, 0.9, 0.9)$, compute B_1 ?

(ii)   Fact 2: x is A_2, $A_2 = (0.1, 0.3, 0.5, 0.7, 0.8, 0.9, 1)$, compute B_2 ?

Compute the results according to the NSM:

(i) $SM_W = SM(A, A_1) = \dfrac{1}{2} * \dfrac{4.2}{4.9} = 0.429 > \lambda = 0.3$,

$B_1 = \min[1, \dfrac{B}{2 * 0.429}] = (0.35, 0.47, 0.58, 0.7, 0.82, 1, 1)$.

(ii) $SM_W = SM(A, A_2) = \dfrac{1}{2} * (1 + 0.124) = 0.562 > \lambda = 0.3$,

$B_2 = 2B * (1 - 0.562) = (0.26, 0.35, 0.44, 0.53, 0.61, 0.79, 0.88)$.

We observe the consequent B_1 and B_2 has larger difference, but the antecedent A_1 and A_2 has a little difference. This is not consistent with our understanding of reality.

2.3   An Improved Algorithm for NSM

When $0 < SM_W < \dfrac{1}{2}$, they compute B' according to the following formula in Yeung and Tsang's NSM:

$$B' = \min[1, \dfrac{B}{2 * SM_W}] = \min[1, \dfrac{B}{2}(\sum_{i=1}^{n} \dfrac{LW_i}{\sum_{j=1}^{n} LW_j} SM_i)^{-1}].$$

We improve the above formula according to the following formula:

$$B' = \min[1, B * \sum_{i=1}^{n} \dfrac{LW_i}{\sum_{j=1}^{n} LW_j} \dfrac{M(A_i')}{M(A_i)}].$$

That is, we compute $B \dfrac{M(A_i')}{M(A_i)}$ corresponding to every antecedent given fact A_i,

$(i = 1, 2, \cdots n)$, and then weight average.

Step3': Compute B'

$$B' = \begin{cases} \min\left[B\sum_{i=1}^{n}\dfrac{LW_i}{\sum_{j=1}^{n}LW_j}\cdot\dfrac{M(A'_i)}{M(A_i)},1\right], & \text{if } 0 < SM_w < \dfrac{1}{2}, \\ 2B*(1-SM_w), & \text{if } \dfrac{1}{2} \le SM_w < 1. \end{cases}$$

We compute the Example 1 according to the improved algorithm.

(i) $\quad SM_w = SM(A,A_1) = \dfrac{1}{2}*\dfrac{4.2}{4.9} = 0.429 > \lambda = 0.3$,

$B_1 = \min[1,0.857B] = (0.26,0.34,0.43,0.51,0.6,0.77,0.86)$.

(ii) $\quad SM_w = SM(A,A_2) = \dfrac{1}{2}*(1+0.124) = 0.562 > \lambda = 0.3$,

$B_2 = 2B*(1-0.562) = (0.26,0.35,0.44,0.53,0.61,0.79,0.88)$.

The antecedent A_1 and A_2 has a little difference, the consequent B_1 and B_2 also has a little difference. It is reasonable. This is consistent with our understanding of reality.

3 Conclusion

This paper indicates the limitation of NSM algorithm, and proposes an improved algorithm for NSM, also gives an example to show the reasonability of the improved algorithm.

Acknowledgements. Thanks to the support by Provincial Natural Science Foundation of Hainan (No. 109002) and Project of Young Teachers Start Research of Hainan Normal University (QN1121).

References

1. Yeung, D.S., Tsang, E.C.C.: Weighted fuzzy production rules. Fuzzy Set Systems 88(3), 299–313 (1997)
2. Zadeh, L.A.: Outline of a new approach to the analysis of complex systems and decision processes. IEEE Trans. System, Man, Cybernetics 3, 28–44 (1973)
3. Turksen, I.B., Zhao, Z.: An approximate analogical reasoning scheme based on similarity measures and interval valued fuzzy sets. Fuzzy Sets and Systems 34(3), 323–346 (1990)
4. Turksen, I.B., Zhao, Z.: An approximate analogical reasoning approach based on similarity measures. IEEE Trans. System, Man, Cybernetics 18, 1049–1051 (1989)

5. Yeung, D.S., Tsang, E.C.C.: A weighted fuzzy production rule evaluation methods. In: Proceedings of Fourth IEEE International Conference on Fuzzy Systems, pp. 461–468 (1995)
6. Yeung, D.S., Tsang, E.C.C.: A comparative study on similarity-based fuzzy reasoning methods. IEEE Transaction on Systems Man Cybernetics 27(2), 216–226 (1997)
7. Yeung, D.S., Tsang, E.C.C.: A multi-level weighted fuzzy reasoning algorithm for expert systems. IEEE Transaction on Systems Man Cybernetics 28(2), 149–158 (1998)
8. Ha, M.H., Liu, Y., Li, H.J.: A similarity-based weighted fuzzy production rule evaluation method. Journal of Hebei University (Natural Science Edition) 25(6), 659–663 (2005)
9. Ha, M.H., Li, H.J.: Two improved similarity measures and their fuzzy reasoning methods. Computer Engineering and Applications 41(35), 31–34 (2005)
10. Ha, M.H., Li, H.J.: Three similarity-based weighted fuzzy reasoning methods. Computer Engineering and Applications 28, 34–37 (2006)

Design of Optimal Cost Fuzzy Controller
for Spatial Double Inverted Pendulum System

Zhi-hong Miao[1], Zhi-hui Li[1], Yong-li Zhang[2], and Hong-xing Li[2]

[1] Department of Fire Protection Engineering, The Chinese People's Armed Police Force
Academy, Langfang, Hebei, 065000, China
[2] Faculty of Electronic Information and Electrical Engineering, Dalian University of
Technology, Dalian 116024, China
miaozhh@21cn.com, zylzhang@126.com, lihx@dlut.edu.cn

Abstract. For a spatial double inverted pendulum system, a design method of
optimal cost fuzzy controller is developed via the parallel distributed compensa-
tion(PDC) approach. Firstly, by using the Lagrange equation, the mathematical
model of the spatial double inverted pendulum is derived. Then, a sufficient condition for the existence of optimal cost fuzzy controller is presented with taking into
account the ratio between the lengths of two pendulums, and it is formed in terms of
linear matrix inequalities. Under a certain cost function, the best ratio between the
lengths of two pendulums is solved by this method.

Keywords: T-S fuzzy model, cost function, spatial double inverted pendulum, linear matrix inequalities.

1 Introduction

Usually, a nonlinear system, if its mathematical model can not be obtained, can
be approximated by a Fuzzy T-S model. In this kind of T-S model, several local
linear models are blended together to form a complete fuzzy model. On this basis,
the typical approaches for control design are carried out via the so-called parallel
distributed compensation (PDC) method [1]. Therefore, the mature linear system
theory can be employed to guide designing a fuzzy controller. In general, stability
conditions and performance of a T-S control system can be formulated into linear
matrix inequalities (LMIs)[1]. The feasible solutions of LMIs can be acquired by
interior-point methods.

On the other hand, an inverted pendulum system is a type of nonlinear, multivariable and have highly unstable characteristics. Usually,it is a ideal apparatus for
demonstrating and motivating various theories and control schemes. Moreover, the
problem of balancing an inverted pendulum is also representative of some other
well-known control problem, and has strong engineering application background,
such as the problem of control a rocket or robot. In recent years, various modern
control theories have been used for controlling the inverted pendulum, for instance,

B.-Y. Cao and X.-J. Xie (Eds.): Fuzzy Engineering and Operations Research, AISC 147, pp. 129–138.
springerlink.com

the nonlinear control theory [2, 3], fuzzy control approaches [4, 5], especially, the variable universe adaptive control technique, which has been successfully used in the hardware implement of the quadruple inverted pendulum [6]. Compared with the single inverted pendulum system, the spatial double inverted pendulum system is more complex in both the establishment of model and experiment design.

Although there have been some literatures [7, 8, 9] about the control of spherical and multi-level inverted pendulum, Researching on the problem of the relationship between lengths of pendulums, especially the ratio of lengths, and the control performance has not be found. Currently, the optimal guaranteed cost control approach has obtained wide concern and achieved great progress [10, 11, 12]. The advantage of that approach is which providing an upper bound of a cost function or performance index, and thus the system performance can be guaranteed to be less than this bound. In this paper,this principle is also employed to design an optimal cost fuzzy controller for spatial double inverted pendulum system.

2   Modeling of Spatial Double Inverted Pendulum

The goal of this section is to present the details of establishing a mathematic model for spatial double inverted pendulum system. The spatial double inverted pendulum system (SDIP) in this paper consists of two rods and a cart, and a space Cartesian coordinate system is also attached in this system. The schematic diagram is shown in Figure 1.

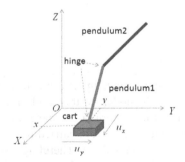

Fig. 1. The schematic diagram of spatial double inverted pendulum system

The main symbols adopted in this system are listed as follows:

- m_c: Mass of the cart; m_1: Mass of the first pendulum
- m_2: Mass of the second pendulum
- l_1: Length from the centroid of the first pendulum to linked point(hinge) on the cart
- l_2: Length from the centroid of the second pendulum to linked point(hinge) on the first pendulum.

In this paper, l_1 and l_2 are called as "centroid length".

For the i-th pendulum, let ψ_i and θ_i be the Euler angles of the pendulum, where ψ_i is precession angle, θ_i is nutational angle.

Let $(x, y, 0)$ be the position of the cart, then the centroid position of the first pendulum is

$$x_{1c} = x + l_1 \sin\theta_1; \; y_{1c} = y - l_1 \sin\psi_1 \cos\theta_1; \; z_{1c} = l_1 \cos\psi_1 \cos\theta_1. \quad (1)$$

and the centroid position of the second pendulum is

$$x_{2c} = x + L_1 \sin\theta_1 + l_2 \sin\theta_2; \; y_{2c} = y - L_1 \sin\psi_1 \cos\theta_1 - l_2 \sin\psi_2 \cos\theta_2;$$
$$z_{2c} = L_1 \cos\psi_1 \cos\theta_1 + l_2 \cos\psi_2 \cos\theta_i. \quad (2)$$

and the kinetic energy of the spatial inverted pendulum can be constituted as

$$T = T_{cart} + T_{pendulum1} + T_{pendulum2} \quad (3)$$

where T_{cart} and $T_{pendulumi}$ represent the kinetic energy of the cart and the i-th rod respectively. Firstly, the the kinetic energy of the cart can be obtained as

$$T_{cart} = \frac{1}{2}m_c(\dot{x}^2 + \dot{y}^2). \quad (4)$$

Then, using the Konig's theorem [13], the kinetic energy of the i-th pendulum

$$T_{pendulumi} = \frac{1}{2}m_i v_{ic}^2 + \frac{1}{2}\omega_i^{(0)T} J_{ic}^{(0)} \omega_i^{(0)}, \quad (5)$$

where $J_{ic}^{(0)}$ is the inertia matrix of the i-th pendulum relative to the centroid. For simplicity, we assume the rod linear densities in each sides with respect to the centroid is uniform, expressed as μ_1 and μ_2. So the inertia matrix with respect to the axis, which passes through the centroid and is perpendicular to the pendulum, will be

$$J_{ic}^{(0)} = \begin{bmatrix} \frac{1}{6}m_i(L_i^2 - 2L_i l_i + 2l_i^2) & 0 & 0 \\ 0 & \frac{1}{6}m_i(L_i^2 - 2L_i l_i + 2l_i^2) & 0 \\ 0 & 0 & 0 \end{bmatrix} (i = 1, 2).$$

The projection of the angular velocity on coordinate system is

$$\omega_i^{(0)} = [\dot{\psi}_i \cos\theta_i, \dot{\theta}_i, \dot{\psi}_i \sin\theta_i]^T, (i = 1, 2).$$

On the other hand, the gravitational potential energy of the SDIP System can be expressed as

$$V = V_{cart} + V_{pendulum1} + V_{pendulum2}, \quad (6)$$

where V_{cart}, $V_{pendulum1}$ and $V_{pendulum2}$ are the potential energies of the cart, the first pendulum and the second pendulum, respectively. Then

$$V_{cart} = 0, \quad V_{pendulum1} = mgz_{1c} = mgl_1 \cos\psi_1 \cos\theta_1.$$

$$V_{pendulum2} = mgz_{2c} = mg(L_1 \cos\psi_1 \cos\theta_1 + l_2 \cos\psi_2 \cos\theta_2). \tag{7}$$

Now, introducing a the Lagrangian function as follows

$$L = T - V, \tag{8}$$

Then, the Lagrangian equations can be expressed as

$$\frac{d}{dt}\left(\frac{\partial L}{\partial \dot{q}_j}\right) - \frac{\partial L}{\partial q_j} = Q_j, \quad (j = 1, 2, \cdots, 6), \tag{9}$$

where $q_1 = x, q_2 = y, q_3 = \psi_1, q_4 = \theta_1, q_5 = \psi_2, q_6 = \theta_2$. After some simple derivations, we have

$$M(q, \dot{q})\ddot{q} + C(q, \dot{q})\dot{q} + G(q) = F, \tag{10}$$

where

$$M(q, \dot{q}) = \begin{bmatrix} h_0 & 0 & 0 & u_{01} & 0 & u_{02} \\ 0 & h_0 & -f_{01} & -\bar{g}_{01} & -f_{02} & -\bar{g}_{02} \\ 0 & -f_{01} & f_{11} & 0 & f_{12} & \bar{g}_{12} \\ u_{01} & -\bar{g}_{01} & 0 & h_3 & -g_{12} & u_{12} \\ 0 & -f_{02} & f_{12} & -g_{12} & f_{22} & 0 \\ u_{02} & -\bar{g}_{02} & \bar{g}_{12} & u_{12} & 0 & h_4 \end{bmatrix},$$

$$C(q, \dot{q}) =$$

$$\begin{bmatrix} 0 & 0 & 0 & 0 & 0 & 0 \\ 0 & 0 & -a_{01} & b_{01} & -a_{02} & b_{02} \\ 0 & 0 & 0 & -b_{11} & a_{12} & -b_{12} \\ 0 & 0 & b_{11} & 0 & \bar{b}_{12} & c_{12} \\ 0 & 0 & -a_{12} & -\bar{b}_{12} & 0 & -b_{22} \\ 0 & 0 & b_{12} & -c_{12} & b_{22} & 0 \end{bmatrix} \begin{bmatrix} 0 \\ 0 \\ \dot{\psi}_1 \\ \dot{\psi}_1 \\ \dot{\psi}_2 \\ \dot{\psi}_2 \end{bmatrix} + \begin{bmatrix} 0 & 0 & 0 & -v_{01} & 0 & -v_{02} \\ 0 & 0 & b_{01} & -a_{01} & b_{02} & -a_{02} \\ 0 & 0 & -b_{11} & 0 & -b_{12} & a_{12} \\ 0 & 0 & 0 & 0 & c_{12} & v_{12} \\ 0 & 0 & -\bar{b}_{12} & -a_{12} & -b_{22} & 0 \\ 0 & 0 & -c_{12} & \bar{v}_{12} & 0 & 0 \end{bmatrix} \begin{bmatrix} 0 \\ 0 \\ \dot{\theta}_1 \\ \dot{\theta}_1 \\ \dot{\theta}_2 \\ \dot{\theta}_2 \end{bmatrix},$$

$$G(q) = \begin{bmatrix} 0, 0, a_{01}g, -b_{01}g, a_{02}g, -b_{02}g \end{bmatrix}^T, \quad F = \begin{bmatrix} u_x, u_y, 0, 0, 0, 0 \end{bmatrix}^T.$$

Some symbols used above expressions are listed as follows:

$$h_0 = m_c + m_1 + m_2; \quad h_1 = m_1 l_1 + m_2 L_1; \quad h_2 = m_2 l_2;$$

$$h_3 = \frac{1}{6}(8m_1 l_1^2 + m_1 L_1^2 - 2m_1 L_1 l_1 + 6m_2 L_1^2); \quad h_4 = \frac{1}{6}m_2(8l_2^2 + L_2^2 - 2L_2 l_2).$$

$$a_{ij} = \rho_{ij} \cos\theta_i \cos\theta_j \sin(\psi_i - \psi_j); \quad b_{ij} = \rho_{ij} \cos\theta_i \sin\theta_j \cos(\psi_i - \psi_j);$$

$$\bar{b}_{ij} = \rho_{ij} \cos\theta_j \sin\theta_i \cos(\psi_i - \psi_j); \quad c_{ij} = \rho_{ij} \sin\theta_i \sin\theta_j \sin(\psi_i - \psi_j);$$

$$f_{ij} = \rho_{ij} \cos\theta_i \cos\theta_j \cos(\psi_i - \psi_j); \quad g_{ij} = \rho_{ij} \sin\theta_i \cos\theta_j \sin(\psi_i - \psi_j);$$

$$\bar{g}_{ij} = \rho_{ij} \sin\theta_j \cos\theta_i \sin(\psi_i - \psi_j);$$

$$u_{ij} = \rho_{ij}[\cos\theta_i \cos\theta_j + \sin\theta_i \sin\theta_j \cos(\psi_i - \psi_j)];$$
$$v_{ij} = \rho_{ij}[-\cos\theta_i \sin\theta_j + \sin\theta_i \cos\theta_j \cos(\psi_i - \psi_j)];$$
$$\bar{v}_{ij} = \rho_{ij}[-\cos\theta_j \sin\theta_i + \sin\theta_j \cos\theta_i \cos(\psi_i - \psi_j)];$$

here $\rho_{ij} = \begin{cases} h_j L_i, & i \neq j; \\ h_{i+2}, & i = j. \end{cases}$ $L_0 = 1, \psi_0 = 0.$

In general, the different choices for the lengths of the pendulums (the centroid length) would influence the control performance of the SDIP system. For convenience, we introduce a parameter $\lambda = \frac{l_2}{l_1}$ to show the ratio of the centroid length of the second pendulum to the centroid length of the first pendulum. If $l_1 = l_2$, then $\lambda = 1$. Therefore, the matrixes $M(q, \dot{q})$, $C(q, \dot{q})$ and $G(q)$ appeared above equation are relevant to the parameter λ, namely $M = M(q, q, \lambda)$, $C = C(q, q, \lambda)$ and $G = G(q, \lambda)$.

Let the state vector

$$\mathbf{x} = [x_1, x_2, \cdots, x_{12}]^T = [x, \dot{x}, y, \dot{y}, \psi_1, \dot{\psi}_1, \theta_1, \dot{\theta}_1, \psi_2, \dot{\psi}_2, \theta_2, \dot{\theta}]^T. \quad (11)$$

Based on the linear method [14], for each operating points, linearization of the dynamic system can be conducted. Then the T-S fuzzy model of the SDIP System is obtained. The fuzzy rules of the T-S fuzzy model are listed:

IF z_1 is $F_1^{(i)}$ and $\cdots$ and z_n is $F_n^{(i)}$ Then $\dot{\mathbf{x}} = A_i(\lambda)\mathbf{x} + B_i(\lambda)\mathbf{u}$,
$\mathbf{x}(0) = \mathbf{x}_0, (i = 1, 2, \cdots, r),$

where $z_1, \cdots, z_n$ are measurable variables. $F_j^{(i)}$, $j = 1, 2, \cdots, n$ is the j-th fuzzy set of the i-th fuzzy rule, $\mathbf{u}(t) = (u_x, u_y)^T$ is the control input and $\mu_j^i(x_j)$ is the membership function of fuzzy set $F_j^{(i)}$.

Similarly, the system matrix and input matrix $A(\lambda)$, $B_i(\lambda)$ are relevant to λ. If $l_1 = l_2$, then the system matrix and input matrix become $A(1)$, $B(1)$. Usually, $l_1 \neq l_2$(i.e.$\lambda \neq 1$), in this case, by using the Taylor formula, we have

$$A_i(\lambda) \approx A_i(1) + \frac{\partial A_i}{\partial \lambda}|_1(\lambda - 1), B_i(\lambda) \approx B_i(1) + \frac{\partial B_i}{\partial \lambda}|_1(\lambda - 1). \quad (12)$$

Denote $E_{Ai} = \frac{\partial A_i}{\partial \lambda}|_1$, $E_{Bi} = \frac{\partial B_i}{\partial \lambda}|_1$. The model (12) can be simplified as

IF z_1 is $F_1^{(i)}$ and $\cdots$ and z_n is $F_n^{(i)}$ Then
$\dot{\mathbf{x}} = (A_i(1) + (\lambda - 1)E_{Ai})\mathbf{x} + (B_i(1) + (\lambda - 1)E_{Bi})\mathbf{u}$,
$\mathbf{x}(0) = \mathbf{x}_0, (i = 1, 2, \cdots, r).$ $\quad (13)$

In the following, $A_i(1)$, $B_i(1)$ are replaced as A_i, B_i respectively.

By using weighted average method, we have the state equation of the SDIP system

$$\dot{\mathbf{x}} = \left[\sum_{i=1}^r \alpha_i(A_i + (\lambda - 1)E_{Ai})\right]\mathbf{x} + \left[\sum_{i=1}^r \alpha_i(B_i + (\lambda - 1)E_{Bi})\right]\mathbf{u}, \quad (14)$$

where $\alpha_i = \frac{\omega^i}{\sum_{i=1}^r \omega^i}$, $\omega^i = \Pi_{j=1}^n \mu_j^i(x_j)$.

3 Problem Description

In this paper, the controller is designed according to the model (14) for the spatial double inverted pendulum. The goal of controlling is to make the both pendulums maintaining in the vertical direction, the car returning to its original position and the closed-loop system maintaining a certain performance index. In addition, the relationship between a certain performance index and the parameter λ is discussed, and give a design method which can attains optimal performance index.

First of all, for fuzzy system (14), we use a fuzzy controller via the parallel distributed compensation approach [1] as follows:

$$\mathbf{u}(t) = \sum_{i=1}^{r} \alpha_i K_i \mathbf{x}(t), \tag{1}$$

where K_i is the feedback gain for the i-th subsystem. By substituting (1) to (14), we obtain the closed-loop system:

$$\dot{\mathbf{x}}(t) = \sum_{i=1}^{r} \sum_{j=1}^{r} \alpha_i \alpha_j [A_i + (\lambda - 1)E_{1_i} + (B_i + (\lambda - 1)E_{2_i})K_j]\mathbf{x}(t). \tag{2}$$

In this paper, the quadratic cost function is appointed for the fuzzy control system as the following

$$J = \int_0^\infty \left[x^T(t)Qx(t) + u^T(t)Ru(t) \right] dt, \tag{3}$$

where Q and R are the given positive definite matrices.

In rest of this paper, the main goal is to design the appropriate state feedback gain matrix K_i for the fuzzy control system (14) such that the closed-loop system is asymptotically stable, in addition, to seek a solution for the parameter λ such that the upper bound of the cost function J reaches minimum. If there exists a fuzzy controller $u^\star(t)$ such that the close-loop system (2) is asymptotically stable and the upper bound of the cost function J reaches minimum, then $u^\star(t)$ is called an optimal cost fuzzy controller.

4 Design of Optimal Cost Fuzzy Controller

In order to design an optimal cost fuzzy controller for the SDIP system, we define a candidate Lyapunov function as follows

$$V(\mathbf{x}) = \mathbf{x}^T P \mathbf{x}, \tag{1}$$

where P is a positive definite matrix. Then the following sufficient condition for the existence of optimal cost fuzzy controller can be easily obtained.

Theorem 1. *For the fuzzy system (14) and the cost function J (3), if there are positive definite matrices P and K_i, and a set of matrices Y_{ij} (where Y_{ii} is symmetrical and $Y_{ij} = Y_{ji}^T$), such that the following matrix inequalities are satisfied*

1) $P[(A_i + B_i K_i) + (\lambda - 1)(E_{Ai} + E_{Bi} K_i)] +$
   $[(A_i + B_i K_i) + (\lambda - 1)(E_{Ai} + E_{Bi} K_i)]^T P + Q + K_i^T R K_i < Y_{ii},$
   $(i = 1, 2, \cdots, r);$

2) $P[(A_i + A_j + B_i K_j + B_j K_i) + (\lambda - 1)(E_{Ai} + E_{Aj} + E_{Bi} K_j + E_{Bj} K_i)]$
   $+[(A_i + A_j + B_i K_j + B_j K_i) + (\lambda - 1)(E_{Ai} + E_{Aj} + E_{Bi} K_j + E_{Bj} K_i)]^T P$
   $+2Q + K_i^T R K_i + K_j^T R K_j < Y_{ij} + Y_{ji}, (i < j);$

3) $\begin{pmatrix} Y_{11} & \cdots & Y_{1r} \\ \vdots & \ddots & \vdots \\ Y_{r1} & \cdots & Y_{rr} \end{pmatrix} < 0,$

then the closed-loop system is asymptotically stable by using fuzzy controller $\mathbf{u}(t) = \sum_{i=1}^{r} \alpha_i K_i \mathbf{x}(t),$ *and* $J_0 = \mathbf{x}_0^T P \mathbf{x}_0$ *is an upper bound of J.*

Proof: The time derivative of $V((\mathbf{x}))$ is:

$$\dot{V}(\mathbf{x}) = \sum_{i=1}^{r} \alpha_i^2 \mathbf{x}^T \{P[(A_i + B_i K_i) + (\lambda - 1)(E_{Ai} + E_{Bi} K_i)]$$

$$+[(A_i + B_i K_i) + (\lambda - 1)(E_{Ai} + E_{Bi} K_i)]^T P\}\mathbf{x} + \sum_{i<j}^{r} \alpha_i \alpha_j$$

$$\mathbf{x}^T \{P[(A_i + A_j + B_i K_j + B_j K_i) + (\lambda - 1)(E_{Ai} + E_{Aj} + E_{Bi} K_j + E_{Bj} K_i)]$$
$$+[(A_i + A_j + B_i K_j + B_j K_i) + (\lambda - 1)(E_{Ai} + E_{Aj} + E_{Bi} K_j + E_{Bj} K_i)]^T P\}\mathbf{x}$$

$$= \sum_{i=1}^{r} \alpha_i^2 \mathbf{x}^T \{P[(A_i + B_i K_i) + (\lambda - 1)(E_{Ai} + E_{Bi} K_i)]$$

$$+[(A_i + B_i K_i) + (\lambda - 1)(E_{Ai} + E_{Bi} K_i)]^T P + Q + K_i^T R K_i\}\mathbf{x} + \sum_{i<j}^{r} \alpha_i \alpha_j$$

$$\mathbf{x}^T \{P[(A_i + A_j + B_i K_j + B_j K_i) + (\lambda - 1)(E_{Ai} + E_{Aj} + E_{Bi} K_j + E_{Bj} K_i)]$$
$$+[(A_i + A_j + B_i K_j + B_j K_i) + (\lambda - 1)(E_{Ai} + E_{Aj}$$
$$+E_{Bi} K_j + E_{Bj} K_i)]^T P + 2Q + K_i^T R K_i + K_j^T R K_j\}\mathbf{x}$$

$$-\mathbf{x}^T [\sum_{i=1}^{r} \alpha_i^2 (Q + K_i R K_i) + \sum_{i<j}^{r} \alpha_i \alpha_j (2Q + K_i^T R K_i + K_j^T R K_j)]\mathbf{x}.$$

Applying the conditions (1) and (2) to (2), the inequality (2) becomes

$$\dot{V}(\mathbf{x}) \leq \sum_{i=1}^{r} \alpha_i^2 \mathbf{x}^T Y_{ii} \mathbf{x} + \sum_{i<j}^{r} \alpha_i \alpha_j \mathbf{x}^T (Y_{ij} + Y_{ji})\mathbf{x}$$

$$-\mathbf{x}^T [\sum_{i=1}^{r} \alpha_i^2 (Q + K_i R K_i) + \sum_{i<j}^{r} \alpha_i \alpha_j (2Q + K_i^T R K_i + K_j^T R K_j)].$$

According to the condition (3), we have

$$\dot{V}(\mathbf{x}) \leq -\mathbf{x}^T [\sum_{i=1}^{r} \alpha_i^2 (Q + K_i R K_i) + \sum_{i<j} \alpha_i \alpha_j (2Q + K_i^T R K_i + K_j^T R K_j)] \mathbf{x}.$$

Since $2X^T SY \leq \inf_{S>0}\{X^T SX + Y^T SY\}$[15], then $K_i^T R K_j + K_j^T R K_i \leq K_i^T R K_i + K_j^T R K_j$. Above inequality changes into the following form:

$$\dot{V}(\mathbf{x}) < -\mathbf{x}^T \sum_{i=1}^{r} \sum_{j=1}^{r} \alpha_i \alpha_j (Q + K_i^T R K_j) \mathbf{x} < 0.$$

Then the closed-loop system is asymptotically stable at its equilibrium.

Moreover, integrating both sides of above equation with respect to t on the interval from 0 to ∞, we have

$$J = \int_0^{\infty} [\mathbf{x}^T(t) Q \mathbf{x}(t) + \mathbf{u}^T(t) R \mathbf{u}(t)] dt \leq V(\mathbf{x}(0)) = \mathbf{x}_0^T P \mathbf{x}_0.$$

The proof is completed.

Since the conditions of Theorem 1 are not linear matrix inequalities of the P and K_i, It's necessary to transform into linear matrix inequalities.

By using Schur Complement [16], the conditions of the Theorem 1 can be change into LMIs. In the following theorem, let $\rho = \epsilon\lambda$.

Theorem 2. *For the fuzzy system (14) and the cost function J (3), if there are positive definite matrices Z and W_i, a set of matrices X_{ij} (where X_{ii} is symmetrical and $X_{ij} = X_{ji}^T$) and constants $\epsilon > 0$, $\rho > 0$, such that the following matrix inequalities are satisfied:*

1) $$\begin{bmatrix} \Omega_{ii} - X_{ii} & (E_{1_i}X + E_{2_i}W_i)^T & W_i^T & Z & (\rho - \epsilon)I \\ E_{1_i}X + E_{2_i}W_i & -\epsilon I & 0 & 0 & 0 \\ W_i & 0 & -R^{-1} & 0 & 0 \\ Z & 0 & 0 & -Q^{-1} & 0 \\ (\rho - \epsilon)I & 0 & 0 & 0 & -\epsilon^{-1}I \end{bmatrix} < 0,$$

$(i = 1, 2 \cdots, r)$

2) $$\begin{bmatrix} \Omega_{ij} - X_{ij} - X_{ji} & \Gamma_{ij}^T & W_i^T & W_j^T & Z & (\rho - \epsilon)I \\ \Gamma_{ij} & -\epsilon I & 0 & 0 & 0 & 0 \\ W_i & 0 & -R^{-1} & 0 & 0 & 0 \\ W_j & 0 & 0 & -R^{-1} & 0 & 0 \\ Z & 0 & 0 & 0 & -\frac{1}{2}Q^{-1} & 0 \\ (\rho - \epsilon)I & 0 & 0 & 0 & 0 & -\epsilon^{-1}I \end{bmatrix} < 0, \ (i < j);$$

3) $$\begin{pmatrix} X_{11} & \cdots & X_{1r} \\ \vdots & \ddots & \vdots \\ X_{r1} & \cdots & X_{rr} \end{pmatrix} < 0,$$

where $\Omega_{ij} = A_i Z + Z A_i^T + B_i W_j + W_j^T B_i^T, \Gamma_{ij} = (E_{1_i} + E_{1_j}) Z + E_{2_i} W_j + E_{2_j} W_i$, the closed-loop system is asymptotically stable by using fuzzy controller $\mathbf{u}(t) = \sum_{i=1}^r \alpha_i K_i \mathbf{x}(t)$, and $J_0 = \mathbf{x}_0^T P \mathbf{x}_0$ is an upper bound of J.

Based on the results of above theorem, the next goal is to minimize the upper bound of the cost function J.

From the results of Theorem 2, we have

$$J \le \mathbf{x}(0)^T P \mathbf{x}(0), \tag{2}$$

Based on the optimal guaranteed cost fuzzy control method [11], the following theorem about the optimal cost fuzzy controller can be derived for the fuzzy system (14).

Theorem 3. *For the fuzzy system (14), the optimal cost fuzzy controller can be designed by solving the following linear matrix inequalities:*

$$min_{Z,\rho,\epsilon,W_i} \ Trace(M) \tag{3}$$

subject to

$$(i) \ Conditions \ 1) \ 2) \ 3) \ of \ the \ Theorem \ 2; \ (ii) \ \begin{bmatrix} M & I \\ I & Z \end{bmatrix} > 0. \tag{4}$$

If the solution of LMIs in Theorem 3 is $(Z^\star, \lambda^\star, \epsilon^\star, W_i^\star)$ then $\mathbf{u}(t) = \sum_{i=1}^r \alpha_i K_i \mathbf{x}(t)$ is a optimal cost fuzzy controller for the fuzzy system (14), where $K_i = W_i^\star Z^{\star^{-1}}$. Moreover, the best ratio between the centroid lengths of two pendulums is $\lambda^\star = \frac{\rho^\star}{\epsilon^\star}$.

5 Conclusion

In this paper, the mathematical model of the spatial double inverted pendulum system(SDIP system) is obtained by using the Lagrange equation. For the SDIP system, a new optimal cost fuzzy controller design method has been provided. Moveover, we consider the ratio between the centroid lengths of two pendulums in the PDIP system as a design parameter. The sufficient condition for the existence of optimal cost fuzzy controller has been presented, which can be solved effectively using the LMI tools in Matlab.

References

1. Tanaka, K., Ikeda, T., Wang, H.O.: Robust stabilization of a class of uncertain nonlinear systems via fuzzy control: Quadratic stabilizability. H control theory and linear matrix inequalities. IEEE Trans. Fuzzy Syst. 4(1), 1–13 (1996)

2. Chung, C.C., Hauser, J.: Nonlinear control of a swinging pendulum. Automatica 31(6), 851–862 (1995)
3. Xue, A.K., Guan, B.L., Shang, Q.L., Wang, J.Z.: Modeling of triple inverted pendulum and H_∞ robust optimal guaranteed cost control. Journal of Zhejiang University (Engineering Science) 38(12), 1637–1641 (2004)
4. Yurkovich, S., Widjaja, M.: Fuzzy controller synthesis for an inverted pendulum system. Control Engineering Practice 4(4), 445–469 (1996)
5. Yi, J., Yubazki, N.: Stabilization fuzzy control of inverted pendulum system. Artifical Intelligence in Engineering 14(2), 153–163 (2000)
6. Li, H.X., Miao, Z.H., Wang, J.Y.: Variable universe adaptive fuzzy control on the quadruple inverted pendulum. Science in China (Series E) 32(1), 65–75 (2002)
7. Liu, G., Nesic, D., Mareels, I.: Non-local stabilization of a spherical inverted pendulum. International Journal of Control 81(7), 1035–1053 (2008)
8. Duan, X.C., Qiu, Y.Y., Duan, B.Y.: Adaptive sliding mode fuzzy control of planar inverted pendulum. Control and Design 22(7), 774–782 (2007)
9. Yuan, P.G., Zhang, G.Y., Qin, Z.Q., Li, Z.X.: Analysis and controller design of plane double inverted pendulum. Control Engineering of China 11(6), 517–520 (2004)
10. Chang, S.S.L., Peng, T.K.C.: Adaptive guaranteed cost control of systems with uncertain parameters. IEEE Trans. Automat. Contr. 17(4), 474–483 (2004)
11. Yu, L.: Optimal guaranteed cost control of linear uncertain system: an LMI approach. Control Theory and Applications 17(3), 423–428 (2000)
12. Zheng, K., Xu, J.M., Yu, L.: Takiga-Sugeno model-based optimal guarenteed cost fuzzy control for inverted pendulum. Control Theory & Applications 21(5), 703–708 (2004)
13. Liu, Y.Z., Yang, H.X., Zhu, B.H.: Theoretical Mechanics. High Education Press (2001)
14. Marcelo, C.M., Teixeira, Stanislaw, H.Z.: Stabilizing controller design for uncertain nonlinear system using fuzzy models. IEEE Trans. on Fuzzy System 7(2), 133–142 (1999)
15. Guan, X.P., Chen, C.L.: Delay-dependent guaranteed cost control for T-S Fuzzy systems with time delays. IEEE Trans. Fuzzy Systems 12(2), 236–249 (2004)
16. Boyd, S., Ghaoui, L.E., Feron, E., Balakrishnan, V.: Linear Matrix Inequalities in System and Control Theory. SIAM, Philadelphia (1994)

New Absolute Stability Criteria for T-S Fuzzy Lurie Control Systems with Time-Delay and Nonlinearities

Xiao-xu Xia[1], Yan-ju Luo[2], and Yong Wang[1,3]

[1] School of Sciences, Southwest Petroleum University, Chengdu, Sichuan, 610500, China
xiaxiaoxuswpi@sohu.com
[2] Department of Computing Technology, Southwest Electric Power Design
Institute of China, Chengdu, Sichuan, 610021, China
[3] State Key Laboratory of Oil and Gas Reservoir Geology and Exploitation,
Southwest Petroleum University, Chengdu, Sichuan, 610500, China

Abstract. In this paper, the problem of absolute stability for a new class of T-S fuzzy Lurie control systems with time-delay and nonlinearities is considered. By utilizing the Lyapunov stability theory and novel techniques, new Lyapunov functions are defined and a new delay-dependent absolute stability condition is derived. Finally, by using LMI Control Toolbox and Simulink toolbox in MATLAB, a numerical example is presented to illustrate feasibility and effectiveness of the proposed results.

Keywords: Absolute stability, Takagi-Sugeno (T-S) fuzzy Lurie systems, time-delay, Lyapunov-Krasovskii functional (LKF), linear matrix inequality (LMI).

1 Introduction

It is well known that the time-delay phenomenon is frequently encountered in various of engineering systems and the existence of time-delays is often the main source of instability and poor performance. Therefore, the studies of stabilization of Lurie systems with time-delay have more important significance than the ones of model with not time delay. However, the stability conditions mentioned above are all delay-independent, which are often conservative when time-delay is small. In order to improve these conditions , some delay-dependent absolute stability conditions for Lurie control systems have been proposed in [1, 2].

On the other hand, The Takagi-Sugeno (T-S) fuzzy model which was described in [3] for the first time has been proven to be a powerful tool for providing an effective representation of complex nonlinear systems. In [4, 5], the T-S fuzzy systems with time-delay have been extensively studied in recent years.

However, to the authors' knowledge, the problem of the delay-dependent condition for absolute stability of T-S fuzzy Lurie control systems with time-delay and

B.-Y. Cao and X.-J. Xie (Eds.): Fuzzy Engineering and Operations Research, AISC 147, pp. 139–145.
springerlink.com

nonlinearities has not been investigated up to now. So the motivation of the present study is to extend the ordinary T-S fuzzy models to describe the Lurie control systems with time-delay and nonlinearities in order to obtain a more general absolute stability criteria. In this paper, based on Lyapunov functional, Linear matrix inequality (LMI) approach and novel techniques, a new delay-dependent absolutely stable condition for such systems is derived. Furthermore, a simulation example will be provided to demonstrate the proposed results.

2 Problem Formulation

In this section , we consider a class of T-S fuzzy Lurie control systems as following form:

$$\dot{x}(t) = \sum_{i=1}^{r} h_i(s(t)) \left[A_i x(t) + B_i x(t - \tau) + \rho f(\sigma(t - \tau)) + b f(\sigma(t)) \right], t \geq 0,$$

(1)

$$\sigma(t) = c^T x(t),$$

(2)

$$x(\theta) = \varphi(\theta), \qquad \theta \in [-\tau, 0\,],$$

(3)

where $x(t) \in R^n$ denotes the state vector; A_i, B_i $(i = 1, 2, \cdots, r)$ are the coefficient matrices with appropriate dimensions; $b, \rho \in R^n$ are the coefficient of the nonlinearities; $c \in R^n$; $\tau \geq 0$ is the time-delay; $\varphi(\cdot) \in C([-\tau, 0\,], R^n)$ is a continuous vector valued initial function; the nonlinearity functions $f(\cdot)$ satisfy the following sector condition:

$$f(\cdot) \in K[0, \infty] = \{ f(\cdot) \,|\, f(0) = 0, 0 < \sigma f(\sigma(t)) < \infty, \sigma \neq 0 \}.$$

(4)

The fuzzy basis functions are described by:

$$h_i(s(t)) = \frac{\omega_i(s(t))}{\sum\limits_{j=1}^{r} \omega_j(s(t))},$$

(5)

$$\omega_i(s(t)) = \prod_{j=1}^{g} \mu_{ij}(s_j(t)), s(t) = [s_1(t), s_2(t), \cdots, s_g(t)]^T, \quad i = 1, \cdots, r,$$

where $s_1(t), s_2(t), \cdots, s_g(t)$ are the premise variables, and each μ_{ij} $(j = 1, 2, \cdots, g)$ is a fuzzy set. $\mu_{ij}(s_j(t))$ is the grade of membership of $s_j(t)$ in μ_{ij}. It is assumed in this paper that

$$\omega_i(s(t)) \geq 0, \quad \sum_{j=1}^{r} \omega_j(s(t)) > 0, \quad i = 1, 2, \cdots, r, \ \forall \, t \geq 0.$$

Therefore, the fuzzy basis functions satisfy

$$\sum_{i=1}^{r} h_i(s(t)) = 1 \ \text{ with } \ h_i(s(t)) \geq 0, \ i = 1, 2, \cdots, r, \ \forall \, t \geq 0.$$

3 Main Results

Theorem 1. *The system descried by (1) in section 2 is absolutely stable, if there are symmetric positive definite matrices* P_1, S_{22}, Q, R, M, *scalars* $r > 0$, $\alpha > 0$, $\beta > 0$, *and any matrices* $S_{12}, P_j (j = 2, \cdots, 15)$, *such that the following LMIs hold:*

$$\begin{bmatrix} P_1 & S_{12} \\ * & S_{22} \end{bmatrix} \geq 0, \tag{1}$$

$$\Omega_i = \begin{bmatrix} \Gamma_{(1,1)} & \Gamma_{(1,2)} & \Gamma_{(1,3)} & \Gamma_{(1,4)} & \Gamma_{(1,5)} & \Gamma_{(1,6)} & \Gamma_{(1,7)} \\ * & \Gamma_{(2,2)} & \Gamma_{(2,3)} & \Gamma_{(2,4)} & \Gamma_{(2,5)} & \Gamma_{(2,6)} & \Gamma_{(2,7)} \\ * & * & \Gamma_{(3,3)} & \Gamma_{(3,4)} & \Gamma_{(3,5)} & \Gamma_{(3,6)} & \Gamma_{(3,7)} \\ * & * & * & \Gamma_{(4,4)} & \Gamma_{(4,5)} & \Gamma_{(4,6)} & \Gamma_{(4,7)} \\ * & * & * & * & \Gamma_{(5,5)} & \Gamma_{(5,6)} & \Gamma_{(5,7)} \\ * & * & * & * & * & \Gamma_{(6,6)} & \Gamma_{(6,7)} \\ * & * & * & * & * & * & \Gamma_{(7,7)} \end{bmatrix} < 0, \tag{2}$$

where

$$\Gamma_{(1,1)} = P_2^T A_i + A_i^T P_2 - P_9^T - P_9 + S_{12}^T + S_{12} + Q + \tau^2 R;$$

$$\Gamma_{(1,2)} = P_1 - P_2^T + A_i^T P_3 - P_{10};$$

$$\Gamma_{(1,3)} = P_2^T B_i + P_9^T + A_i^T P_4 - P_{11} - S_{12};$$

$$\Gamma_{(1,4)} = P_2^T b + A_i^T P_5 - P_{12} + \alpha c; \Gamma_{(1,5)} = P_2^T \rho + A_i^T P_6 - P_{13};$$

$$\Gamma_{(1,6)} = P_9^T + A_i^T P_7 - P_{14}; \Gamma_{(1,7)} = A_i^T P_8 - P_{15} + S_{22};$$

$$\Gamma_{(2,2)} = -P_3^T - P_3 + \tau^2 M; \Gamma_{(2,3)} = P_3^T B_i + P_{10}^T - P_4;$$

$$\Gamma_{(2,4)} = P_3^T b - P_5 + \beta c; \Gamma_{(2,5)} = P_3^T \rho - P_6; \Gamma_{(2,6)} = P_{10}^T - P_7;$$

$$\Gamma_{(2,7)} = -P_8 + S_{12}; \Gamma_{(3,3)} = P_4^T B_i + B_i^T P_4 + P_{11}^T + P_{11} - Q;$$

$$\Gamma_{(3,4)} = P_4^T b + B_i^T P_5 + P_{12}; \Gamma_{(3,5)} = P_4^T \rho + B_i^T P_6 + P_{13};$$

$$\Gamma_{(3,6)} = P_{11}^T + B_i^T P_7 + P_{14}; \Gamma_{(3,7)} = B_i^T P_8 + P_{15} - S_{22};$$

$$\Gamma_{(4,4)} = P_5^T b + b^T P_5 + r; \Gamma_{(4,5)} = P_5^T \rho + b^T P_6;$$

$$\Gamma_{(4,6)} = P_{12}^T + b^T P_7; \Gamma_{(4,7)} = b^T P_8;$$

$$\Gamma_{(5,5)} = P_6^T \rho + \rho^T P_6 - r; \Gamma_{(5,6)} = P_{13}^T + \rho^T P_7;$$

$$\Gamma_{(5,7)} = \rho^T P_8; \Gamma_{(6,6)} = P_{14}^T + P_{14} - M;$$

$$\Gamma_{(6,7)} = P_{15}; \Gamma_{(7,7)} = -R(i = 1, 2, \cdots, r).$$

Proof. Define a new class of Lyapunov functional candidate in the following form:

$$V(t) = V_1(t) + V_2(t) + V_3(t) + V_4(t) + V_5(t) + V_6(t), \tag{3}$$

with

$$V_1(t) = \begin{bmatrix} x(t) \\ \int_{t-\tau}^t x(s)ds \end{bmatrix}^T \begin{bmatrix} P_1 & S_{12} \\ S_{12}^T & S_{22} \end{bmatrix} \begin{bmatrix} x(t) \\ \int_{t-\tau}^t x(s)ds \end{bmatrix},$$

$$V_2(t) = \int_{t-\tau}^t x^T(s)Qx(s)ds, \quad V_3(t) = \tau \int_{-\tau}^0 d\xi \int_{t+\xi}^t x^T(s)Rx(s)ds,$$

$$V_4(t) = \tau \int_{-\tau}^0 \int_{t+\xi}^t \dot{x}^T(s)M\dot{x}(s)dsd\xi, \quad V_5(t) = 2\beta \int_0^{\sigma(t)} f(\sigma)d\sigma,$$

$$V_6(t) = r \int_{t-\tau}^t f^2(\sigma(s))ds.$$

$$X^T(t) = \begin{bmatrix} x^T(t) & \dot{x}^T(t) & x^T(t-\tau) & f^T(\sigma(t)) & f^T(\sigma(t-\tau)) \end{bmatrix}$$

$$\int_{t-\tau}^t \dot{x}^T(s)ds \quad \int_{t-\tau}^t x^T(s)ds \Bigr]. \tag{4}$$

Then, the time derivative of $V(t)$ along the trajectory of system (1) in section 2 is given by

$$\dot{V}(t) = \dot{V}_1(t) + \dot{V}_2(t) + \dot{V}_3(t) + \dot{V}_4(t) + \dot{V}_5(t) + \dot{V}_6(t), \tag{5}$$

where

$$\dot{V}_1(t) = 2x^T(t)P_1\dot{x}(t) + 2 \begin{bmatrix} x(t) \\ \int_{t-\tau}^t x(s)ds \end{bmatrix}^T \begin{bmatrix} 0 & S_{12} \\ S_{12}^T & S_{22} \end{bmatrix} \begin{bmatrix} \dot{x}(t) \\ x(t) - x(t-\tau) \end{bmatrix}$$

$$= 2x^T(t)P_1\dot{x}(t) + 2 \begin{bmatrix} x^T(t) & \int_{t-\tau}^t x^T(s)ds \end{bmatrix}$$

$$\begin{bmatrix} S_{12}x(t) - S_{12}x(t-\tau) \\ S_{12}^T\dot{x}(t) + S_{22}x(t) - S_{22}x(t-\tau) \end{bmatrix}$$

$$= 2x^T(t)P_1\dot{x}(t) + 2x^T(t)S_{12}x(t) - 2x^T(t)S_{12}x(t-\tau).$$

$$+2 \int_{t-\tau}^t x^T(s)ds \left(S_{12}^T\dot{x}(t) + S_{22}x(t) - S_{22}x(t-\tau) \right). \tag{6}$$

According to Lemma 1 in [1], the following inequalities can been obtained

$$\dot{V}_3(t) \leq \tau^2 x^T(t)Rx(t) - \int_{t-\tau}^t x^T(s)dsR \int_{t-\tau}^t x(s)ds. \tag{7}$$

$$\dot{V}_4(t) \leq \tau^2 \dot{x}^T(t)M\dot{x}(t) - \int_{t-\tau}^t \dot{x}^T(s)dsM \int_{t-\tau}^t \dot{x}(s)ds. \tag{8}$$

Furthermore, using (4) in section 2, there holds

$$\dot{V}_5(t) \le 2\beta c^T \dot{x}(t) f(\sigma(t)) + 2\alpha c^T x(t) f(\sigma(t)).$$ (9)

By using (1) in section 2 and fact $x(t-h) = x(t) - \int_{t-h}^{t} \dot{x}(s)ds$, we have:

$$\sum_{i=1}^{r} h_i(s(t)) \left[x^T(t) P_2^T + \dot{x}^T(t) P_3^T + x^T(t-\tau) P_4^T + f^T(\sigma(t)) P_5^T \right.$$

$$\left. + f^T(\sigma(t-\tau)) P_6^T + \int_{t-\tau}^{t} \dot{x}^T(s)ds P_7^T + \int_{t-\tau}^{t} x^T(s)ds P_8^T \right] [A_i x(t) + B_i x(t-\tau)$$

$$+ \rho f(\sigma(t-\tau)) + bf(\sigma(t)) - \dot{x}(t)] = 0$$ (10)

and

$$\sum_{i=1}^{r} h_i(s(t)) \left[x^T(t) P_9^T + \dot{x}^T(t) P_{10}^T + x^T(t-\tau) P_{11}^T + f^T(\sigma(t)) P_{12}^T \right.$$

$$\left. + f^T(\sigma(t-\tau)) P_{13}^T + \int_{t-\tau}^{t} \dot{x}^T(s)ds P_{14}^T + \int_{t-\tau}^{t} x^T(s)ds P_{15}^T \right]$$

$$\left[x(t-\tau) - x(t) + \int_{t-\tau}^{t} \dot{x}(s)ds \right] = 0.$$ (11)

Adding up (10) and (11) to (5), we can see that

$$\dot{V}(t) \le \sum_{i=1}^{r} h_i(s(t)) X^T(t) \Omega_i X(t),$$ (12)

where $X(t)$ is the same as the corresponding item in (4). If $\Omega_i < 0 (i = 1, 2, \cdots, r)$, then there is a scalar $a > 0$, such that $\dot{V}(t) \le -a \|x(t)\|^2$ for $x(t) \ne 0$, which shows that the system described by (1) is absolutely stable. This completes the proof.

4 Simulation Example

The system considered in this example is with two rules, $f(\sigma(t)) = tan(\sigma(t))$, and $\tau = 2.5$. The fuzzy basis functions are $h_1(s_1(t)) = \sin^2(\pi s_1(t))$, $h_2(s_1(t)) = \cos^2(\pi s_1(t))$. And we let

$$A_1 = \begin{bmatrix} -0.2 & -0.2 \\ 0.4 & -0.2 \end{bmatrix}, A_2 = \begin{bmatrix} -0.4 & -0.2 \\ 0.3 & -0.15 \end{bmatrix}, B_1 = \begin{bmatrix} -0.2 & 0.3 \\ 0.5 & -0.2 \end{bmatrix},$$

$$B_2 = \begin{bmatrix} -0.15 & 0.4 \\ 0.4 & -0.1 \end{bmatrix}, b = \begin{bmatrix} -0.2 \\ -0.3 \end{bmatrix}, c = \begin{bmatrix} 0.4 \\ 0.6 \end{bmatrix}, \rho = \begin{bmatrix} 0.2 \\ -0.3 \end{bmatrix}.$$

Now, we use the LMI Control Toolbox in MATLAB, and we can obtain feasible solutions as follows:

$$P_1 = \begin{bmatrix} 219.3246 & -77.7073 \\ -77.7073 & 181.2928 \end{bmatrix}, S_{22} = \begin{bmatrix} 18.2872 & 8.6944 \\ 8.6944 & 18.7736 \end{bmatrix}, Q = \begin{bmatrix} 245.7597 & 2.6214 \\ 2.6214 & 34.4028 \end{bmatrix},$$

$$R = \begin{bmatrix} 14.0150 & 5.0294 \\ 5.0294 & 13.1675 \end{bmatrix}, M = \begin{bmatrix} 66.5865 & 104.2300 \\ 104.2300 & 179.0560 \end{bmatrix}, r = 174.6868,$$

$$\alpha = 53.3308, \beta = 903.0572.$$

By using the Simulink Toolbox in MATLAB, the state response of the system is shown in Fig.1.

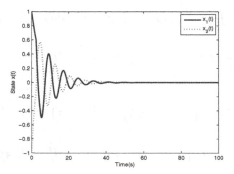

Fig. 1. Response of the state $x(t)$.

From the numerical and simulated results , we can see that the above T-S fuzzy Lurie control system with time-delay and nonlinearities is absolutely stable.

5 Conclusion

The absolute stability for a new class of T-S fuzzy Lurie control systems with time-delay and nonlinearities is investigated in this paper. A new system is created, and a new delay-dependent condition for such system is obtained by using novel techniques. Finally, the results of the presented simulation example have shown that the proposed results in Theorem1 are feasible and effective.

Acknowledgement. This work was supported by the Sichuan Youth Science and Technology Fund (No. 2011JQ0044), the National Program on Key Basic Research Project (973 Program, Grant No. 2011CB201005), the Open Fund (PLN1003) of State Key Laboratory of Oil and Gas Reservoir Geology and Exploitation (Southwest Petroleum University), the Scientific Research Fund (10ZB113) of Sichuan Provincial Educational Department and the Petroleum Technology Innovation Fund of CNPC. The authors are grateful to the editor and referee for their careful reading of the manuscript and helpful suggestions on this work.

References

1. Cao, J.W., Zhong, S.M., Hu, Y.Y.: Delay-dependent condition for absolute stability of Lurie control systems with multiple time delays and nonlinearities. J. Math. Anal. Appl. 338, 497–504 (2008)
2. Gao, J.F., Pan, H.P., Ji, X.F.: A new delay-dependent absolute stability criterion for Lurie systems with time-varying delay. Acta Automatica Sinica 36, 845–850 (2010)
3. Takagi, T., Sugeno, M.: Fuzzy identification of systems and its application to modeling and control. IEEE Trans. Systems Man Cybernet. 15, 116–132 (1985)
4. Sheng, L., Gao, M., Yang, H.Z.: Delay-dependent robust stability for uncertain stochastic fuzzy Hopfield neural networks with time-varying delays. Fuzzy Sets and Systems 160, 3503–3517 (2009)
5. Li, Y.M., Xu, S.Y., Zhang, B.Y., Chu, Y.M.: Robust stabilization and H_∞ control for uncertain fuzzy neutral systems with mixed time delays. Fuzzy Sets and Systems 159, 2730–2748 (2008)

An Improved Method for Ranking Fuzzy Numbers by Distance Minimization

Ruo-ning Xu[1] and Xiao-yan Zhai[2],⋆

[1] School of Mathematics and Information Sciences, Guangzhou University,
Guangzhou 510006 China
rnxu@163.com
[2] School of Management, Guangdong University of Foreign Studies,
Guangzhou 510420 China
zhaixiaoy@163.com

Abstract. In this paper, the problem of ranking fuzzy number is discussed. An improved ranking method is proposed for finding the correct order of fuzzy numbers. Firstly, Based on the distance in fuzzy number space, an index for characterizing uncertain degree is defined and its properties are discussed. Then, by this index the ordering rules are given. Finally, three numerical examples are presented to illustrate the efficiency of the proposed model.

Keywords: Fuzzy number, ranking function, distance.

1 Introduction

For the complex decision making problems, we frequently cannot get its exact numerical data since the complexity of problems themselves, the vagueness in decision makers' thinking and judgment. For this situation, fuzzy numbers are regarded as one of the most suitable instrument for dealing with this type of information. When the information is expressed as fuzzy number, it is necessary to evaluate and compare fuzzy data before a decision is made. Therefore, ranking fuzzy numbers becomes an important problem in the decision making process.

There have been quite a few methods proposed for ranking fuzzy numbers [3, 4, 5, 6, 7]. These researches are based on methods including distance between fuzzy numbers, weighted mean value, centroid point and coefficient of variation. Because each of these methods exists some deficiency, for instance, counter-intuitive or no discriminating, there is yet no method that can always give a satisfactory solution to every situation. Recently, Asady and Zendehnam [2] proposed a new method for ranking fuzzy numbers by distance minimization. The drawback of this method is that for all triangular fuzzy number $\tilde{A} = (a, \alpha, \beta)$ with $a = \frac{\alpha - \beta}{4}$ and also for trapezoidal fuzzy numbers $\tilde{A} = (a_l, a_r, \alpha, \beta)$ with $a_l + a_r = \frac{\alpha - \beta}{2}$ gives the same results. But this is not so reasonable because these fuzzy numbers do not place in

---

⋆ Corresponding author.

B.-Y. Cao and X.-J. Xie (Eds.): Fuzzy Engineering and Operations Research, AISC 147, pp. 147–153.
springerlink.com © Springer-Verlag Berlin Heidelberg 2012

an equivalence class. In order to overcome this problem, Abbasbandy and Hajjari [1] presented a magnitude method for ranking trapezoidal fuzzy numbers based on the left and right spreads at some λ−levels of trapezoidal fuzzy numbers. Unfortunately, this method cannot resolve above problem since it cannot discriminate these trapezoidal fuzzy numbers $\tilde{A} = (a_l, a_r, \alpha, \beta)$ when $a_l + a_r = \frac{\alpha - \beta}{6}$.

To remove the flaws mentioned above, we propose an improved method for ranking fuzzy numbers. An index for characterizing uncertain degree is defined, and then the order of fuzzy number is derived based on the distance in fuzzy number space. In next section, we review some basic concepts and results about fuzzy number. Section 3 presents a improved method for ranking fuzzy numbers by distance minimization. Section 4 presents several numerical examples to illustrate the proposed model. The paper ends with some concluding remarks.

2 Preliminaries

In this section, some basis concepts and results on fuzzy numbers are reviewed, which are necessary for developing our improved method.

Definition 1. Let R denote the set of real number. A fuzzy number $\tilde{A}$ is a fuzzy subset of R, and its membership function $\tilde{A}(x)$ satisfies the following criteria:

1) The λ -level set of $\tilde{A}$, denoted as $A_\lambda = \{x : \tilde{A}(x) \geq \lambda\}$, is a closed convex interval; and
2) $\exists x \in R$ such that $\tilde{A}(x) = 1$.

The set of all fuzzy numbers is denoted as $\tilde{F}(R)$. From the Definition 1, the λ-level set of fuzzy number $\tilde{A}$ can be represented as $A_\lambda = [l(\lambda), r(\lambda)]$ for all $\lambda \in (0, 1]$.

Trapezoidal fuzzy number is a type of popular fuzzy number used in many real-world problems. A fuzzy number $\tilde{A}$ is called a trapezoidal fuzzy number if its membership function has the following form:

$$\tilde{A}(x) = \begin{cases} 1 - \frac{a_l - x}{\alpha}, & \text{if} \quad a_l - \alpha < x \leq a_l, \\ 1, & \text{if} \quad a_l < x \leq a_r, \\ 1 - \frac{x - a_r}{\beta}, & \text{if} \quad a_r < x \leq a_r + \beta, \\ 0, & \text{otherwise}, \end{cases} \tag{1}$$

where $[a_l, a_r]$ is the most possible value of fuzzy number $\tilde{A}$, and α, β are the lower and upper spread bound of $\tilde{A}$ respectively. We denote it as $\tilde{A} = (a_l, a_r, \alpha, \beta)$. When $a_l = a_r$, $\tilde{A}$ is called a triangular fuzzy number, and denoted as $\tilde{A} = (a, \alpha, \beta)$, where $a = a_l$. Obviously, the λ-level set of trapezoidal fuzzy number $\tilde{A}$ can be represented as

$$A_\lambda = [l(\lambda), r(\lambda)] = [a_l - \alpha(1 - \lambda), a_r + \beta(1 - \lambda)] \tag{2}$$

for all $\lambda \in (0, 1]$.

In order to measure the near degree between two fuzzy numbers, paper [8] presented the following distance.

Definition 2. Let $\tilde{A}$ and $\tilde{B}$ be fuzzy numbers with $A_\lambda = [a_l(\lambda), a_r(\lambda)]$ and $B_\lambda = [b_l(\lambda), b_r(\lambda)]$, the quantity

$$\tilde{d}(\tilde{A}, \tilde{B}) = \left(\int_0^1 d^2(A_\lambda, B_\lambda) \, d\lambda \right)^{\frac{1}{2}} \tag{3}$$

is the distance between $\tilde{A}$ and $\tilde{B}$, where

$$d^2(A_\lambda, B_\lambda) = (a_l(\lambda) - b_l(\lambda))^2 + (a_r(\lambda) - b_r(\lambda))^2. \tag{4}$$

The distance defined here synthetically reflects the information on every membership degree, and the meaning of this distance is visual and natural.

Using this distance, Asady and Zendehnam [2] showed the following conclusions.

Theorem 1. *Let $\tilde{A}$ be a fuzzy number. Then*

$$C(\tilde{A}) = \frac{1}{2} \int_0^1 (a_l(\lambda) + a_r(\lambda)) \, d\lambda \tag{5}$$

is the nearest point to the fuzzy number $\tilde{A}$ with respect to distance $\tilde{d}(*, *)$.

Theorem 2. *If $\tilde{A} = (a_l, a_r, \alpha, \beta)$ is a trapezoidal fuzzy number, then the nearest point to $\tilde{A}$ is*

$$C(\tilde{A}) = \frac{a_l + a_r}{2} + \frac{\beta - \alpha}{4}. \tag{6}$$

If $\tilde{A} = (a, \alpha, \beta)$ is a triangular fuzzy number, then the nearest point to $\tilde{A}$ is

$$C(\tilde{A}) = a + \frac{\beta - \alpha}{4}. \tag{7}$$

3   Improvement of Distance Minimization Method

In the paper [2], Asady and Zendehnam used $C(\tilde{A})$, the nearest point to the fuzzy number $\tilde{A}$, as a ranking function, which mapped each fuzzy number into the real line, and can be used to define the order for the fuzzy number set by the natural order existed on R. Nevertheless, this method can not rank fuzzy numbers correctly since $C(\tilde{A})$ cannot discriminate some kind of fuzzy numbers mentioned ahead. In this section, we present a improved method for ranking fuzzy numbers to overcome this drawback.

For $\tilde{A} \in \tilde{F}(A)$, the distance of fuzzy number $\tilde{A}$ to its nearest point $C(\tilde{A})$ can be expressed as

$$
\tilde{d}(\tilde{A}, C(\tilde{A})) = \left(\int_0^1 d^2(A_\lambda, C(\tilde{A})) \, d\lambda \right)^{\frac{1}{2}}
$$

$$
= \left(\int_0^1 (a_l(\lambda) - C(\tilde{A}))^2 + (a_r(\lambda) - C(\tilde{A}))^2 \, d\lambda \right)^{\frac{1}{2}}, \qquad (8)
$$

which can be rewritten as

$$
\tilde{d}(\tilde{A}, C(\tilde{A})) = \left(\int_0^1 [a_l(\lambda)^2 + a_r(\lambda)^2] \, d\lambda - 2C(\tilde{A})^2 \right)^{\frac{1}{2}}. \qquad (9)
$$

Obviously, $\tilde{d}(\tilde{A}, C(\tilde{A}))$ is determined by $\int_0^1 [a_l(\lambda)^2 + a_r(\lambda)^2] \, d\lambda$ and $C(\tilde{A})$. Therefore, we denote

$$
D(\tilde{A}) = \int_0^1 [a_l(\lambda)^2 + a_r(\lambda)^2] \, d\lambda, \ \forall \tilde{A} \in \tilde{F}(R), \qquad (10)
$$

and use it as an index for ranking fuzzy numbers when two fuzzy numbers have same nearest point. Now, we present a improved method for Asady and Zendehnam's distance minimization method in the following.

Definition 3. *For any two fuzzy numbers $\tilde{A}$ and $\tilde{B}$, define the ranking of $\tilde{A}$ and $\tilde{B}$ as follows:*

1) *If $C(\tilde{A}) < C(\tilde{B})$, then $\tilde{A} \prec \tilde{B}$;*
2) *If $C(\tilde{A}) > C(\tilde{B})$, then $\tilde{A} \succ \tilde{B}$;*
3) *If $C(\tilde{A}) = C(\tilde{B})$, then*
   if $D(\tilde{A}) < D(\tilde{B})$, then $\tilde{A} \succ \tilde{B}$, and
   if $D(\tilde{A}) > D(\tilde{B})$, then $\tilde{A} \prec \tilde{B}$;
4) *Else $\tilde{A} \approx \tilde{B}$.*

Theorem 3. *If $\tilde{A} = (a_l, a_r, \alpha, \beta)$ is a trapezoidal fuzzy number, then*

$$
D(\tilde{A}) = 2C(\tilde{A})^2 + \frac{5}{24}(\alpha - \beta)^2 + \frac{2}{3}\alpha\beta + \frac{1}{2}(a_l - a_r)^2 + \frac{1}{2}(\alpha + \beta)(a_r - a_l). \qquad (11)
$$

Proof. The left and right points of A_λ can be represented as

$$
l(\lambda) = a_l - \alpha(1 - \lambda), \ r(\lambda) = a_r + \beta(1 - \lambda),
$$

respectively, and thus

$$
a_l(\lambda)^2 + a_r(\lambda)^2 = a_l^2 + a_r^2 + (\alpha^2 + \beta^2)(\lambda - 1)^2 + (\lambda - 1)(\alpha a_l - \beta a_r).
$$

According to equation (10), we can derive $D(\tilde{A})$ as follows

$$
\begin{aligned}
D(\tilde{A}) &= \int_0^1 [a_l(\lambda)^2 + a_r(\lambda)^2]\, d\lambda \\
&= \int_0^1 [a_l^2 + a_r^2 + (\alpha^2 + \beta^2)(\lambda - 1)^2 + 2(\lambda - 1)(\alpha a_l - \beta a_r)]\, d\lambda \\
&= a_l^2 + a_r^2 + \frac{1}{3}(\alpha^2 + \beta^2) + \beta a_r - \alpha a_l \\
&= 2\left(\frac{a_l + a_r}{2} + \frac{\beta - \alpha}{4}\right)^2 \\
&\quad + \frac{5}{24}(\alpha - \beta)^2 + \frac{2}{3}\alpha\beta + \frac{1}{2}(a_l - a_r)^2 + \frac{1}{2}(\alpha + \beta)(a_r - a_l) \\
&= 2C(\tilde{A})^2 + \frac{5}{24}(\alpha - \beta)^2 + \frac{2}{3}\alpha\beta + \frac{1}{2}(a_l - a_r)^2 + \frac{1}{2}(\alpha + \beta)(a_r - a_l).
\end{aligned}
$$

When $\tilde{A}$ is a triangular fuzzy number, we can easily get the following conclusion.

Theorem 4. If $\tilde{A} = (a, \alpha, \beta)$ is a triangular fuzzy number, then

$$
D(\tilde{A}) = 2C(\tilde{A})^2 + \frac{5}{24}(\alpha - \beta)^2 + \frac{2}{3}\alpha\beta. \tag{12}
$$

4   Numerical Examples

In this section, we use three examples to illustrate this improved method. The first two are taken from Abbasbandy and Hajjari [1], and the third is proposed to compare with Abbasbandy and Hajjari's method.

Example 1. Consider triangular fuzzy numbers

$$
\tilde{A} = (0, 1, 1), \ \tilde{B} = (1, 5, 1).
$$

By Asady and Zendehnam's distance minimization method,

$$
C(\tilde{A}) = 0, \text{ and } C(\tilde{B}) = 0,
$$

therefore the ranking order is

$$
\tilde{A} \sim \tilde{B},
$$

which is an unreasonable result. By the proposed method, we have

$$
D(\tilde{A}) = 2C(\tilde{A})^2 + 5/24(\alpha - \beta)^2 + 2/3\alpha\beta = 2/3, \text{ and } D(\tilde{B}) = 20/3,
$$

therefore the ranking order is

$$
\tilde{A} \succ \tilde{B}.
$$

So, our method overcome the shortcoming of Asady and Zendehnam's distance minimization method.

Example 2. Consider the four fuzzy numbers

$$\tilde{A} = (1, 5, 1), \ \tilde{B} = (1/4, 2, 1), \ \tilde{S} = (2, 9, 1), \ \tilde{T} = (0, 1, 2, 0).$$

We have

$$C(\tilde{A}) = C(\tilde{B}) = C(\tilde{S}) = C(\tilde{T}) = 0.$$

Thus, the ranking order, by Asady and Zendehnam's distance minimization method, is

$$\tilde{A} \sim \tilde{B} \sim \tilde{S} \sim \tilde{T}.$$

On the other hand, according to the proposed improved method, we have

$$D(\tilde{A}) = 20/3, \ D(\tilde{B}) = 37/24, \ D(\tilde{S}) = 58/3, \ D(\tilde{T}) = 7/3,$$

which gives the ranking order

$$\tilde{B} \succ \tilde{T} \succ \tilde{A} \succ \tilde{S}.$$

Example 3. In the paper [1], Abbasbandy and Hajjari defined the magnitude of the trapezoidal fuzzy number as

$$Mag(\tilde{A}) = \int_0^1 [a_l(\lambda) + a_r(\lambda) + x_0 + y_0] f(\lambda) \, d\lambda, \tag{13}$$

where $f(\lambda) = \lambda$, $[x_0, y_0]$ is the most possible values and $a_l(\lambda), a_r(\lambda)$ are the left and right points of A_λ, respectively.

Consider trapezoidal fuzzy numbers

$$\tilde{A} = (0, 0.5, 4, 1), \ \tilde{B} = (-0.5, 0.5, 1, 1).$$

By (13), we have

$$Mag(\tilde{A}) = 0, \ \text{and} \ Mag(\tilde{B}) = 0,$$

therefore Abbasbandy and Hajjari's method gives that the ranking order is

$$\tilde{A} \sim \tilde{B},$$

which means that Abbasbandy and Hajjari's method can not discriminate these two fuzzy numbers. By the proposed method, we have

$$C(\tilde{A}) = -1/2, \ \text{and} \ C(\tilde{B}) = 0,$$

which gives the ranking order as

$$\tilde{A} \prec \tilde{B}.$$

5   Conclusion

This paper concerns with the problem of ranking fuzzy numbers. The shortcomings of distance minimization, proposed by Asady and Zendehnam, are discussed and an improved method is put forward based on the distance in fuzzy number space. By this method, the shortcomings existed in Asady and Zendehnam, and Abbasbandy and Hajjari's methods can be resolved. Compared to Abbasbandy and Hajjari's ranking method for trapezoidal fuzzy numbers, the applicable scope of the proposed method is more extensive, and the method is suitable to general fuzzy numbers. Numerical examples show that the proposed method can discriminate those fuzzy numbers which Asady and Zendehnam, and Abbasbandy and Hajjari's methods can not. Since the proposed method is based on the distance between fuzzy numbers, it is easy to rank fuzzy numbers in a way similar to the original method and just need a little more calculation.

References

1. Abbasbandy, S., Hajjari, T.: A new approach for ranking of trapezoidal fuzzy numbers. Computers and Mathematics with Applications 57, 413–419 (2009)
2. Asady, B., Zendehnam, M.: Ranking fuzzy numbers by distance minimization. Applied Mathematical Modelling 31, 2589–2598 (2007)
3. Bortolan, G., Degani, R.: A review of some methods for ranking fuzzy numbers. Fuzzy Sets and Systems 15, 1–19 (1985)
4. Choobineh, F., Li, H.: An index for ordering fuzzy numbers. Fuzzy Sets and Systems 54, 287–294 (1993)
5. Lee, K.M., Cho, C.H., Lee-Kwang, H.: Ranking fuzzy values with satisfaction function. Fuzzy Sets and Systems 64, 295–311 (1994)
6. Fortemps, P., Roubens, M.: Ranking and defuzzification methods based on area compensation. Fuzzy Sets and Systems 82, 319–330 (1996)
7. Cheng, C.H.: A new approach for ranking fuzzy numbers by distance method. Fuzzy Sets and Systems 95, 307–317 (1998)
8. Xu, R.: A linear regression model in fuzzy environment. Advances in Modeling & Simulation 27, 31–40 (1991)

The Reductivity of Fuzzy Inference

Chun-ling Zhang, Xue-hai Yuan, and Yun-tao Wang

Faculty of Electronic Information and Electrical Engineering,
Dalian University of Technology, Dalian 116024, China
hzy_dahua2@163.com

Abstract. This paper proposed a new method of fuzzy inference which being called generalized CRI Algorithm, it is discussed on the basis of the common CRI Algorithms. In this study, we also considered the reductivity of 21 common fuzzy implication operators together with CRI algorithm and generalized CRI Algorithm, along with the results in the form of tables.

Keywords: Fuzzy inference, CRI algorithm, generalized CRI algorithm, reductivity.

1 Introduction

1.1 The Methods of Fuzzy Inference

Fuzzy inference plays a crucial role during the process of modeling the fuzzy system. Now, the most commonly used methods of fuzzy inference are the CRI Algorithm and the Triple I Method.

In 1973, Zadeh proposed the CRI Algorithm, the basic idea of it is to transform the fuzzy rule base to the fuzzy relation R in $X \times Y$ by fuzzy implication combining with the fuzzy input A^* and R, and then we get the output:

$$B^*(y) = \bigvee_{x \in X} (A^*(x) \wedge R(A(x), B(y))) \, [1].$$

In 1999, professor Guo-Jun Wang proposed Triple I Method, which has sophisticated logical fundamental. The basic idea is that if we have got B^*, then $A(x) \to B(y)$ and $A^*(x) \to B^*(y)$ should have certain logical relationship during the process of fuzzy inference. In general,

$$(A(x) \to B(y)) \to (A^*(x) \to B^*(y)) \tag{1}$$

The B^* is the minimum fuzzy set which assuring formula (1) gets the maximum value [2-3].

B.-Y. Cao and X.-J. Xie (Eds.): Fuzzy Engineering and Operations Research, AISC 147, pp. 155–163.
springerlink.com   © Springer-Verlag Berlin Heidelberg 2012

1.2   Definition of the Reductivity [2]

We suppose that A is a normal fuzzy set (there exists $x_0 \in X$ such that $A(x_0) = 1$), with regard to the problem of fuzzy inference, if when $A^* = A$, there has $B^* = B$, then we could take this method as having the reductivity.

When the input $A^* = A$ is given, what we desire is to get $B^* = B$, which means we wish the inference method having reductivity. So whether the method has reductivity or not is an essential standard to judge its reasonableness and effectiveness. In this paper, we considered the reductivity of 21 common fuzzy implication operators together with the CRI Algorithm and the Generalized CRI Algorithm.

2   Discussion on the Reductivity of CRI Algorithm and Generalized CRI Algorithm

On the basis of studying CRI Algorithm, a new method of fuzzy inference that is being called Generalized CRI Algorithm is proposed. During the process of fuzzy inference, we choose $\overset{\rightarrow}{}$ as R_{KD} implication operator, then

$$R(x, y) = A(x) \rightarrow B(y) = (1 - A(x)) \vee B(y) \qquad (2)$$

So formula (2) can be written as:

$$B^*(y) = \underset{x \in X}{\wedge} (A^*(x) \rightarrow (A(x) \rightarrow B(y))) \qquad (3)$$

In formula (3), when we choose different implication operators, then:

$$B^*(y) = \underset{x \in X}{\wedge} (A^*(x) \rightarrow_1 (A(x) \rightarrow_2 B(y))) \qquad (4)$$

The method of getting B^* through formula (3) and (4) is called Generalized CRI Algorithm.

The common 21 implication operators [4]:

$\theta_1(a,b) = (1-a) \vee b$ (*Kleene − Dienes operator*);

$\theta_2(a,b) = 1 - a + ab$ (*Reichenbach operator*);

$\theta_3(a,b) = (1 - a + b) \wedge 1$ (*Lukasiewicz operator*);

$$\theta_4(a,b) = \begin{cases} 1, & a \leq b, \\ \dfrac{b}{a}, & a > b, \end{cases} \text{(\textit{Goguen operator})};$$

$$\theta_5(a,b) = \begin{cases} 1, & a \le b, \\ b, & a > b, \end{cases} \quad (\text{Gödel operator}) \, ;$$

$$\theta_6(a,b) = \begin{cases} (1-a) \vee b, & (1-a) \wedge b = 0, \\ 1, & else, \end{cases} ;$$

$$\theta_7(a,b) = \begin{cases} 1, & a < 1, \\ b, & a = 1, \end{cases} ;$$

$$\theta_8(a,b) = (1-a) \vee (a \wedge b) \; (\text{Zadeh operator});$$

$$\theta_9(a,b) = (1 - a + 2ab - a^2 b) \wedge 1;$$

$$\theta_{10}(a,b) = \begin{cases} 1, & a < 1 \text{ or } b = 1 \\ 0, & else, \end{cases} ;$$

$$\theta_{11}(a,b) = \begin{cases} 1, & a \le b, \\ 0, & a > b, \end{cases} (\text{Gaines} - \text{Re}cher \ operator);$$

$$\theta_{12}(a,b) = \begin{cases} 1, & a = 0 \text{ and } b = 0, \\ b^a, & else, \end{cases} (\text{Yager operator});$$

$$\theta_{13}(a,b) = a \wedge b \; (\text{Mamdani operator});$$

$$\theta_{14}(a,b) = ab \; (\text{Larsen operator});$$

$$\theta_{15}(a,b) = 0 \vee (a + b - 1) \, ;$$

$$\theta_{16}(a,b) = \begin{cases} a \wedge b, & a \vee b = 1, \\ 0, & a \vee b < 1, \end{cases} ;$$

$$\theta_{17}(a,b) = a \vee b;$$

$$\theta_{18}(a,b) = a + b - ab;$$

$$\theta_{19}(a,b) = 1 \wedge (a + b);$$

$$\theta_{20}(a,b) = \begin{cases} a \vee b, & a \wedge b = 0, \\ 1, & a \wedge b > 0, \end{cases} .$$

2.1 Discussion on the Reductivity of CRI Algorithm

In this paper, we rule:

$$T_1(a,b) = a \wedge b, \; T_2(a,b) = (a + b - 1) \vee 0, \; T_3(a,b) = ab$$

Theorem 2.1.1 *Using the CRI Algorithm and "$\vee - T_1$" synthetic operation, when we choose separately implication operators $\rightarrow$ as $\theta_5, \theta_{12}, \theta_{13}, \theta_{14}, \theta_{15}$, the corresponding CRI Algorithm has reductivity; when we choose separately implication operators $\rightarrow$ as $\theta_0, \theta_1, \theta_2, \theta_3, \theta_4, \theta_5, \theta_6, \theta_7, \theta_8,$ $\theta_9, \theta_{10}, \theta_{11}, \theta_{12},$ $\theta_{13}, \theta_{14}, \theta_{15}, \theta_{16},$ $\theta_{17}, \theta_{18}, \theta_{19}, \theta_{20}$, the corresponding CRI Algorithm doesn't have reductivity.*

The proofs of some chosen implication operators have been given in paper [1]. In this paper, we just add the reductivity when choosing some other implication operators. The proof process is similar, so we won't give them any longer.

Theorem 2.1.2 *Using the CRI Algorithm and "$\vee - T_2$" synthetic operation, when we choose separately implication operators $\rightarrow$ as $\theta_0, \theta_1, \theta_2, \theta_3, \theta_4, \theta_5, \theta_6, \theta_7, \theta_8, \theta_9, \theta_{10}, \theta_{11}, \theta_{12}, \theta_{13}, \theta_{14}, \theta_{15}$, the corresponding CRI Algorithm has the reductivity; when we choose separately implication operators $\rightarrow$ as $\theta_6, \theta_7, \theta_9, \theta_{10}, \theta_{16}, \theta_{17}, \theta_{18}, \theta_{19}, \theta_{20}$, the corresponding CRI Algorithm doesn't have the reductivity.*

Proof. Taking θ_2 implication operator as an example, similar to others' proof. $\theta_2(a,b) = 1 - a + ab$. According to the definition of reductivity, let $A^* = A$ and A is a normal fuzzy set. Then

$$B^*(y) = \bigvee_{x \in X} ((A^*(x) + A(x) \rightarrow B(y) - 1) \vee 0)$$
$$= \bigvee_{x \in X} ((A(x) + (1 - A(x) + A(x)B(y)) - 1) \vee 0)$$
$$= \bigvee_{x \in X} A(x)B(y) = B(y)$$

So when we use "$\vee - T_2$" synthetic operation and choose implication operator $\rightarrow$ as θ_2, the fuzzy inference has reductivity.

Theorem 2.1.3 *Using the CRI algorithm and "$\vee - T_3$" synthetic operation, when we choose separately implication operators $\rightarrow$ as $\theta_4, \theta_5, \theta_{11}, \theta_{12}, \theta_{13}, \theta_{14}$, the corresponding CRI Algorithm has the reductivity; when choose separately implication operators $\rightarrow$ as $\theta_0, \theta_1, \theta_2, \theta_3, \theta_6, \theta_7, \theta_8, \theta_9, \theta_{10}, \theta_{15}, \theta_{16}, \theta_{17}, \theta_{18}, \theta_{19}, \theta_{20}$, the corresponding CRI Algorithm doesn't have the reductivity.*

Proof. Taking θ_4 as an example, similar to others' proof.

$$\theta_4(a,b) = \begin{cases} 1, & a \le b, \\ \dfrac{b}{a}, & a > b, \end{cases} \quad (Goguen)$$

according to the definition of reductivity, let $A^* = A$ and A be a normal fuzzy set. Then

$$B^*(y) = \underset{x \in X}{\vee} (A^*(x)(A(x) \to B(y)))$$

$$= (\underset{A(x) \le B(y)}{\vee} (A(x)*1)) \vee (\underset{A(x) > B(y)}{\vee} A(x)*\frac{B(y)}{A(y)}) = B(y).$$

So when we use "$\vee - T_3$" synthetic operation and choose implication operator $\to$ as θ_4, the fuzzy inference has reductivity.

2.2 Discussion on the Reductivity of Generalized CRI Algorithm

Theorem 2.2.1 *Using the generalized CRI Algorithm, and we choose implication operator $\to_1$ as $\theta_1(R_{KD})$ in the formula (4). When we choose separately implication operators $\to_2$ as $\theta_0, \theta_1, \theta_2, \theta_3, \theta_4, \theta_5, \theta_7, \theta_9, \theta_{12}$, the corresponding generalized CRI Algorithm has the reductivity. When we choose separately implication operators $\to_2$ as $\theta_6, \theta_8, \theta_9, \theta_{10}, \theta_{11}, \theta_{13}, \theta_{14}, \theta_{15}, \theta_{16}, \theta_{17}, \theta_{18}, \theta_{19}, \theta_{20}$, the corresponding generalized CRI Algorithm doesn't have the reductivity.*

Proof. We choose $\to_1$ as $\theta_1(R_{KD})$ and $\to_2$ as

$$\theta_5(a,b) = \begin{cases} 1, & a \le b, \\ b, & a > b, \end{cases} \quad (\ G\ddot{o}del\ )$$

and others are similar. According to the definition of reductivity, let $A^* = A$ and A is a normal base set.
If $B(y) < 1$, then

$$B^*(y) = (A^*(x) \to_1 (A(x) \to_2 B(y)))$$

$$= \underset{x \in X}{\wedge} ((1 - A(x)) \vee (A(x) \to B(y)))$$

$$= [\underset{A(x) \le B(y)}{\wedge} ((1 - A(x)) \vee 1)] \wedge [\underset{A(x) > B(y)}{\wedge} ((1 - A(x)) \vee B(y))]$$

$$= \underset{A(x) > B(y)}{\wedge} ((1 - A(x)) \vee B(y)) = 0 \vee B(y) = B(y).$$

If $B(y)=1$, then $B^*(y)= \bigwedge_{x \in X}((1-A(x)) \vee 1)=1$.

So $B^*(y)=B(y)$, that is to say, the fuzzy inference has reductivity.

Theorem 2.2.2 *Using the generalized CRI Algorithm and in the formula (4), we choose separately implication operator* $\to_1$ *as* $\theta_3(R_{LU})$ *and* $\to_2$ *as* $\theta_0, \theta_1, \theta_2, \theta_3, \theta_4, \theta_5, \theta_7, \theta_8, \theta_9, \theta_{12}, \theta_{13}, \theta_{14}, \theta_{15}$, *the corresponding generalized CRI Algorithm has the reductivity; when we choose separately implication operators* $\to_2$ *as* $\theta_6, \theta_{10}, \theta_{11}, \theta_{16}, \theta_{17}, \theta_{18}, \theta_{19}, \theta_{20}$, *the corresponding generalized CRI Algorithm doesn't have the reductivity.*

Proof. We choose $\to_1$ as $\theta_3(Lukasiewicz)$ and $\to_2$ as

$$\theta_0(a,b)=\begin{cases}1, & a \le b, \\ (1-a) \vee b, & a > b,\end{cases} (R_0)$$

and similar to others' proof.

According to the definition of reductivity, let $A^*=A$ and A be a normal fuzzy set.

If $B(y)<1$, let $A(a)=1$, then

$$B^*(y)=(A^*(x) \to_1 (A(x) \to_2 B(y)))$$
$$= \bigwedge_{x \in X}(1-A(x)+(A(x) \to B(y)) \wedge 1$$
$$= \bigwedge_{A(x)>B(y)}(1-A(x)+(1-A(x)) \vee B(y))$$
$$=1-A(a)+(1-A(a)) \vee B(y)=B(y).$$

If $B(y)=1$, it's obvious that $B^*(y)=1$.

So $B^*(y)=B(y)$, that is to say, the fuzzy inference has the reductivity.

Theorem 2.2.3 *Using the generalized CRI Algorithm, and we choose* $\to_1$ *as* θ_7 *in the formula (4). When we choose separately implication operators* $\to_2$ *as* $\theta_0, \theta_1, \theta_2, \theta_3, \theta_4, \theta_5, \theta_6, \theta_7, \theta_8, \theta_9, \theta_{12}, \theta_{13}, \theta_{14}, \theta_{15}, \theta_{16}, \theta_{18}$, *the corresponding generalized CRI Algorithm has the reductivity; when we choose separately implication operators* $\to_2$ *as* $\theta_{10}, \theta_{11}, \theta_{17}, \theta_{19}, \theta_{20}$, *the corresponding generalized CRI Algorithm doesn't have the reductivity.*

Proof. We choose $\to_1$ as and $\to_2$ as $\theta_3(a,b)=(1-a+b) \wedge 1(Lukasiewicz)$ and similar to others' proof. According to the definition of the reductivity, let $A^*=A$ and A be a normal fuzzy set. Then

$$B^*(y) = (A^*(x) \rightarrow_1 (A(x) \rightarrow_2 B(y)))$$
$$= [\underset{A(x)<1}{\wedge} 1] \wedge [\underset{A(x)=1}{\wedge} ((1-A(x)+B(y)) \wedge 1)]$$
$$= 1 \wedge B(y) = B(y).$$

So $B^*(y) = B(y)$, that is to say, the fuzzy inference has reductivity.

There are some results in paper [1], and in this paper we show the theorems within the following forms, in which "√"means the method has the reductivity and "×"means the method doesn't have it.

Table 1. CRI Algorithm:"$\vee - T_1$"synthetic operation

θ_0	θ_1	θ_2	θ_3	θ_4	θ_5	θ_6
×	×	×	×	×	√	×
θ_7	θ_8	θ_9	θ_{10}	θ_{11}	θ_{12}	θ_{13}
×	×	×	×	×	√	√
θ_{14}	θ_{15}	θ_{16}	θ_{17}	θ_{18}	θ_{19}	θ_{20}
√	√	×	×	×	×	×

Table 2. CRI Algorithm:"$\vee - T_2$"synthetic operation

θ_0	θ_1	θ_2	θ_3	θ_4	θ_5	θ_6
√	√	√	√	√	√	×
θ_7	θ_8	θ_9	θ_{10}	θ_{11}	θ_{12}	θ_{13}
×	√	×	×	√	√	√
θ_{14}	θ_{15}	θ_{16}	θ_{17}	θ_{18}	θ_{19}	θ_{20}
√	√	×	×	×	×	×

Table 3. CRI Algorithm : "$\vee - T_3$"synthetic operation

θ_0	θ_1	θ_2	θ_3	θ_4	θ_5	θ_6
×	×	×	×	√	√	×
θ_7	θ_8	θ_9	θ_{10}	θ_{11}	θ_{12}	θ_{13}
×	×	×	×	√	√	√
θ_{14}	θ_{15}	θ_{16}	θ_{17}	θ_{18}	θ_{19}	θ_{20}
√	×	×	×	×	×	×

Table 4. Generalized CRI Algorithm : $\rightarrow_1$ is chosen as R_{KD} and $\rightarrow_2$ is chosen as the following

θ_0	θ_1	θ_2	θ_3	θ_4	θ_5	θ_6
√	√	√	√	√	√	×
θ_7	θ_8	θ_9	θ_{10}	θ_{11}	θ_{12}	θ_{13}
√	×	√	×	×	√	×
θ_{14}	θ_{15}	θ_{16}	θ_{17}	θ_{18}	θ_{19}	θ_{20}
×	×	×	×	×	×	×

Table 5. Generalized CRI Algorithm : $\rightarrow_1$ is chosen as R_{Lu} and $\rightarrow_2$ is chosen as the following

θ_0	θ_1	θ_2	θ_3	θ_4	θ_5	θ_6
√	√	√	√	√	√	×
θ_7	θ_8	θ_9	θ_{10}	θ_{11}	θ_{12}	θ_{13}
√	√	√	×	×	√	√
θ_{14}	θ_{15}	θ_{16}	θ_{17}	θ_{18}	θ_{19}	θ_{20}
√	√	×	×	×	×	×

Table 6. Generalized CRI Algorithm : $\rightarrow_1$ is chosen as θ_7 and $\rightarrow_2$ is chosen as the following

θ_0	θ_1	θ_2	θ_3	θ_4	θ_5	θ_6
√	√	√	√	√	√	√
θ_7	θ_8	θ_9	θ_{10}	θ_{11}	θ_{12}	θ_{13}
√	√	√	×	×	√	√
θ_{14}	θ_{15}	θ_{16}	θ_{17}	θ_{18}	θ_{19}	θ_{20}
√	√	√	×	√	×	×

3 Conclusion

Generalized CRI Algorithm is a new fuzzy inference method and we can get the following conclusions through the discussion on the reductivity when different implication operators combining with the two inference methods.

(1) Considering the reductivity of CRI Algorithm, we have proved the reductivity of some other implication operators based on other paper's conclusions.

(2) Considering the reductivity of generalized CRI Algorithm, the θ_7 implication operator has commendable property, which means when it combined with many other implication operators, the inference method has the reductivity.

Acknowledgements. Thanks to the support by National Natural Science Foundation of China (No. 90818025 and No. 61074044) and Faculty of Electronic Information and Electrical Engineering of Dalian University of Technology.

References

1. Hou, J., Li, H.X.: Reductivity of some fuzzy inference methods. Fuzzy Systems and Mathematics 4, 57–62 (2005)
2. Wang, G.J.: Non-classical logic and approximate reasoning. Science Press, Beijing (2000) (Chinese)
3. Wang, G.J.: The triple method with total inference rules of fuzzy reasoning. Science in China E 29, 43–53 (1999)
4. You, F., Feng, Y.B., Li, H.X.: Fuzzy implication operators and their construction: fuzzy implication operators and their properties. Journal of Beijing Normal University (Natural Science) 39, 52–56 (2003)

Information Entropy of Discrete Quasi-random Variables and Its Properties

Yang Yang[1], Chao Wang[2], and Ming-hu Ha[3,*]

[1] College of Mathematics & Computer Sciences, Hebei University, Baoding,
071002, P.R. China
[2] College of Physics Science & Technology, Hebei University, Baoding, 071002,
P.R. China
[3] College of Mathematics & Computer Sciences, Hebei University, Baoding, 071002,
P.R. China, and College of Science, Hebei University of Engineering,
Handan, 056038, P.R. China
yy429041@126.com, wangchaohbu@yahoo.com.cn, mhha@hbu.edu.cn

Abstract. In order to discuss the information entropy on quasi-probability space, combining Shannon entropy on probability space with quasi-probability on quasi-probability space, the definitions of self-information and information entropy on quasi-probability space are proposed, and basic properties of information entropy of discrete quasi-random variables on quasi-probability space are also given.

Keywords: Information entropy, quasi-random variables, quasi-probability space, self-information.

1   Introduction

Shannon entropy is an important measure of the uncertainty related to random variables on probability space, which was firstly introduced by Shannon [1] in 1948. This concept has played a significant role in information theory, economic decision, meteorology, cybernetics, statistical inference, and so on [2].

However, Shannon entropy is based on probability and established on probability space. As we all known, probability is a set function which satisfies additivity. The additivity is so strict that sometimes it cannot be satisfied in some practical applications [3]. Zhang and Yang [4] have paid attention to this problem in information entropy and given the concept of information entropy of discrete information source with g_λ distribution on Sugeno measure space, which is a kind of non-probability space and is an interesting extension of probability space. In this paper, we propose the concept of information entropy on quasi-probability space (another non-probability space) and discuss its basic properties.

2   Preliminaries

Let X be a nonempty set and $\mathcal{F}$ be an σ-algebra of subsets of X.

---

[*] Corresponding author.

B.-Y. Cao and X.-J. Xie (Eds.): Fuzzy Engineering and Operations Research, AISC 147, pp. 165–172.
springerlink.com   © Springer-Verlag Berlin Heidelberg 2012

Definition 2.1[5]. *Let* $a \in (0, \infty]$. *An extended real function* $\theta : [0, a] \to [0, \infty]$ *is called a* T *-function if and only if* θ *is continuous, strictly increasing, and such that* $\theta(0) = 0$ *and* $\theta^{-1}(\{\infty\}) = \emptyset$ *or* $\{\infty\}$, *according to* a *being finite or not.*

Obviously, θ^{-1} *is continuous, strictly increasing when* θ *is continuous, strictly increasing.*

Definition 2.2[5]. *The function* μ *is called quasi-additive if and only if there exists a* T *-function* θ, *whose domain of definition contains the range of* μ, *such that the set function* $\theta \circ \mu$ *defined on* $\mathcal{F}$ *by*

$$(\theta \circ \mu(E)) = (\theta(\mu(E))), \forall E \in \mathcal{F}, \tag{1}$$

is additive.

Definition 2.3[5]. *The function* μ *is called quasi-probability if and only if there exists a* T *-function* θ, *such that* $\theta(0) = 0$, $\theta(1) = 1$ *and the set function* $\theta \circ \mu$ *is a classical probability on* $\mathcal{F}$. *The triple* $(X, \mathcal{F}, \mu)$ *is called a quasi-probability space.*

Example 2.1[5]. Assume μ is a classical probability measure. If we take T -function $\theta(y) = y$, then we have $\theta \circ \mu = \mu$, so $\theta \circ \mu$ is a classical probability measure. According to the definition 2.3, μ is a quasi-probability measure. Obviously quasi-probability measure is an extension of classical probability measure.

Example 2.2[5]. Assume μ is a Sugeno measure. If we take T -function

$$\theta(y) = \begin{cases} \dfrac{\ln(1 + \lambda y)}{\ln(1 + \lambda)}, & \text{if } \lambda \neq 0, \\ y, & \text{if } \lambda = 0, \end{cases}$$

then we have $\theta \circ \mu = \mu$, so $\theta \circ \mu$ is a classical probability measure. According to the definition 2.3, μ is a quasi-probability measure.

More specifically, let $X = \{a, b\}, \mathcal{F} = \mathcal{P}(X)$, g_λ be defined by

$$g_\lambda(E) = \begin{cases} 0, & \text{if } E = \emptyset, \\ 0.2, & \text{if } E = \{a\}, \\ 0.4, & \text{if } E = \{b\}, \\ 1, & \text{if } E = X. \end{cases}$$

Then g_λ is a Sugeno measure with a parameter $\lambda = 5$. If we take

$$\theta(y) = \frac{\ln(1+\lambda y)}{\ln(1+\lambda)} = \frac{\ln(1+5y)}{\ln 6},$$

then we have

$$(\theta \circ g_\lambda)(E) = \begin{cases} 0, & \text{if } E = \varnothing, \\ 0.387, & \text{if } E = \{a\}, \\ 0.613, & \text{if } E = \{b\}, \\ 1, & \text{if } E = X. \end{cases}$$

It is clear to see that $\theta \circ g_\lambda$ is a probability measure. According to the definition 2.3, g_λ is a quasi-probability measure.

Obviously quasi-probability measure is also an extension of Sugeno measure.

Definition 2.4[6,7]. *Let* $(X, \mathcal{F}, \mu)$ *be a quasi-probability space. A real function* $\xi = \xi(\omega), \omega \in X$ *is called a quasi-random variable if*

$$\{\omega | \xi(\omega) \le x\} \in \mathcal{F}, \forall x \in R. \tag{2}$$

Definition 2.5[6,7]. *A quasi-random variable* ξ *is discrete if its range is finite or countable.*

Definition 2.6[8]. *Let* ξ *be a discrete quasi-random variable on quasi-probability space* $(X, \mathcal{F}, \mu)$. *Then the expected value of* ξ *is defined by*

$$E_\mu[\xi] = \sum_{i=1}^{\infty} x_i \mu_i, \tag{3}$$

if $\sum_{i=1}^{\infty} |x_i| \mu_i < \infty$.

3 Information Entropy of Discrete Quasi-random Variables and Its Properties

3.1 Self-Information of Discrete Quasi-random Variables

Firstly, we propose the definition of the self-information of discrete quasi-random variables on quasi-probability space. Consider a discrete quasi-random variable ξ which follows the quasi-probability distribution:

$$\begin{bmatrix} \xi \\ \mu \end{bmatrix} = \begin{bmatrix} x_1 & \cdots & x_i & \cdots & x_n \\ \mu(x_1) & \cdots & \mu(x_i) & \cdots & \mu(x_n) \end{bmatrix}, \ i = 1, 2, \cdots, n, \ n \in N^+. \tag{4}$$

Definition 3.1. *The self-information of the event* $\xi = x_i$ *is defined as*

$$I(x_i) = -\log_a(\theta \circ \mu(x_i)), \ a > 1, \ i = 1, 2, \cdots, n, \ n \in N^+. \tag{5}$$

The base of the logarithm a is not specified in the definition. The most common bases of the logarithm are 2, Euler number e and 10, and the unit of I is bit for $b = 2$, nat for $b = e$, and dit for $b = 10$.

Theorem 3.1. *Suppose that* ξ *is a discrete quasi-random variable which satisfies (4) on quasi-probability space* $(X, \ \mathcal{F}, \ \mu)$. *It is obtained that if* $\mu(x_1) < \mu(x_2)$, *then* $I(x_1) > I(x_2) \geq 0$.

Proof. According to the definition 2.3 and the definition 3.1, we have

$$I(x_1) = -\log_a(\theta \circ \mu(x_1)) \geq 0, I(x_2) = -\log_a(\theta \circ \mu(x_2)) \geq 0.$$

Since θ is strictly increasing, it follows from the condition $\mu(x_1) < \mu(x_2)$ that

$$\theta \circ \mu(x_1) < \theta \circ \mu(x_2).$$

$-\log_a x \ (a > 1)$ is decreasing, so we have

$$I(x_1) = -\log_a(\theta \circ \mu(x_1)) > -\log_a(\theta \circ \mu(x_2)) = I(x_2) \geq 0.$$

The theorem is proved.

Corollary 3.1. *Suppose that* ξ *is a random variable on probability space. It is obtained that if* $\mu(x_1) < \mu(x_2)$, *then* $I(x_1) > I(x_2) \geq 0$.

Proof. Since the random variable ξ is a quasi-random variable with $\theta(y) = y$, it is easy to have the conclusion according to the theorem 3.1.

Corollary 3.2. *Suppose that* ξ *is a* g_λ *random variable on Sugeno measure space. It is obtained that if* $\mu(x_1) < \mu(x_2)$, *then* $I(x_1) > I(x_2) \geq 0$.

Proof. Since the g_λ random variable ξ is a quasi-random variable with

$$\theta(y) = \begin{cases} \dfrac{\ln(1 + \lambda y)}{\ln(1 + \lambda)}, & \text{if } \lambda \neq 0, \\ y, & \text{if } \lambda = 0, \end{cases}$$

It is easy to have the conclusion according to the theorem 3.1.

3.2   Information Entropy of Discrete Quasi-random Variables

Next, the definition of information entropy of quasi-random variables on quasi-probability space is given on the basis of the definition of self-information. Consider a discrete quasi-random variable ξ which follows the quasi-probability distribution:

$$\begin{bmatrix} \xi \\ \mu \end{bmatrix} = \begin{bmatrix} x_1 & \cdots & x_i & \cdots & x_n \\ \mu(x_1) & \cdots & \mu(x_i) & \cdots & \mu(x_n) \end{bmatrix}, \ i=1,2,\cdots,n,\ n\in N^+.$$

For simplicity, we shall take μ_i instead of $\mu(x_i)$, that is

$$\mu = \big(\mu(x_1), \mu(x_2), \cdots, \mu(x_n) \big) = \big(\mu_1, \mu_2, \cdots, \mu_n \big).$$

Then the quasi-probability distribution of discrete quasi-random variable ξ can be denoted by

$$\begin{bmatrix} \xi \\ \mu \end{bmatrix} = \begin{bmatrix} x_1 & \cdots & x_i & \cdots & x_n \\ \mu_1 & \cdots & \mu_i & \cdots & \mu_n \end{bmatrix}, i=1,2,\cdots,n,\ n\in N^+. \tag{6}$$

Definition 3.2. *The information entropy of a quasi-random variable ξ is defined as*

$$H(\xi) = E\big[I(x_i)\big] = \sum_{i=1}^{n} I(x_i) \cdot \big(\theta \circ \mu(x_i) \big), \ n\in N^+. \tag{7}$$

Obviously, $H(\xi)$ is a function of the quasi-probability $\mu(x_i)$, so the information entropy of ξ can be denoted by

$$H(\mu) = H\big(\mu(x_1), \mu(x_2), \cdots, \mu(x_n) \big) = H\big(\mu_1, \mu_2, \cdots, \mu_n \big), n\in N^+.$$

Theorem 3.2. *Suppose that ξ is a discrete quasi-random variable which satisfies (6) on quasi-probability space $(X,\ \mathcal{F},\ \mu)$, and $H(\mu)$ is the information entropy of ξ. Then the following properties of $H(\mu)$ is obtained:*

(i) $H(\mu) = H(\mu_1,\ \mu_2,\ \cdots,\ \mu_n) \geq 0,\ n\in N^+$;

(ii) $H(c,\ 0) = H(c,\ 0,0) = H(c,\ 0,0,0) = \cdots = H(c,\ 0,\cdots,\ 0) = 0,\ c > 0$,
   $n\in N^+$;

(iii) $H(\mu_1,\ \mu_2,\ \cdots,\ \mu_n) = H(\mu_2,\ \mu_1,\ \cdots,\ \mu_n) = \cdots = H(\mu_n,\ \mu_1,\ \cdots,\ \mu_{n-1})$,
   $n\in N^+$;

(iv) $\lim_{\varepsilon\to 0} H_{n+1}(\mu_1,\ \mu_2,\ \cdots,\ \mu_n - \varepsilon, \varepsilon) = H_n(\mu_1,\ \mu_2,\ \cdots,\ \mu_n),\ n\in N^+$;

(v) $\lim_{\varepsilon\to 0} H(\mu_1,\ \mu_2,\ \cdots,\ \mu_{n-1} - \varepsilon, \mu_n + \varepsilon) = H(\mu_1,\ \mu_2,\ \cdots,\ \mu_n),\ n\in N^+.$

Proof. (i) According to the definition 3.2, we have

$$H(\mu) = H(\mu_1, \mu_2, \cdots, \mu_n) = \sum_{i=1}^{n} I(x_i) \cdot (\theta \circ \mu_i).$$

It follows from the theorem 3.1 that $I(x_i) \geq 0$. According to the definition 2.3, we obtain $\theta \circ \mu$ is a classical probability, so $\theta \circ \mu_i \geq 0$. It is clear that

$$H(\mu) = H(\mu_1, \mu_2, \cdots, \mu_n) = \sum_{i=1}^{n} I(x_i) \cdot (\theta \circ \mu_i) \geq 0.$$

(ii) Let T-function $\theta(x) = \dfrac{x}{c}$. According to the definition 3.2, we have

$$H(c, 0) = I(x_1) \cdot (\theta \circ \mu_1) + I(x_2) \cdot (\theta \circ \mu_2),$$

where $c > 0, \mu_1 = c, \mu_2 = 0$.

It follows from the properties of logarithmic function that $H(c, 0) = 0$. Similarly, we can obtain that

$$H(c, 0, 0) = H(c, 0, 0, 0) = \cdots = H(c, 0, \cdots, 0) = 0, c > 0.$$

(iii) According to the definition 3.2, it is obvious that

$$H(\mu_1, \mu_2, \cdots, \mu_n) = H(\mu_2, \mu_1, \cdots, \mu_n)$$
$$= \cdots = H(\mu_n, \mu_1, \cdots, \mu_{n-1})$$
$$= \sum_{i=1}^{n} I(x_i) \cdot (\theta \circ \mu_i).$$

(iv) Since θ is continuous, according to the definition 3.2, we have

$$\lim_{\varepsilon \to 0} H_{n+1}(\mu_1, \mu_2, \cdots, \mu_n - \varepsilon, \varepsilon) = \lim_{\varepsilon \to 0} \left(\sum_{i=1}^{n-1} I(x_i) \cdot (\theta \circ \mu_i) \right)$$
$$- \lim_{\varepsilon \to 0} \left(\theta(\mu_n - \varepsilon) \cdot \log_a \theta(\mu_n - \varepsilon) + \theta(\varepsilon) \cdot \log_a \theta(\varepsilon) \right)$$
$$= \sum_{i=1}^{n-1} I(x_i) \cdot (\theta \circ \mu_i) - \theta(\mu_n) \cdot \log_a \theta(\mu_n)$$
$$= \sum_{i=1}^{n} I(x_i) \cdot (\theta \circ \mu_i)$$
$$= H_n(\mu_1, \mu_2, \cdots, \mu_n).$$

(v) Since θ is continuous, according to the definition 3.2, we have

$$\lim_{\varepsilon \to 0} H\left(\mu_1,\ \mu_2,\ \cdots,\ \mu_{n-1}-\varepsilon,\mu_n+\varepsilon\right)=\lim_{\varepsilon \to 0}\left(\sum_{i=1}^{n-2}I\left(x_i\right)\cdot\left(\theta\circ\mu_i\right)\right)$$

$$-\lim_{\varepsilon \to 0}\left(\theta\left(\mu_{n-1}-\varepsilon\right)\cdot\log_a\theta\left(\mu_{n-1}-\varepsilon\right)+\theta\left(\mu_n+\varepsilon\right)\cdot\log_a\theta\left(\mu_n+\varepsilon\right)\right)$$

$$=\sum_{i=1}^{n-2}I\left(x_i\right)\cdot\left(\theta\circ\mu_i\right)-\theta\left(\mu_{n-1}\right)\cdot\log_a\theta\left(\mu_{n-1}\right)-\theta\left(\mu_n\right)\cdot\log_a\theta\left(\mu_n\right)$$

$$=\sum_{i=1}^{n}I\left(x_i\right)\cdot\left(\theta\circ\mu_i\right)$$

$$=H\left(\mu_1,\ \mu_2,\ \cdots,\ \mu_n\right).$$

Lemma 3.1[9]. *Suppose that a continuous function $f(x)$ is convex up. If*
$$\sum_{i=1}^{n}a_i=1,\ and\ 0\le a_i\le 1,\ 1\le i\le n,\ then\ \sum_{i=1}^{n}a_i f\left(x_i\right)\le f\left(\sum_{i=1}^{n}a_i x_i\right).$$

Theorem 3.3. *Suppose that ξ is a discrete quasi-random variable which satisfies (6) on quasi-probability space $(X,\ \mathcal{F},\ \mu)$. Then*
$$H\left(\mu\right)=H\left(\mu_1,\ \mu_2,\cdots,\ \mu_n\right)\le\log n,\ n\in N^{+}.$$

Proof. According to the definition 3.2, we have
$$H\left(\mu\right)=H\left(\mu_1,\ \mu_2,\ \cdots,\ \mu_n\right)$$
$$=\sum_{i=1}^{n}I\left(x_i\right)\cdot\left(\theta\circ\mu_i\right)$$
$$=\sum_{i=1}^{n}\left(\theta\circ\mu_i\right)\cdot\log\frac{1}{\theta\circ\mu_i}\ .$$

It follows from the definition 2.3 that $\sum_{i=1}^{n}\left(\theta\circ\mu_i\right)=1$. According to the lemma 3.1, we have

$$H\left(\mu\right)=\sum_{i=1}^{n}\left(\theta\circ\mu_i\right)\cdot\log\frac{1}{\theta\circ\mu_i}$$
$$\le\log\left(\sum_{i=1}^{n}\left(\theta\circ\mu_i\right)\cdot\frac{1}{\theta\circ\mu_i}\right)$$
$$=\log n\ .$$

The theorem is proved.

Corollary 3.3[9]. *Suppose that ξ is a discrete random variable on probability space. Then $H\left(\mu\right)=H\left(\mu_1,\ \mu_2,\cdots,\ \mu_n\right)\le\log n,\ n\in N^{+}.$*

Proof. Since the random variable ξ is a quasi-random variable with $\theta(y)=y$, it is easy to have the conclusion according to the theorem 3.3.

Corollary 3.4[4]. *Suppose that ξ is a discrete g_λ random variable on Sugeno measure space. Then $H(\mu) = H(\mu_1, \mu_2, \cdots, \mu_n) \leq \log n$, $n \in N^+$.*

Proof. Since the g_λ random variable ξ is a quasi-random variable with

$$\theta(y) = \begin{cases} \dfrac{\ln(1+\lambda y)}{\ln(1+\lambda)}, & \text{if } \lambda \neq 0, \\ y, & \text{if } \lambda = 0, \end{cases}$$

It is easy to have the conclusion according to the theorem 3.3.

4 Conclusion

In this paper, we firstly propose the concepts of self-information, information entropy on quasi-probability space and discuss their basic properties. Since Sugeno measure space is an important extension of probability space and the quasi-probability space is also an interesting extension of Sugeno measure space, the main results of the paper generalize the corresponding conclusions on the probability space and Sugeno measure space. Furthermore, the paper provides the foundation for establishing information theory on quasi-probability space.

Acknowledgements. Thanks to the support by National Natural Science Foundation of China (No. 60773062 and No. 61073121).

References

1. Shannon, C.E.: A mathematical theory of communication. Bell System Technical Journal 27, 379–423 (1948)
2. Bai, X.M., Zhao, S.S.: The cognition and application of entropy. Statistics & Information Tribune 16(3), 7–12 (2001)
3. Ha, M.H., Li, Y., Li, J., Tian, D.Z.: The key theorem and the bounds on the rate of uniform covergence of learning theory on Sugeno measure space. Science in China: Series F, Information Sciences 49(3), 372–385 (2006)
4. Zhang, J.L., Yang, Q.: Information entropy of discrete information source with g_λ distribution and their application. Fuzzy Systems and Mathematics 23(1), 52–58 (2009)
5. Wang, Z.Y., Klir, G.J.: Fuzzy Measure Theory. Plenum Press, New York (1992)
6. Ha, M.H., Feng, Z.F., Song, S.J., Gao, L.Q.: The key theorem and the bounds on the rate of uniform convergence of statistical learning theory on quasi-probability spaces. Chinese Journal of Computers 31(3), 476–485 (2008)
7. Ha, M.H., Yang, L.Z., Wu, C.X.: Introduction to Generalized Fuzzy Set-Valued Measure. Science Press, Beijing (2009)
8. Ha, M.H., Wang, C., Zhang, Z.M., Tian, D.Z.: Uncertainty Statistical Learning Theory. Science Press, Beijing (2010)
9. Fu, Z.Y.: Information Theory Principles and Applications (The Third Edition). Publishing House of Electronics Industry, Beijing (2011)

Fuzzy Diagnosis of High Capacity Channel

Ming Ni, Ci-yuan Xiao, and Ya-xun Yang

Southwest Petroleum University, Chengdu 610500, China
86nm189@163.com

Abstract. The concept of High Capacity Channel was originally presented by oil field of Shengli oil field in 1980s. Oilfield development researchers pointed out that High Capacity Channel is a relative concept, it referred to the channel which was formed after the water flooding in reservoir, so that the injection water is easier to go through or enter into the channel, and it should not be only restricted to the size of pore throat. Based on analyzing the form mechanism of High Capacity Channel, This paper selects the evaluation index, and ensures the membership function. It also proposes using the method of Fuzzy Synthetic Evaluation, constructs a new fuzzy identification model, which has been tested by examples. The results have showed this method can accurately identify the High Capacity Channel .The identification rate can meet with the request of precision.

Keywords: High capacity channel, membership function, fuzzy diagnosis, oil-field development.

1   Introduction

A great change of pore configuration will take place in the later period of oilfield development and high permeability zones and extra-high permeability zones will be easily formed, which are regarded as High Capacity Channel. As injection water flows through formation by High Capacity Channel, it greatly reduces the recovery ratio of oil reservoir, and influences the improvement of development efficiency. How to solve the problem of plugging, increase the conformance efficiency of injected water, improve development efficiency and enhance the development benefit, these issues are to be eagerly solved in the development of high water-cut stage sandstone reservoirs. So the research of High Capital Cannel identification is extremely significant to the development of sandstone when the oil enters into the High water-cut stage.

2   The Forming Mechanism of High Capacity Channel

The forming of high capacity channel is influenced by the development feature of reservoir and its producing process. For the medium or high permeability sandstone reservoirs, which are mainly formed by muddy cementation, pore structure of reservoir will change significantly for a long period of water erosion, and small

B.-Y. Cao and X.-J. Xie (Eds.): Fuzzy Engineering and Operations Research, AISC 147, pp. 173–180.

size of argillaceous were taken away, which makes rock particles easily migrated. Intergranule pore diameter expands apparently. Reservoir layer, interlayer and plane heterogeneity performance obviously After long-term water-flooding, the heterogeneity of reservoir rises quickly. Because of the scour of injected water, particles between pore passages have been scoured away, and passageway as well as permeability both increases gradually. Therefore, high-capacity channels will be easily formed. Experimental study has showed that the production rate will deeply affect the formation of high-capacity channels when the irrational production methods are used, such as forced recovery and injection, low pressure water flooding and High-speed development, etc.

3 Diagnosis Method of High-Capacity Channels

Diagnosis method of high-capacity channels mostly can be found between quantitative and semi-quantitative. The current methods mainly contain interwell tracer test, Water conservancy detection method, log parameter method, comprehensive evaluation approach and so on. This paper uses the method of fuzzy mathematics and presentsanew way of identification method of high-capacity channels.

3.1 Establish the Hierarchical Structure Model (Table 1)

As shown in the table1, the influence factors can be divided into static development element and dynamic development element. The former one is also called Geofactor and the latter one is also called Effect Factors of development.

Table 1.

Destination Layer	Rule layer	
Comprehensive evaluation	Geofactor	Heterogeneity
		Degree of cementation
		Whether the oil field is heavy crude
		Porosity
		Whether it is sandstone reservoir
		Permeability
	Effect factors of develment	Injector producer differential pressure
		Intensity of anomaly of entry profile
		The increase level of apparent water infectivity.
		Water content
		The level of sand entry
		The increase level of fluid productivity

3.2   Index Selection

3.2.1   Selection of Geofacto Index

(1) Permeability. It is an important aspect of confirming the existence of high capacity channel. The high capacity channel can be easily formed in high permeability zone.

(2) Heterogeneity. Because of the formation heterogeneity, the onrush of injected water forms secondary high-capacity channels.

(3) Degree of cementation. The higher degree of cementation is more difficult to form high capacity channel.

(4) Whether the oil field is heavy crude. The larger the viscosity is, the easier the high capacity channel will form. In addition, it is more helpful to form high capacity channel when the large distinction of viscosity between water and oil.

(5) Porosity. Porosity and permeability share the same effect. The larger the porosity is, the easier the high capacity channel will form.

(6) Whether it is sandstone reservoir. Sandstone reservoir is one of the essential conditions to form high capacity channel. Under the influence of both weakly cemented sands and thicken oil, wormhole appears in reservoir stratum caused by mass inflow of sands.

3.2.2   Selection of Dynamic Factors Index

(1) Injection-production differential pressure index. Under practical production conditions, injection-production ratio of well group equals to the ratio of well fluid and water injection of Related Wells. In the same flooding unit, the injection-production ratio of the well which is connected to High Capacity Channel is greatly larger than others. Take the ratio of Injection-production differential pressure under present condition to normal condition (no high capacity channel)

(2) Intensity of anomaly of entry profile. It is necessary to focus on the information of entry profile if water sucking level wants to be studied.

(3) The increase level of apparent water injectivity   index. Apparent water injectivity index refers to the ratio of single well injection to water injection pressure. When the High Capacity Channel has formed, water injectivity index will raise suddenly. Take the ratio of apparent water injectivity index under normal situation to observation time.

(4) The increase level of fluid productivity index. Fluid productivity index refers to the ratio of fluid productivity per well to drawdown pressure.

(5) Water content. Another obvious representation of the form of High Capacity Channel is water ratio. If the water ratio changes suddenly, some abnormal condition will be certain to appear underground.

(6) The level of sand entry. The quantity of sand entry indirectly decides the volume of high capacity channel. If the degree of sand entry is different, the development degree of high capacity channel will be different too.

3.3 Membership Functions of Every Evaluation Index

Suppose the membership function of permeability is $u_1(x)$ $(10^{-3} \mu m^2)$

$$u_1(x) = \begin{cases} 0, & x < 50, \\ \dfrac{x-50}{250}, & 50 < x < 300, \\ 1, & x > 300. \end{cases} \tag{1}$$

The membership function of heterogeneity is $u_2(x)$

$$u_2(x) = \begin{cases} 0.5, & \text{ordinary,} \\ 1.0, & \text{serious.} \end{cases} \tag{2}$$

The membership function of degree of cementation is $u_3(x)$

$$u_3(x) = \begin{cases} 0, & \text{strong,} \\ 0.5, & \text{ordinary,} \\ 1, & \text{weak.} \end{cases} \tag{3}$$

The membership function of whether the oil field is heavy crude is $u_4(x)$

$$u_4(x) = \begin{cases} 0.75, & \text{heavy crude,} \\ 0.25, & \text{else.} \end{cases} \tag{4}$$

The membership function of porosity is $u_5(x)$

$$u_5(x) = \begin{cases} 0, & x < 0.15, \\ \dfrac{x-0.15}{0.1}, & 0.15 < x < 0.25, \\ 1, & x > 0.25. \end{cases} \tag{5}$$

The membership function of whether it is sandstone reservoir is $u_6(x)$

$$u_6(x) \begin{cases} 1, & \textbf{sandstone} \\ 0, & \textbf{else} \end{cases} \tag{6}$$

The membership function of injection-production differential pressure index is $u_7(x)$

$$u_7(x) \begin{cases} 1, & x < 0.4 \\ 1 - \dfrac{x-0.4}{0.3}, & 0.4 < x < 0.7 \\ 0, & x > 0.7 \end{cases} \qquad (7)$$

The membership function of intensity of anomaly of entry profile is $u_8(x)$

$$u_8(x) = \begin{cases} 0, & \text{strong,} \\ 0.5, & \text{ordinary,} \\ 1, & \text{weak.} \end{cases} \qquad (8)$$

The membership function of the increase level of apparent water injectivity index is $u_9(x)$

$$u_9(x) = \begin{cases} 1, & x < 0.4, \\ 1 - \dfrac{x-0.4}{0.3}, & 0.4 < x < 0.7, \\ 0, & x > 0.7. \end{cases} \qquad (9)$$

The membership function of the increase level of fluid productivity index is $u_{10}(x)$

$$u_{10}(x) = \begin{cases} 1, & x < 0.4, \\ 1 - \dfrac{x-0.4}{0.3}, & 0.4 < x < 0.7, \\ 0, & x > 0.7. \end{cases} \qquad (10)$$

The membership function of water content is $u_{11}(x)$.

$$u_{11}(x) = \begin{cases} 0, & x < 0.6, \\ \dfrac{x-0.6}{0.3}, & 0.6 < x < 0.9, \\ 1, & x > 0.9. \end{cases} \qquad (11)$$

The membership function of the level of sand entry is $u_{12}(x)$.

$$u_{12}(x) \begin{cases} 0, & \textbf{weak} \\ 0.5, & \textbf{strong} \\ 1, & \textbf{serious} \end{cases} \qquad (12)$$

3.4   The Method for Confirm the Weight of Index

Take all index compared with one another. Applied nine-demarcation matrix method (Table 2) to construct judgment matrix.

Table 2.

scale	meaning
1	Have equal importance
3	One is a little important than the other one.
5	One is Apparently important than the other one.
7	One is Much more important than the other one.
9	One is Extremely important than the other one.
2,4,6,8	Judgment mid-value of two adjacent factors
reprocal	Suppose the Judgment value which compare facor i to facor j is b_{ij} ,then, the Judgment value which compare facor j to facor i is $\dfrac{1}{b_{ij}}$

Calculation results is shown in table 3.

Table 3.

facor	T_1	T_2	T_3	T_4	T_5	T_6	T_7	T_8	T_9	T_{10}	T_{11}	T_{12}
weight	0.06	0.03	0.06	0.03	0.06	0.06	0.14	0.14	0.14	0.07	0.07	0.14

(The ordinal notes for heterogeneity, porosity ,permeability, whether the oil field is heavy crude, whether it is sandstone reservoir, degree of cementation, intensity of anomaly of entry profile, injector producer differential pressure, the increase level of apparent water infectivity , water content, the increase level of fluid productivity, water content and the level of sand entry as T_1, T_2, T_3, T_4, T_5, T_6, T_7, T_8, T_9, T_{10}, T_{11} and T_{12}.)

If some data are missed, the missing index weight may be taken as 0 and the weight shall be readjusted in accordance with it.

3.5   Computing Method

Suppose the Comprehensive discriminated subordinate function is u ,

$$u=0.06u_1+0.06u_2+0.06u_3+0.03u_4+0.03u_5+0.06u_6+0.14u_7+0.14u_8+0.14u_9+0.14u_{10}+0.07u_{11}+0.07u_{12} \tag{13}$$

Calculate membership of objects. Calculate memberships of 43 oil wells which have been distinguished. According to the practical data, we regard the ideal cutoff as 0.4.Take another 17 oil wells, Calculation results can be seen in table 4.

Table 4.

	T_1 ($10^{-3}\ \mu m^2$)	T_2	T_3	T_4	T_5 %	T_6	T_7 %	T_8	T_9 %	T_{10} %	T_{11} %	T_{12}	u	result	reality
1	51.06	strong	weak	no	16.18	no	85.67	weak	109.5	62	<60	weak	0.17	not exist	not exist
2	126.43	strong	weak	no	14.81	no	54.53	weak	109.5	74	<60	weak	0.24	not exist	not exist
3	85.20	strong	weak	no	16.74	no	68.55	weak	109.5	84	<60	weak	0.29	not exist	not exist
4	30.79	strong	weak	no	17.71	no	124.30	weak	120.5	95	<60	weak	0.16	not exist	not exist
5	72.39	ordinary	weak	no	17.21	yes	102.82	ordinary	101.5	93	<60	weak	0.25	not exist	not exist
6	33.52	ordinary	weak	no	16.78	yes	59.03	ordinary	101.5	59	<60	weak	0.41	exist	not exist
7	264.91	ordinary	weak	no	17.87	yes	103.2	ordinary	85.2	75	<60	weak	0.30	not exist	not exist
8	194.84	ordinary	weak	no	17.32	yes	105.66	weak	85.2	83	<60	weak	0.21	not exist	not exist
9	46.62	ordinary	weak	no	16.72	yes	60.13	strong	85.2	91	<60	weak	0.36	not exist	not exist
10	93.42	ordinary	strong	no	17.20	yes	138.70	ordinary	78.1	67	<60	weak	0.20	not exist	not exist
11	97.05	ordinary	strong	no	17.54	yes	145.39	weak	6.2	46	<60	weak	0.13	not exist	not exist
12	214.7	ordinary	ordinary	yes	10.21	yes	0.241	ordinary	56.5	83	<60	weak	0.53	exist	exist
13	278.21	ordinary	ordinary	yes	11.52	yes	0.289	strong	78.2	55	<60	weak	0.48	exist	exist
14	184.83	ordinary	ordinary	yes	10.1	yes	0.433	ordinary	93.2	71	<60	weak	0.38	not exist	exist
15	137.5	ordinary	ordinary	yes	10.17	yes	0.075	weak	49.3	39	<60	weak	0.54	exist	exist
16	325.04	strong	weak	no	23.74	yes	1.71	strong	37.9	67	98.1	weak	0.92	exist	exist
17	739.89	strong	weak	no	22.77	yes	2.43	strong	55.1	43	89.2	weak	0.84	exist	exist

(The ordinal notes for permeability, heterogeneity, degree of cementation, whether the oil field is heavy crude, porosity. whether it is sandstone reservoir, injection-production differential pressure index, intensity of anomaly of entry profile, the increase level of apparent water infectivity , water content, the increase level of fluid productivity, water content and the level of sand entry as T_1, T_2, T_3, T_4, T_5, T_6, T_7, T_8, T_9, T_{10}, T_{11} and T_{12}.).

The experiment has proved that the judgment precision is more than 80%.

4 Conclusion

The paper has mainly analyzed the formation mechanism and influence factors of the high capacity channel, sets up fuzzy identification method, decides whether the high capacity channel exists or not. Compared with traditional methods, the method mentioned in this paper is more brief and fast. To some extent, it is meaningful to guide the profile control and water shutoff as well as improving recovery ratio for oil fields.

Acknowledgements. Thanks to Professor Hengsheng Yao from Southwest Petroleum University for his aid in sample collection. I am particularly grateful to Professor Ci-yuan Xiao for much helpful guidance.

References

1. Xiao, C.Y.: Engineering fuzzy system, pp. 140–143. Science Press, Beijing (2005)
2. Zeng, L.F.: The macroscopic throats forming mechanism of unsolidated sand-reservior and their identyfying method. Journal of Basic Science and Engineering 3, 56–64 (2002)
3. He, C., Li, P., Wang, Z.Y.: The characteristics of larger pore path in gudao oilfied and its profile control or plugging method. Oil Drilling & Production Technology 22(5), 63–66 (2000)
4. She, Y.M., Xiong, H.: Research on method of fuzzy pattern recognition. Journal of Yunnan Institute of the Nationalites(Natral Sciences Edition) 7(1), 14–20 (1998)
5. Dou, Z.L., Zeng, L.F., Zhang, Z.H.: Research on the diagnosis and description of wormhole. Petroleum Exploration and Development 28(1), 75–77 (2001)
6. Wu, S.Y., Li, Z.A., Yao, F.: The study on identifying macropore path and water plugging-profiling technoogy. Journal of East China Institute of Technology 29(3), 245–247 (2006)
7. Liu, Y.T., Sun, B.L., Yu, Y.S.: Fuzzy identification and quantative calculation method for big pore throat. Oil Drilling & Production Technology 25(5), 54–59 (2003)

Part II
The Mathematical Theory
of Fuzzy Systems

Fuzzy Tolerance Region and Process Capability Analysis

Bahram Sadeghpour Gildeh and Vahab Moradi

Department of Statistics, Faculty of Mathematical Science, University of Mazandaran,
Postal Code 47416-95447, Babolsar, Iran
sadeghpour@umz.ac.ir, Vahab.Moradi25@gmail.com

Abstract. Process capability analysis is designed to estimate the proportion of parts that do not meet engineering requirements in a stable production process. When the quality of products is related to two or more variables, the analysis should be based on a multivariate statistical technique. In this paper, we review and criticize the capability indices typically used in industry for this purpose, and propose a general multivariate process capability index based on fuzzy tolerance region which has not the restriction of conventional process capability indices. We further derive its analytical expression under standard assumptions, discuss numerical approximations, and compare three traditional process capability indices with the new index based on fuzzy approach.

Keywords: Fuzzy number, fuzzy tolerance, process capability, multivariate normal distribution.

1 Introduction

In manufacturing industry, process capability analysis is used to flag high values of the proportion of parts being produced that do not meet engineering requirements, in order to prevent further production of unacceptable output. This analysis assumes that the process is stable, that is, that any random sample of observations from the process may be regarded as a random sample from the same underlying distribution, a situation often described in the engineering literature as that of a process in statistical control. Capability analysis is typically performed by evaluating capability indices which relate the allowable spread of the process, defined by the engineering specifications, to the natural spread of the process, represented by a multiple of the standard deviation of the output. Assuming that the output is normally distributed, the expected proportion of non-conforming parts, i.e., those which will lie outside the engineering specification limits, may be estimated from the capability index. However, the abundance of outputs from skewed distributions and the censoring effects induced by the finite precision of actual measurements, make often

B.-Y. Cao and X.-J. Xie (Eds.): Fuzzy Engineering and Operations Research, AISC 147, pp. 183–193.
springerlink.com   © Springer-Verlag Berlin Heidelberg 2012

rather unreasonable the normality assumption on which traditional capability indices are intuitively based. Moreover, the sampling distributions of the estimators of the capability indices are often intractable, even under normality assumptions. As a consequence, point estimators of the capability indices, with no reference to their precision, are usually quoted; this is a misleading practice, for even relatively large samples may produce rather unreliable estimators.

In this paper, we propose a fuzzy approach to evaluating process capability which, within a decision-theoretical framework, directly assesses the proportion of future parts which may be expected to lie outside the tolerance region. This results in a new multivariate capability index which (i) has a solid, decision theoretical foundation, (ii) does not require the process to be normal, (iii) may accommodate measurements with error, and (iv) contains the conventional index as a limiting case. The proposed capability index is a direct function of the data, whose value is sufficient to solve the relevant decision problem. Nevertheless, robustness considerations make it desirable to assess the precision of the capability index thus obtained; within the Fuzzy framework this is achieved by deriving its Gaussian shape fuzzy number.

In Section 2, we review and criticize the indices traditionally used to evaluate process capability. In Section 3, we recall some notions of fuzzy sets theory, probability of fuzzy event and process capability index defined by Yongting (1996). In Section 4, we propose a new multivariate capability index motivated by the definition of probability of fuzzy event and discuss some of its properties. Section 5 contains two numerical examples to compare three traditional process capability indices with the new index.

2 Traditional Process Capability Indices

In the literature, one of the proposed definitions on process capability index consider that as the ratio of the real performance of process to requested performance (see Pillet (1995)), that is,

$$Capability\ index = \frac{The\ width\ of\ tolerance\ interval}{The\ width\ of\ process\ dispersion}. \tag{1}$$

But, just as we recalled in the introduction section, for a given tolerance interval $[LSL, USL]$ and a risque α, a process with the quality characteristic X is said to be capable if,

$$P(X \in [LSL, USL]) \geq 1 - \alpha. \tag{2}$$

If the center of the distribution of X ($\mu = E(X)$) and the mid-point of the tolerance interval (standard center) be equal ($\mu = M = \frac{(LSL+USL)}{2}$), then we have

$$P(X \in [LSL, USL]) = P\left(|X - \mu| \leq \frac{ULS - LSL}{2}\right). \tag{3}$$

Let
$$r = \min\{c : P(|X - \mu| \le c) \ge 1 - \alpha\}. \tag{4}$$

Since $\{|X - \mu| \le c\}$ is monotone with respect to c, we can say that r is the unique solution of the equation $P(|X - \mu| \le c) = 1 - \alpha$, and the process will be capable if,
$$\frac{USL - LSL}{2} \ge r, \tag{5}$$
or
$$\frac{USL - LSL}{2r} \ge 1. \tag{6}$$

The ratio $\frac{USL - LSL}{2r}$, is called capability index.

If X be normally distributed, that is,
$$X \longrightarrow N(\mu, \sigma^2), \ \mu \in R \ and \ \sigma \in R^+, \tag{7}$$

then, we have
$$P(|X - \mu| \le 3\sigma) = 0.9927. \tag{8}$$

Therefore, with a risk $\alpha = 0.0027$, the simplest capability index which is called the process potential index defined by
$$C_p = \frac{USL - LSL}{6\sigma}. \tag{9}$$

For more information see Kane (1986) or Montgomery (2001).

In the case $\mu = M$ (that is perfect location of the process) a C_p value of at least 1 assures that at most 0.27% of the produced items fall outside the tolerance interval. But if $\mu \ne M$ a C_p value of 1 does not all guarantee a small fraction of non confirming items because the C_p index does not take into consideration the location μ of the distribution. In order to reflect departures from the target value as well as changes in the process variation several order indices have been proposed such as C_{pk} (Sullivan, 1984 & 1985, Kane, 1986) and C_{pm} (Chan, Chen and Spiring, 1988). A through discussion of these and other indices can be found in the book of Kotz and Johnson (1993).

The two indices
$$CPU = \frac{USL - \mu}{3\sigma} \ and \ CPL = \frac{\mu - LSL}{3\sigma}, \tag{10}$$

measure the performance of the process with respect to the upper and lower specification limits. They are used in unilateral tolerance situations where only one single specification limit is given. Again a process e.g. with upper specification limit is considered as capable if $CPU \ge 1$.

From these definition we get for the specification case the C_{pk} index,
$$C_{pk} = \min(CPU, CPL). \tag{11}$$

C_{pk} measures the distance between the process mean and the closest specification limit relation to the one-side actual process spread 3σ. Kane (1986) gave an equivalent representation:

$$C_{pk} = C_p(1 - k), \quad k = \frac{2\,|\mu - M|}{USL - LSL}. \tag{12}$$

For fixed μ, C_{pk} decreases with increasing σ and for fixed σ, C_{pk} decrease with increasing difference $|\mu - M|$. Because $k \geq 0$ it holds that $C_{pk} \leq C_p$, where equality is given if and only if $\mu = M$. Though C_{pk} take into account deviation from the target as change in variations, a high value of C_{pk} does not assure that the process is located near to target. There is also no exact relation between C_{pk} and the fraction p of non conforming items.

Departures from the target value carry more weight with the other well-known capability index C_{pm} defined by

$$C_{pm} = \frac{USL - LSL}{6\sigma'}, \quad \sigma' = \sqrt{\sigma^2 + (\mu - M)^2}. \tag{13}$$

In principal, C_{pm} behaved like C_{pk} but C_{pm} is bounded above as $\sigma \to 0$ and $\mu \neq M$. It holds

$$C_{pm} \leq \frac{USL - LSL}{6\,|\mu - M|}, \tag{14}$$

which implies that if μ falls outside the middle third of the tolerance interval (LSL, USL), then C_{pm} is smaller than 1 in spite of how small σ might be. For $\mu = M$ it holds $Cp = CPU = CPL = C_{pk} = C_{pm}$.

Because the capability indices depend on the unknown parameters μ and σ. These parameters have to be estimated from random samples $(X_1, ..., X_n)$. Common practice is to use as estimators for μ, σ and σ' respectively:

$$\widehat{\mu} = \overline{X}, \quad \widehat{\sigma} = S = \sqrt{\frac{1}{n-1} \sum_{i=1}^{n}(X_i - \overline{X})^2} \tag{15}$$

and

$$\widehat{\sigma'} = \sqrt{\frac{1}{n} \sum_{i=1}^{n}(X_i - \overline{X})^2 + (\overline{X} - M)^2}. \tag{16}$$

3 Process Capability Index Based on Fuzzy Tolerance Interval

In this section, we recall some notions of fuzzy sets theory and the probability of a fuzzy event. Then, we try to define a process capability index as the probability of a fuzzy event which present the tolerance interval.

Let X be a given universal set, a set can be defined by a function, usually called a characteristic function, that declares which elements of X are members of the set and which are not. Set A is defined by its characteristic function, χ_A, as follows:

$$\chi_A(x) = \begin{cases} 1, & if \quad x \in A, \\ 0, & if \quad x \in X - A. \end{cases} \tag{17}$$

That is, the characteristic function maps elements of X to element of the set $\{0, 1\}$, which is formally expressed by $\chi_A : X \to \{0, 1\}$.

The concept of characteristic function can be generalized such that the values assigned to the elements of the universal set fall within a specified range and indicate the membership grade of these elements in the set in question. Larger values denote higher degrees of membership. Such a function is called a membership function, and the set defined by it a fuzzy set. The most commonly used range of values of membership functions is the unit interval $[0, 1]$. In this case, each membership function maps elements of a given universal set X, which is always a crisp set, into real numbers in $[0, 1]$. Two distinct notations are most commonly employed in the literature to denote membership functions. In one of them, the membership function of a fuzzy set $\widetilde{A}$ is denoted by $\mu_{\widetilde{A}}$: that is, $\mu_{\widetilde{A}} : X \to [0, 1]$. In the other one, the function is denoted by $\widetilde{A}$ and has, of course, the same form: that is $\widetilde{A} : X \to [0, 1]$. In this paper, we use the second notation.

Now, we recall the definition of the probability of a fuzzy event which has been given by Zadeh (1968). Let $(R^n, \mathcal{B}^n, P)$ be a probability space, in which $\mathcal{B}^n$ is the σ-field of Borel set in R^n and P is the probability measure over R^n. Then, a fuzzy event in R^n is a fuzzy set $\widetilde{A}$ ($\widetilde{A} : R^n \to [0, 1]$, is Borel measurable).

The probability of a fuzzy event $\widetilde{A}$ is defined by the Lebesgue-Stieltjes integral:

$$P(\widetilde{A}) = \int_{R^n} \widetilde{A}(x) dP. \tag{18}$$

The existence of the Lebesgue-Stieltjes integral is insured by the assumption that $\widetilde{A}$ is Borel measurable.

According to the traditional definition of process capability index (C_p), a process is said to be capable, if for a given risque α ($\alpha \in (0, 1)$):

$$P(X \in [LSL, USL]) = \int_R \chi_{[LSL,USL]}(x) dP \geq 1 - \alpha. \tag{19}$$

To modify the ambiguity which exist in the definition of tolerance interval (especially in the limits) and for the mentioned reasons in the introduction section, in the following we realize a relation between the characteristic function of a tolerance interval and the membership function of a fuzzy set which employ to present the tolerance interval. Then, we define the fuzzy process capability index as the probability of fuzzy tolerance interval.

Now, the fuzzy process capability index can be defined as the probability of fuzzy up-to-standard products turned out in the process of production it is labelled as $C_{\widetilde{p}}$ (for more information see Yongting (1996)).

When the quality index of process products is a continuous random variable,

$$C_{\widetilde{p}} = \int_{-\infty}^{+\infty} \widetilde{A}(x) f(x) dx, \tag{20}$$

where $f(x)$ is the probability density function of quality characteristic X, $\widetilde{A}(x)$ is the membership function of fuzzy up-to-standard products.

When the quality characteristic of process is a discrete random variable X and the values of X is x_i $(i = 1,,n)$,

$$C_{\widetilde{p}} = \sum_{i=1}^{n} \widetilde{A}(x_i)P_X(x_i), \tag{21}$$

where $P_X(x_i)$ is the probability when $X = x_i$, $\widetilde{A}(x_i)$ is the membership function of fuzzy up-to-standard products.

Obviously, we have to know the value of the probability density of quality characteristic and the membership function of fuzzy up-to-standard products, before we can work out $C_{\widetilde{p}}$.

The membership function of fuzzy up-to-standard products generally can help to realize the fuzzy nature of clear technical standard boundaries.

Example 3.1. *Let the distribution of quality characteristic X be normal (i.e., $X \longrightarrow N(\mu, \sigma^2)$). The degree of membership of fuzzy up-to-standard products can be work out through the formula $C_{\widetilde{p}}$. In the case of trapezoidal membership function, we have*

$$
\begin{aligned}
C_{\widetilde{p}(trap)} = &-\frac{\sigma}{\sqrt{2\pi}(a-b)}\left[\exp\left(\frac{-(a-\mu)^2}{2\sigma^2}\right) - \exp\left(\frac{-(b-\mu)^2}{2\sigma^2}\right)\right] \\
&+\frac{\sigma}{\sqrt{2\pi}(c-d)}\left[\exp\left(\frac{-(c-\mu)^2}{2\sigma^2}\right) - \exp\left(\frac{-(d-\mu)^2}{2\sigma^2}\right)\right] \\
&+\frac{\mu-a}{b-a}\left[\Phi(\frac{b-\mu}{\sigma}) - \Phi(\frac{a-\mu}{\sigma})\right] \\
&+\frac{\mu-d}{c-d}\left[\Phi(\frac{d-\mu}{\sigma}) - \Phi(\frac{c-\mu}{\sigma})\right] \\
&+\left[\Phi(\frac{c-\mu}{\sigma}) - \Phi(\frac{b-\mu}{\sigma})\right].
\end{aligned}
\tag{22}
$$

In the case of $b = c = \mu$ (i.e., in the case of triangular membership function),

$$
\begin{aligned}
C_{\widetilde{p}(tri)} = &-\frac{\sigma}{\sqrt{2\pi}(a-\mu)}\left[e^{-(a-\mu)^2/2\sigma^2} - 1\right] \\
&+\frac{\sigma}{\sqrt{2\pi}(\mu-d)}\left[1 - e^{-(d-\mu)^2/2\sigma^2}\right] \\
&+\left[\Phi(\frac{d-\mu}{\sigma}) - \Phi(\frac{a-\mu}{\sigma})\right].
\end{aligned}
\tag{23}
$$

And when the membership function is exponential (i.e., $\widetilde{A}(x) = \exp\left[-\frac{(x-m)^2}{2\lambda^2}\right]$), we have

$$C_{\tilde{p}(\exp)} = \int_{-\infty}^{+\infty} \exp\left(\frac{-(x-m)^2}{2\lambda^2}\right) \frac{1}{\sqrt{2\pi}\sigma} \exp\left(\frac{-(x-\mu)^2}{2\sigma^2}\right) dx$$

$$= \left(\frac{\lambda}{\sqrt{\sigma^2 + \lambda^2}}\right) \exp\left(\frac{-(\mu-m)^2}{2(\sigma^2 + \lambda^2)}\right), \tag{24}$$

where for fixed μ, m and σ, it is a function of modification parameter of tolerance interval $(\lambda > 0)$.

4 Process Capability Index Based on Fuzzy Tolerance Region

Multivariate capability indices appeared in the literature during the early 1990s. Most of them assumed multivariate normal data, a stable process, and were generalizations of their univariate counterparts. Wang et al. (2001) reviewed three multivariate methods (Taam et al. 1993, Chen 1994, and Shahriari et al. 1995) in detail and computed capability for four problems. In general, Taam et al. (1993) presented both a multivariate Cp and Cpm (MCp and MCpm). Given the elliptical equiprobability contours of the multivariate normal distribution, elliptical specifications were assumed. They addressed a hole-diameter application that required elliptical specifications. Shahriari et al. (1995) proposed a multivariate capability index using hyper-rectangular process boxes rather than ellipses as the specification area. They (1995) recognized the need for multiple measures of multivariate capability and proposed a three-component multivariate capability vector. The first component of the vector is a multivariate capability index analogous to the ratio of lengths of the univariate Cp index. The numerator is the area (two-dimensional case) or the volume (three or more dimensions) defined by the engineering tolerance region. The denominator is the area or volume of a "modified process region," defined as the smallest region similar in shape to the engineering tolerance region, circumscribed about a specified probability contour. The number of dimensions of the process data is captured by taking the vth root of the ratio. That is, the first component, labeled C_{pM}, is defined as

$$C_{pM} = \left[\frac{vol.\ of\ engineering\ tolerance\ region}{vol.\ of\ modified\ process\ region}\right]^{\frac{1}{v}}. \tag{25}$$

The second component of the vector of this method is based on the assumption that the center of the engineering specifications is considered to be the true underlying mean of the process. The third component of the vector summarizes a comparison of the location of the modified process region and the tolerance region. It indicates whether any part of the modified process region falls outside the engineering specifications. It has a value of 1, if the entire modified process region is contained within the tolerance region and, otherwise, a value of 0.

Here we extend the definition of $C_{\tilde{p}}$ to multivariate case which has not the restriction of traditional methods.

Let $\underline{X}$ be a v-vector of quality characteristic, if $f(\underline{x})$ be the probability density function of $\underline{X}$ and $\widetilde{A}(\underline{x})$ be the membership function of fuzzy tolerance region, then we denote the multivariate process capability index based on fuzzy tolerance region by $MC_{\tilde{p}}$ and define it as follows

$$MC_{\tilde{p}} = E\left(\tilde{A}(\underline{X})\right) = \int_{R^\nu} \tilde{A}(\underline{x}) f(\underline{x}) d\underline{x}. \tag{26}$$

We say the process is capable if for a given α, $MC_{\tilde{p}} \geq 1 - \alpha$.

If the quality characteristic random vector normally distributed and the membership function of fuzzy tolerance region be a Gaussian fuzzy number i.e.,

$$f(\underline{x}) = |2\pi\Sigma|^{-\frac{1}{2}} \exp\left[-\frac{1}{2}\left(\underline{x} - \underline{\mu}\right)' \Sigma^{-1}\left(\underline{x} - \underline{\mu}\right)\right], \tag{27}$$

and

$$\tilde{A}(\underline{x}) = \exp\left[-\frac{1}{2}\left(\underline{x} - \underline{m}\right)' \Lambda^{-1}\left(\underline{x} - \underline{m}\right)\right]. \tag{28}$$

Then we have

$$MC_{\tilde{p}} = E\left(\tilde{A}(\underline{X})\right) \tag{29}$$

$$= \int_{R^\nu} \exp\left[-\frac{1}{2}\left(\underline{x} - \underline{m}\right)' \Lambda^{-1}\left(\underline{x} - \underline{m}\right)\right] |2\pi\Sigma|^{-\frac{1}{2}} \exp\left[-\frac{1}{2}\left(\underline{x} - \underline{\mu}\right)' \Sigma^{-1}\left(\underline{x} - \underline{\mu}\right)\right] d\underline{x}.$$

Suppose that $\underline{u} = \left(\underline{x} - \underline{m}\right)' \Lambda^{-1}\left(\underline{x} - \underline{m}\right) + \left(\underline{x} - \underline{\mu}\right)' \Sigma^{-1}\left(\underline{x} - \underline{\mu}\right)$, then we have

$$\underline{u} = \left(\underline{x} - \underline{\mu}\right)' (\Lambda^{-1} + \Sigma^{-1})\left(\underline{x} - \underline{\mu}\right) + 2\left(\underline{x} - \underline{\mu}\right)' \Lambda^{-1}\left(\underline{\mu} - \underline{m}\right) + \left(\underline{\mu} - \underline{m}\right)' \Lambda^{-1}\left(\underline{\mu} - \underline{m}\right)$$

$$= \underline{y}\underline{y}' + 2\underline{y}'(\Lambda^{-1} + \Sigma^{-1})^{-\frac{1}{2}} \Lambda^{-1}\left(\underline{\mu} - \underline{m}\right) + \left(\underline{\mu} - \underline{m}\right)' \Lambda^{-1}\left(\underline{\mu} - \underline{m}\right)$$

$$= \left[\underline{y} + (\Lambda^{-1} + \Sigma^{-1})^{-\frac{1}{2}} \Lambda^{-1}\left(\underline{\mu} - \underline{m}\right)\right]' \left[\underline{y} + (\Lambda^{-1} + \Sigma^{-1})^{-\frac{1}{2}} \Lambda^{-1}\left(\underline{\mu} - \underline{m}\right)\right]$$

$$- \left(\underline{\mu} - \underline{m}\right)' \left[\Lambda^{-1}(\Lambda^{-1} + \Sigma^{-1})^{-1} \Lambda^{-1} - \Lambda^{-1}\right]\left(\underline{\mu} - \underline{m}\right). \tag{30}$$

and

$$MC_{\tilde{p}} = \frac{\exp\left\{\frac{1}{2}\left(\underline{\mu} - \underline{m}\right)' \left[\Lambda^{-1}(\Lambda^{-1} + \Sigma^{-1})^{-1} \Lambda^{-1} - \Lambda^{-1}\right]\left(\underline{\mu} - \underline{m}\right)\right\}}{\sqrt{|\Sigma\Lambda^{-1} + I|}}, \tag{31}$$

where $\underline{y} = (\Lambda^{-1} + \Sigma^{-1})^{\frac{1}{2}}\left(\underline{x} - \underline{\mu}\right)$, and μ is mean vector, Σ covariance matrix of $\underline{X}$, $\underline{m}$ target vector and $\Lambda = diag\left(\lambda_1^2, \lambda_2^2, ..., \lambda_\nu^2\right)$, $\lambda_i \in R^+$; $i = 1, 2, ..., \nu$, is the diagonal matrix of tolerance region parameters.

In the case $\upsilon = 1$, we have

$$C_{\tilde{p}} = \frac{\exp\left\{\frac{1}{2}(\mu - m)^2 \left[\lambda^{-2}\left(\lambda^{-2} + \sigma^{-2}\right)^{-1} - \lambda^{-2}\right]\right\}}{\sqrt{(\sigma^2\lambda^{-2} + 1)}}$$

$$= \frac{\lambda}{\sqrt{\lambda^2 + \sigma^2}} \exp\left[-\frac{(\mu - m)^2}{2\left(\lambda^2 + \sigma^2\right)}\right]. \tag{32}$$

Because the capability indices depend on the unknown parameters μ and Σ. These parameters have to be estimated from random samples or data matrix $X = (\underline{X}_1, ..., \underline{X}_n)$. We estimate μ and Σ by $\overline{X}$ (sample mean vector) and S (sample covariance matrix). We propose two estimators for $MC_{\widetilde{p}}$ as follows:

$$\widehat{MC_{\widetilde{p}_1}} = \frac{\exp\left\{\frac{1}{2}\left(\overline{X} - \underline{m}\right)' \left[\Lambda^{-1}\left(\Lambda^{-1} + S^{-1}\right)^{-1}\Lambda^{-1} - \Lambda^{-1}\right]\left(\overline{X} - \underline{m}\right)\right\}}{\sqrt{|S\Lambda^{-1} + I|}}, \quad (33)$$

$$\widehat{MC_{\widetilde{p}_2}} = E\left(\widetilde{A}(\underline{X})\right) = \frac{1}{n}\sum_{i=1}^{n}\exp\left[-\frac{1}{2}\left(\underline{X}_i - \underline{m}\right)'\Lambda^{-1}\left(\underline{X}_i - \underline{m}\right)\right]. \quad (34)$$

5 Numerical Examples

Example 5.1. Jackson (1956) discusses a bivariate quality control example involving the joint control of Elon (E) and Hydroguinone (H), the two components of a process used to make a film-developing solution. Seventy-five successive samples were taken; however, only summary statistics are provided. The specifications were defined as an interval for each variate $(235, 295)$ and $(440, 500)$ respectively, and together form a rectangular tolerance region. The center of the specifications was $(265, 470)$ and was assumed to be the target of the process. The sample mean vector was $(264.32, 471.48)$ and the sample covariance matrix was $\begin{pmatrix} 102.65 & 68.87 \\ 68.87 & 107.96 \end{pmatrix}$. Three multivariate process capability indices were calculated in [14] and Table 5.1 shows the results of them and the output of fuzzy approach.

Table 5.1.

Method	Shahriari et al.	Taam and Subbiah	Chen	Fuzzy $\left(\widehat{MC_{\widetilde{p}_1}}\right)$
$Index$	$[0.85, 0.1, 0]$	0.93	0.91	0.7

The first component of method 1, as well as the index of methods 2 and 3, has value less than 1, implying that variability of the process is more than allowed by engineering drawing (probability level is 99.73%). However the second component of method 1 and denominator of method 2 indicate that the mean process is not on the target vector. Fuzzy approach $(\widehat{MC_{\widetilde{p}_1}})$ shows that the process is capable (with cut point 0.7).

Example 5.2. In this example we just change the target vector of Example 5.1 $((265, 470)$ to $(250, 460))$, and the sensitivity our fuzzy approach with respect to target vector is given in the Table 5.2.

Table 5.2.

Method	Shahriari et al.	Taam and Subbiah	Chen	Fuzzy $\left(\widehat{MC_{\tilde{p}_1}}\right)$
Index	$[0.85, 0, 0]$	0.18	0.91	0.32

Results show that the second component of method 1, method 2 and fuzzy approach are sensitive to difference between mean vector and target vector. But the method 3 (Chen's method) is not sensitive to this difference. In the following table, we summarized the sensitivity of $\widehat{MC_{\tilde{p}_1}}$ with respect to target vector, mean vector and λ vector changes.

Table 5.3.

Target vector	$\widehat{MC_{\tilde{p}_1}}$	Mean vector	$\widehat{MC_{\tilde{p}_1}}$	λ vector	$\widehat{MC_{\tilde{p}_1}}$
$[265, 470]$	0.7	$[265, 470]$	0.7	$[14.7, 14.7]$	0.7
$[230, 420]$	0.18	$[250, 450]$	0.32	$[9.6, 9.6]$	0.5
$[290, 480]$	0.49	$[290, 490]$	0.15	$[20, 20]$	0.8

6 Conclusion

A new approach and procedure for performing multivariate process capability presented. The procedure is based on the probability of fuzzy event defined by Zadeh (1968) by considering a fuzzy tolerance region. The new approach appears to be able to correctly establish process capability index for given fuzzy information of tolerance region, according to results of the case study presented. It is sensitive to target value, process variation and consideration loss function. Three recently proposed multivariate methodologies for assessing capability are compared with this new approach.

Acknowledgement. The authors wish to gratefully acknowledge the support of the research by the Research Center of Algebraic Hyperstructures and Fuzzy Mathematics (RCHFM), University of Mazandaran.

References

1. Chan, L.K., Chen, S.W., Spiring, F.A.: A new measure of process capability: Cpm. Journal of Quality Technology 20, 162–175 (1988)
2. Chen, H.: A Multivariate Process Capability Index Over a Rectangular Solid Tolerance Zone. Statistica Sinica 4, 749–758 (1994)
3. Kane, V.E.: Process capability indices. Journal of Quality Technology 18, 41–52 (1986)
4. Klir, G.J., Bo, Y.: Fuzzy set and fuzzy logic, theory and applications. Prentice-Hall Inc., New Jersey (1995)
5. Kotz, S., Johnson, N.L.: Process Capability Indices. Chapman and Hall, London (1993)

6. Pearn, W.L., Kotz, S., Johnson, N.L.: Distribution and inferential properties of capability indices. Journal of Quality Technology 24, 216–231 (1992)
7. Pillet, M.: Appliquer la maîtrise statistique de procédés (MSP/SPC), d'organisation, Paris (1995)
8. Montgomery, D.C.: Introduction to Statistical Quality Control. John Wiley & Sons, New York (2001)
9. Rodringuez, R.N.: Recent development in process capability analysis. Journal of Quality Technology 24, 176–182 (1992)
10. Sadeghpour, G.B., Gien, D.: Measurement error effects on the performance of the fuzzy capability index. In: Actes du Quatrième Congrès International Pluridisciplinaire Qualité et S ûreté de Fonctionnement, Annecy, France, pp. 222–230 (2001)
11. Sadeghpour, G.B.: Comparison of Cp, Cpk and Cp-tilde process capability indices in the case of measurement error occurrence. In: Proceeding of the 10th IFSA World Congress, Istanbul, Turkey, pp. 563–567 (2003)
12. Sadeghpour, G.B.: Multivariate Process Capability Index Based on Fuzzy Tolerance. In: Proceeding of 37th Annual Iranian International Mathematics Conference, Tabriz, Iran, pp. 607–610 (2006)
13. Shahriari, H., Hubele, N.F., Lawrence, F.P.: A Multivariate Process Capability Vector. In: Proceedings of the 4th Industrial Engineering Research Conference, Nashville, TN, pp. 303–308 (1995)
14. Sullivan, L.P.: Reducing variability: a new approach to quality. Quality Progress 17, 15–21 (1984)
15. Sullivan, L.P.: Letters. Quality Progress 18, 7–8 (1995)
16. Taam, W., Subbaiah, P., Liddy, J.W.: A Note on Multivariate Capability Indices. Journal of Applied Statistics 20, 339–351 (1993)
17. Wang, F.K., Hubele, N.F., Lawrence, F.P., Miskulin, J.O., Shariari, H.: Comparison of Three Multivariate Process Capability Indices. Journal of Quality Technology 32, 263–275 (2002)
18. Yong, T.C.: Fuzzy quality and analysis on fuzzy probability. Fuzzy Sets and Systems 83, 283–290 (1996)
19. Zadeh, L.A.: Fuzzy set. Inform. Control 8, 338–353 (1967)
20. Zadeh, L.A.: The concept of a linguistic variable and its application to approximate reasoning. Information Sciences 9, 43–80 (1975)

Complex Fuzzy Measurable Function

Fu-chuan Chen[1], Mei-qin Chen[2], Yi-nan Huang[2], and Sheng-quan Ma[3,*]

[1] QiongTai Teachers College, Haikou, Hainan, 571100, China
  fuchuan9688@163.com
[2] Department of Mathematics, Hainan Normal University, Haikou, Hainan, 571158, P.R. China
[3] School of information Science and Technology, Hainan Normal University, Haikou, Hainan, 571158, P.R. China
  mashengquan@163.com

Abstract. The paper [2] defined the measurable functions of complex fuzzy, with its definition researching that it is the complex fuzzy decision theorem of measurable functions, and introduced the various convergence's definition of its sequence, according to the definition we further studied the various convergence theorem of its sequence, It builds the certain foundation for the complex fuzzy integral.

Keywords: Complex fuzzy measurable function, convergence in complex fuzzy measure, convergence almost uniformly.

1990-1991, Buckley [1] proposed the concept of fuzzy complex numbers and fuzzy complex valued functions. The theory of complex fuzzy measure continued the fuzzy measure theory and classical theory of complex measure. In 1997, Qiu Jiqing [2-3] and other researchers referred to the classical measure theory's method, firstly proposing the concept of complex fuzzy measure, later studying the relationship of the convergence of its sequence. Since 2000, Ma S.Q [5-7]has done some exploratory work from the problems, and has made a sequence of outcome studies in this regard. The Theory of fuzzy complex measure is an important part of the fuzzy complex analysis, it has a strong practical background and important applications in the fuzzy system identification, fuzzy control, multi-classifier system design and so on. Because of it being much more complex than the fuzzy real value measure, the theory research and develop slowly.

1   Preparation

Definition 1.1[2] *Suppose* (X, F) *is complex fuzzy measurable space, for arbitrary* $A \in \mathrm{F}$, *the mapping* $f : X \rightarrow C$ *is called complex fuzzy measurable function on* A, *if for arbitrary* $\alpha = a + bi \in C$, *then* $A \cap F_\alpha \in \mathrm{F}$, *where*

---

* Correspondence author.

B.-Y. Cao and X.-J. Xie (Eds.): Fuzzy Engineering and Operations Research, AISC 147, pp. 195–206.

$$\{x \in X \,|\, \mathrm{Re}[f(x)] \geq a, \mathrm{Im}[f(x)] \geq b\}. \; \alpha = a + bi, \beta = a' + b'i \in C,$$

If $-\infty \leq a \leq +\infty, -\infty \leq b \leq +\infty$, then we denote $-\infty \leq \alpha \leq +\infty$.

Denote: $F_{\alpha^-} = \{x \in X \,|\, \mathrm{Re}[f(x)] \leq a, \mathrm{Im}[f(x)] \leq b\}$;

$F_{.\alpha} = \{x \in X \,|\, \mathrm{Re}[f(x)] > a, \mathrm{Im}[f(x)] > b\}$;

$F_{\underline{\infty}} = \{x \in X \,|\, \mathrm{Re}[f(x)] = -\infty, \mathrm{Im}[f(x)] = -\infty\}$;

$F_{\overline{\infty}} = \{x \in X \,|\, \mathrm{Re}[f(x)] = +\infty, \mathrm{Im}[f(x)] = +\infty\}$;

$F_{<\alpha,\beta>} = \{x \in X \,|\, a \leq \mathrm{Re}[f(x)] \leq a', b \leq \mathrm{Im}[f(x)] \leq b'\}$.

2 Complex Fuzzy Measurable Functions

Theorem 2.1. *Suppose* (X, F) *is complex fuzzy measurable space, the mapping* $f : X \to C$ *is function on* $A \in F$, *if* $A \cap F_{\underline{\infty}} \in F$, *then* f *is complex fuzzy measurable function if and only if for arbitrary* $\alpha \in C$, *then* $A \cap F_{.\alpha} \in F$.

Proof: (Necessity) If $\alpha = +\infty$, then

$$F_{\overline{\infty}} = \{x \in X \,|\, \mathrm{Re}[f(x)] > +\infty, \mathrm{Im}[f(x)] > +\infty\} = \varphi,$$

so $A \cap F_{.+\infty} = \phi \in F$; If $-\infty < \alpha < +\infty$, then

$$A \cap F_{.\alpha} = \bigcup_{n=1}^{\infty} (A \cap F_{.\alpha + \frac{1}{n}}), \text{ so } A \cap F_{.\alpha} \in F;$$

If $\alpha = -\infty$, then $A \cap F_{.-\infty} = \bigcup_{n=1}^{\infty}(A \cap F_{-n})$, so $A \cap F_{.-\infty} \in F$. Therefore for arbitrary $\alpha \in C$, then $A \cap F_{.\alpha} \in F$.

(Sufficiency): If $\alpha = -\infty$, then $A \cap F_{\underline{\infty}}$, $A \cap F_{.-\infty} \in F$ and $A \cap F_{.-\infty} = A = (A \cap F_{\underline{\infty}}) \cup (A \cap F_{.-\infty})$, so $A \cap F_{.-\infty} \in F$;

If $-\infty < \alpha < +\infty$, then $A \cap F_{\alpha} = \bigcap_{n=1}^{\infty}(A \cap F_{.(\alpha - \frac{1}{n})})$, so $A \cap F_{\alpha} \in F$;

If $\alpha = +\infty$, then $A \cap F_{.+\infty} = A \cap F_{\overline{\infty}} = \bigcap_{n=1}^{\infty}(A \cap F_n)$, so $A \cap F_{.+\infty} \in F$.

Therefore f is complex fuzzy measurable function.

Theorem 2.2. *Suppose* (X,F) *is complex fuzzy measurable space, the mapping* $f: X \to C$ *is complex fuzzy measurable function on* $A \in F$ *if and only if for arbitrary* $\alpha, \beta \in C$ *,then* $A \cap F_{<\alpha,\cdot\beta>} \in F$ *and* $A \cap F_{\overline{\infty}} \in F$.

Proof: (Necessity) $A \cap F_{+\infty} = A \cap F_{\overline{-}}$, so $A \cap F_{\overline{\infty}} \in F$. For arbitrary $\alpha, \beta \in C$, $A \cap F_{\alpha}$, $A \cap F_{\beta} \in F$,due to $A \cap F_{<\alpha,\cdot\beta>} = A \cap (F_{\alpha} \cap F_{\beta}^{c}) = (A \cap F_{\alpha}) \cap (A \cap F_{\beta}^{c})$.

So $A \cap F_{<\alpha,\cdot\beta>} \in F$.

(Sufficiency): For arbitrary $-\infty \leq \alpha < +\infty$, then

$$A \cap F_{\alpha} = (A \cap F_{<\alpha,\cdot\beta>}) \cup (A \cap F_{\overline{\infty}}),$$

so $A \cap F_{\alpha} \in F$ .If $\alpha = +\infty$,then $A \cap F_{+\infty} = A \cap F_{\overline{-}}$, so $A \cap F_{+\infty} \in F$.

Theorem 2.3. *Suppose* (X,F) *is complex fuzzy measurable space, the mapping* $f: X \to C$ *is complex fuzzy measurable function on* $A \in F$ *if and only if for arbitrary* $\alpha \in C$, *then* $A \cap F_{\alpha^{-}} \in F$.

Proof: (Necessity) If $\alpha = -\infty$, then $A \cap F_{-\infty^{-}} = A \cap F_{\overline{\infty}}$. From the Theorem 2.1 ,we can know $A \cap F_{\overline{\infty}} \in F$; If $-\infty < \alpha < +\infty$ then $F_{\alpha^{-}} = F_{\alpha}^{c}$, so $A \cap F_{\alpha^{-}} = A \cap F_{\cdot\alpha}^{c}$, And $A \cap F_{\alpha^{-}} \in F$;If $\alpha = +\infty$, due to $A \cap F_{+\infty^{-}} = A$, then $A \cap F_{+\infty^{-}} \in F$. Therefore for arbitrary $\alpha \in C$, then $A \cap F_{\alpha^{-}} \in F$.

(Sufficiency) If $\alpha = -\infty$, then $A \cap F_{\overline{\infty}} = A \cap F_{-\infty^{-}} \in F$.

If $\alpha > -\infty$, then $A \cap F_{\alpha} = A \cap F_{\alpha^{-}} \in F$. From the Theorem 2.1, f is complex fuzzy measurable function on A.

Theorem 2.4. *Suppose* (X,F) *is complex fuzzy measurable space, the mapping* $f: X \to C$ *is complex fuzzy measurable function on* $A \in F$ *if and only if for arbitrary* $\alpha \in C$, *then* $A \cap F_{\alpha^{-}} \in F$ *and* $A \cap F_{\overline{\infty}} \in F$.

Theorem 2.5. *Suppose* (X,F) *is complex fuzzy measurable space, for arbitrary* $A \in F$, *the mapping* $f: X \to C$, *then*

(1) *If* f *is measurable,* f *is measurable function on the subset* A_1 *of* A.

(2) *If* $A_1, A_2 \in F$, $A_1 \cap A_2 = \varphi$ *and* $A = A_1 \cup A_2$, *then* f *is measurable on* A *if and only if* f *is both measurable on* A_1 *and* A_2.

Proof. (1) For arbitrary $\alpha \in C$, due to

$$A_1 \cap F_\alpha = (A_1 \cap A) \cap F_\alpha = A_1 \cap (A \cap F_\alpha),$$

A_1 and $A \cap F_\alpha$ are measurable sets, so $A_1 \cap F_\alpha \in F$. Therefore f is meas-urable on A_1.

(2) Suppose f is measurable on A, from (1), we can know f is both measurable on A_1 and A_2. On the contrary, if f is both measurable on A_1 and A_2, then for arbitrary $\alpha \in C$, due to

$$A \cap F_\alpha = (A_1 \cup A_2) \cap F_\alpha = (A_1 \cap F_\alpha) \cup (A_2 \cap F_\alpha),$$

so $A \cap F_\alpha \in F$. Therefore f is complex fuzzy measurable function on A.

Theorem 2.6. *Suppose* (X, F) *is complex fuzzy measurable space, for arbitrary* $A \in F, f f$ *and* g *are both complex fuzzy measurable function on* A, *then*

(1) *For arbitrary complex number* $\lambda = \lambda_1 + i\lambda_2$, *if* $\lambda \cdot f$ *is meaning, then* λf *is complex fuzzy measurable function on* A.
(2) *If* $f + g$ *is meaning, then* $f + g$ *is complex fuzzy measurable function on* A.
(3) *If* $f \cdot g, f / g$ *are both meaning, then* $f \cdot g, f / g$ *are both complex fuzzy measurable function on* A.
(4) $\max(f, g), \min(f, g)$ *are both complex fuzzy measurable function on* A.

Proof: They are similar to the proof process of real fuzzy measurable function. Omitted.

3   Complex Fuzzy Measurable Function Sequences

Theorem 3.1. *Suppose* (X, F) *is complex fuzzy measurable space, for arbitrary* $A \in F, \{f_n\}$ *is complex fuzzy measurable function sequence on* A, *so the sup function, infimum function, upper limit function, lower limit function of* $\{f_n\}$ *are all complex fuzzy measurable function on* A.

Proof. First, we can prove the sup function g of $\{f_n\}$ is measurable. We can know

$$g(x) = \lim_{n \to \infty} \max\{f_1(x), f_2(x), \cdots f_n(x)\} \cdot g(x) = \max\{f_1(x), f_2(x), \cdots f_n(x)\}$$

is complex fuzzy measurable function and $\{g_n(x)\}$ is monotone increasing sequence,

$$g(x) = \lim_{n \to \infty} g_n(x), \text{ so for arbitrary } \alpha \in C \text{, then } A \cap g_{\cdot\alpha} = \bigcup_{n=1}^{\infty} [A \cap (g_n)_{\cdot\alpha}],$$

therefore $A \cap g_{\cdot\alpha}$ is measurable set, $A \cap g_{\underline{\infty}} \in F$ is obvious, so g is complex fuzzy measurable function on A.

Similarly, we can prove the infimum function g' is measurable.

Now, we will prove upper limit function $\overline{\lim_{n \to \infty}} f_n$ is measurable .Denote G_n is the sup function of the sequence $f_n, f_{n+1}, \cdots, f_{n+m}, \cdots$.According to the above proof, G_n is measurable. Due to $\overline{\lim_{n \to \infty}} f_n = \lim_{n \to \infty} G_n$ and $G_1 \geq G_2 \geq \cdots \geq G_n \geq \cdots$,so $\overline{\lim_{n \to \infty}} f_n$ is the infimum function of the measurable function sequence $\{G_n\}$, so $\overline{\lim_{n \to \infty}} f_n$ is complex fuzzy measurable function on A.

Similarly, we can prove $\underline{\lim_{n \to \infty}} f_n$ is complex fuzzy measurable function on A.

Corollary 3.1. *Suppose* (X, F) *is complex fuzzy measurable space, for arbitrary* $A \in F$, $\{f_n\}$ *is complex fuzzy measurable function sequence on* A, *so* $f = \lim_{n \to \infty} f_n$ *is complex fuzzy measurable function on* A.

If for arbitrary $\varepsilon = \varepsilon_1 + i\varepsilon_2$, $\varepsilon_1, \varepsilon_2 > 0$,*then* $\left| \mathrm{Re}[f_n(x)] - \mathrm{Re}[f(x)] \right| \geq \varepsilon_1$, *and*

$$\left| \mathrm{Im}[f_n(x)] - \mathrm{Im}[f(x)] \right| \geq \varepsilon_2 \text{ ,we denote } \left| f_n(x) - f(x) \right| \geq \varepsilon .$$

Definition 3.1. *Suppose* (X, F, μ) *is complex Fuzzy measure space,* $f_n (n = 1, 2, \cdots), f$ *is complex fuzzy measurable function, for arbitrary* $A \in F$, $\varepsilon_1, \varepsilon_2 > 0$

(1) $\{f_n\}$ *converge in complex fuzzy measure* μ *to* f *on* A ,*denote* $f_n \xrightarrow{\ u. \ } f$,*,if for arbitrary* $\varepsilon = \varepsilon_1 + i\varepsilon_2$, *such that*

$$\lim_{n \to \infty} \mu(\{x | \mathrm{Re} | f_n - f | \geq \varepsilon_1, \mathrm{Im} | f_n - f | \geq \varepsilon_2\} \cap A) = 0.$$

(2) $\{f_n\}$ *converge in pseudo complex fuzzy measure* μ *to* f *on* A ,*denote* $f_n \xrightarrow{\ p.u. \ } f$, *if for arbitrary* $\varepsilon = \varepsilon_1 + i\varepsilon_2$, *such that*

$$\lim_{n \to \infty} \mu(\{x | \mathrm{Re} | f_n - f | < \varepsilon_1, \mathrm{Im} | f_n - f | < \varepsilon_2\} \cap A) = \mu(A).$$

(3) $\{f_n\}$ *is basic in complex fuzzy measure, if for arbitrary* $\varepsilon = \varepsilon_1 + i\varepsilon_2$,*such that*

$$\lim_{\substack{n \to \infty \\ m \to \infty}} \mu\{x | \mathrm{Re} | f_n - \tilde{f}_m | \geq \varepsilon_1, \mathrm{Im} | f_n - f_m | \geq \varepsilon_2\} = 0 .$$

(4) $\{f_n\}$ is basic in pseudo complex fuzzy measure, if for arbitrary $\varepsilon = \varepsilon_1 + i\varepsilon_2$, such that

$$\lim_{\substack{n\to\infty \\ m\to\infty}} \mu(\{x | \mathrm{Re} | f_n - f_m | \geq \varepsilon_1, \mathrm{Im} | f_n - f_m | \geq \varepsilon_2\} \cap A) = \mu(A).$$

(5) $\{f_n\}$ almost uniformly converge to f, if for arbitrary $\varepsilon = \varepsilon_1 + i\varepsilon_2$, there exists a measurable set E, such that $\mu(E) < \varepsilon$, and $\{f_n\}$ uniformly converge to the measurable function f on E^c.

(6) $\{f_n\}$ pseudo-almost uniformly converge to f, if for arbitrary $\varepsilon = \varepsilon_1 + i\varepsilon_2$, there exists a measurable set $E \subset A$ ($A \in F$), such that $\mu(E) > \mu(A) - \varepsilon$, and $\{f_n\}$ uniformly converge to the measurable function f on E^c.

Theorem 3.2. Suppose (X,F,μ) is complex fuzzy measure space, $\{f_n\}$ and $\{g_n\}$ are complex fuzzy measurable function sequence. f and g are complex fuzzy measurable function. μ is self-continuous. If $\{f_n\}$ converge in complex fuzzy measure μ to f on A, then $\{f_n\}$ is basic in complex fuzzy measure.

Proof.   For arbitrary $\varepsilon = \varepsilon_1 + \varepsilon_2, \varepsilon_1, \varepsilon_2 > 0$, then

$$\{x \big| | f_n(x) - f_m(x)| \geq \varepsilon\} \subset \{x \| f_n(x) - f(x) | \geq \frac{\varepsilon}{2}\} \cup \{x \| f_m(x) - f(x) | \geq \frac{\varepsilon}{2}\},$$

so

$$A \cap \{x \| f_n(x) - f_m(x) | \geq \varepsilon\} \subset (A \cap \{x \| f_n(x) - f(x) | \geq \frac{\varepsilon}{2}\}) \cup (A \cap \{x \| f_m(x) - f(x) | \geq \frac{\varepsilon}{2}\}).$$

Due to $\{f_n\}$ converge in complex fuzzy measure μ to f on A, denote

$$\lim_{n\to\infty} \mu(A \cap \{x \| f_n(x) - f(x) | \geq \varepsilon\}) = 0, \text{so } \lim_{\substack{n\to\infty \\ m\to\infty}} \mu(A \cap \{x \| f_n(x) - f_m(x) | \geq \varepsilon\}) = 0.$$

Therefore $\{f_n\}$ is basic in complex fuzzy measure.

Theorem 3.3. Suppose (X,F,μ) is complex fuzzy measure space, $A \in F$, $\mu(A) < +\infty$. $\{f_n\}$ and $\{g_n\}$ are complex fuzzy measurable function sequence. f and g are complex fuzzy measurable function. μ is pseudo-self -continuous.

(1) If $\{f_n\}$ converge in pseudo complex fuzzy measure μ to f on A, then $\{f_n\}$ is basic in pseudo complex fuzzy measure.

(2) If $\{f_n\}, \{g_n\}$ separately converge in pseudo complex fuzzy measure μ to f and g on A, then for arbitrary $\alpha, \beta \in C$, $\{\alpha f_n + \beta g_n\}$ converge in pseudo complex fuzzy measure to $\alpha f + \beta g$.

Proof. for arbitrary $\varepsilon = \varepsilon_1 + \varepsilon_2, \varepsilon_1, \varepsilon_2 > 0$, then

$$\{x \| \tilde{f}_n(x) - \tilde{f}_m(x) \| < \varepsilon\} \supset \{x \| \tilde{f}_n(x) - \tilde{f}(x) \| < \frac{\varepsilon}{2}\} \cap \{x \| \tilde{f}_m(x) - \tilde{f}(x) \| < \frac{\varepsilon}{2}\},$$

so

$$A \cap \{x \| f_n(x) - f_m(x) \| < \varepsilon\} \supset (A \cap \{x \| f_n(x) - f(x) \| < \frac{\varepsilon}{2}\}) \cap (A \cap \{x \| f_m(x) - f(x) \| < \frac{\varepsilon}{2}\}),$$

$\{f_n\}$ converge in pseudo complex fuzzy measure μ to f on A, then

$$\lim_{n \to \infty} \mu(A \cap \{x \| f_n(x) - f(x) \| < \varepsilon\}) = \mu(A), \text{ so } \lim_{\substack{n \to \infty \\ m \to \infty}} \mu(A \cap \{x \| f_n(x) - f_m(x) \| < \varepsilon\}) = \mu(A).$$

Therefore $\{f_n\}$ is basic in pseudo complex fuzzy measure.

(2) For arbitrary $\varepsilon = \varepsilon_1 + \varepsilon_2, \varepsilon_1, \varepsilon_2 > 0$, if $\alpha \neq 0, \beta \neq 0$, then

$$\{x \| (\alpha f_n(x) + \beta g_n(x)) - (\alpha f(x) + \beta g(x)) \| < \varepsilon\}$$

$$= \{x \| \alpha(f_n(x) - f(x)) + \beta(g_n(x) - g(x)) \| < \varepsilon\}$$

$$\supset \{x \| \alpha(f_n(x) - f(x)) \| < \frac{\varepsilon}{2}\} \cap \{x \| \beta(g_n(x) - g(x)) \| < \frac{\varepsilon}{2}\}$$

$$= \{x \| f_n(x) - f(x) \| < \frac{\varepsilon}{2\alpha}\} \cap \{x \| g_n(x) - g(x) \| < \frac{\varepsilon}{2\beta}\},$$

so

$$A \cap \{x \| (\alpha f_n(x) + \beta g_n(x)) - (\alpha f(x) + \beta g(x)) \| < \varepsilon\}$$

$$\supset A \cap \{x \| f_n(x) - f(x) \| < \frac{\varepsilon}{2\alpha}\} \cap \{x \| g_n(x) - g(x) \| < \frac{\varepsilon}{2\beta}\}.$$

$$\lim_{n \to \infty} \mu(A \cap \{x \| f_n(x) - f(x) \| < \frac{\varepsilon}{2\alpha}\}) = \mu(A), \lim_{n \to \infty} \mu(A \cap \{x \| g_n(x) - g(x) \| < \frac{\varepsilon}{2\beta}\}) = \mu(A).$$

So $\mu(A \cap \{x \| (\alpha f_n(x) + \beta g_n(x)) - (\alpha f(x) + \beta g(x)) \| < \varepsilon\}) = \mu(A) \backslash.$

If $\alpha = 0$ or $\beta = 0$, then

$$\mu(A \cap \{x \| (\alpha f_n(x) + \beta g_n(x)) - (\alpha f(x) + \beta g(x)) \| < \varepsilon\}) = \mu(A).$$

So, $\{\alpha f_n + \beta g_n\}$ converge in pseudo complex fuzzy measure to $\alpha f + \beta g$.

Theorem 3.4 *Suppose* (X, F, μ) *is complex fuzzy measure space,* $\mu(X) < \infty$. $\{f_n\}$ *and* $\{g_n\}$ *are complex fuzzy measurable function sequence.* f *and* g *are complex fuzzy measurable function,*

(1) *If* μ *is upper-self-continuous,* $\{f_n\}$, $\{g_n\}$ *separately converge in complex fuzzy measure* μ *to*

f and g, $\mu\{x \mid \text{Re}[g(x)] = 0, \text{Im}[g(x)] = 0\} = 0$, $\mu\{x \mid \text{Re}[g_n(x)] = 0, \text{Im}[g_n(x)] = 0\} = 0$

($n = 1, 2, \cdots$), so $\{f_n / g_n\}$ *converge in complex fuzzy measure to* f / g.

(2) *If μ is pseudo lower-self-continuous, $\{f_n\}$, $\{g_n\}$ separately converge in pseudo complex fuzzy measure μ to f and g, and*

$$\mu\{x \mid \mathrm{Re}[g(x)] \neq 0, \mathrm{Im}[g(x)] \neq 0\} = \mu(X),$$
$$\mu\{x \mid \mathrm{Re}[g_n(x)] \neq 0, \mathrm{Im}[g_n(x)] \neq 0\} = \mu(X) \ n = 1, 2, \cdots, so \ \{f_n / g_n\}$$

converge in pseudo complex fuzzy measure to f / g.

Proof. (1) μ is upper-self-continuous, so for arbitrary $x \in X$, $g(x) \neq 0$, $g_n(x) \neq 0, n = 1, 2, \cdots$. Suppose $\{f_{n_k} / g_{n_k}\}$ is arbitrary subsequence of $\{f_n / g_n\}$. Due to $\{g_n\}$ converge in complex fuzzy measure to g, so there exists a subsequence $\{g_{n_{k_s}}\}$ almost anywhere converge to g of $\{g_{n_k}\}$. Similarly there exists a subsequence $\{f_{n_{k_s}}\}$ almost anywhere converge to f of $\{f_{n_k}\}$, so $\{f_{n_{k_s}} / g_{n_{k_s}}\}$ almost anywhere converge to f / g, therefore $\{f_n / g_n\}$ converge in complex fuzzy measure to f / g.

(2) Similarly, we can prove $\{f_n / g_n\}$ converge in pseudo complex fuzzy measure to f / g.

Theorem 3.5. *Suppose (X, F, μ) is complex fuzzy measure space, $\mu(X) < \infty$. $\{f_n\}$ is complex fuzzy measurable function sequence,*

(1) *If μ is uniform-upper-self-continuous, $\{f_n\}$ is basic in complex fuzzy measure, then there exists a subsequence $\{f_{n_k}\}$ almost uniformly converge of $\{f_n\}$.*

(2) *If μ is uniform-pseudo- lower -self-continuous, $\{f_n\}$ is basic in pseudo complex fuzzy measure, then there exists a subsequence $\{f_{n_k}\}$ pseudo almost uniformly converge of $\{f_n\}$.*

Proof. (1) $\{f_n\}$ is basic in pseudo complex fuzzy measure, then for arbitrary positive integer k, exists n'_k, such that $m, n \geq n'_k$, so

$$\mu(\{x \mid \| f_m(x) - f_n(x) \| \geq \frac{1}{2^k}\}) < \frac{1}{2^k}.$$

Let $n_2 = (n_1 + 1) \vee n'_2, \cdots$, $n_k = (n_{k-1} + 1) \vee n'_k, \cdots$. Then $n_1 < n_2 < \cdots$, so $\{f_{n_k}\}$ is an infinite subsequence of $\{f_n\}$.

Let $E_k = \mu(\{x \mid \| f_{n_{k+1}}(x) - f_{n_k}(x) \| \geq \frac{1}{2^k}\})$. Due to μ is uniform-upper-self-continuous, so $\lim\limits_{n \to \infty} \mu(\bigcup\limits_{k=s}^{n} E_k) = 0$. We will prove $\{f_{n_k}\}$ uniform convergence on $(\bigcup\limits_{k=s}^{\infty} E_k)^c$, so we can prove $\{f_{n_k}\}$ uniform basic on $(\bigcup\limits_{k=s}^{\infty} E_k)^c$.

$$(\bigcup_{k=s}^{\infty} E_k)^c = \bigcap_{k=s}^{\infty} E_k^c = \bigcap_{k=s}^{\infty} (\{x \| f_{n_{k+1}}(x) - f_{n_k}(x) \ge \frac{1}{2^k}\})^c = \bigcap_{k=s}^{\infty} \{x \| f_{n_{k+1}}(x) - f_{n_k}(x) < \frac{1}{2^k}\}.$$

So for arbitrary $x \in (\bigcup_{k=s}^{\infty} E_k)^c$ and $k \ge s$, then $\| f_{n_{k+1}}(x) - f_{n_k}(x) \| < \frac{1}{2^k}$ is true, so for

arbitrary $j \ge i \ge s$ and $x \in (\bigcup_{k=s}^{\infty} E_k)^c$

$$\| f_{n_j}(x) - f_{n_i}(x) \| \le \sum_{k=i}^{\infty} \| f_{n_k}(x) - f_{n_{k+1}}(x) \| \le \frac{1}{2^{i-1}} \le \frac{1}{2^{s-1}} < \varepsilon,$$

therefore $\{f_{n_k}\}$ almost uniformly converge.

(2) $\{f_n\}$ is basic in pseudo complex fuzzy measure, then for arbitrary positive

integer k, exists n_k', such that $m, n \ge n_k'$, so

$$\mu(\{x \| f_m(x) - f_n(x) \| < \frac{1}{2^k}\}) > \mu(X) - \frac{1}{2^k}.$$

Let $n_2 = (n_1 + 1) \vee n_2', \cdots$, $n_k = (n_{k-1} + 1) \vee n_k', \cdots$. Then $n_1 < n_2 < \cdots$, so $\{f_{n_k}\}$ is an

infinite subsequence of $\{f_n\}$. Let $E_k = \mu(\{x \| f_{n_{k+1}}(x) - f_{n_k}(x) \| < \frac{1}{2^k}\})$. Due to

μ is uniform- pseudo-lower-self-continuous, so $\lim_{n \to \infty} \mu(\bigcap_{k=s}^{n} E_k) = \mu(X)$. We will

prove $\{f_{n_k}\}$ uniform convergence on $\bigcap_{k=s}^{\infty} E_k$, so we can prove $\{f_{n_k}\}$ uniform basic

on $\bigcap_{k=s}^{\infty} E_k$. Due to $\bigcap_{k=s}^{\infty} E_k = \bigcap_{k=s}^{\infty} E_k = \bigcap_{k=s}^{\infty} (\{x \| f_{n_{k+1}}(x) - f_{n_k}(x) \| < \frac{1}{2^k}\})$, and prove

$\{f_{n_k}\}$ almost uniformly converge from above. So we prove it.

Theorem 3.6. *Suppose* (X,F,μ) *is complex fuzzy measure space.* $\mu(X) < \infty$. $\{f_n\}$
is complex fuzzy measurable function sequence, f *is complex fuzzy measurable*
function, then

(1) *If* μ *is uniform-upper-self-continuous,* $\{f_n\}$ *is basic in complex fuzzy*
measure, and a subsequence $\{f_{n_k}\}$ *converge in complex fuzzy measure to* f, *then*
$\{f_n\}$ *converge in complex fuzzy measure to* f.

(2) *If* μ *is uniform-pseudo- lower -self-continuous,* $\{f_n\}$ *is basic in pseudo*
complex fuzzy measure, and a subsequence $\{f_{n_k}\}$ *converge in pseudo complex*
fuzzy measure to f, *then* $\{f_n\}$ *converge in pseudo complex fuzzy measure to* f.

Proof. For arbitrary $\varepsilon_i > 0 (i = 1,2)$, $\delta > 0$, exists $N > 0$, such that $n, m > N$, $n_k > N$,

then $\mu\{x | \text{Re} | f_n(x) - f_m(x) | \geq \frac{\varepsilon_1}{2}, \text{Im} | f_n(x) - f_m(x) | \geq \frac{\varepsilon_2}{2}\} < \frac{\delta}{2}$,

$\mu\{x | \text{Re} | f_{n_k}(x) - f(x) | \geq \frac{\varepsilon_1}{2}, \text{Im} | f_{n_k}(x) - f(x) | \geq \frac{\varepsilon_2}{2}\} < \frac{\delta}{2}$. For arbitrary $x \in X$, then

$$\text{Re} | f_n(x) - f(x) | \leq \text{Re} | f_n(x) - f_{n_k}(x) | + \text{Re} | f_{n_k}(x) - f(x) |$$

$$\text{Im} | f_n(x) - f(x) | \leq \text{Im} | f_n(x) - f_{n_k}(x) | + \text{Im} | f_{n_k}(x) - f(x) |$$

If $\text{Re} | f_n(x) - f(x) | \geq \varepsilon_1$, $\text{Im} | f_n(x) - f(x) | \geq \varepsilon_2$, then $\text{Re} | f_n(x) - f_{n_k}(x) | \geq \frac{\varepsilon_1}{2}$,

$\text{Im} | f_n(x) - f_{n_k}(x) | \geq \frac{\varepsilon_2}{2}$ or $\text{Re} | f_{n_k}(x) - f(x) | \geq \frac{\varepsilon_1}{2}$, $\text{Im} | f_{n_k}(x) - f(x) | \geq \frac{\varepsilon_2}{2}$.

so

$$\{x | \text{Re} | f_n(x) - f(x) | \geq \varepsilon_1, \text{Im} | f_n(x) - f(x) | \geq \varepsilon_2\}$$

$$\subset \{x | \text{Re} | f_n(x) - f_{n_k}(x) | \geq \frac{\varepsilon_1}{2}, \text{Im} | f_n(x) - f_{n_k}(x) | \geq \frac{\varepsilon_2}{2}\}$$

$$\cup \{x | \text{Re} | f_{n_k}(x) - f(x) | \geq \frac{\varepsilon_1}{2}, \text{Im} | f_{n_k}(x) - f(x) | \geq \frac{\varepsilon_2}{2}\}.$$

Due to μ is uniform-upper-self-continuous, let $m = n_k$, $n > N$, then

$$0 \leq \mu\{x | \text{Re} | f_n(x) - f(x) | \geq \varepsilon_1, Im | f_n(x) - f(x) | \geq \varepsilon_2\}$$

$$\leq \mu(\{x | \text{Re} | f_n(x) - f_{n_k}(x) | \geq \frac{\varepsilon_1}{2}, \text{Im} | f_n(x) - f_{n_k}(x) | \geq \frac{\varepsilon_2}{2}\}$$

$$\cup \{x | \text{Re} | f_{n_k}(x) - f(x) | \geq \frac{\varepsilon_1}{2}, \text{Im} | f_{n_k}(x) - f(x) | \geq \frac{\varepsilon_2}{2}\})$$

$$\leq \mu(\{x | \text{Re} | f_n(x) - f_{n_k}(x) | \geq \frac{\varepsilon_1}{2}, \text{Im} | f_n(x) - f_{n_k}(x) | \geq \frac{\varepsilon_2}{2}\}) + \frac{\delta}{2}$$

$$\leq \frac{\delta}{2} + \frac{\delta}{2} = \delta.$$

So $\{f_n\}$ converge in complex fuzzy measure to f.

(2) Similarly, we can prove $\{f_n\}$ converge in pseudo complex fuzzy measure to f.

Theorem 3.7. *Suppose* (X, F, μ) *is complex fuzzy measure space,* $\mu(X) < \infty$. $\{f_n\}$ *is complex fuzzy measurable function sequence, then*

(1) *If* μ *is uniform-upper-self-continuous,* $\{f_n\}$ *is basic in complex fuzzy measure, then exists a complex fuzzy measurable function* f *such that* $\{f_n\}$ *converge in com- plex fuzzy measure to* f.

(2) *If μ is uniform-pseudo- lower -self-continuous, $\{f_n\}$ is basic in pseudo complex fuzzy measure, then exists a complex fuzzy measurable function f such that $\{f_n\}$ converge in pseudo complex fuzzy measure to f.*

Proof. (1) If μ is uniform-upper-self-continuous, $\{f_n\}$ is basic in complex fuzzy measure, according to the Theorem 3.5 (1),there exists a almost uniformly basic subse- quence $\{f_{n_k}\}$,so $\{f_{n_k}\}$ is almost anywhere basic. For arbitrary x,such that there exists $\lim\limits_{k\to\infty} f_{n_k}(x)$ .Let $f(x) = \lim\limits_{k\to\infty} f_{n_k}(x)$ .If exists x ,such that $\lim\limits_{k\to\infty} f_{n_k}(x)$ don't exist, we let $f(x) = 0$ .so f is measurable function, and $\{f_{n_k}\}$ almost anywhere converge to f .According to Lebesgue theorem, we can know $\{f_{n_k}\}$ converge in complex fuzzy measure to f ,again according to the Theorem 3.6 (1), $\{f_n\}$ converge in complex fuzzy measure to f .

(2) Similarly, we can prove $\{f_n\}$ converge in pseudo complex fuzzy measure to f .

Theorem 3.8 *Suppose* (X,F,μ) *is complex fuzzy measure space,* $\mu(X) < \infty$. $\{f_n\}$ *is complex fuzzy measurable function sequence, then*

(1) *If μ is uniform-upper-self-continuous, $\{f_n\}$ is basic in complex fuzzy measure if and only if there exists a complex fuzzy measurable function f such that $\{f_n\}$ converge in complex fuzzy measure to f.*

(2) *If μ is uniform-pseudo- lower -self-continuous, $\{f_n\}$ is basic in pseudo com- plex fuzzy measure if and only if there exists a complex fuzzy measurable function f such that $\{f_n\}$ converge in pseudo complex fuzzy measure to f.*

Proof. The necessity can be proved by the Theorem 3.7.

The sufficiency can be proved by the Theorem 3.2 and 3.3(1).

Acknowledgements. Thanks to the support by Natural Science Foundation of Hainan Province (No.111007) and Hainan International cooperation Key Project (No.GJXM201105).

References

1. Buckley, J.J.: Fuzzy complex numbers. FSS 33, 333–345 (1989)
2. Qiu, J.Q., Liu, F.C., Su, L.Q.: Complex fuzzy measure and complex fuzzy integral. Journal of Hebei Institute of Chemical Technology and Light Industry 1, 1–5 (1997)
3. Qiu, J.Q.: Relation on convergence of measurable functions sequence on fuzzy complex measure space. Fuzzy Sets and Systems 21, 92–96 (2007)
4. Zhang, G.Q.: Fuzzy measure theory. Guizhou Science and Technology Press (1994)

5. Ma, S.Q.: Fuzzy complex number and that of several properties. Journal of Lanzhou University (Natural Science Edition) 32, 643–645 (1996)
6. Ma, S.Q.: The fuzzy complex numbers based on the form Z=re$^{i\theta}$ and some operation properties. Journal of Northwest for Nationalities (Natural Science) 1, 8–9 (1997)
7. Ma, S.Q.: Some properties of Rectangular fuzzy complex numbers. Journal of Northwest for Nationalities (Natural Science), 62–63 (1998)
8. Ha, M.H., Wu, C.X.: Fuzzy measures and fuzzy integral theory. Science Press (1998)

Fuzzy Complex-Valued Fuzzy Measure Base on Fuzzy Complex Sets

Zhi-qing Zhao[1] and Sheng-quan Ma[2,*]

[1] Department of Mathematics, Hainan Normal University, Haikou, Hainan,
571158, P.R. China
[2] School of information science and Technology, Hainan Normal University,
Haikou, Hainan, 571158, P.R. China
mashengquan@163.com

Abstract. Based on the conception of fuzzy complex-valued measure, Firstly, the definition of fuzzy complex-valued fuzzy measure on the fuzzy complex sets is introduced, the properties of them is discussed, and gain some better results; Secondly, we introduce the conception of fuzzy complex-valued fuzzy measurable function on fuzzy complex sets , and finally discuss its convergence. Establish the foundation for further research.

Keywords: Fuzzy complex-valued fuzzy measure, fuzzy complex σ- algebraic, measurable.

1 Introduction

Fuzzy complex-valued measure is the important part of fuzzy complex analysis, which is the further advancement of fuzzy measure, in recent decades, the fuzzy complex-valued measure theory has developed very quickly, Ma, S.-Q ,Qiu, J.-Q, etc, have take a large study for it, In this paper, fuzzy complex-valued measure is expanded ,we introduce the definition of fuzzy complex-valued fuzzy measure on fuzzy complex sets, and discuss its properties in detail, then we introduce the definition of fuzzy complex-valued fuzzy measurable function on fuzzy complex sets, and discuss the basic properties of them, establish the foundation for further investigation.

2 Preliminary

Let $F^\cdot(R)$ denotes the set of all fuzzy real number on R, $F^\cdot(C)$ denotes the set of all fuzzy complex number on C, and $F_+^\cdot(C)$ denotes the non-negative fuzzy complex number, which shows that the real part and the imaginary part are non-negative.

---

* Corresponding author.

B.-Y. Cao and X.-J. Xie (Eds.): Fuzzy Engineering and Operations Research, AISC 147, pp. 207–212.
springerlink.com
© Springer-Verlag Berlin Heidelberg 2012

Definition 2.1[1]. Let $\hat{c}_1, \hat{c}_2 \in F^*(C)$, the order of $\hat{c}_1$ and $\hat{c}_2$ is defined as following:

1) $\hat{c}_1 \le \hat{c}_2 \Leftrightarrow \operatorname{Re}\hat{c}_1 \le \operatorname{Re}\hat{c}_2, \operatorname{Im}\hat{c}_1 \le \operatorname{Im}\hat{c}_2$;

2) $\hat{c}_1 = \hat{c}_2 \Leftrightarrow \hat{c}_1 \le \hat{c}_2, \hat{c}_2 \le \hat{c}_1$.

Definition 2.2[1] Let $\hat{a}, \hat{b} \in F^*(C)$, the fuzzy distance of fuzzy complex numbers on $F^*(C)$ is defined as following:

$$\hat{\rho}: F^*(C) \times F^*(C) \to F_+^*(C)$$

$$\hat{\rho}(\hat{a}, \hat{b}) = \left(\hat{\rho}(\operatorname{Re}\hat{a}, \operatorname{Re}\hat{b}), \hat{\rho}(\operatorname{Im}\hat{a}, \operatorname{Im}\hat{b})\right),$$

where

$$\hat{\rho}(\operatorname{Re}\hat{a}, \operatorname{Re}\hat{b}) = \bigcup_{\lambda \in [0,1]} \lambda \left[\left|(\operatorname{Re}\hat{a})_1^- - (\operatorname{Re}\hat{b})_1^-\right|, \sup_{\lambda \le \eta \le 1} \left|(\operatorname{Re}\hat{a})_\eta^- - (\operatorname{Re}\hat{b})_\eta^-\right| \vee \left|(\operatorname{Re}\hat{a})_\eta^+ - (\operatorname{Re}\hat{b})_\eta^+\right|\right],$$

$$\hat{\rho}(\operatorname{Re}\hat{a}, \operatorname{Re}\hat{b}) = \bigcup_{\lambda \in [0,1]} \lambda \left[\left|(\operatorname{Im}a')_1^- - (\operatorname{Im}b')_1^-\right|, \sup_{\lambda \le \eta \le 1} \left|(\operatorname{Im}\hat{a})_\eta^- - (\operatorname{Im}\hat{b})_\eta^-\right| \vee \left|(\operatorname{Im}\hat{a})_\eta^+ - (\operatorname{Im}\hat{b})_\eta^+\right|\right].$$

Definition 2.3[2]. Let $\{\hat{c}_n\} \subset F(C)$ and $\hat{c} \in F(C)$, $\{\hat{c}_n\}$ is called converges to $\hat{c}$, if $\operatorname{Re}\hat{c}_n \to \operatorname{Re}\hat{c}$, $\operatorname{Im}\hat{c}_n \to \operatorname{Im}\hat{c}(n \to \infty)$. Notes $\lim_{n \to \infty} \hat{c}_n = \hat{c}$.

Definition 2.4. Let $\hat{F} \subset F^*(C)$, $\hat{F}$ is called a fuzzy complex σ-algebra. If the following conditions hold:

FA1 : $\varphi, X \in \hat{F}$;

FA2 : If $\hat{A} \in \hat{F}$, then $\hat{A}^c \in \hat{F}$;

FA3 : If $\{\hat{A}_n\} \subset \hat{F}$, then $\bigcup_{n=1}^{\infty} \hat{A}_n \in \hat{F}$.

3   Fuzzy Complex-Valued Fuzzy Measures on Fuzzy Complex Sets

Definition 3.1. The mapping $\hat{\mu}: \hat{F} \to F_+^*(C)$ is called a fuzzy complex-valued fuzzy measure on $\hat{F}$, where $\hat{F}$ is a fuzzy complex σ-algebra, if $\hat{\mu}$ have the following properties:

FCFM1 : If $\varphi \in \hat{F}$, then $\hat{\mu}(\varphi) = \hat{0}$;

FCFM2 : If $\hat{A}, \hat{B} \in \hat{F}, \hat{A} \subset \hat{B}$ then $\hat{\mu}(\hat{A}) \le \hat{\mu}(\hat{B})$;

FCFM3 : If $\{\hat{A}_n\} \subset \hat{F}, \hat{A}_n \subset \hat{A}_{n+1}$, then $\hat{\mu}\left(\bigcup_{n=1}^{\infty} \hat{A}_n\right) = (\hat{\rho})\lim_{n \to \infty} \hat{\mu}(\hat{A}_n)$;

FCFM4: If $\{\hat{A}_n\} \subset \hat{F}, \hat{A}_n \supset \hat{A}_{n+1},$ *and existing* n *with* $n \geq 1$ *such that* $\hat{\mu}(\hat{A}_{n_0}) \neq \infty$,

then $\hat{\mu}\left(\bigcap\limits_{n=1}^{\infty} \hat{A}_n\right) = (\hat{\rho})\lim\limits_{n \to \infty}\hat{\mu}(\hat{A}_n)$.

$\hat{\mu}$ is also called upper semi-continuous (lower semi-continuous) fuzzy complex value fuzzy measure, if and only if it satisfy (FCFM1), (FCFM2) and (FCFM3) ((FCFM1) (FCFM2) and (FCFM4)).

Definition 3.2. *Fuzzy complex-valued fuzzy functions* $\hat{\mu}$ *on* $\hat{F}$ *is called null-additive (null-subtractive), if the following conditions hold:*

$$\hat{\mu}(\hat{A}) = \hat{0} \Rightarrow \hat{\mu}(\hat{A} \cup \hat{B}) = \hat{\mu}(\hat{B}) \ or \ \hat{\mu}(\hat{B}) = \hat{0} \Rightarrow \hat{\mu}(\hat{A} \cup \hat{B}) = \hat{\mu}(\hat{A}),$$

$$(\ \hat{\mu}(\hat{A}) = \hat{0} \Rightarrow \hat{\mu}(\hat{A}^c \cap \hat{B}) = \hat{\mu}(\hat{B}) \ or \ \hat{\mu}(\hat{B}) = \hat{0} \Rightarrow \hat{\mu}(\hat{A} \cap \hat{B}^c) = \hat{\mu}(\hat{A}) \ )$$

for arbitrary $\hat{A}, \hat{B} \in \hat{F}$.

Definition 3.3. *Fuzzy complex-valued fuzzy functions* $\hat{\mu}$ *on* $\hat{F}$ *is called pseudo null-additive (pseudo null-subtractive), if the following conditions hold:*

$$\hat{\mu}(\hat{A} \cap \hat{C}) = \hat{\mu}(\hat{A}) \Rightarrow \hat{\mu}((\hat{A} \cap \hat{B}^c) \cup \hat{C}) = \hat{\mu}(\hat{C}),$$

$$(\ \hat{\mu}(\hat{A} \cap \hat{C}^c) = \hat{\mu}(\hat{A}) \Rightarrow \hat{\mu}((\hat{A} \cap \hat{B}^c) \cup \hat{C}) = \hat{\mu}(\hat{C}) \ )$$

for arbitrary $\hat{A}, \hat{B} \in \hat{F}$ *and arbitrary* $\hat{C} \in \hat{A} \cap \hat{F} = \{\hat{A} \cap \hat{D} \mid \hat{D} \in \hat{F}\}$.

Definition 3.4. *Fuzzy complex-valued fuzzy function* $\hat{\mu}: \hat{F} \to F_+^*(C)$ *on* $\hat{F}$ *is called Auto-continuity from above (below), if* $\{\hat{A} \cup \hat{B}_n\} \subset \hat{F}$ *and the following conditions hold:*

$$\lim\limits_{n \to \infty}\hat{\mu}(\hat{B}_n) = \hat{0} \Rightarrow \lim\limits_{n \to \infty}\hat{\mu}(\hat{A} \cup \hat{B}_n) = \hat{\mu}(\hat{A}),$$

$$(\ \lim\limits_{n \to \infty}\hat{\mu}(\hat{B}_n) = \hat{0} \Rightarrow \lim\limits_{n \to \infty}\hat{\mu}(\hat{A} \cap \hat{B}_n^c) = \hat{\mu}(\hat{A}) \ )$$

for arbitrary $\hat{A} \in \hat{F}$ *and* $\{\hat{B}_n\} \subset \hat{F}$.

Theorem 3.1. *Let* $(X, \hat{F}, \hat{\mu})$ *be fuzzy complex-valued fuzzy measure space. Necessary and sufficient conditions that* $\hat{\mu}$ *is null-additive is that* $\hat{\mu}(\hat{A} \cup \hat{B}) = \hat{\mu}(\hat{B})$ *for arbitrary* $\hat{A}, \hat{B} \in \hat{F}$ *with* $\hat{\mu}(\hat{A}) = \hat{0}$.

Proof. *Necessary: Suppose* $\hat{\mu}(\hat{A}) = \hat{0}$, *then*

$$\hat{0} \leq \text{Re}\,\hat{\mu}(\hat{A} - \hat{B}) \leq \text{Re}\,\hat{\mu}(\hat{A}) = \hat{0},$$

$$\hat{0} \leq \text{Im}\,\hat{\mu}(\hat{A} - \hat{B}) \leq \text{Im}\,\hat{\mu}(\hat{A}) = \hat{0},$$

such that

$$\hat{0} \le \hat{\mu}(\hat{A} - \hat{B}) \le \hat{\mu}(\hat{A}) = \hat{0},$$

$$(\hat{A} - \hat{B}) \cap \hat{B} = \phi,$$

then

$$\operatorname{Re} \hat{\mu}(\hat{A} \cup \hat{B}) = \operatorname{Re} \hat{\mu}((\hat{A} - \hat{B}) \cup \hat{B}) = \operatorname{Re} \hat{\mu}(\hat{B}),$$

$$\operatorname{Im} \hat{\mu}(\hat{A} \cup \hat{B}) = \operatorname{Im} \hat{\mu}((\hat{A} - \hat{B}) \cup \hat{B}) = \operatorname{Im} \hat{\mu}(\hat{B}),$$

so

$$\hat{\mu}(\hat{A} \cup \hat{B}) = \hat{\mu}((\hat{A} - \hat{B}) \cup \hat{B}) = \hat{\mu}(\hat{B}).$$

Sufficient: It is clear.

Theorem 3.2. *Let* $(X, \hat{F}, \hat{\mu})$ *is fuzzy complex-valued fuzzy measure space. Then the following are equivalent:*

1) $\hat{\mu}$ *is null-additive ;*

2) $\hat{\mu}(\hat{A}) = \hat{0}$ *and* $\hat{A} \subset \hat{B} \Rightarrow \hat{\mu}(\hat{B} - \hat{A}) = \hat{\mu}(\hat{B}), \forall \hat{A}, \hat{B} \in \hat{F}$;

3) $\hat{\mu}(\hat{A}) = \hat{0} \Rightarrow \hat{\mu}(\hat{B} - \hat{A}) = \hat{\mu}(\hat{B}), \forall \hat{A}, \hat{B} \in \hat{F}$.

Proof.

(1) $\Rightarrow$ (2) we know it by $\hat{\mu}(\hat{A}) = \hat{\mu}((\hat{A} \cap \hat{B}^c) \cup \hat{B})$. So this proof is very easy

(2) $\Rightarrow$ (3) since $\hat{\mu}(\hat{A} - \hat{B}) = \hat{\mu}(\hat{A} - (\hat{B} - \hat{A}))$ and $\hat{A} \cap \hat{B} \subset \hat{A}$, we have $\hat{0} \le \hat{\mu}(\hat{A} \cap \hat{B}) \le \hat{\mu}(\hat{A}) = \hat{0}$, so it is clear and we omit it

(3) $\Rightarrow$ (1) If $\hat{A} \subset \hat{B}$, then we know it by (3);

If $\hat{A} \not\subset \hat{B}$, we know it by $\hat{\mu}(\hat{A}) = \mu((\hat{A} \cup \hat{B}) - \hat{B})$.

Theorem 3.3. *Necessary and sufficient condition that* $\hat{\mu}$ *is called null-additive on* $\hat{F}$ *is that if* $\hat{\mu}(\hat{B}_n) \to \hat{0}$ *, then* $\hat{\mu}(\hat{A} - \hat{B}_n) \to \hat{\mu}(\hat{A})$ *for arbitrary reduction sequence* $\{\hat{B}_n\} \subset \hat{F}$ ·

Proof. Necessity: clearly.

Sufficiency: Since $\hat{\mu}(\hat{B}) = \hat{0}$ for arbitrary $\hat{B} \in \hat{F}$, Let $\hat{B}_n \equiv \hat{B}, n = 1, 2, \cdots$, then $\hat{B}_n \downarrow \hat{B}$ and $\hat{\mu}(\hat{B}) \equiv \hat{0}, n = 1, 2 \cdots$, so $\hat{A} - \hat{B}_n \to \hat{A} - \hat{B}$ for arbitrary $\hat{A} \in \hat{F}$, hence

$$\hat{\mu}(\hat{A} - \hat{B}) = \hat{\mu}(\hat{A} - \hat{B}_n) \to \hat{\mu}(\hat{A}),$$

therefore $\hat{\mu}(\hat{A} - \hat{B}) = \hat{\mu}(\hat{A})$,We know $\hat{\mu}$ is null-additive by theorem3.2. The proof is completed.

4 Fuzzy Complex-Valued Fuzzy Measurable Function on $\hat{F}$

Definition 4.1. Let $(X, \hat{F}, \hat{\mu})$ be a fuzzy complex-valued fuzzy measure space, the mapping $\hat{f} : \hat{F} \to F^*(C)$ is called fuzzy complex-valued fuzzy measurable function on $\hat{F}$, where A is membership functions, if the following conditions hold:

$$\left\{ x \in X \mid A(\hat{f}_\alpha(x)) \geq \alpha \right\} \in \hat{F}$$

for arbitrary $\alpha \in [0,1]$.

Definition 4.2. Let $(X, \hat{F}, \hat{\mu})$ be a fuzzy complex-valued fuzzy measure space, Let $\hat{f}_n$ $(n=1,2,\cdots)$ and $\hat{f} : C \to F^*(C)$ denote fuzzy complex-valued fuzzy measurable function, and arbitrary $\hat{A} \in \hat{F}$, then

1) $\left\{ \hat{f}_n \right\}$ is called almost everywhere converges to $\hat{f}$ on $\hat{A}$, if $\hat{\mu}(\hat{E}) = \hat{0}$ for arbitrary $\hat{E} \in \hat{F}$ and such that $\left\{ \hat{f}_n \right\}$ converges to $\hat{f}$ point by point on $\hat{A} - \hat{E}$. Note $\hat{f}_n \underline{a.e.} \hat{f}$.

2) $\left\{ \hat{f}_n \right\}$ is called almost uniform converges to $\hat{f}$ on $\hat{A}$, if $\exists \hat{E} \in \hat{F}$, for arbitrary $\varepsilon > 0$, $\left| \hat{\mu}_\alpha(\hat{E}) \right| < \varepsilon$, such that $\left\{ \hat{f}_n \right\}$ uniform converges to $\hat{f}$ point by point. Note $\hat{f}_n \underline{a.u.} \hat{f}$.

3) $\left\{ \hat{f}_n \right\}$ is called pseudo almost everywhere converges to $\hat{f}$, if $\hat{\mu}(\hat{A} - \hat{E}) = \hat{\mu}(\hat{E})$ for arbitrary $\hat{E} \in \hat{F}$ such that $\left\{ \hat{f}_n \right\}$ converges to $\hat{f}$ point by point on $\hat{A} - \hat{E}$. Note $\hat{f}_n \underline{p.a.e} \hat{f}$.

4) $\left\{ \hat{f}_n \right\}$ is called pseudo almost uniform converges to $\hat{f}$, if $\lim_{n \to \infty} \hat{\mu}(\hat{A} - \hat{E}_k) = \hat{\mu}(\hat{A})$ for arbitrary $\left\{ \hat{E}_k \right\} \subset \hat{F}$ such that $\left\{ \hat{f}_n \right\}$ uniform converges to $\hat{f}$ point by point on $\hat{A} - \hat{E}$ for arbitrary fixed point $k = 1, 2, 3 \cdots$. Note $\hat{f}_n \underline{p.a.u} \hat{f}$.

5) $\left\{ \hat{f}_n \right\}$ is called converges in measure to $\hat{f}$, if

$$\lim_{n \to \infty} \hat{\mu}\left(\left\{ x \mid \left\{ \hat{f}_n(x) - \hat{f}(x) \right\} > \varepsilon \right\} \cap \hat{A} \right) = 0$$

for arbitrary $\varepsilon > 0$. Note $\hat{f}_n \underline{\hat{\mu}} \hat{f}$.

6) $\left\{ \hat{f}_n \right\}$ is called pseudo converges in measure to $\hat{f}$, if

$$\lim_{n \to \infty} \hat{\mu}\left(\left\{ x \mid \left\{ \hat{f}_n(x) - \hat{f}(x) \right\} > \varepsilon \right\} \cap \hat{A} \right) = \hat{\mu}(\hat{A}),$$

Note $\hat{f}_n \underline{p.\hat{\mu}} \hat{f}$.

Theorem 4.1. *Fuzzy complex-valued fuzzy measurable function* $\bar{\mu}$ *is called auto-continuity from above (below) on* $\bar{F}$, *if it is null-additive.*

Proof. It is clear by the definition of auto-continuity and null-additive.

5 Conclusion

Fuzzy complex-valued fuzzy measure on fuzzy complex sets has good application background, Especially in Comprehensive Evaluation, Engineering, Artificial intelligence, Machine learning, Pattern recognition and Information fusion, the work of this paper provides a powerful guarantee for its future application.

Acknowledgements. Thanks to the support by Natural Science Foundation of Hainan Province (No.111007) and Hainan International cooperation Key Project (No.GJXM201105).

References

1. Ma, S.Q.: Theoretical basis of fuzzy complex analysis, pp. 79–169. Science Press, Beijing (2010)
2. Xiao, Y., Wang, G.J.: Fuzzy complex-valued measures and Fuzzy complex-valued integral, vol. 3, p. 201. Tianjin University
3. Ha, M.H., Wu, C.X.: Fuzzy measure and fuzzy integral theory, pp. 6–40. Science press, Beijing (1998)
4. Chen, M.Q., Ma, S.Q., Huang, Y.N.: Expansion of complex fuzzy measure. Journal of Hainan Normal University (Nature Science) 23, 358–360 (2010)
5. Qiu, J.Q., Sun, T.: Relation on convergence of measurable function sequence on fuzzy complex measure space. Fuzzy Systems and Mathematices 21, 92–96 (2007)
6. Zhang, G.Q.: Fuzzy measure theory. Guizhou Science and Technology Press (1994)
7. Wang, X.Z.: The application of fuzzy measure and fuzzy integral in classification technology. Science Press (2008)

An FLP Complementary Slackness Theorem Based on Fuzzy Relationship

Liu Xin[*]

School of Mathematics and Quantitative Economics
Dongbei University of Finance and Economics, Dalian, 116025, China
liuxin060@dufe.edu.cn

Abstract. In order to improve and spread FLP (fuzzy linear programming) duality theorem, this paper applies fuzzy relationship and fuzzy number theorem to the study of FLP duality theorem which is based on fuzzy relationship. It indicates that the important results of the fundamental concept and nature of the FLP duality problems and that classical LP problems can be spread in fuzzy linear programming based on fuzzy relationship; and it advances and proves the symmetry theorem and the complementary slackness theorem of DFLP, which provides theoretical basis of a lot of fuzzy optimization issues in the reality.

Keywords: Fuzzy relationship, fuzzy number, fuzzy linear programming, fuzzy duality.

1 Introduction

Linear Programming is applied widely in the real world. Due to the uncertainty of collecting samples, the coefficients of the LP equations are not always be exact numbers in many optimization problems. The coefficients of the LP equations could not always be exact numbers. Hence, in such cases, linear programming with fuzzy coefficients is capable of explaining the problem in more detail and is more applicable [1].

Let's review the concepts and characteristics of optimum solutions of Fuzzy FLP.

Let $M = \{1, 2, ..., m\}$, $N = \{1, 2, ..., n\}$, m, n are positive integers, consider the following LP scenario:

$$\begin{cases} \max Z(\min Z) = \sum_{j=1}^{n} c_j x_j \\ \sum_{j=1}^{n} a_{ij} x_j \leq b_i, i \in M, \\ x_j \geq 0, j \notin N. \end{cases} \qquad (1)$$

---

[*] Liu Xin (1961−),female(Fuxin, Liaoning), professor, mainly work on the study of fuzzy logic and economy optimization.

B.-Y. Cao and X.-J. Xie (Eds.): Fuzzy Engineering and Operations Research, AISC 147, pp. 213–228.
springerlink.com   © Springer-Verlag Berlin Heidelberg 2012

This is a classical LP problem, coefficients $a_{ij}, b_i, c_j, i \in M, j \in N$ of Equation (1) are all constants, if we change $a_{ij}, b_i, c_j, i \in M, j \in N$ into fuzzy numbers $\tilde{a}_{ij}, \tilde{b}_i, \tilde{c}_j$, then the original LP becomes the following FLP problem:

$$\begin{cases} \widetilde{\max} Z(\widetilde{\min} Z) = \sum_{j=1}^{n} \tilde{c}_j x_j \\ \sum_{j=1}^{n} \tilde{a}_{ij} x_j \le \tilde{b}_i, i \in M, \\ x_j \ge 0, j \notin N. \end{cases} \tag{2}$$

Let the membership function of fuzzy numbers $\tilde{a}_{ij}, \tilde{b}_i, \tilde{c}_j$ be:

$$\mu_{\tilde{a}_{ij}} : R \to [0,1],$$
$$\mu_{\tilde{b}_i} : R \to [0,1],$$
$$\mu_{\tilde{c}_j} : R \to [0,1], i \in M, j \in N.$$

There exist the following extension theorems:

Theorem 1[2]: *Let* $\tilde{c}_j, \tilde{a}_{ij} \in F_0(R), x_j \ge 0, i \in M, j \in N$. *Then*

$$\tilde{c}_1 x_1 \oplus ... \oplus \tilde{c}_n x_n, \tilde{a}_{i1} x_1 \oplus ... \oplus \tilde{a}_{in} \tilde{x}_n$$

are also fuzzy numbers.

Let $\tilde{P}$ be a fuzzy relationship on R, which is also the fuzzy extension of duality relationship $\le$ on R. Then FLP Problem (2) can be expressed as:

$$\begin{cases} \widetilde{\max} Z(\widetilde{\min} Z) = \tilde{c}_1 x_1 \oplus ... \oplus \tilde{c}_n x_n \\ \tilde{a}_{i1} x_1 \oplus ... \oplus \tilde{a}_{in} x_n \tilde{P} \tilde{b}_i \quad i \in M, \\ x_j \ge 0 \quad j \in N. \end{cases} \tag{3}$$

Definition 1[3]: *Let* $\mu_{\tilde{a}_{ij}} : R \to [0,1], \mu_{\tilde{b}_i} : R \to [0,1], i \in M, j \in N$ *be the membership function of fuzzy number* $\tilde{a}_{ij}, \tilde{b}_i$. *Let* $\tilde{P}$ *be normal fuzzy extension of relationship* $P(\le)$ *on* R. *For fuzzy set* $\tilde{X}$, *define its membership function* $\mu_{\tilde{X}}$ *as* $\forall x \in R^n$.

$$\mu_{\tilde{X}}(x) = \begin{cases} if \quad x_j \geq 0, \forall j \in N \\ \min\{\mu_{\tilde{P}}(\sum_{j=1}^{n}\tilde{a}_{1j}x_j\tilde{b}_1)...\mu_{\tilde{P}}(\sum_{j=1}^{n}\tilde{a}_{mj}x_j\tilde{b}_m), \\ otherwise, \quad 0. \end{cases}$$

Then $\tilde{X}$ is the fuzzy set of feasible solution, or simplified as FLP feasible region, of FLP (3).

For $\beta \in [0,1]$, vector $x \in [\tilde{X}]_\beta$ is FLP's β − feasible solution.

Note: In FLP, $\tilde{X}$ is a fuzzy set. Also, all β − feasible solutions are vectors, and they are members of fuzzy set $\tilde{X}$'s β − cut set. Obviously, if all coefficients $\tilde{a}_{ij}, \tilde{b}_i$ are classical fuzzy numbers, which are also called "clear numbers", then the fuzzy feasible region is equivalent to set of all feasible solutions in classical LP.

Definition 2: *Let $\tilde{P}$ be a fuzzy relationship on R, $\alpha \in [0,1], \tilde{a}, \tilde{b}$ are fuzzy numbers. If $\mu_{\tilde{P}}(\tilde{a},\tilde{b}) \leq \alpha$, which is denoted as $\tilde{a}\tilde{P}_\alpha\tilde{b}$, $\tilde{P}_\alpha$ is α − relationship of $\tilde{P}$ on R; if $\tilde{a}\tilde{P}_\alpha\tilde{b}$, and $\mu_{\tilde{P}}(\tilde{a},\tilde{b}) < \alpha$, which is denoted as $\tilde{a}\tilde{P}_\alpha^*\tilde{b}$, $\tilde{P}_\alpha^*$ is strong α − relationship of $\tilde{P}$ on R.*

Note: $\tilde{P}_\alpha$ and $\tilde{P}_\alpha^*$ are duality relationships on fuzzy relationship set $F_0(R)$ established by fuzzy set $\tilde{P}$ on $\alpha \in [0,1]$. If $\tilde{a}, \tilde{b}$ are classical fuzzy numbers corresponding real numbers a, b, and $\tilde{P}$ is fuzzy extension of relationship $\leq$, then

$$a\tilde{P}b \Leftrightarrow a \leq b.$$

Hence, $\forall \alpha \in (0,1), a\tilde{P}_\alpha^*b \Leftrightarrow a < b$.

Apply the results above on the special fuzzy relationship

$$\tilde{P} \in \{\precsim^{Pos}, \prec^{Nes}, \lesssim^M, \lesssim_M\},$$

we have the following conclusion:

Proposition 1:[4] *Let $\tilde{a}$ and $\tilde{b}$ be fuzzy numbers, $\alpha \in (0,1]$.*

(1) *Let* $\tilde{P} \in \{ \preceq^{Pos}, \tilde{\lesssim}^M \}$,   *then* $\tilde{a}\tilde{P}_\alpha\tilde{b} \Leftrightarrow \tilde{a}^L(\alpha) \leq \tilde{b}^R(\alpha)$,

$\tilde{a}\tilde{P}_\alpha^*\tilde{b} \Leftrightarrow \tilde{a}^R(\alpha) < \tilde{b}^L(\alpha)$.

(2) *Let $\tilde{P} \in \{ \prec^{Nes}, \tilde{\lesssim}_M \}$, then*

$\tilde{a}\tilde{P}_\alpha\tilde{b} \Leftrightarrow \tilde{a}^R(1-\alpha) \leq \tilde{b}^L(1-\alpha)$, $\begin{aligned} &\tilde{a}\tilde{P}_\alpha^*\tilde{b} \Leftrightarrow \tilde{a}^R(1-\alpha) \leq \tilde{b}^L(1-\alpha), \\ &\text{and } \tilde{a}^L(1-\alpha) < \tilde{b}^R(1-\alpha). \end{aligned}$

Definition 3. *Let $a_{ij}, b_i, c_j, i \in M, j \in N$ be the fuzzy numbers on R, $\tilde{P}$ be the fuzzy relationship on R, $\alpha \in (0,1]$, $X = (x_1,...,x_n)^T$ be $\alpha-$feasible solution of FLP (3), $\tilde{c}X = \tilde{c}_1x_1 \oplus ... \oplus c_nx_n$. If there does not exist any $X^{'} \in [\tilde{X}]_\alpha$ satisfying $\tilde{c}X\tilde{P}_\alpha\tilde{c}X^{'}$, vector $X \in R^n$ is called $(\alpha, \alpha)-$ maximum solution of maximum problem of the object function in FLP (3). If there does not exist any $X^{'} \in [\tilde{X}]_\alpha$ satisfying $\tilde{c}X^{'}\tilde{P}_\alpha\tilde{c}X$, vector $X \in R^n$ is called $(\alpha, \alpha)-$minimum solution of minimum problem of the object function in FLP (3).*

Definition 4. *Let $a_{ij}, b_i, c_j, i \in M, j \in N$ be fuzzy numbers on R, $\tilde{P}$ be a fuzzy relationship on R, which is also fuzzy extension of a normal duality relationship on R. Let $\alpha, \beta \in (0,1]$. If there does not exist $X^{'} \in [\tilde{X}]_\beta$, $X \neq X^{'}$ satisfying $\tilde{c}X\tilde{P}_\alpha^*\tilde{c}X^{'}$, then $\beta-$feasible solution $X \in [\tilde{X}]_\beta$ of (3) is called $(\alpha, \beta)-$maximum solution.*

Note: All $(\alpha, \beta)-$ maximum solution are $\beta-$ feasible solutions, with some special characteristics, of the object function in FLP (3). Obviously, when all coefficients of FLP (3) are classical fuzzy numbers, $(\alpha, \beta)-$ feasible solutions in FLP model are the same as classical solutions of LP (1).

2   Concepts and Characteristics of the Feasible Solutions in Duality Fuzzy Linear Programming (DFLP)

Consider the following FLP problem:

$$(P) \begin{cases} \mathrm{m\tilde{a}x}\, Z = \tilde{c}_1 x_1 \oplus ... \oplus \tilde{c}_n x_n \\ \tilde{a}_{i1} x_1 \oplus ... \oplus \tilde{a}_{in} x_n \tilde{P} \tilde{b}_i \quad i \in M , \\ x_j \geq 0 \quad j \in N, \end{cases} \quad (4)$$

where $\tilde{c}_j, \tilde{a}_{ij}, \tilde{b}_i$ are standard fuzzy numbers, and there membership functions are

$$\mu_{\tilde{c}_j} : R \to [0,1], \mu_{\tilde{a}_{ij}} : R \to [0,1], \mu_{\tilde{b}_i} : R \to [0,1], \quad i \in M, j \in N.$$

FLP (4) is called original (P). FLP duality problem (D) is defined as:

$$(D) \begin{cases} \mathrm{m\tilde{i}n}\, \tilde{W} = \tilde{b}_1 \tilde{y}_1 \oplus ... \oplus \tilde{b}_m y_m \\ \tilde{c}_j \tilde{Q} (\tilde{a}_{1j} y_1 \oplus ... \oplus \tilde{a}_{mj} y_m), j \in N, \\ y_i \geq 0, i \in M, \end{cases} \quad (5)$$

where $\tilde{P}$ and $\tilde{Q}$ are fuzzy duality relationship, especially,

$$\tilde{P} = \tilde{\leq}^M \text{ and } \tilde{Q} = \tilde{\leq}_M \text{ or } \tilde{P} = \prec^{Pos} \text{ and } \tilde{Q} = \prec^{Nes} \text{ or } \tilde{P} = \prec^{Nes} \text{ and } \tilde{Q} \prec^{Pos} .$$

Definition 5[5]: *Let* $\mu_{\tilde{a}_{ij}} : R \to [0,1], \mu_{\tilde{c}_j} : R \to [0,1], i \in M, j \in N$ *be the membership functions of fuzzy numbers* $\tilde{a}_{ij}, \tilde{c}_j$, $\tilde{Q}$ *be the fuzzy extension of duality relationship* P *on* R, $\tilde{Y}$ *be fuzzy set and its membership function is defined as* $\forall y \in R^n$ *there exists:*

$$\mu_{\tilde{Y}}(y) = \begin{cases} \text{if } y_i \geq 0, \forall i \in M \\ \min \{ \mu_{\tilde{Q}}(\tilde{c}_1, \tilde{a}_{11} y_1 \oplus ... \oplus \tilde{a}_{m1} y_m)... \\ \mu_{\tilde{Q}}(c_n, \tilde{a}_{1n} y_1 \oplus .. \tilde{a}_{mn} y_m), \\ \quad Otherwise \quad 0, \end{cases} \quad (6)$$

$\tilde{Y}$ is called fuzzy feasible solution set of (D), or feasible solution set of duality FLP (D). $\forall \beta \in (0,1)$, vector $y \in [\tilde{Y}]_\beta$ is called $\beta -$ feasible solution of FLP (D).

Definition 6. *Let* $\tilde{a}_{ij}, \tilde{c}_j, b_i, i \in M, j \in N$ *be fuzzy numbers on* R, $\tilde{Q}$ *be fuzzy extension of normal duality relationship* $\leq$ *on* R. *Let* $\alpha, \beta \in (0,1]$. *If there does not exist* $y' \in [\tilde{Y}]_\beta$, $y' \neq y$ *satisfying* $\tilde{b} y' \tilde{Q}_\alpha^* \tilde{b} y$, *then* $\beta -$ *feasible solution*

$y \in [\tilde{Y}]_\beta$ of (5) is called $(\alpha, \beta) - minimum$ solution, where $\tilde{Q}_\alpha^*$ is $\alpha - strong$ relationship of $\tilde{Q}$ on R.

Let $\tilde{X}$ be the feasible solution set of FLP (P), $\tilde{Y}$ be the feasible solution set of FLP duality (D). Obviously, $\tilde{X}$ is a fuzzy subset of R^n; $\tilde{Y}$ is a fuzzy subset of R^n. Note that in classical scenario, $\tilde{a}_{ij}, \tilde{c}_j, b_i, i \in M, j \in N$ are classical numbers, relationship $\preceq^{Pos}, \prec^{Nes} (\tilde{\leqq}^M, \tilde{\leqq}_M)$ corresponds to $\leq$. Hence, classical scenario (P) and (D) becomes LP problem's original problems and its duality problem.

Proposition 2: [6] Let $\tilde{a}_{ij}, \tilde{c}_j, i \in M, j \in N, \alpha \in (0,1), \preceq^{Pos}, \prec^{Nes} (\tilde{\leqq}^M, \tilde{\leqq}_M)$ are fuzzy relationships defined on R. Thus, $\forall j \in N$, there exists:

1)
$$\mu_{\preceq^{Pos}} (\tilde{c}_j, \tilde{a}_{1j} y_1 \oplus ... \oplus \tilde{a}_{mj} y_m) \geq \alpha$$
$$\Leftrightarrow \sum_{i \in M} \tilde{a}_{ij}^R (\alpha) y_i \geq \tilde{c}_j^L (\alpha) \qquad (7)$$

2)
$$\mu_{\prec^{Nes}} (\tilde{c}_j, \tilde{a}_{1j} y_1 \oplus ... \oplus \tilde{a}_{mj} y_m) \geq \alpha$$
$$\Leftrightarrow \sum_{i \in M} \tilde{a}_{ij}^L (1-\alpha) y_i \geq \tilde{c}_j^R (1-\alpha) \qquad (8)$$

3)
$$\mu_{\tilde{\leqq}^M} (\tilde{c}_j, \tilde{a}_{1j} y_1 \oplus ... \oplus \tilde{a}_{mj} y_m) \geq \alpha$$
$$\Leftrightarrow \sum_{i \in M} \tilde{a}_{ij}^R (\alpha) y_i \geq \tilde{c}_j^L (\alpha)$$

4)
$$\mu_{\tilde{\leqq}_M} (\tilde{c}_j, \tilde{a}_{1j} y_1 \oplus ... \oplus \tilde{a}_{mj} y_m) \geq \alpha$$
$$\Leftrightarrow \sum_{i \in M} \tilde{a}_{ij}^L (1-\alpha) y_i \geq \tilde{c}_j^R (1-\alpha)$$

Corollary 1: 1) Let $\tilde{P} = \preceq^{Pos} (\tilde{P} = \tilde{\leqq}^M)$, vector $y = (y_1, ..., y_m)$ is the $\beta - feasible$ solution of FLP (5) $\Leftrightarrow y = (y_1, ..., y_m)$ is the non-negative solution of inequality $\sum_{i \in M} \tilde{a}_{ij}^R (\beta) y_i \geq \tilde{c}_j^L (\beta), j \in N$.

2) Let $\tilde{P} = \prec^{Nes} (\tilde{P} = \tilde{\leqq}_M)$, vector $y = (y_1, ..., y_m)$ be $\beta - feasible$ solution $\Leftrightarrow y = (y_1, ..., y_m)$ is the non-negative solution of inequality

$$\sum_{i \in M} \tilde{a}_{ij}^L (1-\beta) y_i \geq \tilde{c}_j^R (1-\beta), j \in N.$$

3   Duality Theorem of Fuzzy Linear Programming

Theorem 2 (Symmetric theorem): *In FLP (P), the dual (D)'s dual is the original (P).*

Proof. Let the original problem be (P)

$$\begin{cases} \tilde{\max} Z = \tilde{c}_1 x_1 \oplus \dots \oplus \tilde{c}_n x_n \\ \tilde{a}_{i1} x_1 \oplus \dots \oplus \tilde{a}_{in} x_n \tilde{P} \tilde{b}_i, \quad i \in M, \\ x_j \geq 0, \quad j \in N. \end{cases}$$

Its duality problem be (D)

$$\begin{cases} \tilde{\min} W = \tilde{b}_1 \tilde{y}_1 \oplus \dots \oplus \tilde{b}_m y_m \\ \tilde{c}_j \tilde{Q}(\tilde{a}_{1j} y_1 \oplus \dots \oplus \tilde{a}_{mj} y_m), j \in N, . \\ y_i \geq 0, i \in M. \end{cases}$$

Multiply (–1) on both sides, also, since $\tilde{\min}(W) = -\tilde{\max}(-W)$, thus

$$\begin{cases} \tilde{\max}(-W) = -(\tilde{b}_1 y_1 \oplus \dots \oplus \tilde{b}_m y_m) \\ -(\tilde{a}_{1j} y_1 \oplus \dots \oplus \tilde{a}_{mj} y_m) \tilde{Q}(-\tilde{c}_j), j \in N, \\ y_i \geq 0, i \in M. \end{cases}$$

According to symmetric transformation relationship, the duality problem of the equations above is:

$$\begin{cases} \tilde{\min}(-W^{'}) = -(\tilde{c}_1 x_1 \oplus \dots \oplus \tilde{c}_n x_n) \\ (-\tilde{b}_i) \tilde{P}[-(\tilde{a}_{i1} x_1 \oplus \dots \oplus \tilde{a}_{in} x_n)], \quad i \in M, \\ x_j \geq 0, \quad j \in N. \end{cases}$$

Also, since $\tilde{\min}(-W^{'}) = -\tilde{\max}(W^{'})$, let $Z = W^{'}$, we have the following:

$$\begin{cases} -\tilde{\max}(W^{'}) = -\tilde{\max} Z = -(\tilde{c}_1 x_1 \oplus \dots \oplus \tilde{c}_n x_n) \\ (\tilde{a}_{i1} x_1 \oplus \dots \tilde{a}_{in} x_n) \tilde{P} \tilde{b}_i, i \in M, \\ x_j \geq 0, j \in N. \end{cases}$$

Which can be also expressed as:

$$\begin{cases} \tilde{\max} Z = (\tilde{c}_1 x_1 \oplus ... \oplus \tilde{c}_n x_n) \\ (\tilde{a}_{i1} x_1 \oplus ... \tilde{a}_{in} x_n) \tilde{P} \tilde{b}_i, i \in M, \\ \qquad x_j \geq 0, j \in N. \end{cases}$$

Which is the same as the original problem.

Theorem 3[7]: Let $\tilde{a}_{ij}, \tilde{c}_j, \tilde{b}_i, i \in M, j \in N$ be fuzzy numbers, $\alpha \in (0,1)$, $\tilde{X}$ be the feasible solution set with fuzzy relation $\tilde{P} = \tilde{\leq}^M$ of FLP (4), be the feasible solution set with fuzzy relation $\tilde{Q} = \tilde{\leq}_M$ of FLP (5). If vector $X = (x_1, ..., x_n)^T \geq 0, Y = (y_1, ..., y_m) \geq 0, X \in [\tilde{X}]_\alpha, Y \in [\tilde{Y}]_{1-\alpha}$, then

$$\sum_{j \in N} \tilde{c}_j^R(\alpha) x_j \leq \sum_{i \in M} \tilde{b}_i^R(\alpha) y_i \qquad (9)$$

Corollary 2: Let $\tilde{a}_{ij}, \tilde{c}_j, \tilde{b}_i, i \in M, j \in N$ be fuzzy numbers, $\alpha \in (0,1)$, $\tilde{X}$ be the feasible solution set with fuzzy relation $\tilde{P} = \preceq^{Pos}$ of FLP (4), $\tilde{Y}$ be the feasible solution set with fuzzy relation $\tilde{Q} = \prec^{Nes}$ of FLP (5). If vector $X = (x_1, ..., x_n)^T \geq 0, Y = (y_1, ..., y_m) \geq 0, X \in [\tilde{X}]_\alpha, Y \in [\tilde{Y}]_{1-\alpha}$, then

$$\sum_{j \in N} \tilde{c}_j^R(\alpha) x_j \leq \sum_{i \in M} \tilde{b}_i^R(\alpha) y_i . \qquad (10)$$

Corollary 3: Let $\tilde{a}_{ij}, \tilde{c}_j, \tilde{b}_i, i \in M, j \in N$ be fuzzy numbers, $\alpha \in (0,1)$, $\tilde{X}$ be the feasible solution set with fuzzy relation $\tilde{P} = \prec Nes$ of FLP (4), $\tilde{Y}$ be the feasible solution set with fuzzy relation $\tilde{Q} = \preceq^{Pos}$ of FLP (5). If vector $X = (x_1, ..., x_n)^T \geq 0, Y = (y_1, ..., y_m) \geq 0, X \in [\tilde{X}]_{1-\alpha}, Y \in [\tilde{Y}]_\alpha$, then

$$\sum_{j \in N} \tilde{c}_j^L(\alpha) x_j \leq \sum_{i \in M} \tilde{b}_i^L(\alpha) y_i . \qquad (11)$$

Theorem 4[8]: (1st Weak duality theorem): Let $\tilde{a}_{ij}, \tilde{c}_j, \tilde{b}_i, i \in M, j \in N$ be fuzzy numbers, $\alpha \in (0,1)$, $\tilde{X}$ be the feasible solution set with fuzzy relation $\tilde{P}$ of FLP (4), $\tilde{Y}$ be the feasible solution set with fuzzy relation $\tilde{Q}$ of FLP (5). If vector

$$X = (x_1,...,x_n)^T \geq 0, Y = (y_1,...,y_m) \geq 0, X \in [\tilde{X}]_\alpha,$$

$$Y \in [\tilde{Y}]_\beta, \alpha + \beta = 1 \; \tilde{P}$$

And $\tilde{Q}$ are a set of dual relationship, denote as $(\tilde{P}, \tilde{Q})$

1) If $(\tilde{P}, \tilde{Q}) \in \{(\tilde{\leq}^M, \tilde{\leq}_M), (\preceq^{Pos}, \prec^{Nes})\}$, then

$$\sum_{j \in N} \tilde{c}_j^{\,R}(\alpha) x_j \leq \sum_{i \in M} \tilde{b}_i^{\,R}(\alpha) y_i. \tag{12}$$

2) If $(\tilde{P}, \tilde{Q}) \in \{(\tilde{\leq}_M, \tilde{\leq}^M), (\prec^{Nes}, \preceq^{Pos})\}$, then

$$\sum_{j \in N} \tilde{c}_j^{\,L}(\beta) x_j \leq \sum_{i \in M} \tilde{b}_i^{\,L}(\beta) y_i. \tag{13}$$

Theorem 5: 1) Let $\tilde{a}_{ij}, \tilde{c}_j, \tilde{b}_i, i \in M, j \in N$ be fuzzy numbers, $\alpha \in (0,1)$, $\tilde{X}$ be the feasible solution set with fuzzy relation $\tilde{P} = \tilde{\leq}^M$ of FLP (4), $\tilde{Y}$ be the feasible solution set with fuzzy relation $\tilde{Q} = \tilde{\leq}_M$ of FLP (5). If vector $X = (x_1,...,x_n)^T \geq 0, Y = (y_1,...,y_m) \geq 0, X \in [\tilde{X}]_\alpha, Y \in [\tilde{Y}]_{1-\alpha}$, satisfying

$$\sum_{j \in N} \tilde{c}_j^{\,R}(\alpha) x_j = \sum_{i \in M} \tilde{b}_i^{\,R}(\alpha) y_i. \tag{14}$$

Then X is the $(\alpha, \alpha) -$ maximum solution of FLP (4), Y is the $(1-\alpha, 1-\alpha) -$ minimum solution of FLP (5).

2) Let $\tilde{a}_{ij}, \tilde{c}_j, \tilde{b}_i, i \in M, j \in N$ be fuzzy numbers, $\alpha \in (0,1)$, $\tilde{X}$ be the feasible solution set with fuzzy relation $\tilde{P} = \tilde{\leq}_M$ of FLP (4), $\tilde{Y}$ be the feasible solution set with fuzzy relation $\tilde{Q} = \tilde{\leq}^M$ of FLP (5). If vector $X = (x_1,...,x_n)^T \geq 0, Y = (y_1,...,y_m) \geq 0, X \in [\tilde{X}]_{1-\alpha}, Y \in [\tilde{Y}]_\alpha$, satisfying

$$\sum_{j \in N} \tilde{c}_j^{\,L}(1-\alpha) x_j = \sum_{i \in M} \tilde{b}_i^{\,L}(1-\alpha) y_i. \tag{15}$$

Then X is the $(1-\alpha, 1-\alpha) -$ maximum solution of FLP (4). Y is the $(\alpha, \alpha) -$ minimum solution of FLP (5).

Corollary 4: *Let* $\tilde{a}_{ij}, \tilde{c}_j, \tilde{b}_i, i \in M$, $j \in N$ *be fuzzy numbers.* $\alpha \in (0,1)$, $\tilde{X}$ *be the feasible solution set with fuzzy relation* $\tilde{P} = \preceq^{Pos}$ *of FLP (4),* $\tilde{Y}$ *be the feasible solution set with fuzzy relation* $\tilde{Q} = \prec^{Nes}$ *of FLP (5). If vector* $X = (x_1, ..., x_n)^T \geq 0, Y = (y_1, ..., y_m) \geq 0, X \in [\tilde{X}]_\alpha, Y \in [\tilde{Y}]_{1-\alpha}$, *satisfying*

$$\sum_{j \in N} \tilde{c}_j^{R}(\alpha) x_j = \sum_{i \in M} \tilde{b}_i^{R}(\alpha) y_i. \tag{16}$$

Then X *is the* $(\alpha, \alpha)-$ *maximum solution of FLP (4),* Y *is the* $(1-\alpha, 1-\alpha)-minimum$ *solution of FLP (5).*

Corollary 5: *Let* $\tilde{a}_{ij}, \tilde{c}_j, \tilde{b}_i, i \in M$, $j \in N$ *be fuzzy numbers,* $\alpha \in (0,1)$, $\tilde{X}$ *be the feasible solution set with fuzzy relation* $\tilde{P} = \prec^{Nes}$ *of FLP (4),* $\tilde{Y}$ *be the feasible solution set with fuzzy relation* $\tilde{Q} = \preceq^{Pos}$ *of FLP (5). If vector* $X = (x_1, ..., x_n)^T \geq 0, Y = (y_1, ..., y_m) \geq 0, X \in [\tilde{X}]_{1-\alpha}, Y \in [\tilde{Y}]_\alpha$, *satisfying*

$$\sum_{j \in N} \tilde{c}_j^{L}(1-\alpha) x_j = \sum_{i \in M} \tilde{b}_i^{L}(1-\alpha) y_i. \tag{17}$$

Then X *is the* $(1-\alpha, 1-\alpha)-$ *maximum solution of FLP (4),* Y *is the* $(\alpha, \alpha)-minimum$ *solution of FLP (5).*

Theorem 6[8]: (2nd weak duality theorem): *Let* $\tilde{a}_{ij}, \tilde{c}_j, \tilde{b}_i, i \in M$, $j \in N$ *be fuzzy numbers,* $\alpha \in (0,1)$, $\tilde{X}$ *be the feasible solution set with fuzzy relation* $\tilde{P}$ *of FLP (4),* $\tilde{Y}$ *be the feasible solution set with fuzzy relation* $\tilde{Q}$ *of FLP (5). If some vector*

$$X = (x_1, ..., x_n)^T \geq 0, Y = (y_1, ..., y_m) \geq 0, X \in [\tilde{X}]_\alpha,$$

$$Y \in [\tilde{Y}]_\beta, \alpha + \beta = 1,$$

$\tilde{P}$ *and* $\tilde{Q}$ *are a pair of fuzzy dual,*

$$(\tilde{P}, \tilde{Q}) \in \{(\preceq^{Pos}, \prec^{Nes}), (\tilde{\leqq}^M, \tilde{\leqq}_M),$$
$$(\prec^{Nes}, \preceq^{Pos}), (\tilde{\leqq}_M, \tilde{\leqq}^M)\}$$

When $(\tilde{P}, \tilde{Q}) \in \{(\preceq^{Pos}, \prec^{Nes}), (\lesssim^M, \lesssim_M)\}$, $\sum_{j \in N} \tilde{c}_j^R(\alpha) x_j = \sum_{i \in M} \tilde{b}_i^R(\alpha) y_i$

is satisfied, or
When

$(\tilde{P}, \tilde{Q}) \in \{(\prec^{Nes}, \preceq^{Pos}), (\lesssim_M, \lesssim^M)\}$, $\sum_{j \in N} \tilde{c}_j^L(\beta) x_j = \sum_{i \in M} \tilde{b}_i^L(\beta) y_i$

is satisfied,

Then X is the (α, α) − maximum solution of FLP (4). Y is the (β, β) − minimum solution of FLP (5).

Now, consider the original problem of LP (P) and its duality problem (D) when there exists the fuzzy relationships

$$\tilde{P} = \preceq^{Pos}, \tilde{Q} = \prec^{Nes} \qquad (\tilde{P} = \lesssim^M, \tilde{Q} = \lesssim_M), \alpha = \beta,$$

$$(\text{P}_1) \begin{cases} \max Z = \sum_{j \in N} \tilde{c}_j^R(\alpha) x_j \\ \sum_{j \in N} \tilde{a}_{ij}^L(\alpha) x_j \leq \tilde{b}_i^R(\alpha), i \in M, \\ x_j \geq 0, j \in N, \end{cases} \qquad (18)$$

$$(\text{D}_1) \begin{cases} \min W = \sum_{i \in M} \tilde{b}_i^R(\alpha) y_i \\ \sum_{j \in N} \tilde{a}_{ij}^L(\alpha) y_i \geq \tilde{c}_j^R(\alpha), j \in N, \\ y_i \geq 0, i \in M. \end{cases} \qquad (19)$$

Also, consider the duality problem of LP when there exists the fuzzy relationships

$$\tilde{P} = \prec^{Nes}, \tilde{Q} = \preceq^{Pos} \quad (\tilde{P} = \lesssim_M, \tilde{Q} = \lesssim^M),$$

$$(\text{P}_2) \begin{cases} \max Z = \sum_{j \in N} \tilde{c}_j^L(\alpha) x_j \\ \sum_{j \in N} \tilde{a}_{ij}^R(\alpha) x_j \leq \tilde{b}_i^L(\alpha), i \in M, \\ x_j \geq 0, j \in N \end{cases} \qquad (20)$$

$$(D_2) \begin{cases} \min W = \sum_{i \in M} \tilde{b}_i^L(\alpha) y_i \\ \sum_{j \in N} \tilde{a}_{ij}^R(\alpha) y_i \geq \tilde{c}_j^L(\alpha), j \in N, \\ y_i \geq 0, i \in M. \end{cases} \qquad (21)$$

Note that (P_1) and (D_1), (P_2) and (D_2) are all classical duality LP problem.

Theorem 7[9]: Let $\tilde{a}_{ij}, \tilde{c}_j, \tilde{b}_i, i \in M, j \in N$ be fuzzy numbers.

1) If $\tilde{X}$ be the feasible solution set with fuzzy relation $\tilde{P} = \preceq^{Pos}$ of FLP (4), $\tilde{Y}$ be the feasible solution set with fuzzy relation $\tilde{Q} = \prec^{Nes}$ of FLP (5), if, for some $\alpha \in (0,1)$, $[\tilde{X}]_\alpha$ and $[\tilde{Y}]_{1-\alpha}$ are not empty, then there exists X^* as the $(\alpha, \alpha)-maximum$ solution of FLP (4), and Y^* as the $(1-\alpha, 1-\alpha)-minimum$ solution of FLP (5). And they satisfy the following relationship:

$$\sum_{j \in N} \tilde{c}_j^R(\alpha) x_j^* = \sum_{i \in M} \tilde{b}_i^R(\alpha) y_i^* . \qquad (22)$$

2) If $\tilde{X}$ be the feasible solution set with fuzzy relation $\tilde{P} = \prec^{Nes}$ of FLP (4), $\tilde{Y}$ be the feasible solution set with fuzzy relation $\tilde{Q} = \preceq^{Pos}$ of FLP (5), if, for some $\alpha \in (0,1)$, $[\tilde{X}]_{1-\alpha}$ and $[\tilde{Y}]_\alpha$ are not empty, then there exists X^* as the $(1-\alpha, 1-\alpha)-maximum$ solution of FLP (4), and Y^* as the $(\alpha, \alpha)-minimum$ solution of FLP (5). And they satisfy the following relationship:

$$\sum_{j \in N} \tilde{c}_j^L(1-\alpha) x_j^* = \sum_{i \in M} \tilde{b}_i^L(1-\alpha) y_i^* . \qquad (23)$$

Theorem 8: Let $\tilde{a}_{ij}, \tilde{c}_j, \tilde{b}_i, i \in M, j \in N$ be fuzzy number. $\alpha \in (0,1)$, $\tilde{X}$ be the feasible solution set with fuzzy relation $\tilde{P}$ of FLP (4), $\tilde{Y}$ be the feasible solution set with fuzzy relation $\tilde{Q}$ of FLP (5), if, for some $\alpha, \beta \in (0,1)$, $[\tilde{X}]_\alpha$ and $[\tilde{Y}]_\beta$ are not empty, $\alpha + \beta = 1$, $\tilde{P}$ and $\tilde{Q}$ are a pair of fuzzy dual relationship.

1) When $(\tilde{P}, \tilde{Q}) \in \{(\preceq^{Pos}, \prec^{Nes}), (\lesssim^M, \lesssim_M)\}$, then there exists X^* as the $(\alpha, \alpha)-maximum$ solution of FLP (P) (4), and Y^* as the $(\beta, \beta)-minimum$ solution of FLP (D). And they satisfy the following relationship:

$$\sum_{j\in N}\tilde{c}_j^{\,R}(\alpha)x_j^{\,*}=\sum_{i\in M}\tilde{b}_i^{\,R}(\alpha)y_i^{\,*}\ .\qquad(24)$$

2) When $(\tilde{P},\tilde{Q})\in\{(\prec^{Nes},\preceq^{Pos}),(\tilde{\leqq}_M,\tilde{\leqq}^M)\}$, then there exists X^* as the $(\alpha,\alpha)-maximum$ solution of FLP (P) (4), and Y^* as the $(\beta,\beta)-minimum$ solution of FLP (D). And they satisfy the following relationship:

$$\sum_{j\in N}\tilde{c}_j^{\,L}(\beta)x_j^{\,*}=\sum_{i\in M}\tilde{b}_i^{\,L}(\beta)y_i^{\,*}\ .\qquad(25)$$

Theorem 9 (FLP complementary slackness theorem).

Let $\tilde{a}_{ij},\tilde{c}_j,\tilde{b}_i, i\in M$, $j\in N$ be fuzzy numbers, $\alpha\in(0,1)$, $\tilde{X}$ be the feasible solution set with fuzzy relation $\tilde{P}$ of FLP (4), $\tilde{Y}$ be the feasible solution set with fuzzy relation $\tilde{Q}$ of FLP (5), if, for some $\alpha,\beta\in(0,1)$, $[\tilde{X}]_\alpha$ and $[\tilde{Y}]_\beta$ are not empty, $\alpha+\beta=1$, $\tilde{P}$ and $\tilde{Q}$ are a pair of fuzzy dual relationship.

1) When $(\tilde{P},\tilde{Q})\in\{(\preceq^{Pos},\prec^{Nes}),(\tilde{\leqq}^M,\tilde{\leqq}_M)\}$, then the necessary and sufficient condition of " X^* is the $(\alpha,\alpha)-maximum$ solution of FLP (P) (4), and Y^* is the $(\beta,\beta)-minimum$ solution of FLP (D)" is:

$$Y^*X_T=0\ 和\ Y_S X^*=0,$$

Where $X_T=\{x_{n+1},...x_{n+t}\}^{\mathrm{T}},Y_S=\{y_{m+1},...,y_{m+s}\}$ are slackness variables.

2) When $(\tilde{P},\tilde{Q})\in\{(\prec^{Nes},\preceq^{Pos}),(\tilde{\leqq}_M,\tilde{\leqq}^M)\}$, then the necessary and sufficient condition of " X^* is the $(\alpha,\alpha)-maximum$ solution of FLP (P) (4), and Y^* is the $(\beta,\beta)-minimum$ solution of FLP (D)" is:

$$X^*Y_S=0\ 和\ X_T Y^*=0,$$

where $X_T=\{x_{n+1},...x_{n+t}\}^{\mathrm{T}},Y_S=\{y_{m+1},...,y_{m+s}\}$ are slackness variables.

Proof: 1) When $(\tilde{P},\tilde{Q})\in\{(\preceq^{Pos},\prec^{Nes}),(\tilde{\leqq}^M,\tilde{\leqq}_M)\}$, obviously, $[\tilde{X}]_\alpha$ is the set containing all feasible solutions of equation (18) in LP (P_1), and $[\tilde{Y}]_\beta$ is the set containing all feasible solutions of equation (19) in LP (D_1). If $[\tilde{X}]_\alpha$, $[\tilde{Y}]_\beta$ are not empty, since (P_1) and (D_1) are commonly defined LP duality problem, the original problem and its duality problem are:

Original Problem:

$$
\begin{cases}
\max Z = \displaystyle\sum_{j \in N} \tilde{c}_j^R(\alpha) x_j \\
\displaystyle\sum_{j \in N} \tilde{a}_{ij}^L(\alpha) x_j + \sum_{j \in N'} x_j = \tilde{b}_i^R(\alpha), i \in M, N' = \{n+1,\dots,n+t\}, \\
\qquad\qquad x_j \geq 0, j \in N \cup N'.
\end{cases}
$$

Duality Problem:

$$
\begin{cases}
\min W = \displaystyle\sum_{j \in N} \tilde{b}_j^R(\alpha) y_i \\
\displaystyle\sum_{j \in N} \tilde{a}_{ij}^L(\alpha) y_i - \sum_{i \in M} y_i = \tilde{c}_j^R(\alpha), j \in N, M' = \{m+1,\dots,m+s\}, \\
\qquad\qquad y_i \geq 0, i \in M \cup M'.
\end{cases}
$$

Thus,

$$
\begin{aligned}
Z = \sum_{j \in N} \tilde{c}_j^R(\alpha) x_j &= \sum_{j \in N} \Big(\sum_{i \in M} \tilde{a}_{ij}^L(\alpha) y_i - \sum_{i \in M'} y_i \Big) x_j \\
&= \sum_{j \in N} \sum_{i \in M} \tilde{a}_{ij}^L(\alpha) y_i x_j - \sum_{j \in N} \sum_{i \in M'} y_i x_j \quad , \qquad (26)
\end{aligned}
$$

$$
\begin{aligned}
W = \sum_{i \in M} \tilde{b}_i^R(\alpha) y_i &= \sum_{i \in M} \Big(\sum_{j \in N} \tilde{a}_{ij}^L(\alpha) x_j + \sum_{j \in N'} x_j \Big) y_i \\
&= \sum_{i \in M} \sum_{j \in N} \tilde{a}_{ij}^L(\alpha) x_j y_i + \sum_{i \in M} \sum_{j \in N'} x_j y_i \quad . \qquad (27)
\end{aligned}
$$

Necessity: If X, Y are optimum solutions, based on Theorem 8 (1):

$$
\sum_{j \in N} \tilde{c}_j^R(\alpha) x_j^* = \sum_{i \in M} \tilde{b}_i^R(\alpha) y_i^* .
$$

Furthermore, from (26), (27)

$$
\sum_{j \in N} \sum_{i \in M} \tilde{a}_{ij}^L(\alpha) y_i^* x_j^* - \sum_{j \in N} \sum_{i \in M'} y_i x_j^* = \sum_{i \in M} \sum_{j \in N} \tilde{a}_{ij}^L(\alpha) x_j^* y_i^* + \sum_{i \in M} \sum_{j \in N'} x_j y_i^* .
$$

Therefore,

$$-\sum_{j\in N'}\sum_{i\in M'}y_i x_j^* = \sum_{i\in M'}\sum_{j\in N'}x_j y_i^*.$$

Since $\forall i, j, j \in N \bigcup N', i \in M \bigcup M'$, and $x_j \geq 0, y_i \geq 0$,

$$\sum_{j\in N'}\sum_{i\in M'}y_i x_j^* = 0 \text{ 和} \sum_{i\in M'}\sum_{j\in N'}x_j y_i^* = 0,$$

which is $X^* Y_s = 0$ 和 $X_T Y^* = 0$.

Sufficiency: Let $X^* Y_s = 0$ and $X_T Y^* = 0$. So

$$\sum_{j\in N'}\sum_{i\in M'}y_i x_j^* = 0 \text{ 和} \sum_{i\in M'}\sum_{j\in N'}x_j y_i^* = 0.$$

Thus, from (24) and (25)

$$\sum_{j\in N}\tilde{c}_j^R(\alpha)x_j^* = \sum_{j\in N}\sum_{i\in M}\tilde{a}_{ij}^L(\alpha)y_i^* x_j^* = \sum_{i\in M}\tilde{b}_i^R(\alpha)y_i^*.$$

From Theorem 6, X^* is the (α, α) − maximum solution of FLP (P) (4), and Y^* is the (β, β) − minimum solution of FLP (D).

Part (2) could be proved in a similar way.

Acknowledgements. Thanks to the support by research projects of institutions of higher education of The Education Department of Liaoning Province (2009S034), The key task item in the field of innovative approach particularly supported by basic science and technology work of National Science and Technology Department (2009IM010400-1-39).

References

1. Cao, B.Y., M.: Applied fuzzy mathematics and system, pp. 138–169. Science Press, Beijing (2005)
2. Chen, S.L., L.J.G., Wang, X.G.: Fuzzy set theory and its applications. Science Press, Beijing (2004)
3. Li, A.G., Zhang, Z.H., et al.: Fuzzy mathematics and its applications. Metal-lurgical Industry Press, Beijing (1994)
4. Inuiguchi, M., Ramik, J., Tanino, T., et al.: Satisfying solutions and duality in interval and fuzzy linear programming. Fuzzy Sets and Systems 135, 151–177 (2003)
5. Bactor, C.R., Chandra, S.: On duality in linear programming under fuzzy environment. Fuzzy Sets and Systems 125, 317–325 (2002)

6. Wu, H.-C.: Duality theorems in fuzzy mathematical programming problems based on the concept of necessity. Fuzzy Sets and Systems 139, 363–377 (2003)
7. Ramik, J.: Duality in fuzzy linear programming: some new concepts and results. Fuzzy Optimization and Decision-Making 4, 25–39(2005)
8. Xiong, H. B.: A study of FLP and its algorithms, Nanchang, China, Jiangxi Normal University(2006)
9. Wu, H.-C.: Fuzzy optimization problem based on the embeddingtheorem and possibility measures. Mathematical and Computer Modeling 40, 329–336 (2004)

Choquet Type Fuzzy Complex-Valued Integral and Its Application in Classification

Sheng-quan Ma[1], Fu-chuan Chen[2,*], and Zhi-qing Zhao[3]

[1] School of Information Science and Technology, Hainan Normal University, 571158,
Hainan, Haikou, P.P. China
mashengquan@163.com
[2] QiongTai Teachers College, Haikou, Hainan, 571100, P.R. China
fuchuan9688@163.com
[3] Department of Mathematics, Hainan Normal University, Haikou,
Hainan, 571158, PR. China

Abstract. Fuzzy complex-value integral has a wide range of applications in Compre- hensive Evaluation, Engineering, Artificial intelligence, Machine learning, Pattern recognition, Information fusion and so on. In this paper, we first introduce the concept of Choquet type fuzzy complex-value integral, then discuss its basic properties and provide a design method of fuzzy complex-value integral classifier. Finally, this classification algorithm is proved to be very effective by some examples.

Keywords: Fuzzy complex-value measure, fuzzy complex-value integral, choquet type fuzzy complex-value integral, classifier.

1   Introduction

Fuzzy measure is the extension of classical measure. In 1974, sugeno (see [1]) first introduced the non-additive measure (fuzzy measures) by replacing the additivity with weak monotonicity, and. established the theory of real fuzzy integral based on it. This theory has been developed rapidly, especially in its application to Comprehensive Evaluation, Engineering, Artificial intelligence, Machine learning, Pattern recognition, Information fusion and other fields. In 1989, Buckley first introduced the concept of fuzzy complex numbers (see [2]), then scholars began to discuss fuzzy complex numbers and fuzzy complex sets. In 1997, Based on the idea of Buckely, Chou Jiqing et al. (see [3]) introduced the concepts of complex fuzzy measure and complex fuzzy integral, and studied some basic properties of them. In [4, 5], we further discussed the properties of complex fuzzy measure and complex fuzzy integral. In this paper, we will introduce the specific definition of complex fuzzy measure and complex fuzzy integral, which inherited the original basic properties and can make problem solving easier. Then we will introduce the

---

* Corresponding author.

B.-Y. Cao and X.-J. Xie (Eds.): Fuzzy Engineering and Operations Research, AISC 147, pp. 229–237.
springerlink.com   © Springer-Verlag Berlin Heidelberg 2012

concepts of Choquet type fuzzy complex value integral, and discuss some basic properties about them. In order to facilitate the practical application, the discretization form and its calculation method about complex fuzzy integral will be introduced. Finally, using fuzzy complex value integral as an example to the classifier fusion design, we will give a kind of classifier—fuzzy complex value integral classifier. The result of application shows that the integral has very good application prospects.

2　Preliminary

2.1　Real Fuzzy Measure

Definition 2.1.1[6] *We call non-negative generalized real-value set function* $\mu : \mathbb{F} \to [0,\infty]$ *the real-value fuzzy measure if the following conditions hold*

1) *If* $\phi \in \mathbb{F}$, *then* $\mu(\phi) = 0$;

2) *If* $E \in \mathbb{F}, F \in \mathbb{F}, E \subset F$, *then* $\mu(E) \le \mu(F)$;

3) *If* $E_n \in \mathbb{F}(n = 1, 2, \cdots, \infty), E_1 \subset E_2 \subset \cdots, \bigcup_{n=1}^{\infty} E_n \in \mathbb{F}$, *then* $\lim_{n \to \infty} \mu(E_n) = \mu\left(\bigcup_{n=1}^{\infty} E_n\right)$;

4) *If* $E_n \in \mathbb{F}(n = 1, 2, \cdots, \infty), E_1 \supset E_2 \supset \cdots$, *and existing* n_0 *makes* $\mu(E_{n_0}) < \infty, \bigcap_{n=1}^{\infty} E_n \in \mathbb{F}$,

then $\lim_{n \to \infty} \mu(E_n) = \mu\left(\bigcap_{n=1}^{\infty} E_n\right)$.

We call μ *is the regular real-value fuzzy measure, if* $\mu(X) = 1$.

2.2　g_λ *Fuzzy Measures*

Definition 2.2.1[6]. μ *satisfies* λ *law: there are* $\lambda \in \left(\dfrac{1}{\sup \mu}, \infty\right)$ *and* $\sup \mu = \sup_{E \in \mathbb{F}} \mu(E)$, *let* $E \in \mathbb{F}, F \in \mathbb{F}, E \bigcup F \in \mathbb{F}, E \bigcap F = \phi$. *We have*

$$\mu(E \bigcup F) = \mu(E) + \mu(F) + \lambda \mu(E) \mu(F),$$

μ *satisfies finite* λ *law: there are* $\lambda \in \left(-\dfrac{1}{\sup \mu}, \infty\right)$ *and any finite disjoint sequence* $\{E_1, E_2, \cdots, E_n\}$ *on* $\mathbb{F}$, *we have*

$$\mu\left(\bigcup_{i=1}^{n} E_i\right) = \begin{cases} \dfrac{1}{\lambda}\left\{\displaystyle\prod_{i=1}^{n}[1 + \lambda\mu(E_i)] - 1\right\}, & \lambda \neq 0 \\ \displaystyle\sum_{i=1}^{n}\mu(E_i), & \lambda = 0 \end{cases},$$

μ *satisfies* $\sigma-\lambda$ *law:* *there* *are* $\lambda \in \left(\dfrac{1}{\sup\mu}, \infty \right)$ *and* *any* *disjoint* *sequence* $\{E_1, E_2, \cdots, E_n\}$ *on* $\mathbb{F}$, *we have*

$$\mu\left(\bigcup_{i=1}^{\infty} E_i\right) = \begin{cases} \dfrac{1}{\lambda}\left\{\prod_{i=1}^{\infty}\left[1 + \lambda\mu(E_i)\right] - 1\right\}, & \lambda \neq 0 \\ \sum_{i=1}^{\infty} \mu(E_i), & \lambda = 0 \end{cases}.$$

Definition 2.2.2[6]. *We call* μ *is* g_λ *fuzzy measure on* $\mathbb{F}$, *if and only if it satisfies* $\sigma - \lambda$ *law, and we have at least a set* $E \in \mathbb{F}$ *satisfies* $\mu(E) < \infty$.

Definition 2.2.3[6]. *Let* μ *is real-value fuzzy measure and* f *is non-negative* *real-value measureable function on* X. *The definition of choquet real-value fuzzy* *integral is as following:*

$$(c)\int f d\mu = \int_0^\infty \mu(\{x \mid f(x) > \alpha\}) dl,$$

where the right of the equation is the integral of α *function about lebesgue* *measure.*

Let $X = \{x_1, x_2, \cdots, x_n\}$ is finite sets and f is discrete values function, $\{a_1, a_2, \cdots, a_n\}$ is finite function value sets, suppose $a_1 \leq a_2 \leq \cdots \leq a_n$. Then

$$(c)\int f d\mu = \int_0^\infty \mu(\{x \mid f(x) > \alpha\}) dl,$$

$$\int_0^{a_1} \mu(\{x \mid f(x) \geq a_1\}) dl + \int_{a_1}^{a_2} \mu(\{x \mid f(x) \geq a_2\}) dl + \cdots + \int_{a_{n-1}}^{a_n} \mu(\{x \mid f(x) \geq a_n\}) dl,$$

(where each of integrating range is that the left is closed and the right is opened) So *choquet* type real-value fuzzy integral is as following:

$$(c)\int f(x) d\mu = \sum_{i=1}^{n} (a_i - a_{i-1}) \mu(\{x \mid f(x) \geq a_i\}) = \sum_{i=1}^{n} (a_i - a_{i-1}) \mu(A_i),$$

where $a_0 = 0$, $A_i = \{x_i, x_{i+1}, \cdots, x_n\}$ or $(c)\int f(x) d\mu = \sum_{i=1}^{n} a_i [\mu(A_i) - \mu(A_{i+1})]$ where $A_{n+1} = \phi$.

3   Fuzzy Complex Value Integral

3.1   *Fuzzy Complex Value Measures*

We agreed that $\hat{R}^+ = [0, +\infty), \hat{C}^+ = \{x + iy \mid x, y \in \hat{R}^+\}$ and the order of complex numbers is defined by the order relation of real part and imaginary part simultaneously, the non-negative complex number is the real part and the imaginary part of complex are non-negative.

Definition3.1.1[3]. *Let X is non-empty sets,* $\mathcal{A}$ *is* $\sigma-algebra$ *that is composed by the subset of B. We call* $\mu:\mathcal{A}\to\hat{C}^+$ *is fuzzy complex value measure, if the following conditions hold*

(1) $\mu(\phi)=0$;

(2) *If* $A\subset B$ *, then* $|\mu(A)|\le|\mu(B)|$,

(3) *If* $\{A_n\}_1^\infty\uparrow$ *, just like* $A_n\in A(n=1,2,\cdots)$ $A_1\subseteq A_2\subseteq\cdots\subseteq A_n\subseteq\cdots$,

   Then $\mu(\bigcup_{n=1}^\infty A_n)=\lim_{n\to\infty}\mu(A_n)$;

(4) *If* $\{A_n\}_1^\infty\downarrow$ *, just like* $_{A_n\in A}$ $(n=1,2,\cdots)$ $A_1\supseteq A_2\supseteq\cdots\supseteq A_n\supseteq\cdots$ *and*

   $\exists$ n_0 *, we have* $|\mu(A_{n_0})|<+\infty$ *,then* $\mu(\bigcap_{n=1}^\infty A_n)=\lim_{n\to\infty}\mu(A_n)$;

where $A,B,A_n\in\mathcal{A}(n=1,2\cdots)$ *. We call* $(X,\mathcal{A},\mu)$ *is a fuzzy complex value measure spaces.*

In practical application, we usually use the following more specific Definition 3.1.2.

Definition 3.1.2. *Given measurable space* (X,A) *, we call set function* $_{\mu:}A\to\hat{C}^+$ *is fuzzy complex value measure, if the following conditions hold:*

1) $\mu(\Phi)=0$,

2) $A,B\in A$, $A\subseteq B\Rightarrow$ $\text{Re}(\mu(A))\le\text{Re}(\mu(B))$ *and* $\text{Im}(\mu(A))\le\text{Im}(\mu(B))$ *,note that* $\mu(A)\le\mu(B)$,

3) $_{A_n\in A}$ $(n=1,2,\cdots)$ $A_1\subseteq A_2\subseteq\cdots\subseteq A_n\subseteq\cdots\Rightarrow\mu(\bigcup_{n=1}^\infty A_n)=\lim_{n\to\infty}\mu(A_n)$,

4) $_{A_n\in A}$ $(n=1,2,\cdots)$ $A_1\supseteq A_2\supseteq\cdots\supseteq A_n\supseteq\cdots$ *,and existing* n_0 *,we have*

$\text{Re}(\mu(A_{n_0}))<\infty,\text{Im}(\mu(A_{n_0}))<\infty\Rightarrow\mu(\bigcap_{n=1}^\infty A_n)=\lim_{n\to\infty}\mu(A_n)\cdot$

3.2 Fuzzy Complex-Value Measurable Function and Fuzzy Complex Value Integral

Definition 3.2.1[3]. *Given fuzzy complex-value measure space* (X,A,μ) *,we call* $f:X\to C$ *is fuzzy complex-value measurable function, if the following conditions hold:*

$$\{x\in X|\text{Re}[f(x)]\ge a,\text{Im}[f(x)]\ge b\}\in A \text{ for any } a+bi\in C.$$

Definition 3.2.2. *Let f is fuzzy complex-value measurable function that from fuzzy complex-value measurable space (X, A, μ) to $\hat{C}^+$, $A \in A$, then we call*

$$\int_A \overline{f} d\mu \overset{\Delta}{=} \sup_{\alpha \in [0,\infty)} \min\{\alpha, \operatorname{Re}\mu[\overline{f}_\alpha \cap A]\} + i \sup_{\alpha \in [0,\infty)} \min\{\alpha, \operatorname{Im}\mu[\overline{f}_\alpha \cap A]\}$$

Or $\int_A \overline{f} d\mu \overset{\Delta}{=} \underset{\alpha \in [0,\infty)}{\vee} (\alpha \wedge \operatorname{Re}\mu[\overline{f}_\alpha \cap A]) + i \underset{\alpha \in [0,\infty)}{\vee} (\alpha \wedge \operatorname{Im}\mu[\overline{f}_\alpha \cap A])$,

the fuzzy complex-value integral of f about μ on A,

where $\overline{f}_\alpha = \{x | \operatorname{Re}[\overline{f}(x)] > \alpha, I \operatorname{m}[\overline{f}(x)] > \alpha\}$.

Note that $|\overline{f}|_\alpha = \{x \,\big|\, |\operatorname{Re}[\overline{f}(x)]| > \alpha, |I \operatorname{m}[\overline{f}(x)]| > \alpha\}$.

Theorem 3.2.1. *Fuzzy complex-value integral has the following properties:*

1) *If $\overset{\sqcup}{f_1} \leq \overset{\sqcup}{f_2}$, then $\int_A f_1 d\mu \leq \int_A f_2 d\mu$,*

2) *If $\mu(A) = 0$, then $\int_A f d\mu = 0$,*

3) *$\int (f_1 \vee f_2) d\mu \geq \int f_1 d\mu \vee \int f_2 d\mu$,*

4) *If $A \subseteq B$, then $\int_A f d\mu \leq \int_B f d\mu$,*

5) *If "a" is nonnegative constants, we have $\int_A (a \vee f) d\mu = \int_A a d\mu \vee \int_A f d\mu$.*

   (It can be proved in a similar way as show literature [4]).

Definition 3.2.3. *Let μ is fuzzy complex-value measure and $\overline{f}$ is Non-negative complex-value measurable function on X, the Choque type fuzzy complex-value integral$(C) \int \overline{f} d\mu$ of $\overline{f}$ about μ is*

$$(C) \int_A \overline{f} d\mu \overset{\Delta}{=} \int_0^\infty \operatorname{Re}\mu(\{x | \operatorname{Re}\overline{f}(x) > \operatorname{Re}\alpha\}) dl + i \int_0^\infty \operatorname{Im}\mu(\{x | \operatorname{Im}\overline{f}(x) < \operatorname{Im}\alpha\}) dl,$$

where the two integral of the equation on the right is the integral of α-function about lebesgue measure.

Let $X = \{x_1, x_2, \cdots, x_n\}$ is finite sets and f is discrete values function. The function value $\{a_1, a_2, \cdots, a_n\}$ is finite sets, suppose $a_1 \leq a_2 \leq \cdots \leq a_n$, and then Choquet type fuzzy complex value integral is simplified as

$$(C) \int \overline{f} d\mu = \sum_{j=1}^n (\operatorname{Re}a_j - \operatorname{Re}a_{j-1})\operatorname{Re}\mu(\{x | \operatorname{Re}\overline{f}(x) \geq \operatorname{Re}a_j\}) +$$

$$i \sum_{j=1}^n (\operatorname{Im}a_j - \operatorname{Im}a_{j-1})\operatorname{Im}\mu(\{x | \operatorname{Im}\overline{f}(x) \leq \operatorname{Im}a_j\})$$

$$= \sum_{j=1}^n (\operatorname{Re}a_j - \operatorname{Re}a_{j-1})\operatorname{Re}\mu(A_j) + i \sum_{j=1}^n (\operatorname{Im}a_j - \operatorname{Im}a_{j-1})\operatorname{Im}\mu(A'_j),$$

where $a_0 = 0$, $A_j = \{x_j, x_{j+1}, ..., x_n\}$, $A'_j = \{x_1, x_2, ..., x_j\}$.

Theorem 3.2.2. *The Choquet type fuzzy complex value integral has the following properties:*

1) *If* $\overline{f} \leq \overline{g}$, *then* $(C)\int_A \overline{f} d\mu \leq (C)\int_A \overline{g} d\mu$,

2) $(C)\int X_A d\mu = \mu(A)$,

3) *If a is non-negative real numbers, and b is real-numbers, then we have*
$$(C)\int_A (a\overline{f} + b) d\mu = a(C)\int_A \overline{f} d\mu + b\mu(X),$$

4) *If a is real numbers, then*
$$(C)\int_A \overline{f} d(a\mu) = a(C)\int_A \overline{f} d\mu,$$

5) *If N is null sets, and* $\overline{f}(x) = \overline{g}(x)$ *for any* $x \notin N$, *then*
$$(C)\int_A \overline{f} d\mu = (C)\int_A \overline{g} d\mu.$$

4   An Application of Fuzzy Complex Value Integral: Fuzzy Complex Value Integral Classifier Design

There is a c kinds problem, let $X = \{x_1, x_2, \cdots, x_n\}$ is attribute sets, $f = (f(x_1), f(x_2), \cdots, f(x_n))$ is a function on X,

$$f(x_j) = a_j = \text{Re}a_j + i\text{Im}a_j \in K \ (j = 1, 2, \cdots, n) \text{ (K is complex sets)},$$

we have μ_k is fuzzy complex value measure of class K, which is, based on class K, fuzzy measure that is definited in attribute sets and it can be expressed as 2^n dimension column vector, note that:

$$\mu_k = \left[\mu_k(\{\varphi\}), \mu_k(\{x_1\}), \mu_k(\{x_2\}), \cdots, \mu_k(\{x_1, x_2, \cdots, x_n\}) \right]^T \ (1 \leq k \leq c).$$

Let $b_k \triangleq \int_X f d\mu_k$. Then b_k is the Level of trust that belongs to class k (Support degree), $f(x_j)$ is j th Attribute value of Some examples, we have

$$f(x_j) \triangleq a_j = \text{Re}a_j + i\text{Im}a_j,$$

where $\text{Re}a_j$ is the important level of some example for j_{th} attribute, and $\text{Im}a_j$ is the unimportant level of some example for j_{th} attribute, where $\text{Re}a_j, \text{Im}a_j \in [0,1]$ and $\text{Re}a_j + \text{Im}a_j = 1 \ (j = 1, 2, \cdots, n)$, we have fuzzy complex value measure

$$\mu_k \triangleq \text{Re}\mu_k + i\text{Im}\mu_k (i = \sqrt{-1}), (1 \leq k \leq c) ,$$

where $\text{Re}\mu_k$ is the level of trust of class k in subset of attribute set, and $\text{Im}\mu_k$ is the level of distrust of class k in subset of attribute set,

$$b_k = \int_X f d\mu_k \triangleq \text{Re}b_k + i\text{Im}b_k \triangleq (\text{Re}b_k, \text{Im}b_k), (i = \sqrt{-1}), (1 \leq k \leq c),$$

where $\mathrm{Re}b_k$ is the level of trust that some example belongs to class k, and $\mathrm{Im}b_k$ is the level of distrust that some example belongs to class k.

Classification process:

Step 1: Standardize b_k , The method is as follows:

$$b_k = \left(\mathrm{Re}b_k, \mathrm{Im}b_k\right) \underline{\quad\text{standardize}\quad} \bar{b}_k = \left(\frac{\mathrm{Re}b_k}{\mathrm{Re}b_k + \mathrm{Im}b_k}, \frac{\mathrm{Im}b_k}{\mathrm{Re}b_k + \mathrm{Im}b_k}\right) \triangleq \left(\mathrm{Re}\bar{b}_k, \mathrm{Im}\bar{b}_k\right) \cdot$$

Step 2: Classification principles: Get the $\bar{b}_k = \left(\mathrm{Re}\bar{b}_k, \mathrm{Im}\bar{b}_k\right)$, the principle of classification is "the biggest credibility " or " the smallest distrust ".

For example: Suppose we have 4 examples, which needed to be divided 2 kinds, the data is standardized as following table:

Table 4.1. Standardization table

Case	Attribute x_1	Attribute x_2	Attribute x_3	class
1	0.3+0.7i	0.26+0.74i	0.44+0.56i	?
2	0.31+0.69i	0.45+0.55i	0.24+0.76i	?
3	0.23+0.77i	0.45+0.55i	0.32+0.68i	?
4	0.25+0.75i	0.46+0.54i	0.29+0.71i	?

Fuzzy complex-value measure that is definited on $X = \{x_1, x_2, x_3\}$ is as following table:

Table 4.2. Measure value table

Sets	Fuzzy measure η_1	Fuzzy measure η_2
φ	0	0
$\{x_1\}$	0.1+0.9 i	0.2+0.8i
$\{x_2\}$	0.5+0.5i	0.3+0.7i
$\{x_1, x_2\}$	0.7+0.3i	0.8+0.2i
$\{x_3\}$	0.2+0.8i	0.6+0.4i
$\{x_1, x_3\}$	0.3+0.7i	0.9+0.1i
$\{x_2, x_3\}$	0.9+0.1i	0.7+0.3i
$\{x_1, x_2, x_3\}$	1	1

Calculate:

For case 1

$$f = (f(x_1), f(x_2), f(x_3)) = (0.3 + 0.7i, 0.26 + 0.74i, 0.44 + 0.56i),$$

$$b_1 = (c)\int_X f d\mu_1 = \left[(0.44 - 0.3)\operatorname{Re}\mu_1(\{x_3\}) + (0.3 - 0.26)\operatorname{Re}\mu_1(\{x_1, x_3\}) + (0.26 - 0)\operatorname{Re}\mu_1(\{x_1, x_2, x_3\}) \right]$$

$$+ i\left[(0.74 - 0.7)\operatorname{Im}\mu_1(\{x_1, x_2, x_3\}) + (0.7 - 0.56)\operatorname{Im}\mu_1(\{x_1, x_3\}) + (0.56 - 0)\operatorname{Im}\mu_1(\{x_3\}) \right]$$

$$= [0.14 \times 0.2 + 0.04 \times 0.3 + 0.26 \times 1] + i[0.04 \times 0 + 0.14 \times 0.7 + 0.56 \times 0.8] \triangleq [0.552, 0.546],$$

$$b_2 = (c)\int_X f d\mu_2 = \left[(0.44 - 0.3)\operatorname{Re}\mu_2(\{x_3\}) + (0.3 - 0.26)\operatorname{Re}\mu_2(\{x_1, x_3\}) + (0.26 - 0)\operatorname{Re}\mu_2(\{x_1, x_2, x_3\}) \right]$$

$$+ i\left[(0.74 - 0.7)\operatorname{Im}\mu_2(\{x_1, x_2, x_3\}) + (0.7 - 0.56)\operatorname{Im}\mu_2(\{x_1, x_3\}) + (0.56 - 0)\operatorname{Im}\mu_2(\{x_3\}) \right]$$

$$= [0.14 \times 0.6 + 0.04 \times 0.8 + 0.26 \times 1] + i[0.04 \times 0 + 0.14 \times 0.1 + 0.56 \times 0.4] \triangleq [0.376, 0.238].$$

Normalized

$$\bar{b}_1 = (0.503, 0.497), \bar{b}_2 = (0.612, 0.388).$$

For case 2

$$f = (0.31 + i0.69, 0.45 + 0.55i, 0.24 + 0.76i),$$

$$b_1 = (c)\int_X f d\mu_1 = \left[(0.45 - 0.31)\operatorname{Re}\mu_1(\{x_2\}) + (0.31 - 0.24)\operatorname{Re}\mu_1(\{x_1, x_2\}) + (0.24 - 0)\operatorname{Re}\mu_1(\{x_1, x_2, x_3\}) \right]$$

$$+ i\left[(0.76 - 0.69)\operatorname{Im}\mu_1(\{x_1, x_2, x_3\}) + (0.69 - 0.55)\operatorname{Im}\mu_1(\{x_1, x_2\}) + (0.55 - 0)\operatorname{Im}\mu_1(\{x_2\}) \right]$$

$$= [0.14 \times 0.5 + 0.09 \times 0.7 + 0.24 \times 1] + i[0.07 \times 0 + 0.14 \times 0.3 + 0.55 \times 0.5] \triangleq [0.373, 0.317],$$

$$b_2 = (c)\int_X f d\mu_2 = \left[(0.45 - 0.31)\operatorname{Re}\mu_2(\{x_2\}) + (0.31 - 0.24)\operatorname{Re}\mu_2(\{x_1, x_2\}) + (0.24 - 0)\operatorname{Re}\mu_2(\{x_1, x_2, x_3\}) \right]$$

$$+ i\left[(0.76 - 0.69)\operatorname{Im}\mu_2(\{x_1, x_2, x_3\}) + (0.69 - 0.56)\operatorname{Im}\mu_2(\{x_1, x_3\}) + (0.55 - 0)\operatorname{Im}\mu_2(\{x_2\}) \right]$$

$$= [0.14 \times 0.3 + 0.09 \times 0.8 + 0.24 \times 1] + i[0.07 \times 0 + 0.14 \times 0.2 + 0.55 \times 0.7] \triangleq [0.354, 0.413].$$

Normalized

$$\bar{b}_1 = (0.541, 0.459), \bar{b}_2 = (0.462, 0.538).$$

For case 3

$$f = (0.23 + i0.77, 0.45 + 0.55i, 0.32 + 0.68i),$$

$$b_1 = (c)\int_X f d\mu_1 = \left[(0.45 - 0.32)\operatorname{Re}\mu_1(\{x_2\}) + (0.32 - 0.23)\operatorname{Re}\mu_1(\{x_1, x_3\}) + (0.23 - 0)\operatorname{Re}\mu_1(\{x_1, x_2, x_3\}) \right]$$

$$+ i\left[(0.77 - 0.68)\operatorname{Im}\mu_1(\{x_1, x_2, x_3\}) + (0.68 - 0.55)\operatorname{Im}\mu_1(\{x_1, x_3\}) + (0.55 - 0)\operatorname{Im}\mu_1(\{x_2\}) \right]$$

$$= [0.13 \times 0.1 + 0.09 \times 0.9 + 0.23 \times 1] + i[0.09 \times 0 + 0.13 \times 0.7 + 0.55 \times 0.5] \triangleq [0.324, 0.366],$$

$$b_2 = (c)\int_X f d\mu_2 = \left[(0.45 - 0.32)\operatorname{Re}\mu_2(\{x_2\}) + (0.32 - 0.23)\operatorname{Re}\mu_2(\{x_2, x_3\}) + (0.23 - 0)\operatorname{Re}\mu_2(\{x_1, x_2, x_3\}) \right]$$

$$+ i\left[(0.77 - 0.68)\operatorname{Im}\mu_2(\{x_1, x_2, x_3\}) + (0.68 - 0.55)\operatorname{Im}\mu_2(\{x_1, x_3\}) + (0.55 - 0)\operatorname{Im}\mu_2(\{x_2\}) \right]$$

$$= [0.13 \times 0.3 + 0.09 \times 0.7 + 0.23 \times 1] + i[0.09 \times 0 + 0.13 \times 0.1 + 0.55 \times 0.7] \triangleq [0.332, 0.395].$$

Normalized

$$\overline{b_1} = (0.470, 0.530), \overline{b_2} = (0.457, 0.543).$$

For case 4

$$f = (0.25 + i0.75, 0.46 + 0.54i, 0.29 + 0.71i),$$

$$b_1 = (c)\int_X f d\mu_1 = \left[(0.46 - 0.29)\operatorname{Re}\mu_1(\{x_2\}) + (0.29 - 0.25)\operatorname{Re}\mu_1(\{x_2, x_3\}) + (0.25 - 0)\operatorname{Re}\mu_1(\{x_1, x_2, x_3\})\right]$$

$$+ i\left[(0.75 - 0.71)\operatorname{Im}\mu_1(\{x_1, x_2, x_3\}) + (0.71 - 0.54)\operatorname{Im}\mu_1(\{x_2, x_3\}) + (0.54 - 0)\operatorname{Im}\mu_1(\{x_2\})\right]$$

$$= [0.17 \times 0.5 + 0.04 \times 0.9 + 0.25 \times 1] + i[0.04 \times 0 + 0.17 \times 0.1 + 0.54 \times 0.5] \triangleq [0.371, 0.287],$$

$$b_2 = (c)\int_X f d\mu_2 = \left[(0.4 - 0.29)\operatorname{Re}\mu_2(\{x_2\}) + (0.29 - 0.25)\operatorname{Re}\mu_2(\{x_2, x_3\}) + (0.25 - 0)\operatorname{Re}\mu_2(\{x_1, x_2, x_3\})\right]$$

$$+ i\left[(0.75 - 0.71)\operatorname{Im}\mu_2(\{x_1, x_2, x_3\}) + (0.71 - 0.54)\operatorname{Im}\mu_2(\{x_2, x_3\}) + (0.54 - 0)\operatorname{Im}\mu_2(\{x_2\})\right]$$

$$= [0.17 \times 0.3 + 0.04 \times 0.7 + 0.25 \times 1] + i[0.04 \times 0 + 0.17 \times 0.3 + 0.54 \times 0.7] \triangleq [0.329, 0.429],$$

Normalized

$$\overline{b_1} = (0.564, 0.436), \overline{b_2} = (0.434, 0.566).$$

We clearly know case 1 belonging to the class ω_2 and the others are belonging to the class ω_1 by the maximum of the Real part of b_k (or the minimum of the imaginary part of b_k) principle.

5   Conclusion

Fuzzy complex value integral has good application background, especially in Comprehensive Evaluation, Engineering, Artificial intelligence, Machine learning, Pattern recognition and Information fusion. Fuzzy complex value integral classifier is a new classifier fusion technique in this paper, we will continue studying its application in multiple classifier fusion problems.

Acknowledgements. Thanks to the support by Natural Science Foundation of Hainan Province (No.111007) and Hainan International cooperation Key Project (No.GJXM201105).

References

1. Sugeno, M.: Theory of fuzzy integrals and applications. Tokyo Institute of Technology (1974)
2. Buckley, J.J.: Fuzzy complex numbers. Fuzzy Sets and Systems 33, 333–345 (1989)
3. Qiu, J.Q., Li, F.C., Shu, L.Q.: Complex fuzzy measure and complex fuzzy integral. Hebei Light Chemical College Journals 1, 1–4 (1997)
4. Ma, S.Q., Chao, C.: Fuzzy complex analysis. Minorities Press, Beijing (2002)
5. Ma, S.Q.: Fuzzy complex analysis foundation theory. Science Press, Beijing (2010)
6. Wang, X.Z.: The application of fuzzy measure and fuzzy integral in classification technology. Science Press (2008)

Generalized Set-Valued Mixed Variational Inequalities for Fuzzy Mappings

Xiao-min Wang[1] and Li-min Duan[2]

[1] Northeastern University at Qinhuangdao, Qinhuangdao 066004, China
   xmwang0823@163.com
[2] College of Science, Northeastern University, Shenyang 110004, China
   Duanlimin_5566@163.com

Abstract. In this paper, a new class of generalized set-valued mixed variational inequalities for fuzzy mappings is introduced. We prove the existence of solutions of the variational inequality.We also prove the convergence of a new iterative algorithm approximating the solution for this variational inequality.

Keywords: Variational inequality, fuzzy mapping, algorithm.

1 Introduction

In recent years, the variational inequalities problem for fuzzy mappings was studied by many mathematicians, see [1-6] and the references therein. And many useful results in theoretical and applied fields have been given. Motivated and inspired by these research works, we introduce the generalized set-valued mixed variational inequalities for fuzzy mappings. We also construct the interative algorithm and discuss the convergence of interative sequences generated by the algorithm. The results presented in this paper generalize some recent results in this field.

2 Preliminaries

Let H be a real Hilbert space whose inner product and norm are denoted by $\langle \cdot, \cdot \rangle$ and $\|\cdot\|$ respectively. A fuzzy set in H is a function with domain H and values in $[0,1]$. Let $F(H)$ be a collection of all fuzzy sets over H. If F is a fuzzy mapping on H, then $F(x)$ (denote it by F_x, in sequel) is a fuzzy set on H and $F_x(y)$ is the menbership function of y in F_x.

Let $M \in F(H)$, $q \in [0,1]$. Then the set $M_q = \{x \in H : M(x) \ge q\}$ is called a q-cut set of M.

B.-Y. Cao and X.-J. Xie (Eds.): Fuzzy Engineering and Operations Research, AISC 147, pp. 239–245.
springerlink.com   © Springer-Verlag Berlin Heidelberg 2012

Let $T, A, B: H \rightarrow F(H)$ be three fuzzy mappings satisfy the following condition(1):

There exist three mappings $a, b, c: H \rightarrow [0,1]$ such that for all $x \in H$ the sets $(T_x)_{a(x)}, (A_x)_{b(x)}$ and $(B_x)_{c(x)} \in CB(H)$, where $CB(H)$ denotes the family of all nonempty bounded closed subsets of H. By the fuzzy mappings T, A and B, we can define three set-valued mappings $\tilde{T}, \tilde{A}$ and $\tilde{B}$ as follows:

$$\tilde{T}: H \rightarrow CB(H), x \rightarrow (T_x)_{a(x)},$$

$$\tilde{A}: H \rightarrow CB(H), x \rightarrow (A_x)_{b(x)},$$

$$\tilde{B}: H \rightarrow CB(H), x \rightarrow (B_x)_{c(x)}.$$

In the sequel, $\tilde{T}, \tilde{A}$ and $\tilde{B}$ are called the set-valued mappings induced by the fuzzy mapping T, A and B respectively.

Given set-valued mapping $g: H \rightarrow 2^H$, single-valued mappings f, p, q: $H \rightarrow H$, $N: H \times H \times H \rightarrow H$, a continuous bifunction $\varphi(\cdot, \cdot): H \times H \rightarrow R \cup \{+\infty\}$ with $\operatorname{Im} g(x) \cap \operatorname{dom} \partial \varphi(\cdot, \cdot) \neq \varnothing$, we consider the following problem:

Find x, u, v, z, $w \in H$, such that

$$\begin{cases} T_x(u) \geq a(x), A_x(v) \geq b(x), B_x(z) \geq c(x), w \in g(x) \cap \operatorname{dom} \partial \varphi(\cdot, \cdot), \\ \langle N(f(u), p(v), q(z)), y - w \rangle \geq \varphi(w, w) - \varphi(y, w), \forall y \in H, \end{cases} \quad (2.1)$$

the problem (2.1) is called Generalized Set-valued Mixed Variational Inequalities for Fuzzy Mappings.

For $N(f(u), p(v), q(z)) = N(f(u), p(v))$, $\forall u, v, z \in H$, $\varphi(u, v) = \varphi(u)$, $\forall v \in H$, then the problem (2.1) reduces to finding x, u, v, $w \in H$, such that

$$\begin{cases} T_x(u) \geq a(x), A_x(v) \geq b(x), w \in g(x) \cap \operatorname{dom}(\partial \varphi), \\ \langle N(f(u) - p(v)), y - w \rangle \geq \varphi(w) - \varphi(y), \forall y \in H, \end{cases} \quad (2.2)$$

The problem (2.2) is called Completely Generalized Variational Inclusions for Fuzzy Mappings[4].

For and $g: H \rightarrow H$ is single-valued mapping, then problem (2.2) reduces to finding x, u, $v \in H$, such that

$$\begin{cases} T_x(u) \geq a(x), A_x(v) \geq b(x), g(x) \cap \operatorname{dom} \partial \varphi, \\ \langle N(f(u), p(v)), y - g(x) \rangle \geq \varphi(g(x)) - \varphi(y), \forall y \in H, \end{cases} \quad (2.3)$$

The problem (2.3) is called Completely Generalized Nonlinear Variational Inclusions for Fuzzy Mappings[1].

Definition 2.1. *Set-valued mapping* $g : H \to 2^H$ *is said to be strongly* δ - *monotone if there exists a constant* $\delta > 0$, *such that*

$$\langle w_1 - w_2, x_1 - x_2 \rangle \geq \delta \| x_1 - x_2 \|, \quad \forall x_i \in H, \ w_i \in (g(x_i)), \ i = 1, 2.$$

Definition 2.2. *The set-valued mapping* $\tilde{T} : H \to CB(H)$, *induced by the fuzzy mapping* $T : H \to F(H)$, *is* $t_1 - H - Lipschitz$ *continuous if there exists a constant* $t_1 > 0$, *such that* $H(\tilde{T}x, \tilde{T}y) \leq t_1 \| x - y \|, \quad \forall x, y \in H.$

Similarly, we can define that $\tilde{A}$ is $t_2 - H -$ Lipschitz continuous, $\tilde{B}$ is $t_3 - H -$ Lipschitz continuous, and g is $t_4 - H -$ Lipschitz continuous.

Definition 2.3. *Nonlinear mappings* $f, p, q : H \to H$, $N(\cdot, \cdot, \cdot) : H \times H \times H \to H$, N *is* $k_1 - f - Lipschitz$ *continuous for the first variable, if and only if there exists a constant* $k_1 > 0$, *such that* $\forall x, y, z, w \in H$

$$\| N(f(x), p(z), q(w)) - N(f(y), p(z), q(w)) \| \leq k_1 \| x - y \|.$$

Similarly, we can define $k_2 - p -$ Lipschitz continuous for the second variable and $k_3 - q -$ Lipschitz continuous for the third variable.

Definition 2.4. *The subdifferential* $\partial \varphi(\cdot, \cdot)$ *of a convex, proper and lower semi-contin-uous function* $\varphi(\cdot, \cdot) : H \times H \to R \cup \{+\infty\}$ *is a maximal monotone with respect to the first agrument, and we can define its resolvent by*

$$J_{\varphi(u)} = (I + \rho \partial \varphi(\cdot, u))^{-1} \equiv (I + \rho \partial \varphi(u))^{-1},$$

where $\partial \varphi(u) \equiv \partial \varphi(\cdot, u)$, *unless otherwise specified. It is well known that the operator* $J_{\varphi(u)}$ *satisfies*

$$\| J_{\varphi(u)}(x) - J_{\varphi(u)}(y) \| \leq \| x - y \|, \quad \forall x, y, u \in H.$$

3 Main Results

Lemma 3.1. *The problem (2.1) has the solution* x, u, v, z, $w \in H$, *if and only if satisfies* $u \in \tilde{T}_x$, $v \in \tilde{A}_x$, $z \in \tilde{B}_x$, $w \in g(x) \cap dom(\partial \varphi(w))$

$$w = J_{\varphi(w)}(w - \rho N(f(u), p(v), q(z))),$$

where $\rho \geq 0$ *is a constant.*

Proof. If x, u, v, z, $w \in H$, satisfies the relation $u \in \tilde{T}_x$, $v \in \tilde{A}_x$, $z \in \tilde{B}_x$, $w \in g(x) \cap dom(\partial\varphi(w))$, from definition of $J_{\varphi(w)}$ we have

$$w - \rho N(f(u), p(v), q(z)) \in w + \rho \partial\varphi(w),$$

that is, $-N(f(u), p(v), q(z)) \in \partial\varphi(w)$. From definition of subdifferential, we have

$$\langle N(f(u), p(v), q(z)), y - w \rangle \geq \varphi(w, w) - \varphi(y, w), \forall y \in H \cdot$$

Thus x, u, v, z and w are solution of (2.1).

Conversely, if x, u, v, z and w are solution of (2.1), then

$$\begin{cases} T_x(u) \geq a(x), A_x(v) \geq b(x), B_x(z) \geq c(x), w \in g(x) \cap dom\partial\varphi(\cdot, \cdot), \\ \langle N(f(u), p(v), q(z)), y - w \rangle \geq \varphi(w, w) - \varphi(y, w), \forall y \in H, \end{cases}$$

hence $u \in \tilde{T}_x$, $v \in \tilde{A}_x$, $z \in \tilde{B}_x$, $w \in g(x) \cap dom(\partial\varphi(w))$, $-N(f(u), p(v), q(z)) \in \partial\varphi(w) \cdot$

When $\rho \geq 0$, we have $w - \rho N(f(u), p(v), q(z)) \in w + \rho \partial\varphi(w)$. Thus $u \in \tilde{T}_x$, $v \in \tilde{A}_x$, $z \in \tilde{B}_x$, $w \in g(x) \cap dom(\partial\varphi(w))$, $w = J_{\varphi(w)}(w - \rho N(f(u), p(v), q(z)))$.

The proof is completed.

Suppose that $T, A, B : H \to F(H)$ satisfy the condition (1). Let $\tilde{T}$, $\tilde{A}$, $\tilde{B} :$ $H \to CB(H)$ be set-valued mappings induced by T, A and B respectively. For given $x_0 \in H$, $u_0 \in \tilde{T}x_0$, $v_0 \in \tilde{A}x_0$, $z_0 \in \tilde{B}x_0$, and $w_0 \in g(x_0)$

$$x_1 = x_0 - w_0 + J_{\varphi(w_0)}(w_0 - \rho N(f(u_0), p(v_0), q(z_0))),$$

there exist $u_1 \in \tilde{T}x_1$, $v_1 \in \tilde{A}x_1$, $z_1 \in \tilde{B}x_1$, and $w_1 \in g(x_1)$ such that

$$\|u_1 - u_0\| \leq (1+1) H(\tilde{T}x_1 - \tilde{T}x_0), \qquad \|v_1 - v_0\| \leq (1+1) H(\tilde{A}x_1 - \tilde{A}x_0),$$

$$\|z_1 - z_0\| \leq (1+1) H(\tilde{B}x_1 - \tilde{B}x_0), \qquad \|w_1 - w_0\| \leq (1+1) H(g(x_1) - g(x_0)),$$

where H is the Hausdorff metric on $CB(H)$. We can obtain the following algorithm.

Algorithm 3.1. $x_{n+1} = x_n - w_n + J_{\varphi(w_n)}(w_n - \rho N(f(u_n), p(v_n), q(z_n)))$,

$$\|u_{n+1} - u_n\| \leq (1 + (1+n)^{-1}) H(\tilde{T}x_{n+1} - \tilde{T}x_n), \quad u_n \in \tilde{T}x_n,$$

$$\|v_{n+1} - v_n\| \leq (1 + (1+n)^{-1}) H(\tilde{A}x_{n+1} - \tilde{A}x_n), \quad v_n \in \tilde{A}x_n,$$

$$\|z_{n+1} - z_n\| \leq (1 + (1+n)^{-1}) H(\tilde{B}x_{n+1} - \tilde{B}x_n), \quad z_n \in \tilde{B}x_n,$$

$$\|w_{n+1} - w_n\| \leq (1 + (1+n)) H(g(x_{n+1}) - g(x_n)), w_{n+1} \in g(x_n), \qquad n = 0, 1, 2 \ldots$$

Theorem 3.1. *Let set-valued mapping* $g : H \to CB(H)$, *single-valued mappings* $N : H \times H \times H \to H$, $f, p, q : H \to H$, *if the following conditions hold:*

(i) $\tilde{T}$, $\tilde{A}$, $\tilde{B}$ *and* g *are* $t_i - H -$ *Lipschitz continuous respectively* $(i = 1, 2, 3, 4)$

(ii) N *is* $k_1 - f -$ *Lipschitz continuous for the first variable,* $k_2 - p -$ *Lipschitz contin-uous for the second variable and* $k_3 - q -$ *Lipschitz continuous for the third variable.*

(iii) g *is strongly* δ-*monotone.*

(iv) *If there exists a constant* $\rho \geq 0$, *such that*

$$0 < \sqrt{\left(1 - 2\delta + t_4^2\right)} + t_4 + \rho \left(\sum_{i=1}^{3} k_i t_i \right) < 1 \cdot$$

Then there exist $x^* \in H$, $u^* \in \tilde{T}x^*$, $v^* \in \tilde{A}x^*$, $z^* \in \tilde{B}x^*$ *and* $w^* \in g(x^*)$, *which are solution of (2.1). Moreover,* $\{x_n\}$, $\{u_n\}$, $\{v_n\}$, $\{z_n\}$ *and* $\{w_n\}$ *are strong convergence to* x^*, u^*, v^*, z^* *and* w^* *respectively.*

Proof. From Algorithm 3.1 we have

$$\left\| x_{n+1} - x_n \right\| = \left\| x_n - x_{n-1} - (w_n - w_{n-1}) + J_{\varphi(w_n)}\left(h(x_n)\right) - J_{\varphi(w_{n-1})}\left(h(x_{n-1})\right) \right\|$$

$$\leq \left\| x_n - x_{n-1} - (w_n - w_{n-1}) \right\| + \left\| J_{\varphi(w_n)}\left(h(x_n)\right) - J_{\varphi(w_{n-1})}\left(h(x_{n-1})\right) \right\|, \qquad (3.1)$$

where $h(x_n) = w_n - \rho N\left(f(u_n), p(v_n), q(z_n)\right)$. From Lemma 3.1 we have \

$$\left\| J_{\varphi(w_n)}\left(h(x_n)\right) - J_{\varphi(w_{n-1})}\left(h(x_{n-1})\right) \right\| \leq \left\| h(x_n) - h(x_{n-1}) \right\|$$

$$= \left\| w_n - \rho N\left(f(u_n), p(v_n), q(z_n)\right) - w_n + \rho N\left(f(u_{n-1}), p(v_{n-1}), q(z_{n-1})\right) \right\|$$

$$\leq \left\| w_n - w_{n-1} \right\| + \rho \left\| N\left(f(u_n), p(v_n), q(z_n)\right) - N\left(f(u_{n-1}), p(v_{n-1}), q(z_{n-1})\right) \right\|$$

$$\leq \left\| w_n - w_{n-1} \right\| + \rho \left\| N\left(f(u_n), p(v_n), q(z_n)\right) - N\left(f(u_{n-1}), p(v_n), q(z_n)\right) \right\|$$

$$+ \rho \left\| N\left(f(u_{n-1}), p(v_n), q(z_n)\right) - N\left(f(u_{n-1}), p(v_{n-1}), q(z_n)\right) \right\|$$

$$+ \rho \left\| N\left(f(u_{n-1}), p(v_{n-1}), q(z_n)\right) - N\left(f(u_{n-1}), p(v_{n-1}), q(z_{n-1})\right) \right\| \cdot$$

From condition (ii) and Algorithm 3.1 we have

$$\left\| w_n - w_{n-1} \right\| \leq \left(1 + n^{-1}\right) H\left(g(x_n) - g(x_{n-1})\right) \leq t_4\left(1 + n^{-1}\right) \left\| x_n - x_{n-1} \right\|$$

$$\left\| N\left(f(u_n), p(v_n), q(z_n)\right) - N\left(f(u_{n-1}), p(v_n), q(z_n)\right) \right\| \leq k_1 t_1 \left(1 + n^{-1}\right) \left\| x_n - x_{n-1} \right\|$$

$$\left\| N\left(f(u_{n-1}), p(v_n), q(z_n)\right) - N\left(f(u_{n-1}), p(v_{n-1}), q(z_n)\right) \right\| \leq k_2 t_2 \left(1 + n^{-1}\right) \left\| x_n - x_{n-1} \right\|$$

$$\left\| N\left(f(u_{n-1}), p(v_{n-1}), q(z_n)\right) - N\left(f(u_{n-1}), p(v_{n-1}), q(z_{n-1})\right) \right\| \leq k_3 t_3 \left(1 + n^{-1}\right) \left\| x_n - x_{n-1} \right\|$$

Then

$$\left\| J_{\varphi(w_n)}\left(h(x_n)\right) - J_{\varphi(w_{n-1})}\left(h(x_{n-1})\right) \right\| \le \left(t_4 + \rho\left(1+n^{-1}\right)\left(\sum_{i=1}^{3}k_i t_i\right) \right)\left\|x_{n+1}-x_n\right\| . \quad (3.2)$$

From condition (iii), we obtain

$$\left\|x_n - x_{n-1} - \left(w_n - w_{n-1}\right)\right\|^2 = \left\|x_n - x_{n-1}\right\|^2 - 2\left\langle w_n - w_{n-1}, x_n - x_{n-1}\right\rangle + \left\|w_n - w_{n-1}\right\|^2$$

$$\le \left(1 - 2\delta + \left(1+n^{-1}\right)^2 t_4^2\right)\left\|x_n - x_{n-1}\right\|^2 .$$

Then

$$\left\|x_n - x_{n-1} - \left(w_n - w_{n-1}\right)\right\| \le \sqrt{\left(1 - 2\delta + \left(1+n^{-1}\right)^2 t_4^2\right)}\left\|x_n - x_{n-1}\right\| . \quad (3.3)$$

Taking (3.2), (3.3) in (3.1) , we have

$$\left\|x_{n+1} - x_n\right\| \le \varepsilon_n \left\|x_n - x_{n-1}\right\| ,$$

where $\varepsilon_n = \sqrt{\left(1 - 2\delta + t_4^2\left(1+n^{-1}\right)^2\right)} + t_4 + \rho\left(1+n^{-1}\right)\left(\sum_{i=1}^{3}k_i t_i\right).$

Let $\varepsilon = \sqrt{\left(1 - 2\delta + t_4^2\right)} + t_4 + \rho\left(\sum_{i=1}^{3}k_i t_i\right)$, which implies that $\varepsilon_n \to \varepsilon \ (n \to +\infty)$.

From condition (iv), we obtain $0 < \varepsilon_n < 1$ if n is large enough. Thus $\{x_n\}$ is Cauchy sequence. Let $x_n \to x^* \in H$. From

$$\left\|u_n - u_{n-1}\right\| \le \left(1+n^{-1}\right)H\left(\tilde{T}x_n - \tilde{T}x_{n-1}\right) \le t_1\left(1+n^{-1}\right)\left\|x_n - x_{n-1}\right\|,$$

we know $\{u_n\}$ is Cauchy sequence. Moreover , let $u_n \to u^*$. From

$$d\left(u^*, \tilde{T}x^*\right) = \inf\left\{\left\|u^* - z : z \in \tilde{T}x^*\right\|\right\} \le \left\|u^* - u_n\right\| + d\left(u_n, \tilde{T}x^*\right)$$

$$\le \left\|u^* - u_n\right\| + H\left(\tilde{T}x_n, \tilde{T}x^*\right) \le \left\|u^* - u_n\right\| + t_1\left\|x_n - x^*\right\| \to 0\,(n \to +\infty),$$

thus $u^* \in \tilde{T}x^*$. Similarly we may proof $\{v_n\}$, $\{z_n\}$ and $\{w_n\}$ are Cauchy sequence. Let $v_n \to v^*, z_n \to z^*$ and $w_n \to w^*$, we have $v^* \in \tilde{A}x^*$, $z^* \in \tilde{B}x^*$ and $w^* \in g\left(x^*\right)$.

From $x_{n+1} = x_n - w_n + J_{\varphi(w_n)}\left(w_n - \rho N(f(u_n), p(v_n), q(z_n))\right)$ in Algorithm 3.1

$$\lim_{n \to +\infty} x_{n+1} = \lim_{n \to +\infty}\left(x_n - w_n + J_{\varphi(w_n)}\left(w_n - \rho N(f(u_n), p(v_n), q(z_n))\right)\right),$$

we have $w^* = J_{\varphi(w^*)}\left(w^* - \rho N(f(u^*), p(v^*), q(z^*))\right).$

By Lemma 3.1, we get that $x^* \in H$, $u^* \in \tilde{T}x^*$, $v^* \in \tilde{A}x^*$, $z^* \in \tilde{B}x^*$ and $w^* \in g(x^*)$ are solution of (2.1). According to Algorithm 3.1, we obtain that $\{x_n\}$, $\{u_n\}$, $\{v_n\}$, $\{z_n\}$ and $\{w_n\}$ are strong convergence to x^*, u^*, v^*, z^* and w^* respectively. This completes the proof.

4   Conclusion

We have proved the existence of solutions of generalized set-valued mixed variational inequalities for fuzzy mappings. We have also proved the convergence of a new iterative algorithm approximating the solution for this system.

References

1. Huang, N.J.: Completely Generalized Nonlinear Inclusions for Fuzzy Mappings Czechoslovak. Mathematical Journal 49(124), 767–777 (1999)
2. Noor, M.A.: Variational Inequalities for Fuzzy Mappings (I). Fuzzy Sets and Systems 110, 101–108 (2000)
3. Park, J.Y., Lee, S.Y., Jeong, J.U.: Completely generalized strongly quasi-variational inequalities for fuzzy mappings. Fuzzy Sets and Systems 110, 91–99 (2000)
4. Ding, T.M.: Completely Generalized Variational Inclusions for Fuzzy Mappings. Mathematics in Practice and Theory 34(8), 145–149 (2004)
5. Jin, M.M.: Perturbed Proximal Point Algorithm for General Quasi-Variational Inclusions with Fuzzy Set-Valued Mappings. OR Transactions 9(3), 31–38 (2005)
6. Li, H.G., Tian, Y.X.: Interative Algorithm for a New Class of Generalized Mixed Quasi-Variational Inclusions with Fuzzy Set-valued Mappings. J. of Sichuan Normal University (Natural Science) 31(2), 194–197 (2008)
7. Noor, M.A.: Mixed variational inequalities. Appl. Math. Lett. 3, 73–75 (1990)
8. Noor, M.A.: Mixed quasi-variational inequalities. Applied Mathematics and Computation 146, 553–578 (2003)

Minimization of Lattice-Valued Moore Type of Finite Automaton

Kun Zhang and Zhi-wen Mo

College of Mathematics and Software Science,
Sichuan Normal University, Chengdu 610068, China
zk535043994@yahoo.cn

Abstract. The definitions are given about Lattice-valued Moore type of finite automaton and lattice-valued fuzzy Moore type of finite automaton. Some related properties of them are discussed. Moreover, the minimization of fuzzy Moore automaton is also investigated based on state weak equivalence relations. Finally, some significant results concerning of them are given systematically.

Keywords: lattice-valued Moore type of finite automaton, lattice-valued fuzzy Moore type of finite automaton, weak equivalence, minimization.

1 Introduction

The theory of fuzzy sets was introduced by Zadeh in 1965 [1]. The mathematical formulation of a fuzzy automaton was first proposed by Wee in 1967 [2]. Santos defined the so-called maximin automaton in 1968 [3]. Two classes of fuzzy automaton corresponding to the Mealy and Moore type of ordinary automaton are formulated by Asai and Kitajima in 1971 [4]. Many papers related fuzzy automaton are referenced in [5, 6]. Fuzzy finite automaton have widespread applications, including mathematical models, computing with words, learning system, pattern recognition and so on.

The notion of lattice-valued finite automaton took value in lattice-ordered monoids is advanced by Y. M. Li in 2005 [7], thereafter the theory of L-valued automaton was formed. According to the thought of the classification of the fuzzy automaton by W. Cheng and Z. W. Mo [8], L-valued Mizumoto and L-valued Mealy of finite automaton took value in lattice-ordered monoids was defined in paper [9]. In this paper, we defined the formulation of lattice-valued Moore type finite automaton took value in lattice-ordered monoids. We discuss the properties and minimization of L-valued Moore type of finite automaton. Considering that input and output symbols are single symbol or string, we obtain some properties of L-valued Moore type of finite automaton. Based on the thought of minimization of L-valued Mealy and L-valued Mizumoto finite automaton in [10], we advance a so-called weak equivalence ($\approx$) on L-ordered monoid. On the basis of weak equivalence($\approx$), we get minimization of state ($Q/\approx$), and explore its minimization

B.-Y. Cao and X.-J. Xie (Eds.): Fuzzy Engineering and Operations Research, AISC 147, pp. 247–256.
springerlink.com

problem. Meanwhile, we prove and give the minimization algorithm of the L-valued Moore type of finite automaton. It has better worth in the theory and realistic meaning.

2   Characters of L-Valued Moore Type of Finite Automaton

Let us recall the definition of the Moore type of fuzzy automaton.

Definition 1.1.[4] *The Moore type of fuzzy automaton may be expressed as shown in following: M=(S, X, U, F(x), G(s, u)), where S={s_1, s_2, ..., s_v} is the finite states set; X={x_1, x_2, ..., x_μ}is the input symbols set; U={u_1, u_2, ..., u_n}is the output symbols set; F(x)is the fuzzy transition matrix. G(s, u)is the output matrix.*

According to the Moore type of fuzzy automaton, we can define the lattice-valued Moore type of finite automaton as following.

Definition 1.2. *Let (L, •, ∨) be a lattice-ordered monoid. The standard form of lattice-valued Moore type of finite automaton is a five-tuple M=(Q, X, Y, F, G). Where Q={q_1, q_2, ..., q_n}is the finite states set; X={x_1, x_2, ..., x_n}is the input symbols set; Y ={y_1, y_2, ..., y_n}is the output symbols set; F={F(x)|F(x)=(f_{pq} (x)), x∈X, p, q∈Q}is the lattice-valued transition matrix; G={G(x|y)|G(x|y)=(g_p(x|y)), x∈X, y∈Y, p∈Q}is the lattice-valued output matrix. Let f(p, x, q) = f_{pq} (x) and g(p, x, y) = g_p(x|y), that is to say, f, g represent the mapping from Q×X×Q , Q×X×Y to L, respectively. Let X^* be the set of all word of finite length over X, and Y^* be the set of all words of finite length over Y. Let ∧ stand for the empty word, it is the unit on X^*(Y^*),and e stand for supper on L.*

The following conditions hold.

(1) $\forall p \in Q, a \in X, \exists q \in Q,$ s.t $f(p,a,q) > 0 \Rightarrow \exists b \in Y,$ s.t $g(p,a,b) > 0$.

(2) $\forall p \in Q, a \in X, \exists b \in Y,$ s.t $g(p,a,b) > 0 \Rightarrow \exists q \in Q,$ s.t $f(p,a,q) > 0$.

We can define the lattice-valued transition function $f_{pq}(x)$ and the lattice-valued transition function $g_p(x|y)$.

In the case of single input:

(1) $f : Q \times X \times Q \to L$; $(p_i, x_k, q_j) \to f(p_i, x_k, q_j) = f_{p_i q_j}(x_k)$,

$$f(p, \wedge, q) = \begin{cases} e, & p = q \\ 0, & \text{otherwise} \end{cases},$$

(2) $g : Q \times X \times Y \to L$; $(p_i, x_k, y_j) \to g(p_i, x_k, y_j) = g_{p_i}(x_k|y_j)$,

$$g(p, x, y) = \begin{cases} e, & x = y = \wedge \\ 0, & x \neq \wedge, y = \wedge \text{ or } x = \wedge, y \neq \wedge \end{cases}.$$

We can define the f^* and g^* as follows.
In the case of multi-input: :

(1) $f^*: Q \times X^* \times Q \to L$, $f^*(p, \wedge, q) = \begin{cases} e, & p = q \\ 0, & \text{otherwise} \end{cases}$,

$$f^*(p, xa, q) = \bigvee_{r \in Q} [f^*(p, x, r) \bullet f(r, a, q)], x \in X^*, a \in X$$

(2) $g^*: Q \times X^* \times Y^* \to L$, $g^*(p, x, y) = \begin{cases} e, & x = y = \wedge \\ 0, & \text{otherwise} \end{cases}$,

$$g^*(p, xa, yb) = \bigvee_{r \in Q} [g^*(p, x, y) \bullet f^*(p, x, r) \bullet g(r, a, b)].$$

For any $p, r \in Q$, $x \in X^*$, $y \in Y^*$, $a \in X$, $b \in Y$. According to the fuzzy extended theory, we can get the lattice-valued fuzzy transition function f and output function g based on word (fuzzy language). It is obvious that f and g are the mapping from $F(Q) \times F(X) \times F(Q)$, $F(Q) \times F(X) \times F(Y)$ to L, respectively. For any $A, C \in F(Q)$, $B \in F(X)$, $D \in F(Y)$, it has

$$f(A, B, C) = \bigvee_{p \in A, x \in B, q \in C} (A(p) \bullet B(x) \bullet C(q) \bullet f(p, x, q))$$

and

$$g(A, B, D) = \bigvee_{p \in A, x \in B, y \in C} (A(p) \bullet B(x) \bullet D(y) \bullet g(p, x, y)).$$

Hence $M = (F(Q), F(X), F(Y), F, G)$ is a lattice-valued fuzzy Moore type of finite automaton.

Theorem 1.1. *Let L be a lattice-ordered monoid $(L, \bullet, \vee)$, $M = (Q, X, Y, F, G)$ be a L-valued Moore type of finite automaton. Then*

(1) $f = f^*|_{Q \times X \times Q}$.

(2) *If there exists $a \in X$ satisfying $\bigvee_{r \in Q} f(p, a, r) = e$, for any $p \in Q$, then $g = g^*|_{Q \times X \times Y}$.*

Proof. According to Definition 1.1.

(1) Let $x \in X$, according to definition1. 2, then
$$f^*(p, x, q) = f^*(p, \wedge x, q) = \bigvee_{r \in Q} [f^*(p, \wedge, r) \bullet f(r, x, q)].$$

There exists r in Q satisfying $r = p$, s. t
$$\bigvee_{r \in Q} [f^*(p, \wedge, r) \bullet f(r, x, q)]. = f^*(p, \wedge, p) \bullet f(p, x, q) = e \bullet f(p, x, q) = f(p, x, q).$$

Thus $f = f^*|_{Q \times X \times Q}$.

(2) Let $p \in Q$, $x \in X$, $y \in Y$,

$$g^*(p, x, y) = g^*(p, \wedge x, \wedge y) = \bigvee_{r \in Q} [g^*(p, \wedge, \wedge) \cdot f^*(p, \wedge, r) \cdot g(r, x, y)].$$

From the Definition 1.2, $g^*(p, \wedge, \wedge) = e$, $\bigvee_{r \in Q} f^*(p, \wedge, r) = e$, and $f^*(p, \wedge, p) = e$.

$$\bigvee_{r \in Q} [g^*(p, \wedge, \wedge) \cdot f^*(p, \wedge, r) \cdot g(r, x, y)] = e \cdot f^*(p, \wedge, p) \cdot g(p, x, q)$$

$$= e \cdot e \cdot g(p, x, q) = g(p, x, q).$$

Thus $g = g^*|_{Q \times X \times Y}$.

Theorem 1.2. *Let $M = (Q, X, Y, F, G)$ be a L-valued Moore type of finite automaton. For any $p \in Q$, $x \in X^*$ and $Y \in Y^*$, if $|x| \neq |y|$, $g^*(p, x, y) = 0$. ($|x|$ stands for the length of string x).*

Proof. For any $p \in Q$, $x \in X^*$, $y \in Y^*$, if $|x| = |y|$.

(1) Suppose $|x| > |y| = n$, n is a natural number. Now use mathematics inductive method to prove. Basis step: If $n = 0$, then $y = \wedge$ and $|x| > |y| = 0$, so $x \neq \wedge$, By the Definition 1.2, $g^*(p, x, y) = g^*(p, x, \wedge) = 0$, the conclusion establishes. Induction hypothesis: Suppose $|y| = n$, $n \geq 1$, and the result is true for all $p \in Q$, $x \in X^*$, $Y \in Y^*$. Induction step: While $|y| = n+1$, $n \geq 1$, we can get $x = x'a$, $y = y'b$ ($x' \in X^*$, $y' \in Y^*$, $a \in X$, $b \in Y$). Since $|x'| > |y'|$ and $|y| = n$, $n \geq 1$, by the induction hypothesis, for all $r \in Q$, $x' \in X^*$, $y' \in Y^*$, $g^*(p, x', y') = 0$.

Thus $g^*(p, x, y) = g^*(p, x'a, y'b) = \bigvee_{r \in Q} [g^*(p, x', y') \cdot f^*(p, x', r) \cdot g(r, a, b)]$

$$= g^*(p, x', y') \cdot \{\bigvee_{r \in Q} f^*(p, x', r) \cdot g(r, a, b)\} = 0.$$

(2) Analogously it can prove that for all $p \in Q$, $x \in X^*$, $Y \in Y^*$, $|x| < |y|$, $g^*(p, x, y) = 0$ as is to be shown.

This theorem is to say that if $g^*(p, x, y) > 0$ then $|x| = |y|$.

3   Minimization of L-Valued Moore Type of Finite Automaton on the Basis of Weak Equivalence ($\approx$)

Definition 1.3. *Let $(L, \cdot, \vee)$ be a lattice-ordered monoid, $M_i = (Q_i, X, Y, F_i, G_i)$ be a lattice-valued Moore type of finite automaton, in which $q_i \in Q$, $i = 1, 2$. Next we can define three kinds of state equivalence relations of L-valued Moore type of finite automaton, they are weak equivalence, equivalence and strong equivalence.*

(1) q_1 and q_2 are weak equivalent ($q_1 \approx q_2$) if and only if there exists only $z \in X^*$ such that $|z| = |x|$ and $g_1^*(q_1, x, y) = g_2^*(q_2, z, y)$ for any $x \in X^*$, $y \in Y^*$. Meanwhile, there exists only $z' \in X^*$ such that $|z'| < |x|$, and $g_1^*(q_1, z', y) = g_2^*(q_2, x, y)$.

(2) q_1 and q_2 are k-weak equivalent $(q_1 \approx_k q_2)$ if and only if $|x| < k$, k is a positive integer, there exists only $z \in X^*$ such that $|z| = |x|$ and $g_1^*(q_1, x, y) = g_2^*(q_2, z, y)$ for any $x \in X^*$, $y \in Y^*$. Meanwhile, there exists only $z' \in X^*$ such that $|z'| < |x|$ and $g_1^*(q_1, z', y) = g_2^*(q_2, x, y)$.

(3) M_1 and M_2 are weak equivalent $(M_1 \approx M_2)$ if and only if for any $q_1 \in Q_1$, there exists. $q_2 \in Q_2$, such that $q_1 \approx q_2$, and for any $q_2 \in Q_2$, there exists $q_1 \in Q_1$, such that $q_2 \approx q_1$.

(4) M_1 and M_2 are k-weak equivalent $(M_1 \approx_k M_2)$ if and only if for any k, k is a positive integer, any $q_1 \in Q_1$, there exists $q_2 \in Q_2$, such that $q_1 \approx_k q_2$, and for any $q_2 \in Q_2$, there exists $q_1 \in Q_1$, such that $q_2 \approx_k q_1$.

Definition 1.4. q_1 and q_2 are strong equivalence $(q_1 \doteq q_2) \Leftrightarrow$ for any $x \in X^*$, $y \in Y^*$,

$$\bigvee_{p \in Q} f_1^*(q_1, x, p) = \bigvee_{p \in Q} f_2^*(q_2, x, p) \text{ and } g_1^*(q_1, x, y) = g_2^*(q_2, x, y).$$

If $M_1 = M_2 = M$, then weak equivalence $(\approx)$, strong equivalence $(\doteq)$ and their corresponding k-weak equivalence, k-strong equivalence are equivalence relations.

Definition 1.5. Let $(L, \bullet, \vee)$ be a L-ordered monoid and $M = (Q, X, Y, F, G)$ be a L-valued Moore type of finite automaton, any $p, q \in Q$. Then

(1) If $p \approx q$ can induce $p = q$, then M is called a minimal state L-valued type automaton.

(2) If there exists a output string $y \in Y^*$, for any $z \in X^*$, s. t $|x| = |z|$, it has $g^*(p, x, y) \neq g_2^*(q, z, y)$ or $g^*(p, z, y) \neq g_2^*(q, x, y)$ if and only if p and q are said to be distinguishable by $x \in X^*$.

(3) M is in reduced form if and only if $p \approx q$ can induce $p = q$.

Theorem 1.3. Let $(L, \bullet, e)$ be a ordered monoid without the divisor of zero and $M = (Q, X, Y, F, G)$ be a L-valued Moore type of finite automaton. Then 1(a) and 2(a) are equivalent. Especially, when L is a L-ordered monoid $(L, \bullet, \vee)$ (This is to say, multiplication $\bullet$ to finite supermum $\vee$ satisfy distributive law), and for any $p \in Q$, $a \in X$, $\bigvee_{r \in Q} f(p, a, r) = e$, 1(b) and 2(b) are equivalent. Where

1(a) $\forall p \in Q, x \in X, \exists q \in Q, s. t. f(p, x, q) > 0 \Rightarrow \exists y \in Y, s. t. g(p, x, y) > 0;$

1(b) $\forall p \in Q, x \in X, \exists y \in Y, s. t. g(p, x, y) > 0 \Rightarrow \exists q \in Q, s. t. f(p, x, q) > 0;$

2(a) $\forall p \in Q, x \in X^*, \exists q \in Q, s. t. f^*(p, x, q) > 0 \Rightarrow \exists y \in Y^*, s. t. g^*(p, x, y) > 0;$

2(b) $\forall p \in Q, x \in X^*, \exists y \in Y^*, s. t. g^*(p, x, y) > 0 \Rightarrow \exists q \in Q, s. t. f^*(p, x, q) > 0;$

Proof. Now use mathematics inductive method to prove $(1) \Rightarrow (2)$.

1(a) $\Rightarrow$ 2(a):

Suppose for any $p \in Q, x \in X^*$ there exists $q \in Q, |x| = n$, n is a natural number. Next the inductive proof on n can be given. While

n = 0, $x = \wedge$ and p = q, then there exists $\wedge \in Y^*$, $g^*(p, \wedge, \wedge) = e > 0$.

While n=1, by Theorem 1. 2, 2(a) establishes obviously.

Suppose for any x∈X*, |x|=n, the conclusion establishes. While x∈X*, |x|=n+1, x=x'a, x'∈X*, a ∈X, then by the definition 1. 2 and suppose, it has

$$f *(p, x, q)= f *(p, x'a, q)= \underset{r∈Q}{\vee} [f *(p, x', r) \bullet f(r, a, q)]>0.$$

Hence there exists r∈Q, s. t. f *(p, x', r) • f(r, a, q)>0, so there exists r∈Q, s. t f *(p, x', r) >0 and f(r, a, q)>0. Then by Induction hypothesis, f *(p, x', r) >0, we can get g* (p, x', y')>0 for some y'∈Y. In 1(a), p is replaced by r, then g(r, a, b)>0 by f(r, a, q)>0. Let y=y'b. According to (L, •, e) without the divisor of zero. Now g*(p, x, y)=g*(p, x'a, y'b)= $\underset{r∈Q}{\vee}$ [g*(p, x', y')• f*(p, x', r) • g(r, a, b)]≥ g*(p, x', y')• f*(p, x', r) • g(r, a, b)>0. Consequently, the conclusion establishes.

1(b) ⇒ 2(b):
For any p∈Q, x∈X*, there exists y∈Y*, s. t g* (p, x, y)>0, by the Theorem 1.2, it has |x|=|y|=n, n is a nature number. Now use mathematics inductive method to prove.

While n=0, x=y=∧, there exists q=p, then f * (p, x, q)=f* (p, ∧, p)=e>0

While n=1, x∈X , y∈Y , it has g* (p, x, y)= g(p, x, y), from 1(b) there exists r∈ Q, s. t. f(p, x, r)>0, f*(p, x, r)=f(p, x, r)>0, the conclusion establishes.

Suppose for all x∈X* , y∈Y*and |x|=|y|=n, the conclusion establishes.

While |x|=|y|=n+1, x=x'a, y=y'b (x'∈X*, y'∈Y*, a ∈X, b ∈Y), and |x'|=| y'|=n. From the Definition 1.1, g*(p, x, y)= g *(p, x'a, y'b)= $\underset{r∈Q}{\vee}$ [g*(p, x', y')• f*(p, x', r) • g(r, a, b)]>0.

Thus g*(p, x', y') >0. Hence there exists r∈ Q, s. t. f *(p, x', r)>0. From 1(b), we know g(r, a, b)>0, so there exists q∈Q, s. t. f(r, a, q)>0. Thus

$$f *(p, x, q)= f *(p, x'a, q)= \underset{r∈Q}{\vee} [f *(p, x', r) \bullet f(r, a, q)]>0$$

through ordered monoid (L, •, e) without the divisor of zero. So the conclusion establishes.

(2)⇒(1): The conclusion is obvious.

Theorem 1.4. Let (L, •, ∨) be a L-ordered monoid, M=(Q, X, Y, F, G) be a L-valued Moore type of finite automaton. Then there exists a minimal L-valued Moore type of finite automaton M_m weak equivalent to M. Where M_m=(Q/≈, X, Y, F_m, G_m),

$$F_m= (f_{m\,pq}(x)) = (f_m(p,x,q)), \; G_m= (g_{m\,p}(x|y)) = (g_m(p,x,y))$$

$$f_m^{*}([p],x,[q]) = \underset{s,t∈Q,s≈p,t≈q}{\vee} f^{*}(s,x,t)$$

$$g_m^{*}([p],x,y) = \underset{s∈Q,s≈p}{\vee} g^{*}(s,x,y)$$

Proof. We must prove M_m to be a L-valued Moore type of finite automaton at first, next prove M_m is a state-minimal. finally show $M_m \approx M$.

Step1: Let $[p] \in Q/\approx$, $x \in X^*$, there exists $[q] \in Q/\approx$, s. t $f_m^*([p], x, [q]) > 0$, by the definition of M_m, there exists $s, t \in Q$, $s \approx p$, $t \approx q$, s. t $f^*(s, x, t) > 0$. at this time, there exists only $z \in X^*$ and $|z| = |x|$, for any $y \in Y^*$ s. t $g^*(p, x, y) = g^*(s, z, y)$. Since M is a L-valued Moore type of finite automaton, there exists $y \in Y^*$, s. t $g^*(s, x, y) > 0$. So $g_m^*([p], x, y) = \bigvee\limits_{s \in Q, s \approx p} g^*(s, x, y) \geq g^*(s, x, y) > 0$. Thus it satisfies Definition 1.2 (1).

In return, let $[p] \in Q/\approx$, $x \in X^*$, there exists $y \in Y^*$, s.t. $g_m^*([p], x, y) > 0$. By definition of M_m, there exists $s \in Q$, $s \approx p$, $z \in X^*$ s.t. $g^*(s, x, y) > 0$ and $|z| = |x|$. For any $y \in Y^*$, it has $g^*(p, x, y) = g^*(s, z, y)$. By Definition1.2, there exists $t \in Q$, s. t. $f^*(s, z, t) > 0$. Thus $f_m^*([p], x, [q]) = \bigvee\limits_{s, t \in Q, s \approx p, t \approx q} f^*(s, x, t) \geq f^*(s, x, t) > 0$. It satisfy Definition 1.1(2). So the conclusion establishes.

Step2: prove M_m is a state-minimal.

Let $[p], [q] \in Q/\approx$, $[p] \approx [q]$. Let $p \approx s_2$, so for any $x \in X^*$, $y \in Y^*$, there exists only $z_2 \in X^*$ such that $|z_2| = |x|$, it has $g^*(s_2, x, y) = g^*(p, z_2, y)$. By the definition of g_m^*, there exists $s_1 \in Q$, $s_1 \approx p$, s. t $g_m^*([p], z_2, y) = g^*(s_1, \beta_1, y)$, where β_1 is in X^* and $|z_2| = |\beta_1|$, and z_2 is corresponding to β_1 uniquely. On condition that $s_1 \approx p$, by the uniqueness of z_2 corresponding to β_1, it has $g_m^*([p], z_2, y) = g^*(s_1, \beta_1, y) = g^*(s_2, x, y)$. Since $[p], [q] \in Q/\approx$, $[p] \approx [q]$. So there exists only $z_1 \in X^*$, and $|z_1| = |z_2|$, it has $g_m^*([p], z_2, y) = g_m^*([q], z_1, y)$. By the definition of g_m^*, there exists $t_1 \in Q$, $t_1 \approx q$, and $g_m^*([q], z_1, y) = g^*(t_1, \beta, y)$, where β_1 is in X^* and $|z_1| = |\beta_1|$ and z_1 is corresponding to β uniquely, Thus for any $x \in X^*$, $y \in Y^*$, there exists only $\beta \in X^*$, and $|x| = |\beta|$, it has $g^*(s_2, x, y) = g^*(t_1, \beta, y)$. We can show similarly that $g^*(s_2, \beta, y) = g^*(t_1, x, y)$.

By the definition of weak equivalence ($\approx$), we can get $s_2 \approx t_1$. Since $p \approx s_2$, $t_1 \approx q$, so $p \approx q$. By Definition 1.5(1), we can get $p = q$, hence M_m is a state-minimal L-valued Moore type of finite automaton.

Step3: Show $M_m \approx M$.

For any $p \in Q$ $x \in X^*$, $y \in Y^*$, $g^*(p, x, y) = \bigvee\{ g^*(s, x, y) | s \in Q, p \approx s$, there exists only $z \in X^*$, and $|x| = |z|$ s. t $g^*(p, x, y) = g^*(s, z, y)$. Meanwhile, there exists only $\omega \in X^*$, s.t. $|x| = |\omega|$, s.t $g^*(s, x, y) = g^*(p, \omega, y)\} = g_m^*([p], x, y)$. Hence $p \approx [p]$, thus $M_m \approx M$.

The proof is completed.

Theorem 1.5. *Let $(L, \bullet, \vee)$ be a finite L-ordered monoid with $|L| = k$, $M = (Q, X, Y, F, G)$ be a L-valued Moore type of finite automaton. We can get a state-minimal automaton M_m weak equivalent to M, and all states in Q can be distinguished within at most $(k-1)^n$.*

Proof. Since $F = \{F(x) | F(x) = (f_{pq}(x)) = (f(p, x, q)), x \in X, p, q \in Q\}$ is the lattice-valued transition matrix. $G = \{G(x|y) | G(x|y) = (g_p(x|y)) = (g(p, x, y)), x \in X, y \in Y, p \in Q\}$ is the lattice-valued output matrix. We can get F^* and G^* concerning string $x_1 x_2 \ldots x_n \in X$, the matrix F^* and G^*. Since L is a finite L-ordered monoid. It has the supremum and the infimum, by the definition F^* and G^*, the infimum does not appear in G^*. So $G^*$is at most$(k-1)^n$.

Let $M_m=(Q/\approx, X, Y, F_m, G_m)$ and input string $x_1x_2...x_n \in X$. Now it can obtain $Q/\approx$ by the following minimization algorithm of the L-valued Moore type of finite automaton.

Step1: (Initialization) Let $i=1$, we can get $g(x_1), g(x_2),..., g(x_n)$ and $f(x_1), f(x_2),...,$ $f(x_n)$, Thus it can obtain the equivalence classes $Q/\approx_1$concerning $\approx_1$.

Step2: (circulation)   While $i=2,3,...,$by the definition of g^* and f^*

$$g^*(x_1x_2...x_i)= g^*(x_1) \bullet f(x_1) \bullet g(x_2...x_i)=g(x_1) \bullet f(x_1) \bullet g(x_2...x_i)$$

and

$$f^*(x_1x_2...x_i)= f^*(x_1) \bullet f(x_2...x_i)= f(x_1) \bullet f(x_2...x_i),$$

we can calculate the g^* and f^*.It is to say, it can obtain the equivelence classes $Q/\approx_i$from those $Q/\approx_{i-1}$until$|Q/\approx_i|=n$ or $i=(k-1)^n$,run the next step.

Step3: (outcome) $Q/\approx_i= Q/\approx$

Example: Let $M=(Q, X, Y, F, G)$ be a L-valued Moore type of finite automaton.

Here $L = (0, 0.2, 0.4, 0.5, 0.6, 0.7, 0.8, 1, \vee, \wedge)$, $Q = \{q_1, q_2, q_3\}$, $X = \{0\}$, $Y = \{0, 1\}$, M is given by

$$F(0) = \begin{pmatrix} 1 & 0.2 & 0.4 \\ 0.5 & 0.7 & 1 \\ 0 & 0.6 & 0.7 \end{pmatrix}, F(1) = \begin{pmatrix} 0.7 & 1 & 0 \\ 1 & 0.2 & 0.4 \\ 0.6 & 0.5 & 0.2 \end{pmatrix},$$

$$G(0|0) = \begin{pmatrix} 0.8 \\ 0.4 \\ 0.4 \end{pmatrix}, G(0|1) = \begin{pmatrix} 0.5 \\ 1 \\ 1 \end{pmatrix}, (g(q_{i=1,2,3}, 0, 0/1)) = \begin{pmatrix} 1 & 0.8 & 0.5 \\ 1 & 0.4 & 1 \\ 1 & 0.4 & 1 \end{pmatrix}.$$

Step 1 :   It can obtain $Q/\approx_i = \{q_1, \{q_2, q_3\}\}$ by $(g(q_{i=1,2,3}, 0, 0/1))$

Step 2:   Compute $(g(q_{i=1,2,3}, 00, 00/01/10/11))$

$$(g(q_{i=1,2,3}, 00, 00)) = \begin{pmatrix} 0.8 \\ 0.5 \\ 0.4 \end{pmatrix}, (g(q_{i=1,2,3}, 00, 01)) = \begin{pmatrix} 0.5 \\ 1 \\ 0.7 \end{pmatrix},$$

$$(g(q_{i=1,2,3}, 00, 10)) = \begin{pmatrix} 0.7 \\ 0.8 \\ 0.6 \end{pmatrix}, (g(q_{i=1,2,3}, 00, 11)) = \begin{pmatrix} 1 \\ 0.5 \\ 0.5 \end{pmatrix}$$

$$(g(q_{i=1,2,3},00,00/01/10/11)) = \begin{pmatrix} 0.8 & 0.5 & 0.7 & 1 \\ 0.5 & 1 & 0.8 & 0.5 \\ 0.4 & 0.7 & 0.6 & 0.5 \end{pmatrix}.$$

Then $Q/\approx_2 = \{\{q_1\},\{q_2\},\{q_3\}\}$.

Step 3: $|Q/\approx_2| = 3 = |Q|$, stop.

$$Q/\approx = Q/\approx_2 = \{\{q_1\},\{q_2\},\{q_3\}\}$$

The example illuminating the algorithm is given. For a L-valued Moore type of finite automaton M, there exists a minimal L-valued Moore type of finite automaton M_m weak equivalent to M. It can obtain the weak equivalence classes of $\approx_i$ from those $\approx_{i-1}$. And there exists at most $(k-1)^n$ steps to distinguish states in Q.

3   Conclusion

This paper establishes some basic concepts in L-valued Moore type of finite automaton and generalizes some properties of L-valued Moore automaton. Moreover, it studies major char-acterizations of L-valued Moore automaton, In Consideration of input and output symbols are single symbol or string, we get the characters of L-valued Moore type of finite automaton.

Some important results are obtained . And some relations among these characterizations are clarified. From those weak equivalence relations, a minimization algorithm of the L-valued Moore type of finite automaton is obtained. At last, the illustrative example is included.

Acknowledgements. Thanks to the support by National Natural Science Foundation of China (No.11071178).

References

1. Zadeh, L.A.: Fuzzy sets. Inform. Control 8, 338–353 (1965)
2. Wee, W.G.: On generalizations of adaptive algorithm and application of the fuzzy sets concept to pattern classification. Ph. D. Thesis, Purdue University (1967)
3. Santos, E.S.: Maximin automata. Inform. Control 13, 363–377 (1968)
4. Asai, K., Kitajima, S.: A method for optimizing control of multimodal systems using fuzzy automata. Inform. Sci. 3, 343–353 (1971)
5. Dubois, D., Prade, H.: Fuzzy sets and systems: theory and application. Academic Press, New York (1980)
6. Kandel, A., Lee, S.C.: Fuzzy switching and automata: theory and application. Crane & Russak, Edward Arnord, New York, London (1979)

7. Li, Y.M., Pedrycz, W.: Fuzzy finite automata and fuzzy regular expressions with membership values in lattice-ordered monoids. Fuzzy Sets and Systems 156, 68–92 (2005)

8. Cheng, W., Mo, Z.W.: A kind of classification of fuzzy finite automata. Busefal 84, 51–55 (2000)

9. Ying, M.S.: A formal model of computing with words. IEEE Transactions on Fuzzy Systems 10, 642–652 (2002)

10. Wang, Y., Mo, Z.M.: Minimization of mealy lattice finite automata based on fuzzy strings. Fuzzy Systems and Mathematics 23, 50–55 (2009)

11. Wang, Y., Yang, Q.: Minimization of mizumoto lattice finite automata. Journal of Frontiers of Computer Science and Technology 3, 441–446 (2009)

On Robustness of Min-Implication Fuzzy Relation Equation

Jian-hua Jin[1,2], Qing-guo Li[1], and Chun-quan Li[2]

[1] College of Mathematics and Econometrics, Hunan University,
  Changsha, 410082, China
[2] College of Sciences of Southwest Petroleum University,
  Chengdu, 610500, China
  {jjh2006ok,liqingguoli}@yahoo.com.cn

Abstract. The robustness of solutions of min-implication fuzzy relation equation is investigated in this paper. After proposing the definition of perturbation of fuzzy sets based on some logic-oriented equivalence measure, we discuss the relationship for the existence of solutions between fuzzy relation equation and its fuzzy perturbation equation. When the solutions exist, the perturbation issues of the maximal and minimal solutions are presented in terms of $\delta-$equalities, and the maximum of δ with the corresponding $\delta-$equality satisfied is derived.

Keywords: Fuzzy relation equation, solutions, robustness, fuzzy sets, $\delta-$equality.

1 Introduction

The problem of stability and robustness of fuzzy relation equations is one of the most important issues in engineering mathematics. The perturbation method has been extensively used in solving meaningful problems such as fuzzy control systems, fuzzy reasoning and logic systems [4, 9, 14, 15]. Since approximation errors appear in the formulation of fuzzy models of real systems, it is necessary to require the models of fuzzy systems to be tolerant of approximation errors [2]. There will be a corresponding reversal problem: Given the required small perturbation of results of fuzzy reasoning, how to determine the largest errors in premises that ensures the stability of fuzzy systems? As described in [5], the majority of fuzzy reasoning systems can be implemented by using fuzzy relation equations. We only need to consider the sensitivity and robustness of fuzzy relation equations. This forms the main topic of this research.

It is well known that there are different types of fuzzy relation equations corresponding to different fuzzy relation composite operators. Since fuzzy reasoning systems based on Boolean-type implications have good logic foundation and approximate capability [12], the fuzzy relation equation with min-implication composition based upon Boolean-type implication become an important directions in the study of fuzzy relation equations. Li [7] and Yue [13] discussed the

B.-Y. Cao and X.-J. Xie (Eds.): Fuzzy Engineering and Operations Research, AISC 147, pp. 257–264.
springerlink.com

perturbation theory of fuzzy matrix equations with min-implication based on some special implication operators, respectively. However, their results are not derived from the viewpoint of how the solutions of fuzzy relation equation are impacted by all perturbation elements of fuzzy matrix. In the previous work [1, 6, 11], the perturbation of fuzzy sets was expressed based on the notion of the maximum perturbation or δ-equalities of fuzzy sets via the distance measure on the unit interval $[0, 1]$. While, the behavior of a fuzzy logic system is mainly determined by its internal logic structure constituting the fuzzy connectives and fuzzy implication operators. Having this in mind, we introduce another form of perturbation based on logically equivalence measure or similarity of fuzzy sets, and present the perturbation parameters of solutions of fuzzy relation equations by the perturbation of fuzzy sets in this paper.

The rest of this paper is organized in the following way. Min-implication fuzzy relation equations based on R-implication and their decomposition are discussed in Section 2. The relation for the existence of solutions between fuzzy relation equation and its fuzzy perturbation equation is also presented. When the solutions exist, the perturbation issues of the maximal and minimal solutions are studied in terms of $\delta-$equalities in Section 3 and Section 4, respectively. Where, the maximum of δ that ensures the corresponding $\delta-$equality holds is derived. Finally the conclusions are presented in Section 5.

2 Min-implication Fuzzy Relation Equations Based on R-Implication and Their Decomposition

In this section we review some definitions and results already known, give some notations used throughout the paper and discuss the relation for the existence of solutions between fuzzy relation equation and its fuzzy perturbation equation.

Definition 1. Let t be a triangular $t-$norm (or $t-$norm). Define a binary operator $\to: [0, 1] \times [0, 1] \to [0, 1]$ as follows:

$$a \to b = \vee\{c \in [0, 1] | t(a, c) \leq b\}, \qquad \forall a, b \in [0, 1],$$

then $\to$ is called a residuated implication induced by $t-$norm, in short form, $R-$implication.

Example. The *Gödel* implication $a \to b = \begin{cases} 1, & a \leq b \\ b, & a > b \end{cases}$, the *Lukasiewicz* implication $a \to_L b = min\{1, 1 - a + b\}$ and the *Guoguen* implication $a \to_G b = min\{\frac{b}{a}, 1\}$ are all $R-$implication, which were induced by $t-$norms $t(a, b) = min\{a, b\}, t(a, b) = max\{0, a + b - 1\}$ and $t(a, b) = ab$, respectively.

In the paper, suppose that the universe of discourse is a finite set. Let $X = \{x_1, \cdots, x_n\}, Y = \{y_1, \cdots, y_n\}$ and $N_m = \{1, \cdots, m\}$. Suppose $F(X)$ denote the set of all of fuzzy subsets in X, $A \in F(x), B \in F(Y)$ and $R \in F(X \times Y)$. Since X and $X \times Y$ are finite, any element of $F(X)$ and $F(X \times Y)$ could be denoted by a vector and $n \times m$ matrix, respectively.

Definition 2. Let $A = (a_{ij})_{n \times m}$, $Q = (q_{ij})_{m \times k}$ be two fuzzy matrixes and $\to$ be any $R-$implication. Define the min-implication-composition of A and Q, which is denoted by $S = A \odot Q$, as follows, $s_{ij} = \bigwedge_{l \in N_m}(a_{il} \to q_{lj}), \forall i \in N_n, \forall j \in N_k$, where $S = (s_{ij})_{n \times k}$. If one of A and Q is given, to solve the other, then $S = A \odot Q$ is called a min-implication fuzzy relation equation.

Min-implication fuzzy relation equations have the following two types:

(I) Given $A = (a_{ij})_{n \times m}$, $B = (b_{ij})_{n \times k}$, determine $X = (x_{ij})_{m \times k}$ such that

$$A \odot X = B. \tag{1}$$

(II) Given $Q = (q_{ij})_{m \times k}$, $B = (b_{ij})_{n \times k}$, determine $X = (x_{ij})_{n \times m}$ such that

$$X \odot Q = B. \tag{2}$$

In 2001, Stamou et al. gave the decomposition of sup-t composition fuzzy relation equations in [10]. Similarly, min-implication fuzzy relation equations also have the following decomposition.

For instance, equation (1) can be decomposed as a set of k simpler min-implication fuzzy relation equations

$$A \odot \mathbf{r} = \mathbf{b}, \tag{3}$$

where $\mathbf{r}_{m \times 1}$ and $\mathbf{b}_{n \times 1}$ are the column vectors of X and B, respectively.

Eq.(3) can be further decomposed as a system of n min-implication fuzzy relation equations

$$\mathbf{a} \odot \mathbf{r} = b, \tag{4}$$

where $\mathbf{a}$ is a row vector of A and b is a coordinate of $\mathbf{b}$. Let $\mathcal{S}(\mathbf{a}, b)$ be the solution set of Eq.(4), i.e., $\mathcal{S}(\mathbf{a}, b) = \{\mathbf{r} | \mathbf{a} \odot \mathbf{r} = b\}$.

It is easily observed that Eq.(1) has a solution for X iff(if and only if) all the $k-$equations of form (3) have at least one solution for all column vector $\mathbf{b}$ of B, and the Eq.(3) has a solution for an $\mathbf{r}$ iff all the n equations of form (4) have at least one common solution for $\mathbf{r}$. This is a very useful approach to simplify the resolution of fuzzy relation equation.

In the present paper, we only discuss the stability and robustness of solutions of min-implication fuzzy relation equation (4), min-implication fuzzy relation equation (II) will be studied latter.

Definition 3. Let X be a nonempty set. If $A, B \in F(X)$, $0 \le \delta \le 1$ and $\to$ is a fuzzy connective, then we say that A, B are δ-equal based on logically equivalence measure, which denoted by $A \equiv (\delta)B$ in symbols, if the following condition holds [3,6]: $[A \equiv B] = \bigwedge_{x \in X}(A(x) \to B(x)) \wedge (B(x) \to A(x)) \ge \delta$.

Proposition 1. [3] *Let X be a nonempty set. If $A, B \in F(X)$, $0 \le \delta \le 1$ and $\to$ is Lukasiewicz implication(i.e., $x \to y = \min\{1, 1 - x + y\}$, for any $x, y \in [0, 1]$), then the following conditions are equivalent:*

1) $\bigvee\limits_{x\in X} |A(x) - B(x)| \leq 1 - \delta$;

2) $A \equiv (\delta)B$.

From Proposition 1, it shows that the perturbation of fuzzy sets based on logically equivalent measure is more comprehensive than that of fuzzy sets based on the notion of the maximum perturbation via the distance measure on the unit interval $[0,1]$. That is, the perturbation of fuzzy sets which discussed in [1,11] is a special case of Definition 3 which based on logically equivalence measure when we choose the fuzzy connective $\rightarrow$ as *Lukasiewicz* implication operator. Therefore, more general results about the robustness of fuzzy relation equations will be obtained in this paper. Of course, to obtain some nice results about the robustness of fuzzy relation equations based on logically equivalence measure introduced in this paper, we have to give some restrictions on the choice of fuzzy logical connectives, which include the fuzzy logical connectives used frequently in the practical applications.

Lemma 1. *Let* $A, B \in F(X), 0 < \delta \leq 1$. *Suppose* $\rightarrow$ *is the Guoguen implication induced by* $t-norm$ $T(a,b) = ab$. *Then* $A \equiv_G (\delta)B$ *iff* $\delta B(x) \leq A(x) \leq \frac{B(x)}{\delta} \wedge 1, \forall x \in X$.

Proof. $A \equiv_G (\delta)B$ iff $[A \equiv_G B] \geq \delta$ iff $\bigwedge\limits_{x\in X} (A(x) \rightarrow B(x)) \wedge (B(x) \rightarrow A(x)) \geq \delta$ iff $\bigwedge\limits_{x\in X} (1 \wedge \frac{B(x)}{A(x)} \wedge \frac{A(x)}{B(x)}) \geq \delta$ iff $\forall x \in X, 1 \wedge \frac{B(x)}{A(x)} \wedge \frac{A(x)}{B(x)} \geq \delta$ iff $\forall x \in X, \delta B(x) \leq A(x) \leq B(x)$ or $B(x) \leq A(x) \leq \frac{B(x)}{\delta} \wedge 1$, which is equivalent to that $\forall x \in X, \delta B(x) \leq A(x) \leq \frac{B(x)}{\delta} \wedge 1$.

Remark 1. From Lemma 1, clearly it holds that $A \equiv_G (1)B$ if and only if $A = B$, and it turns out that the more the value of δ at the unit $(0,1]$, the more approximating the distance from the fuzzy set A to the fuzzy set B.

Definition 4. Let Fuzzy set $A = (a_{ij})_{m\times n}, A' = (a'_{ij})_{m\times n}$. Then A' is called an upper perturbation of A, denoted by $A' \equiv_G (+\delta)A$, if $a'_{ij} \in [a_{ij}, \frac{a_{ij}}{\delta} \wedge 1]$, $\forall i \in N_m, j \in N_n$; A' is called a lower perturbation of A, denoted by $A' \equiv_G (-\delta)A$ if $a'_{ij} \in [\delta a_{ij}, a_{ij}], \forall i \in N_m, j \in N_n$; and A' is called a full perturbation of A, denoted by $A' \equiv_G (\delta)A$, if $a'_{ij} \in [\delta a_{ij}, \frac{a_{ij}}{\delta} \wedge 1], \forall i \in N_m, j \in N_n$.

Clearly Definition 4 is well defined by Lemma 1. Now we investigate the relationships about the solutions of the following Eq. (5) and that of Eq.(6).

Given fuzzy sets $A = (a_1, \cdots, a_m), A' = (a'_1, \cdots, a'_m), b = (b)_{1\times 1}$ and $b' = (b')_{1\times 1}$, determine $X = (x_j)_{m\times 1}, X' = (x'_j)_{m\times 1}$ such that

$$\bigwedge_{i=1}^{m} (a_i \rightarrow x_i) = b, \tag{5}$$

and

$$\bigwedge_{i=1}^{m} (a'_i \to x'_i) = b'. \tag{6}$$

For Eq.(5), Luo gave the following result in [8]:

Lemma 2. *For Eq. (5), $S(A,b) \neq \emptyset$ iff $G(b) \neq \emptyset$, where $G(b) = \{i \in N_m | a_i \to 0 \leq b\}$.*

Corollary 1. *For Eq. (5), if A is not a zero vector, then $S(A,b) \neq \emptyset$.*

Proof. Let A be a nonzero vector. Then there exists an $i \in N_m$ such that $a_i \neq 0$. Since $a_i \to 0 = \bigvee\{c \in [0,1] | t(a_i, c) \leq 0\} = 0$, $a_i \to 0 \leq b$. It follows that $G(b) \neq \emptyset$, and so $S(A,b) \neq \emptyset$ by Lemma 2.

Proposition 2. *For Eq.(5), Let $A' \equiv_G (\delta)A$ and $A \neq 0$. Then the set of solutions of Eq.(6) is nonempty, i.e., $S(A',b') \neq \emptyset$.*

Proof. Since $A' \equiv_G (\delta)A$ and $A \neq 0$, by Lemma 1, there exits an $i \in N_m$ such that $a_i \neq 0$, and $\delta a_i \leq a'_i \leq \frac{a_i}{\delta} \wedge 1$. Therefore $a'_i \neq 0$, and so $A' \neq 0$. It is followed that Eq.(6) has at least a solution from Corollary 1.

Proposition 2 studies the existence of solutions of fuzzy perturbation equation corresponding to fuzzy relation equation, and the similar results could be obtained through the same method in the following Corollary 2.

Corollary 2. *Suppose that A is nonzero vector for Eq.(5), then the following results hold:*

i) For Eq.(6), if $A' \equiv_G (+\delta)A$, $b' \equiv_G (+\varepsilon)b$, then $S(A',b') \neq \emptyset$;
ii) For Eq.(6), if $A' \equiv_G (-\delta)A$, $b' \equiv_G (-\varepsilon)b$, then $S(A',b') \neq \emptyset$;
iii) For Eq.(6), if $A' \equiv_G (+\delta)A$, $b' \equiv_G (-\varepsilon)b$, then $S(A',b') \neq \emptyset$;
iv) For Eq.(6), if $A' \equiv_G (-\delta)A$, $b' \equiv_G (+\varepsilon)b$, then $S(A',b') \neq \emptyset$.

3 Perturbation of the Least Solution of Fuzzy Relation Equations

In this section, we will mainly discuss the perturbation issue of the least solution of min-implication fuzzy relation equations, in which the results are presented in terms of δ-equality.

Lemma 3. [8] *If Eq.(5) has a solution, then $X_* = (a_1 * b, \cdots, a_m * b)^T$ is the least solution, where $*$ is the induced $t-$norm corresponding to the implication operator.*

For convenience, we will denote X_* and X'_* as the minimal solutions of Eq.(5) and Eq.(6), respectively. And x_{*i} and x'_{*i} represent the $i-th$ element of X_* and X'_*, respectively. In the following we assume that $t-$norm $*$ is the product $t-$norm, i.e., $a * b = ab$, for any $a, b \in [0,1]$.

Theorem 1. *Let $A \neq 0$, $A' \equiv_G (\delta)A$ and $b' \equiv_G (\varepsilon)b$. Then $X'_* \equiv_G (\delta')X_*$, where $\delta' = \bigwedge_{i=1}^{m} \delta'_i$, $x'_{*i} \equiv_G (\delta'_i)x_{*i}$, and for any $i \in N_m$, δ'_i can be calculated as follows:*

$$\delta'_i = \begin{cases} \delta\varepsilon, & \text{if } a_i \neq 0 \text{ and } b \neq 0, \\ 1, & \text{if } a_i = 0 \text{ or } b = 0. \end{cases}$$

Proof. Let fuzzy sets $A = (a_1, \cdots, a_m)$, $A' = (a'_1, \cdots, a'_m)$, $A' \equiv_G (\delta)A$ and $b' \equiv_G (\varepsilon)b$. Then $\delta a_i \leq a'_i \leq \frac{a_i}{\delta} \wedge 1$, and $\varepsilon b \leq b' \leq \frac{b}{\varepsilon} \wedge 1$. It follows that $a_i = 0$ if and only if $a'_i = 0$, and $b = 0$ if and only if $b' = 0$. Assume that $A \neq 0$, then by Corollary 1, Proposition 2 and Lemma 3, both Eq.(5) and Eq.(6) have the least solutions, respectively. On one hand, if $a_i \neq 0$ and $b \neq 0$, then $\delta'_i = [x'_{*i} \equiv x_{*i}] = (x'_{*i} \to x_{*i}) \wedge (x_{*i} \to x'_{*i}) = 1 \wedge \frac{x_{*i}}{x'_{*i}} \wedge \frac{x'_{*i}}{x_{*i}} = 1 \wedge \frac{a_i b}{a'_i b'} \wedge \frac{a'_i b'}{a_i b} \geq 1 \wedge \frac{a_i b}{(\frac{a_i}{\delta} \wedge 1)(\frac{b}{\varepsilon} \wedge 1)} \wedge \frac{a_i b \delta \varepsilon}{a_i b} = \delta\varepsilon \wedge \frac{a_i b \delta \varepsilon}{(a_i \wedge \delta)(b \wedge \varepsilon)} \geq \delta\varepsilon$. On the other hand, if $a_i = 0$ or $b = 0$, then $\delta'_i = [x'_{*i} \equiv x_{*i}] = (0 \to 0) \wedge (0 \to 0) = 1$.

Similarly, some corollaries could be obtained in the same way.

Corollary 3. *Let $A \neq 0$, $A' \equiv_G (+\delta)A$ and $b' \equiv_G (+\varepsilon)b$. Then $X'_* \equiv_G (\delta')X_*$, where $\delta' = \bigwedge_{i=1}^{m} \delta'_i$, $x'_{*i} \equiv_G (\delta'_i)x_{*i}$, and for any $i \in N_m$, δ'_i can be calculated as follows:*

$$\delta'_i = \begin{cases} \frac{a_i b \delta \varepsilon}{(a_i \wedge \delta)(b \wedge \varepsilon)}, & \text{if } a_i \neq 0 \text{ and } b \neq 0, \\ 1, & \text{if } a_i = 0 \text{ or } b = 0. \end{cases}$$

Corollary 4. *Let $A \neq 0$, $A' \equiv_G (-\delta)A$ and $b' \equiv_G (-\varepsilon)b$. Then $X'_* \equiv_G (\delta')X_*$, where $\delta' = \bigwedge_{i=1}^{m} \delta'_i$, $x'_{*i} \equiv_G (\delta'_i)x_{*i}$, and for any $i \in N_m$, δ'_i can be calculated as follows:*

$$\delta'_i = \begin{cases} \delta\varepsilon, & \text{if } a_i \neq 0 \text{ and } b \neq 0, \\ 1, & \text{if } a_i = 0 \text{ or } b = 0. \end{cases}$$

Corollary 5. *Let $A \neq 0$, $A' \equiv_G (+\delta)A$ and $b' \equiv_G (-\varepsilon)b$. Then $X'_* \equiv_G (\delta')X_*$, where $\delta' = \bigwedge_{i=1}^{m} \delta'_i$, $x'_{*i} \equiv_G (\delta'_i)x_{*i}$, and for any $i \in N_m$, δ'_i can be calculated as follows:*

$$\delta'_i = \begin{cases} \varepsilon \wedge \frac{a_i \delta}{(a_i \wedge \delta)}, & \text{if } a_i \neq 0 \text{ and } b \neq 0, \\ 1, & \text{if } a_i = 0 \text{ or } b = 0. \end{cases}$$

Corollary 6. *Let $A \neq 0$, $A' \equiv_G (-\delta)A$ and $b' \equiv_G (+\varepsilon)b$. Then $X'_* \equiv_G (\delta')X_*$, where $\delta' = \bigwedge_{i=1}^{m} \delta'_i$, $x'_{*i} \equiv_G (\delta'_i)x_{*i}$, and for any $i \in N_m$, δ'_i can be calculated as follows:*

$$\delta'_i = \begin{cases} \delta \wedge \frac{b\varepsilon}{(b \wedge \varepsilon)}, & \text{if } a_i \neq 0 \text{ and } b \neq 0, \\ 1, & \text{if } a_i = 0 \text{ or } b = 0. \end{cases}$$

4 Perturbation of the Maximal Solutions of Min-Implication Fuzzy Relation Equations

Lemma 4. [8] *Given* $i_0 \in G(b)$, *Eq.(5) has a maximal solution* $X^* = (x_j^*)_{m \times 1}$, *where*

$$x_j^* = \begin{cases} a_i * b, & \text{if } j = i_0, \\ 1, & \text{if } j \neq i_0. \end{cases}$$

Property 1. The number of the maximal solutions of Eq.(5) is equal to $|G(b)|$, where $|G(b)|$ denotes the cardinal number of the set $G(b)$.

Proof. It is easily concluded by Lemma 4.

From Property 1, one can see that the change of the number of the maximal solutions of Eq.(5) and Eq.(6) depend completely on the sets $G(b)$ and $G(b')$. Therefore, it suffices to consider the perturbation of maximal solutions in the following conditions in Theorem 2.

Theorem 2. *For* $i \in G(b) \cap G(b')$, *let* X^* *and* X'^* *be the maximal solution of Eq.(5) and Eq.(6), respectively. Then* $X'^* \equiv_G (\delta_i) X^*$, *where* δ_i *is concluded as follows:*

$$\delta_i = \begin{cases} \delta\varepsilon, & \text{if } A' \equiv_G (\delta)A, b' \equiv_G (\varepsilon)b, \\ \frac{b\varepsilon}{b \wedge \varepsilon} \wedge \delta, & \text{if } A' \equiv_G (-\delta)A, b' \equiv_G (+\varepsilon)b, \\ \delta\varepsilon, & \text{if } A' \equiv_G (-\delta)A, b' \equiv_G (-\varepsilon)b, \\ \frac{a_i\delta}{a_i \wedge \delta} \wedge \varepsilon, & \text{if } A' \equiv_G (+\delta)A, b' \equiv_G (-\varepsilon)b, \\ \delta\varepsilon, & \text{if } A' \equiv_G (+\delta)A, b' \equiv_G (+\varepsilon)b. \end{cases}$$

Proof. Let $i \in G(b) \cap G(b')$. Then the corresponding maximal solutions of Eq.(5) and Eq.(6) are respectively $x_j^* = \begin{cases} a_i * b, & \text{if } j = i \\ 1, & \text{if } j \neq i \end{cases}$ and $x_j'^* = \begin{cases} a_i' * b', & \text{if } j = i \\ 1, & \text{if } j \neq i \end{cases}$.
If $A' \equiv_G (\delta)A, b' \equiv_G (\varepsilon)b$, then $\delta a_i \leq a_i' \leq \frac{a_i}{\delta} \wedge 1$, and $\varepsilon b \leq b' \leq \frac{b}{\varepsilon} \wedge 1$. So
$[X^* \equiv X'^*] = 1 \wedge (x_i'^* \to x_i^*) \wedge (x_i^* \to x_i'^*) = 1 \wedge \frac{x_i^*}{x_i'^*} \wedge \frac{x_i'^*}{x_i^*} = 1 \wedge \frac{a_i b}{a_i' b'} \wedge \frac{a_i' b'}{a_i b} \geq 1 \wedge \frac{a_i b}{(\frac{a_i}{\delta} \wedge 1)(\frac{b}{\varepsilon} \wedge 1)} \wedge \frac{a_i b \delta\varepsilon}{a_i b} = \delta\varepsilon \wedge \frac{a_i b \delta\varepsilon}{(a_i \wedge \delta)(b \wedge \varepsilon)} \geq \delta\varepsilon$. Similarly, we could prove the other situations.

Noting that if $G(b') = G(b)$, then the perturbation of the maximal solutions of fuzzy relation equation depends only on the value of δ_i.

5 Conclusion

In this work, we have discussed the sensitivity of resolutions of min-implication fuzzy relation equation based on special R-implication. The perturbation method of fuzzy sets is proposed by using logically equivalence measure, and one can see that the perturbation of fuzzy sets based on logically equivalent measure is more comprehensive than that of fuzzy sets based on the notion of the maximum perturbation via the distance measure on the unit interval $[0, 1]$. By using the proposed method, we study the relation for the existence of solutions between fuzzy relation equation and its fuzzy perturbation equation. When the solutions exist, the perturbation issues of the maximal and minimal solutions are derived in terms of δ−equalities.

Acknowledgements. Thanks to the support by National Science Foundation of China (No.10771056) and 973 program (2011CB311808).

References

1. Cai, K.Y.: Robustness of fuzzy reasoning and δ-equalities of fuzzy sets. IEEE Trans. Fuzzy Syst. 9, 738–750 (2001)
2. Cheng, G.S., Fu, Y.X.: Error estimation of perturbations under CRI. IEEE Trans. Fuzzy syst. 14, 709–715 (2006)
3. Jin, J.H., Li, Y.M., Li, C.Q.: Robustness of fuzzy reasoning via logically equivalence measure. Information Sciences 177(22), 5103–5117 (2007)
4. Lu, J., Zhang, G., Ruand, D.: Intelligent multi-criteria fuzzy group decision-making for situation assessment. Soft Computing 12, 289–299 (2008)
5. Klir, G.J., Yuan, B.: Fuzzy Sets and Fuzzy Logic: Theory and Applications. Prentice-Hall, PTR, USA (1995)
6. Li, Y.M., Li, D.C., Pedrycz, W., et al.: An approach to measure the robustness of fuzzy reasoning. International Journal of Intelligent Systems 4(20), 393–413 (2005)
7. Li, Y., Chen, W.X., Zhao, L.L.: On the solutions and perturbation for the fuzzy matrix equations with $\bigwedge - \rightarrow$ composition. Journal of Liaocheng University (Natural Science) 20(2), 9–11 (2007)
8. Luo, Y.B., Li, Y.M.: Decomposition and resolution of θ−fuzzy relation equation based on R−implication. Fuzzy Systems Math. 17(4), 81–87 (2003)
9. Holcapek, M., Turcan, M.: A structure of fuzzy systems for support of decision making. Soft Computing 7, 234–243 (2003)
10. Stamou, G.B., Tzafestas, S.G.: Resolution of composite fuzzy relation equations based on Archimedean triangular norms. Fuzzy Sets and Systems 120, 395–407 (2001)
11. Ying, M.: Perturbation of fuzzy reasoning. IEEE Trans. Fuzzy Syst. 5, 625–629 (1999)
12. Wang, G.J.: On the logic foundation of fuzzy reasoning. Information Sciences 117, 231–251 (1999)
13. Yue, Q., Tan, Y.J.: Perturbation for matrix equations over Brouwerian lattices in minimplication composition. Journal of Fuzhou University (Natural Science) 33(6), 700–703 (2005)
14. Zadeh, L.A.: Outline of a new approach to the analysis of complex systems and decision processes. IEEE Trans. Syst. Man. Cybem. 3, 28–44 (1973)
15. Zadeh, L.A.: The concept of a linguistic variable and its applications to approximate reasoning, I, II, III. Information Sciences 8, 199–249, 301–357 (1975); 9, 43–80

Fuzzy Complex-Valued Integral and Its Convergence

Sheng-quan Ma[1] and Fu-chuan Chen[2]

[1] School of information science and Technology, Hainan Normal University,
  Haikou, Hainan 571158, China
  mashengquan@163.com
[2] QiongTai Teachers College, Haikou, Hainan 571100, China
  fuchuan9688@163.com

Abstract. The Fuzzy complex value measure and the fuzzy complex value integral are the important content of fuzzy complex analysis. This paper give the specific definition of the concept of fuzzy complex value measure, which is not only easy to research, but also inherits the original basic properties, and obtained some new results. Focusing on fuzzy complex value integral convergence theorem, we obtained 7 important convergence theorems, and build the foundation for fuzzy complex value integration theory.

Keywords: Fuzzy complex-valued measure, fuzzy complex-valued integral, conver gence.

1   Introduction

Fuzzy measure is the advancement of classical measure, Sugeno (see [1])first introduced the set functions using monotonic to replace additive in 1974, which was called fuzzy measures, established the theory of real fuzzy integral based on it, this theory has been developed rapidly, especially in application, which has been successfully applied in Comprehensive Evaluation, Engineering, Artificial intelligence, Machine learning, Pattern recognition, Information fusion and other fields. Similarity, When Buckley first introduced the conception (see [2]) of fuzzy complex number in 1989, the scholars began to study fuzzy complex numbers and fuzzy complex sets. Next, Qiu Jiqing (see [3]) introduced the concept of complex fuzzy measure and complex fuzzy integral in 1997, and studied some basic properties. The authors (see [4, 5]) discussed the properties of complex fuzzy measure and complex fuzzy integral. In order to make problems research more easily, the author introduced it specific definition in this paper, which inherits the original basic properties, and get some new results. Establish some convergence theorem of fuzzy complex integral.

B.-Y. Cao and X.-J. Xie (Eds.): Fuzzy Engineering and Operations Research, AISC 147, pp. 265–273.
springerlink.com   © Springer-Verlag Berlin Heidelberg 2012

2   Preliminary

2.1   Real Fuzzy Measure

Definition2.1.1[6] *Non-negative generalized real-valued set function* $\mu : \mathbb{F} \to [0,\infty]$ *is called real-valued fuzzy measure, if the following conditions hold:*

(1) *If* $\phi \in \mathbb{F}$, *,then* $\mu(\phi) = 0$.

(2) $E \in \mathbb{F}, F \in \mathbb{F}, E \subset F$,*then* $\mu(\mathrm{E}) \leq \mu(\mathrm{F})$.

(3) *If* $E_n \in \mathbb{F}(n=1,2,\cdots,\infty), E_1 \subset E_2 \subset \cdots, \bigcup\limits_{n=1}^{\infty} E_n \in \mathbb{F}$, *then* $\lim\limits_{n \to \infty} \mu(E_n) = \mu\left(\bigcup\limits_{n=1}^{\infty} E_n\right)$.

(4) *If* $E_n \in \mathbb{F}(n=1,2,\cdots,\infty), E_1 \supset E_2 \supset \cdots$, *and* $\exists n_0.s.t.\mu\left(E_{n_0}\right) < \infty, \bigcap\limits_{n=1}^{\infty} E_n \in \mathbb{F}$, *then*

$$\lim_{n \to \infty} \mu(E_n) = \mu\left(\bigcap_{n=1}^{\infty} E_n\right),$$

μ *is called regular real-valued fuzzy measure, if* $\mu(X) = 1$.

2.2   Real Fuzzy Integral

The real fuzzy integral is provided by Japanese scholars Sugeno [1] in 1974, it shows as following:

Let $(X, \mathbb{F}, \mu)$ denotes measure space, $X \in \mathbb{F}$, $\mu : \mathbb{F} \to [0,+\infty)$ denotes real fuzzy measure, $\mathbb{F}$ denotes the finite nonnegative measurable function sets in X (f is measurable on X , if $\{x| f(x) \leq a\} \in \mathbb{F}$ for arbitrary a), Note that $F_\alpha = \{x| f(x) \geq \alpha\}, F_{\alpha'} = \{x| f(x) > \alpha\}$ for arbitrary $f \in \mathbb{F}$ and $\alpha \in [0,+\infty), F_\alpha, F_{\alpha'}$ denotes α -Cut sets and α -strong Cut sets.

Definition 2.2.1[6]. *Let* $A \in \mathbb{F}, f \in \mathbb{F}$, *and* μ *denotes real fuzzy measure. Sugeno type real fuzzy integral of* f *about* μ *on* A *is defined as following:*

$$(s)\int_A f d\mu = \sup_{\alpha \in [0,+\infty)} \left[\alpha \wedge \mu(A \cap F_\alpha)\right].$$

If $A = X$, *which be denoted by* $(s)\int f d\mu$.

3   Fuzzy Complex-Valued Integral

3.1   Fuzzy Complex-Valued Measure

We agreed that $\hat{R}^+ = [0,+\infty), \hat{C}^+ = \{x + iy | x, y \in \hat{R}^+\}$ and the order of complex numbers is defined according to the condition that the real part and the imaginary part satisfy the order relation at the same time. Non-negative complex refers to the real and imaginary parts are non-negative.

Definition 3.1.1[3]. *Let X is non-empty sets.* $\mathcal{A}$ *is* $\sigma-$ *algebra that is composed by the subset of B,* $\mu:\mathcal{A}\to\hat{C^+}$ *is called fuzzy complex-valued measure, if the following conditions hold:*

(1) $\mu(\phi)=0$;

(2) *If* $A\subset B$ *, then* $|\mu(A)|\leq|\mu(B)|$;

(3) *If* $\{A_n\}_1^\infty\uparrow$, *where* $A_n\in A(n=1,2,\cdots)$ $A_1\subseteq A_2\subseteq\cdots\subseteq A_n\subseteq\cdots$, *then*

$$\mu(\bigcup_{n=1}^\infty A_n)=\lim_{n\to\infty}\mu(A_n);$$

(4) *If* $\{A_n\}_1^\infty\downarrow$ *, where* $A_n\in A$ $(n=1,2,\cdots)$ $A_1\supseteq A_2\supseteq\cdots\supseteq A_n\supseteq\cdots$ *and exiting* n_0 *such that* $|\mu(A_{n_0})|<+\infty$ *, then*

$$\mu(\bigcap_{n=1}^\infty A_n)=\lim_{n\to\infty}\mu(A_n),$$

where $A,B,A_n\in\mathcal{A}(n=1,2\cdots)$ *. The space* $(X,\mathcal{A},\mu)$ *is called fuzzy complex- valued measure spaces.*

In practical application, we usually use the following more specific Definition 2.1.2.

Definition 3.1.1$'$ *. Let* (X,A) *measurable space. Set function* $\mu:A\to\hat{C^+}$ *is called fuzzy complex-valued measure, if the following conditions hold:*

(1) $\mu(\Phi)=0$.

(2) $A,B\in A$ **,** $A\subseteq B\Rightarrow$ $\mathrm{Re}(\mu(A))\leq\mathrm{Re}(\mu(B))$ *and* $\mathrm{Im}(\mu(A))\leq\mathrm{Im}(\mu(B))$.
Note $\mu(A)\leq\mu(B)$.

(3) $A_n\in A$ $(n=1,2,\cdots)$ $A_1\subseteq A_2\subseteq\cdots\subseteq A_n\subseteq\cdots\Rightarrow\mu(\bigcup_{n=1}^\infty A_n)=\lim_{n\to\infty}\mu(A_n)\cdot$

(4) $A_n\in A$ $(n=1,2,\cdots)$ $A_1\supseteq A_2\supseteq\cdots\supseteq A_n\supseteq\cdots$ *and* $\exists n_0$ *,we have*

$\mathrm{Re}(\mu(A_{n_0}))<\infty,\mathrm{Im}(\mu(A_{n_0}))<\infty\Rightarrow\mu(\bigcap_{n=1}^\infty A_n)=\lim_{n\to\infty}\mu(A_n)\cdot$

3.2 Fuzzy Complex-Valued Measurable Function and Fuzzy Complex Valued Integral

Definition 3.2.1[3]. *Let* (X,A,μ) *be fuzzy complex-valued measure space. The mapping* $f:X\to C$ *is called fuzzy complex-valued measurable function, if the following conditions hold:*

$$\{x\in X|\mathrm{Re}[f(x)]\geq a,\mathrm{Im}[f(x)]\geq b\}\in A$$

for arbitrary $a+bi\in C$.

Definition 3.2.2. *Let the mapping* $f:(X,A,\mu) \to \hat{C}^{+}$ *denotes fuzzy complex-valued measurable function and fuzzy complex-valued integral of* f *about* μ *on* A *is defined as following*

$$\int_{A} \overline{f} d\mu \overset{\Delta}{=} \sup_{\alpha \in [0,\infty)} \min\{\alpha, \operatorname{Re}\mu[\overline{f}_{\alpha} \cap A]\} + i \sup_{\alpha \in [0,\infty)} \min\{\alpha, \operatorname{Im}\mu[\overline{f}_{\alpha} \cap A]\}$$

or

$$\int_{A} \overline{f} d\mu \overset{\Delta}{=} \vee_{\alpha \in [0,\infty)} (\alpha \wedge \operatorname{Re}\mu[\overline{f}_{\alpha} \cap A]) + i \vee_{\alpha \in [0,\infty)} (\alpha \wedge \operatorname{Im}\mu[\overline{f}_{\alpha} \cap A])$$

for arbitrary $A \in A$, *where*

$$\overline{f}_{\alpha} = \{x \mid \operatorname{Re}[\overline{f}(x)] > \alpha, I\operatorname{m}[\overline{f}(x)] > \alpha\}, |\overline{f}|_{\alpha} = \{x \mid \operatorname{Re}[\overline{f}(x)]| \triangleright \alpha, |I\operatorname{m}[\overline{f}(x)]| \triangleright \alpha\}.$$

Theorem 3.2.1. *Fuzzy complex-valued integral has the following properties:*

(1) *If* $f_{1} \leq f_{2}$,*then* $\int_{A} f_{1} d\mu \leq \int_{A} f_{2} d\mu$.

(2) *If* $\mu(A) = 0$,*then* $\int_{A} f d\mu = 0$.

(3) $\int (f_{1} \vee f_{2}) d\mu \geq \int f_{1} d\mu \vee \int f_{2} d\mu$.

(4) *If* $A \subseteq B$,*then* $\int_{A} f d\mu \leq \int_{B} f d\mu$.

(5) *If a is Nonnegative constants, we have* $\int_{A} (a \vee f) d\mu = \int_{A} a d\mu \vee \int_{A} f d\mu$.

(It can be proved in a similar way as show literature [4])

4   The Convergence Theorem of Fuzzy Complex-Valued Integral

Definition 4.3.1. *Let* (X,A,μ) *be fuzzy complex-valued measure space,* $\overline{f}_{n}(n = 1,2,...), \overline{f}$ *denotes fuzzy complex measurable function,* $\{\overline{f}_{n}\}$ *is called fuzzy complex mean convergence on* A, *If* $\lim_{n \to \infty} \int |\overline{f}_{n} - \overline{f}| d\mu = 0$.

Definition 4.3.2. *Let* (X,A,μ) *be fuzzy complex-valued measure space,* $\overline{f}_{n}(n = 1,2,...), \overline{f}$ *denotes fuzzy complex measurable function, and arbitrary* $A \in A$, $\{\overline{f}_{n}\}$ *is called convergence in fuzzy complex measure* $\overline{f}$, *if the following conditions hold:*

$$\mu(\{x \mid |\overline{f}_{n} - \overline{f}| \geq \varepsilon\} \cap A) \to 0 \text{ for every arbitrary } \varepsilon_{i} > 0, (i = 1,2) ,$$

where $\varepsilon = \varepsilon_{1} + i\varepsilon_{2}$.*Note* $\overline{f}_{n} \overset{\mu.}{\longrightarrow} \overline{f}$.

Definition 4.3.3. *Let* (X,A,μ) *be fuzzy complex-valued measure space,* $\overline{f}_{n}(n = 1,2,...), \overline{f}$ *denotes fuzzy complex measurable function, then*

(1) $\{\tilde{f}_{n}\}$ *is called Almost everywhere convergence in* $\overline{f}$, *if existing* $B \in A$ *such that* $\mu(B) = 0$, *and* $\{\overline{f}_{n}\}$ *Converges to* $\overline{f}$ *in* $A \setminus B$.*Note* $\overline{f}_{n} \overset{a.e.}{\longrightarrow} \overline{f}$.

(2) $\{\overline{f_n}\}$ is called almost everywhere uniform converges to $\overline{f}$, if existing $B \in A$ such that $\mu(B) = 0$, and $\{\overline{f_n}\}$ uniform converges to $\overline{f}$ in $A \setminus B$, Note $\overline{f_n} \xrightarrow{a.e.u.} \overline{f}$

(3) $\{\overline{f_n}\}$ is called almost everywhere uniform converges to $\overline{f}$, if existing Set sequence $\{E_k\}$ such that $\tilde{\mu}(E_k) \to 0$, and $\{\overline{f_n}\}$ uniform converges to $\overline{f}$ in $A \setminus E_k$ for arbitrary k.

Definition 4.3.4. Let (X, A, μ) be fuzzy complex-valued measure space, μ is called null-additive, if $\mu(F) = 0, E \cap F = \phi \Rightarrow \mu(E \cup F) = \mu(E)$ for arbitrary $E, F \in A$.

Theorem 4.3.1. Let (X, A, μ) be fuzzy complex-valued measure space, $\overline{f_n}$ fuzzy mean convergence in $\overline{f}$ on A is equivalent with $\overline{f_n} \xrightarrow{\mu} \overline{f}$ on A.

Proof. Suppose $\overline{\mu}(\{x \mid |\overline{f_n} - \overline{f}| \geq \varepsilon\} \cap A) \to 0$ for arbitrary $\varepsilon_i > 0, (i = 1, 2)$ on A, where $\varepsilon = \varepsilon_1 + i\varepsilon_2$, so $\mu(\{x \mid |\text{Re}(\overline{f_n} - \overline{f})| \geq \varepsilon_1, |\text{Im}(\overline{f_n} - \overline{f})| \geq \varepsilon_2\} \cap A) < \varepsilon$

$\int_A |\overline{f_n} - \overline{f}| d\mu = \underset{\alpha \in [0,\infty)}{\vee} (\alpha \wedge \text{Re}\mu[|\overline{f_n} - \overline{f}|_\alpha \cap A]) + i \underset{\alpha \in [0,\infty)}{\vee} (\alpha \wedge \text{Im}\mu[|\overline{f_n} - \overline{f}|_\alpha \cap A])$,

$\text{Re}\mu[|\overline{f_n} - \overline{f}|_\varepsilon \cap A] = \text{Re}\mu[\{x \mid |\text{Re}(\overline{f_n} - \overline{f})| \geq \varepsilon_1, |\text{Im}(\overline{f_n} - \overline{f})| \geq \varepsilon_2\} \cap A]) < \varepsilon$,

$\text{Im}\mu[|\overline{f_n} - \overline{f}|_\varepsilon \cap A] = \text{Im}\mu[\{x \mid |\text{Re}(\overline{f_n} - \overline{f})| \geq \varepsilon_1, |\text{Im}(\overline{f_n} - \overline{f})| \geq \varepsilon_2\} \cap A]) < \varepsilon$

for arbitrary $n > n_0$, so

$\int_A |\overline{f_{n_i}} - \overline{f}| d\mu = \underset{\alpha \in [0,\infty)}{\vee} (\alpha \wedge \text{Re}\mu[|\overline{f_{n_i}} - \overline{f}|_\alpha \cap A]) + i \underset{\alpha \in [0,\infty)}{\vee} (\alpha \wedge \text{Im}\mu[|\overline{f_{n_i}} - \overline{f}|_\alpha \cap A])$

$\geq [\varepsilon_1' \wedge \text{Re}\mu(\{x \mid |\text{Re}(\overline{f_{n_i}} - \overline{f})| \geq \varepsilon_1', |\text{Im}(\overline{f_{n_i}} - \overline{f})| \geq \varepsilon_2'\} \cap A)] + i[\varepsilon_2' \wedge \text{Im}\mu(\{x \mid |\text{Re}(\overline{f_{n_i}} - \overline{f})|$

$\geq \varepsilon_1', |\text{Im}(\overline{f_{n_i}} - \overline{f})| \geq \varepsilon_2'\} \cap A)]$

$\geq (\varepsilon_1' \wedge \delta_1) + i(\varepsilon_2' \wedge \delta_2) > 0$,

contradicted with $\lim_{n \to \infty} \int_A |\overline{f_n} - \overline{f}| d\mu = 0$, so we have $\overline{f_n} \xrightarrow{\mu} \overline{f}$ on A.

Theorem 4.3.2. Let (X, A, μ) be fuzzy complex-valued measure spaces, $\{\overline{f_n}\}$ denotes fuzzy complex measurable functions sequence, and arbitrary $A \in A$, if $\{\overline{f_n}\}$ monotone decreasing converges to $\overline{f}$ and existing n_0 such that

$$\text{Re}(\mu(\{x \mid \overline{f}_{n_0} > \int_A \overline{f} d\mu + \varepsilon\} \cap A)) < \infty, \text{Im}(\mu(\{x \mid \overline{f}_{n_0} > \int_A \overline{f} d\mu + \varepsilon\} \cap A)) < \infty$$

for arbitrary $\varepsilon_i' > 0, (i = 1, 2)$, where $\varepsilon' = \varepsilon_1' + i\varepsilon_2'$

Proof. Since $\overline{f}_1 \geq \overline{f}_2 \geq \ldots$, then

$$\int_A \overline{f}_1 d\mu \geq \int_A \overline{f}_2 d\mu \geq \ldots$$

So

$$\lim_{n\to\infty} \int_A \overline{f}_n d\mu = \bigwedge_{n=1}^{\infty} \int_A \overline{f}_n d\mu$$

(where $\bigwedge_{n=1}^{\infty} \int_A \overline{f}_n d\mu = \bigwedge_{n=1}^{\infty} \mathrm{Re} \int_A \overline{f}_n d\mu + i \bigwedge_{n=1}^{\infty} \mathrm{Im} \int_A \overline{f}_n d\mu$)

Since $\overline{f}_n \geq \overline{f}$ for any n, then

$$\int_A \overline{f}_n d\mu \geq \int_A \overline{f} d\mu, \text{ so } \bigwedge_{n=1}^{\infty} \int_A \overline{f}_n d\mu \geq \int_A \overline{f} d\mu$$

If $\bigwedge_{n=1}^{\infty} \int_A \overline{f}_n d\mu > \int_A \overline{f} d\mu$, then

$$\int_A \overline{f} d\mu = \lambda < \infty \text{ (Where } \lambda = \lambda_1 + i\lambda_2 \text{)}$$

and existing $\gamma_i \in (0,\infty)$ (where $\gamma = \gamma_1 + i\gamma_2$, $(i=1,2)$), we have

$$\bigwedge_{n=1}^{\infty} \int_A \overline{f}_n d\mu > \gamma > \lambda \Rightarrow \forall n, \bigvee_{\alpha\in[0,\infty)} [\alpha \wedge \mathrm{Re}\mu(A\cap(\overline{f}_n)_\alpha)] + i \bigvee_{\alpha\in[0,\infty)} [\alpha \wedge \mathrm{Im}\mu(A\cap(\overline{f}_n)_\alpha)] > \gamma$$

$$\Rightarrow \forall n, \mathrm{Re}\mu(A\cap(\overline{f}_n)_\alpha) + i\mathrm{Im}\mu(A\cap(\overline{f}_n)_\alpha) > \gamma.$$

Let $\varepsilon_i = \dfrac{\gamma_i + \lambda_i}{2}$ $(i=1,2)$. We knows existing n_0 such that

$$\mu(\{x \mid \overline{f}_{n_0} > \int_A \overline{f} d\mu + \varepsilon\} \cap A) < \infty,$$

$$\gamma_i = \lambda_i + 2\varepsilon_i > \lambda_i + \varepsilon_i \Rightarrow \{x \mid \overline{f}_{n_0} \geq \gamma\} \subseteq \{x \mid \overline{f}_{n_0} > \lambda + \varepsilon\}$$

$$\Rightarrow \mu(A\cap(\overline{f}_{n_0})_\gamma) \leq \mu(\{x \mid \overline{f}_{n_0} > \lambda + \varepsilon\} \cap A) < \infty.$$

By the continuity of μ, we have $A\cap(\overline{f}_{n_1})_\gamma \supseteq A\cap(\overline{f}_{n_2})_\gamma \supseteq \ldots$

$$\mu(A\cap(\overline{f})_\gamma) = \mu(\bigcap_{n=1}^{\infty}[A\cap(\overline{f}_n)_\gamma]) = \lim \mu(A\cap(\overline{f}_n)_\gamma) \geq \gamma \geq \lambda$$

$$\Rightarrow \int_A \overline{f} d\mu \overset{\Delta}{=} \bigvee_{\alpha\in[0,\infty)} (\alpha \wedge \mathrm{Re}\mu[f_\alpha \cap A]) + i \bigvee_{\alpha\in[0,\infty)} (\alpha \wedge \mathrm{Im}\mu[f_\alpha \cap A])$$

$$\geq (\gamma_1 \wedge \mathrm{Re}\mu[f_\alpha \cap A]) + i(\gamma_2 \wedge \mathrm{Im}\mu[f_\alpha \cap A]) > \lambda_1 + i\lambda_2 = \lambda$$

contradicted with $\int_A \overline{f} d\mu = \lambda$, so we have $\bigwedge_{n=1}^{\infty} \int_A \overline{f}_n d\mu = \int_A \overline{f} d\mu$, that is

$$\lim_{n\to\infty} \int_A \overline{f}_n d\mu = \int_A \overline{f} d\mu.$$

Theorem 4.3.3. Let (X,A,μ) be fuzzy complex-valued measure spaces, $\{\overline{f}_n\}$ denotes fuzzy complex measurable functions sequence, and arbitrary $A \in \mathrm{A}$. If $\overline{f}_n \to \overline{f}$ on A and existing n_0 such that

$$\mathrm{Re}(\mu(\{x \mid \sup_{n\geq n_0} \overline{f}_n > \int_A \overline{f} d\mu + \varepsilon\} \cap A)) < \infty,$$

$$\mathrm{Im}(\mu(\{x \mid \sup_{n \geq n_0} \overline{f_n} > \int_A \overline{f} d\mu + \varepsilon\} \cap A)) < \infty \quad \text{for} \quad \text{arbitrary} \quad \varepsilon_i' > 0, (i = 1,2) \quad \text{,where}$$

$\varepsilon' = \varepsilon_1' + i\varepsilon_2'$, then $\lim_{n \to \infty} \int_A \overline{f_n} d\mu = \int_A \overline{f} d\mu$.

Proof. Let $\overline{h_n} = \bigvee_{n=1}^{\infty} \overline{f_i}$ and $\overline{g_n} = \bigwedge_{n=1}^{\infty} \overline{f_i}$, then $\{h_n\}$ is monotone decreasing and $\{g_n\}$

is monotone increasing, both of them Converges to $\overline{f}$ and $\overline{g_n} \leq \overline{f_n} \leq \overline{h_n}$, hence,

$\int_A \overline{g_n} d\mu \leq \int_A \overline{f_n} d\mu \leq \int_A \overline{h_n} d\mu$, we know by theorem 3.3.2 that

$$\int_A \overline{f} d\mu = \lim_{n \to \infty} \int_A \overline{g_n} d\mu = \lim_{n \to \infty} \int_A \overline{h_n} d\mu = \lim_{n \to \infty} \int_A \overline{f_n} d\mu.$$

Theorem 4.3.4. *Let (X, A, μ) be fuzzy complex-valued measure space, $\overline{f_n}, \overline{f}$ denotes fuzzy complex measurable function and $\overline{f_n} \xrightarrow{a.e.} \overline{f}$ for each $A \in A$. If $\tilde{\mu}$ is null-additive and existing n_0 such that*

$$\mathrm{Re}(\mu(\{x \mid \sup_{n \geq n_0} \overline{f_n} > \int_A \overline{f} d\mu + \varepsilon\} \cap A)) < \infty,$$

$$\mathrm{Im}(\mu(\{x \mid \sup_{n \geq n_0} \overline{f_n} > \int_A \overline{f} d\mu + \varepsilon\} \cap A)) < \infty$$

for arbitrary $\varepsilon_i' > 0, (i = 1,2)$, where $\varepsilon' = \varepsilon_1' + i\varepsilon_2'$. then

$$\lim_{n \to \infty} \int_A \overline{f_n} d\mu = \int_A \overline{f} d\mu.$$

Proof. If $\overline{f_n} \xrightarrow{a.e.} \overline{f}$ on A, then existing $B \in A$ such that $\mu(B) = 0$, so

$$\int_{A \setminus B} \overline{f} d\mu = \bigvee_{\alpha \in [0,\infty)} (\alpha \wedge \mathrm{Re}\mu[(A \setminus B) \cap \overline{f_\alpha}]) + i \bigvee_{\alpha \in [0,\infty)} (\alpha \wedge \mathrm{Im}\mu[(A \setminus B) \cap \overline{f_\alpha}])$$

$$= \bigvee_{\alpha \in [0,\infty)} (\alpha \wedge \mathrm{Re}\mu[(A \cap \overline{f_\alpha}) \setminus B]) + i \bigvee_{\alpha \in [0,\infty)} (\alpha \wedge \mathrm{Im}\mu[(A \cap \overline{f_\alpha}) \setminus B])$$

$$= \bigvee_{\alpha \in [0,\infty)} (\alpha \wedge \mathrm{Re}\mu(A \cap \overline{f_\alpha})) + i \bigvee_{\alpha \in [0,\infty)} (\alpha \wedge \mathrm{Im}\mu(A \cap \overline{f_\alpha}))$$

$$= \int_A \overline{f} d\mu.$$

Similarly

$$\int_A \overline{f_n} d\mu = \int_{A \setminus B} \overline{f_n} d\mu.$$

Since $\mu(\{x \mid \sup_{n \geq n_0} \overline{f_n} > \int_A \overline{f} d\mu + \varepsilon\} \cap A) < \infty$,

and we know by theorem3.3.3 that $\lim_{n \to \infty} \int_{A \setminus B} \overline{f_n} d\mu = \int_{A \setminus B} \overline{f} d\mu$.

Hence $\lim_{n \to \infty} \int_A \overline{f_n} d\mu = \int_A \overline{f} d\mu$.

Theorem 4.3.5. *Let (X, A, μ) be fuzzy complex-valued measure spaces, $\{\overline{f_n}\}$ denotes a fuzzy complex measurable function sequence. If $\overline{f_n}$ uniform converges to $\overline{f}$ on $A \in A$, then*

$$\lim_{n \to \infty} \int_A \overline{f_n} d\mu = \int_A \overline{f} d\mu.$$

Proof.

(1) If $\int_A \overline{f} d\mu = \infty$, the proof is similar with the proof of Theorem 2.3.3, let

$\overline{g}_n = \overset{\infty}{\underset{n=1}{\wedge}} \overline{f}_i$ Then $\lim_{n\to\infty} \int_A \overline{f}_n d\mu \geq \lim_{n\to\infty} \int_A \overline{g}_n d\mu = \int_A \overline{f} d\mu = \infty$.

(2) If $\int_A \overline{f} d\mu = \underset{\alpha\in[0,\infty)}{\vee}(\alpha \wedge \mathrm{Re}\,\mu(\overline{f}_\alpha \cap A)) + i \underset{\alpha\in[0,\infty)}{\vee}(\alpha \wedge \mathrm{Im}\,\mu(\overline{f}_\alpha \cap A)) = \lambda < \infty$.

Then for any $\alpha \leq \lambda, \mu(\overline{f}_\alpha \cap A) \geq \lambda$, and for arbitrary $\alpha > \lambda, \mu(\overline{f}_\alpha \cap A) \leq \lambda$.

Let α_n monotone decreasing approach to λ . Then $\overline{f}_{\alpha_1} \subseteq \overline{f}_{\alpha_2} \subseteq \cdots$

and $\overset{\infty}{\underset{n=1}{\cup}}\overline{f}_{\alpha_n} = \overline{f}_\lambda$, we know by the lower continuous of μ that

$$\mu(\overline{f}_\lambda \cap A) = \lim_{n\to\infty}\mu(\overline{f}_{\alpha_n} \cap A) \leq \lambda,$$

If $\overline{f}_n \overset{u.}{\longrightarrow} \overline{f}$ on A and existing n_0 such that

$$\Rightarrow \{x|\sup_{n\geq n_0}\overline{f}_n(x) \geq \lambda + \varepsilon\} \cap A \subseteq \{x|f(x) \geq \lambda\} \cap A = \overline{f}_\lambda \cap A$$

$$\Rightarrow \mu(\{x|\sup_{n\geq n_0}\overline{f}_n(x) \geq \lambda + \varepsilon\} \cap A) \leq \lambda < \infty$$

for arbitrary $\varepsilon_i' > 0 (i = 1,2)$ and arbitrary $x \in A$, where $\varepsilon' = \varepsilon_1' + i\varepsilon_2'$.

Hence we know by theorem 3.3.3 that $\lim_{n\to\infty} \int_A \overline{f}_n d\mu = \int_A \overline{f} d\mu$.

Theorem 4.3.6. Let (X, A, μ) be fuzzy complex-valued measure space, $\overline{f}_n, \overline{f}$ denotes fuzzy complex measurable function and $\overline{f}_n \overset{a.e.}{\longrightarrow} \overline{f}$ for each $A \in A$. If $\tilde{\mu}$ is null-additive and existing $\{B_k\} \subseteq A, B_1 \supseteq B_2 \supseteq \cdots, \mu(B_k) \to 0$, we have $\lim_{n\to\infty} \int_{A\backslash B_k} \overline{f}_n d\mu = \int_{A\backslash B_k} \overline{f} d\mu$.

Proof: it is similar with the proof of theorem 3.3.4

Theorem 4.3.7. Let (X, A, μ) be fuzzy complex-valued measure space, $\overline{f}_n, \overline{f}$ denotes fuzzy complex measurable function and $\overline{f}_n \overset{a.e.}{\longrightarrow} \overline{f}$ for each $A \in A$. If $\tilde{\mu}$ is null-additive and existing $\{B_k\} \subseteq A$ such that $B_1 \supseteq B_2 \supseteq \cdots$, and $\mu(B_k) \to 0$, then $\lim_{n\to\infty} \int_{A\backslash B_k} \overline{f}_n d\mu = \int_{A\backslash B_k} \overline{f} d\mu$

Proof. (1) If $\overline{f}_n \overset{a.u.}{\longrightarrow} \overline{f}$ on A, then existing $\{E_k\} \subseteq A,$, we have $\mu(E_k) \to 0$.

If $\overline{f}_n \overset{u.}{\longrightarrow} \overline{f}$ on $A \backslash E_k$, let $B_k = \overset{k}{\underset{i=1}{\cap}} E_i \subseteq E_k$. Then $B_1 \supseteq B_2 \supseteq \cdots, \mu(B_k) \to 0$,

$A \backslash B_k = \overset{k}{\underset{i=1}{\cup}}(A \backslash E_i),$

Since $\overline{f}_n \overset{u.}{\longrightarrow} \overline{f}$ on $A \backslash E_k$ for any k , so we have $\overline{f}_n \overset{u.}{\longrightarrow} \overline{f}$ on $A \backslash B_k$, we know it is clear by Theorem 3.3.5.

5 Conclusion

Fuzzy complex-valued measure and fuzzy complex-valued integral are the important contents of fuzzy complex analytics, which in comprehensive evaluation, engineering technology, artificial intelligence, machine learning, pattern recognition and information fusion fields will have broad application prospect, the author introduce its specific definition, which not only to make problems research more easily, but also to inherits the original basic properties, then get some new results, Established convergence theorem of Complex fuzzy integral in accordance with it.

Acknowledgements. Thanks to the support by Natural Science Foundation of Hainan Province (No.111007) and Hainan International cooperation Key Project (No.GJXM201105).

References

1. Sugeno, M.: Theory of Fuzzy Integrals and Applications. Tokyo Institute of Technology (1974)
2. Buckley, J.J.: Fuzzy complex numbers. Fuzzy Sets and Systems 33, 333–345 (1989)
3. Qiu, J.Q., Li, F.C., Shu, L.Q.: Complex fuzzy measure and complex fuzzy integral. Hebei Light Chemical College Journals 1, 1–4 (1997)
4. Ma, S.Q., Chao, C.: Fuzzy complex analysis. Minorities Press, Beijing (2002)
5. Ma, S.Q.: Fuzzy complex analysis foundation theory. Science Press, Beijing (2010)
6. Wang, X.Z.: The application of Fuzzy measure and fuzzy integral in classification technology. Science Press (2008)

On the Generalized Nonlinear Fuzzy Variational Inclusions

Xiao-min Wang[1] and Jia Liu[2]

[1] Northeastern University at Qinhuangdao, Qinhuangdao 066004, China
xmwang0823@163.com
[2] College of Science, Northeastern University, Shenyang 110004, China
liujia19852006@126.com

Abstract. In this paper, a new generalized nonlinear fuzzy variational inclusion in real Hilbert space is introduced. Using the auxiliary principle technique, a iterative algorithm to compute the approximate solution is constructed, and the convergence criteria is proved under some mild conditions.

Keywords: Variational inclusions, iterative algorithm, fuzzy mapping, existence.

1   Introduction

Variational inclusion problems are among the most interesting and intensively studied classes of mathematical problems and have wide applications in the fields of optimization and control, economics and transportation equilibrium, engineering science. For the past years, many existence results and iterative algorithms for various variational inequality and variational inclusion problems have been studied. For details, please see [1–6] and the references therein. Recently, Wang [1] studied the existence and approximation of solutions on generalized fuzzy varational inclusions with set-valued accretive mappings in Banach space. On the other hand, Xu [2] and Lan [4] respectively introduced and studied a class of nonlinear variational inclusions. Inspired and motivated by the above works, this paper introduces a new type of generalized nonlinear fuzzy variational inclusion.

2   Preliminaries

Let E be a Hilbert space with inner product and norm denoted by $< \cdot, \cdot >$ and $\|\cdot\|$, respectively. Let $CB(E)$ denotes the families of all the nonempty bounded closed subsets of E, $\tilde{H}(\cdot,\cdot)$ denotes the Hausdorff metric on $CB(E)$ defined by

$$\tilde{H}(M,N) = \max\left\{\sup_{a \in M} d(a,N), \sup_{b \in N} d(M,b)\right\}, \quad \forall M, N \in CB(E).$$

where $d(a,N) = \inf_{b \in N} \|a-b\|$, $d(M,b) = \inf_{a \in M} \|a-b\|$.

B.-Y. Cao and X.-J. Xie (Eds.): Fuzzy Engineering and Operations Research, AISC 147, pp. 275–281.
springerlink.com   © Springer-Verlag Berlin Heidelberg 2012

Let $\mu : E \to [0,1]$. Then μ is called a fuzzy set on E. $F(E)$ denotes a collection of all fuzzy sets over E. Let $A : E \to F(E)$. Then A is called a fuzzy mapping on E. $A(x)$ (we denote it by A_x in the sequel) is a fuzzy set on E and $A_x(y)$ is the membership function of y in $A(x)$.

Let $A \in F(E)$, $\alpha \in [0,1]$. Then the set $(A)_\alpha = \{x \in E, A(x) \geq \alpha\}$ is called a α-cut set of A.

Let T, $A : E \to F(E)$ be two fuzzy mappings. There exist two mappings p, $q : E \to [0,1]$, such that for all $x \in E$, $(T_x)_{p(x)} \in CB(E)$, $(A_x)_{q(x)} \in CB(E)$. We can define two set-valued mappings $\tilde{T}$ and $\tilde{A}$ as follows:

$$\tilde{T} : E \to CB(E), \quad x \to (T_x)_{p(x)}; \quad \tilde{A} : E \to CB(E), \quad x \to (A_x)_{q(x)}.$$

$\tilde{T}$ and $\tilde{A}$ are called the set-valued mappings induced by the fuzzy mappings T and A.

For given mappings p, $q : E \to [0,1]$, fuzzy mappings T, $A : E \to F(E)$, two nonlinear single-valued mappings $N(\cdot,\cdot) : E \times E \to E$ and $\eta : E \times E \to E$, for any given $f \in E$, consider the problem of finding x, u, $v \in E$, such that $T_x(u) \geq p(x)$, $A_x(v) \geq q(x)$ and

$$< N(u,v) + f, \eta(y,x) > + b(x,y) - b(x,x) \geq 0, \qquad \forall y \in E. \qquad (2.1)$$

The problem (2.1) is called a generalized nonlinear fuzzy variational inclusion.

Assumptions 2.1

(1) $b(\cdot,\cdot) : E \times E \to R$ satisfies the following conditions:

    (i) $b(x,y)$ is linear in the first argument,

    (ii) $b(x,y)$ is convex in the second argument,

    (iii) $b(x,y)$ is bounded. That is, there exists a constant $r > 0$ such that

$$b(x,y) \leq r\|x\| \cdot \|y\|, \quad \forall x, y \in E.$$

    (iv) $b(x,y) - b(x,z) \leq b(x, y-z), \quad \forall x, y, z \in E.$

(2) $\eta : E \times E \to E$ satisfies the following condition:

$$\eta(y,x) + \eta(x,y) = 0, \quad \forall x, y \in E.$$

(3) There exists a constant $\tau > 0$ satisfying

$$\|N(u,v)\| \leq \tau(\|u\| + \|v\|), \quad \forall x, y \in E.$$

(4) For given f, u, $v \in E$, mapping $x \mapsto < N(u,v) + f, \eta(y,x) >$ is concave and upper semi-continuous.

Definition 2.1[2]. Let $\eta : E \times E \to E$ be a single-valued mapping, η is said to be

(1) δ-Lipschitz continuous if there exists a constant $\delta > 0$ satisfying
$$\|\eta(y,x)\| \le \delta \|y - x\|, \quad \forall x, \ y \in E.$$

(2) σ-strongly monotone if there exists a constant $\sigma > 0$ satisfying
$$< x - y, \eta(x,y) > \ge \sigma \|x - y\|, \quad \forall x, \ y \in E.$$

Definition 2.2[1]. $\tilde{T} : E \to CB(E)$ is said to be

(1) γ-Lipschitz continuous if there exists a constant $\gamma > 0$ satisfying
$$\tilde{H}(\tilde{T}(x), \tilde{T}(y)) \le \gamma \|x - y\|, \quad \forall x, \ y \in E.$$

(2) ξ-strongly monotone if there exists a constant $\xi > 0$ satisfying
$$< u - v, x - y > \ge \xi \|x - y\|, \quad \forall x, \ y \in E, \ u \in \tilde{T}(x), \ v \in \tilde{T}(y).$$

Definition 2.3[2]. Single-valued mapping $N(\cdot, \cdot) : E \times E \to E$ is said to be

(1) α-Lipschitz continuous in the first argument with respect to $\tilde{T}$ if there exists a constant $\alpha > 0$ satisfying
$$\|N(u, \cdot) - N(v, \cdot)\| \le \alpha \|x - y\|, \quad \forall x, \ y \in E, \ u \in \tilde{T}(x), \ v \in \tilde{T}(y).$$

(2) β-strongly monotone in the first argument with respect to $\tilde{T}$ if there exists a constant $\beta > 0$ satisfying
$$< N(u, \cdot) - N(v, \cdot), x - y > \ge \beta \|x - y\|, \quad \forall x, \ y \in E, \ u \in \tilde{T}(x), \ v \in \tilde{T}(y).$$

Lemma 2.1[6]. Let $X \subset E$ be a nonempty subset, and ϕ, $\psi : X \times X \to R$ satisfy the following conditions:

(1) $\psi(x,y) \le \phi(x,y)$, $\forall x, \ y \in X$, and $\psi(x,x) \ge 0$, $\forall x \in X$;

(2) For any $x \in X$, $\phi(x,y)$ is upper semi-continuous in y ;

(3) For any $y \in X$, the set $\{x \in X \mid \psi(x,y) < 0\}$ is convex;

(4) If there exists a nonempty compact set $K \subset X$, and $x_0 \in K$ such that $\psi(x_0, y) < 0$, $\forall y \in X \setminus K$, then there exists y^* such that $\phi(x, y^*) \ge 0, \forall x \in X$.

3   Iterative Algorithm

Algorithm 3.1. *For given* $x_0 \in E$, $u_0 \in \tilde{T}(x_0)$, $v_0 \in \tilde{A}(x_0)$, *we can compute the sequences* $\{x_n\}$, $\{u_n\}$ *and* $\{v_n\}$ *by the following way:*

$$\begin{cases} u_n \in \tilde{T}(x_n), \ \|u_n - u_{n+1}\| \leq (1 + \frac{1}{n+1})\tilde{H}(\tilde{T}(x_n), \tilde{T}(x_{n+1})), \\ v_n \in \tilde{A}(x_n), \ \|v_n - v_{n+1}\| \leq (1 + \frac{1}{n+1})\tilde{H}(\tilde{A}(x_n), \tilde{A}(x_{n+1})), \\ <x_{n+1}, y - x_{n+1}> \geq <x_n, y - x_{n+1}> - \rho <N(u_n, v_n) + f, \eta(y, x_{n+1})> \\ \qquad + \rho b(x_n, x_{n+1}) - \rho b(x_n, y), \quad \forall y \in E. \end{cases}$$

where $\rho > 0$ is a constant.

Theorem 3.1. *Let* E *be a Hilbert space*, $\tilde{T} : E \rightarrow CB(E)$, $x \rightarrow (T_x)_{p(x)}$ *be* γ *-Lipschitz continuous*, $\tilde{A} : E \rightarrow CB(E)$, $x \rightarrow (A_x)_{q(x)}$ *be* μ *-Lipschitz continuous*, $N(\cdot, \cdot) : E \times E \rightarrow E$ *be* β *-Lipschitz continuous*, α *-strongly monotone with respect to* $\tilde{T}$ *in the first argument and* ξ *-Lipschitz continuous in the second argument*, η *be* δ *-Lipschitz continuous and* σ *- strongly monotone, and* $0 < \sqrt{1 - 2\rho\alpha + \rho^2\beta^2\gamma^2} + \rho\beta\gamma\sqrt{1 - 2\sigma + \delta^2} + \rho\xi\mu\delta + \rho r < 1$. *The conditions in Assumptions 2.1 hold, then the sequences* $\{x_n\}$, $\{u_n\}$, $\{v_n\}$ *converge to* x^*, u^*, v^*, *respectively, where* $\{x_n\}$, $\{u_n\}$, $\{v_n\}$ *are the sequences generated by Algorithm 3.1, and* (x^*, u^*, v^*) *is a solution of problem (2.1).*

Proof. For given f, $x \in E$, $u \in \tilde{T}(x)$, $v \in \tilde{A}(x)$, find $\omega \in E$. Let ϕ, $\psi : X \times X \rightarrow R$

$$\phi(y, \omega) = <y, y - \omega> - <x, y - \omega> + \rho <N(u, v) + f, \eta(y, \omega)> \\ - \rho b(x, \omega) + \rho b(x, y),$$
$$\psi(y, \omega) = <w, y - \omega> - <x, y - \omega> + \rho <N(u, v) + f, \eta(y, \omega)> \\ - \rho b(x, \omega) + \rho b(x, y).$$

Using Lemma 2.1, we get there exists $\omega^* \in E$, for all $y \in E$ such that $\phi(y, \omega^*) \geq 0$. Let $y_t = ty + (1 - t)\omega^*$, $t \in (0, 1]$, $y \in E$. And taking $y = y_t$ in $\phi(y, \omega^*) \geq 0$, then let $t \rightarrow 0^+$, we can get

$$< \omega^*, y - \omega^* >\geq< x, y - \omega^* > -\rho < N(u,v) + f, \eta(y, \omega^*) >$$
$$+ \rho b(x, \omega^*) - \rho b(x, y), \quad \forall y \in E.$$

That is, for all $y \in E$, there exists $x_n \in E$, $x_{n+1} \in E$ such that

$$< x_n, y - x_n >\geq< x_{n-1}, y - x_n >$$
$$-\rho < N(u_{n-1}, v_{n-1}) + f, \eta(y, x_n) > + \rho b(x_{n-1}, x_n) - \rho b(x_{n-1}, y), \quad (3.1)$$

$$< x_{n+1}, y - x_{n+1} >\geq< x_n, y - x_{n+1} >$$
$$-\rho < N(u_n, v_n) + f, \eta(y, x_{n+1}) > + \rho b(x_n, x_{n+1}) - \rho b(x_n, y). \quad (3.2)$$

Taking $y = x_{n+1}$ in (3.1), and $y = x_n$ in (3.2) and adding these inequalities, we obain

$$< x_{n+1} - x_n, x_n - x_{n+1} >\geq< x_n - x_{n-1}, x_n - x_{n+1} >$$
$$+ \rho < N(u_{n-1}, v_{n-1}) - N(u_n, v_n), \eta(x_n, x_{n+1}) > + \rho b(x_{n-1} - x_n, x_n)$$
$$+ \rho b(x_n - x_{n-1}, x_{n+1}),$$

$$\|x_n - x_{n+1}\|^2 \leq< x_{n-1} - x_n, x_n - x_{n+1} > -\rho < N(u_{n-1}, v_{n-1}) - N(u_n, v_n),$$
$$\eta(x_n, x_{n+1}) > -\rho b(x_{n-1} - x_n, x_n) - \rho b(x_n - x_{n-1}, x_{n+1}),$$
$$\leq< x_{n-1} - x_n - \rho(N(u_{n-1}, v_{n-1}) - N(u_n, v_{n-1})), x_n - x_{n+1} >$$
$$+ \rho < N(u_{n-1}, v_{n-1}) - N(u_n, v_{n-1}), x_n - x_{n+1} - \eta(x_n, x_{n+1}) >$$
$$+ \rho < N(u_n, v_n) - N(u_n, v_{n-1}), \eta(x_n, x_{n+1}) > + \rho b(x_n - x_{n-1}, x_n - x_{n+1}).$$

Since $b(x, y)$ is bounded, we have

$$\|x_n - x_{n+1}\|^2 \leq \|x_{n-1} - x_n - \rho(N(u_{n-1}, v_{n-1}) - N(u_n, v_{n-1}))\| \cdot \|x_n - x_{n+1}\|$$
$$+ \rho \|N(u_{n-1}, v_{n-1}) - N(u_n, v_{n-1})\| \cdot \|x_n - x_{n+1} - \eta(x_n, x_{n+1})\|$$
$$+ \rho \|N(u_n, v_n) - N(u_n, v_{n-1})\| \cdot \|\eta(x_n, x_{n+1})\| + \rho r \|x_n - x_{n-1}\| \cdot \|x_n - x_{n+1}\|. \quad (3.3)$$

By using the Lipschitz continuity and strong monotonicity for N and η, we obtain

$$\|x_{n-1} - x_n - \rho(N(u_{n-1}, v_{n-1}) - N(u_n, v_{n-1}))\|^2 \leq$$
$$\|x_n - x_{n-1}\|^2 + \rho^2 \|N(u_{n-1}, v_{n-1}) - N(u_n, v_{n-1})\|^2$$
$$-2\rho < N(u_{n-1}, v_{n-1}) - N(u_n, v_{n-1}), x_{n-1} - x_n >$$
$$\leq [1 - 2\rho\alpha + \rho^2 \beta^2 \gamma^2 (1 + \tfrac{1}{n})^2] \cdot \|x_{n-1} - x_n\|^2 \quad (3.4)$$

$$\left\| x_n - x_{n+1} - \eta(x_n, x_{n+1}) \right\|^2 = \left\| x_n - x_{n+1} \right\|^2 - 2 < x_n - x_{n+1}, \eta(x_n, x_{n+1}) >$$

$$+ \left\| \eta(x_n, x_{n+1}) \right\|^2 \leq (1 - 2\sigma + \delta^2) \left\| x_n - x_{n+1} \right\|^2 . \tag{3.5}$$

It follows from (3.3) to (3.5) that

$$\left\| x_n - x_{n+1} \right\| \leq \sqrt{1 - 2\rho\alpha + \rho^2 \beta^2 \gamma^2 (1 + \tfrac{1}{n})^2} \cdot \left\| x_{n-1} - x_n \right\|$$

$$+ \rho\beta\gamma(1 + \tfrac{1}{n})\sqrt{1 - 2\sigma + \delta^2} \left\| x_n - x_{n-1} \right\|$$

$$+ \rho\xi\mu\delta(1 + \tfrac{1}{n}) \left\| x_n - x_{n-1} \right\| + \rho r \left\| x_n - x_{n-1} \right\| .$$

Let $\theta_n = \sqrt{1 - 2\rho\alpha + \rho^2 \beta^2 \gamma^2 (1 + \tfrac{1}{n})^2} + \rho\beta\gamma(1 + \tfrac{1}{n})\sqrt{1 - 2\sigma + \delta^2}$

$$+ \rho\xi\mu\delta(1 + \tfrac{1}{n}) + \rho r ,$$

$$\theta = \sqrt{1 - 2\rho\alpha + \rho^2 \beta^2 \gamma^2} + \rho\beta\gamma\sqrt{1 - 2\sigma + \delta^2} + \rho\xi\mu\delta + \rho r .$$

Then

$$\left\| x_n - x_{n+1} \right\| \leq \theta_n \left\| x_n - x_{n-1} \right\| \tag{3.6}$$

and $\theta_n \to \theta, (n \to \infty)$. Since $0 < \theta < 1$, (3.6) implies that $\{x_n\}$ is a Cauchy sequence. Thus there exists $x^* \in E$ such that $x_n \to x^*$ $(n \to \infty)$. By Algorithm 3.1, we know that $\{u_n\}$, $\{v_n\}$ are all Cauchy sequences, and so there exist $u^* \in E$, $v^* \in E$, i.e., $u_n \to u^*$, $v_n \to v^*$, $(n \to \infty)$. We prove that $u \in \tilde{T}(x)$, $v \in \tilde{A}(x)$. In fact

$$d(u, \tilde{T}(x)) = \inf_{z \in \tilde{T}(x)} \left\| u - z \right\| \leq \left\| u - u_n \right\| + d(u_n, \tilde{T}(x))$$

$$\leq \left\| u - u_n \right\| + \tilde{H}(\tilde{T}(x_n), \tilde{T}(x)) \leq \left\| u - u_n \right\| + r \left\| x_n - x \right\|,$$

that is $d(u, \tilde{T}(x)) = 0$, we can know $u \in \tilde{T}(x)$. Similarly we can easily get $v \in \tilde{A}(x)$. Therefore

$$< x, y - x > \geq < x, y - x > -\rho < N(u, v) + f, \eta(y, x) > + \rho b(x, x) - \rho b(x, y),$$

that is $< N(u, v) + f, \eta(y, x) > + b(x, y) - b(x, x) \geq 0$, $\forall y \in E$.

This completes the proof of Theorem 3.1.

4 Conclusion

The paper presents an iterative algorithm to compute the approximate solution of a generalized nonlinear fuzzy variational inclusion (2.1), and proves the convergence criteria under some mild conditions.

References

1. Wang, J.H., He, Z.Q.: A class of iterative algorithm for generalized set-valued mixed implicit variational inequalities. Journal of Hangzhou Normal University 8(5), 350–353 (2009)
2. Xu, H.L., Guo, X.M.: Auxiliary principle and three-set iterative algorithms for generalized set-valued strongly nonlinear mixed. Applied Mathematics and Mechanics 28(6), 643–649 (2007)
3. Huang, N.J., Cao, S.Y.: A class of variational inclusions for fuzzy mappings. Journal of Xinjiang University 4, 22–28 (1996)
4. Lan, H.Y.: Generalized fuzzy variational inclusions with set-valued accretive mappings in Banach spaces. Journal of Sichuan University 39(6), 1014–1018 (2002)
5. Wan, B., Jiang, X.T., Cao, Y.Z.: Implicit iterative methods for general mixed quasi-variational inequalities. Pure and Applied Mathematics 25(2), 384–389 (2009)
6. Zhang, S.S., Xiang, S.W.: On the existence of solutions for a class of quasi-bilinear variational inequalities. J. Systems Sci. Math. Sci. 16(3), 136–140 (1996)

Part III
Fuzzy Engineering Application and Soft Computing Method

Fuzzy Process and the Application to Option Pricing in Risk Mangement

Shu-xia Liu, Qin-juan Jing, and Dian-yu Zhao

College of Business Administration, Hebei Normal University of Science
& Technology, QinHuangDao, P.R. China
xxglliushuxia@126.com

Abstract. In this paper an option valuation model using fuzzy process is discussed. We demonstrate how fuzzy L process can be successfully applied to the risk neutral option pricing model. Through option pricing theory and fuzzy set theory we get results that allow us to effectively price option in a fuzzy environment.

Keywords: Option pricing, option, fuzzy variables, risk management.

1 Introduction

The concept of fuzzy set was initialized by Zadeh [32] in 1965 via membership function. Fuzzy set theory has been well developed and applied in a wide variety of real problems. Possibility theory was proposed by Zadeh [33] in 1978 and has been developed by many researchers such as Nahmias [21], Dubois and Prade [6]. However, possibility measure is not self-dual. Since a self-dual measure is absolutely needed in both theory and practice, Liu [14] presented a self-dual credibility measure in 2002. From then on, credibility theory has been well developed. An axiomatic foundation was given by Liu [14]. A detailed survey on credibility theory may be found in Liu [12]. While, in real life, there exist much dynamic fuzzy phenomena, which is not enough to be described in one time, its evolutionary process should be described repeatedly or continuously. This prompted the emergence of fuzzy process, which was initiated by Liu [16]. To model fuzzy dynamic systems, fuzzy differential equations are natural choice. The fuzzy differential equation was first introduced by Kandel and Byatt in 1978, for the importance of studying differential equations is well known, many researchers have been interested in fuzzy differential equations such as Puri and Ralescu.However, these fuzzy differential equations are mainly concerning possibilistic uncertainty. In 2007, a special fuzzy process, C process, was proposed by Liu [16], which plays the role of Brownian motion in stochastic process. Based on Liu process, a new kind of fuzzy differential was introduced in this paper.

B.-Y. Cao and X.-J. Xie (Eds.): Fuzzy Engineering and Operations Research, AISC 147, pp. 285–296.
springerlink.com © Springer-Verlag Berlin Heidelberg 2012

Options are among the most important kinds of financial derivatives because of their abundant functions and properties, and they have been bringing profound influence to the global financial market. Good performance of options in financial market must be based on appropriately pricing them. Option pricing theory has become the scientific theory since Black, Scholes [2] and Merton [17] accomplished their great work in 1973. But empirical research finds out that there are systematic error between real market price and the classical model-based price. Namely, there exists mispricing. Black and Scholes, Robinstein, Bates and others have pointed out that phenomena like violation for Call-Put parity, volatility smile, volatility term structure, and it has the unsolvable difficulty for classical option pricing models. Trying to correct these defects, many scholars proposed many new models, such as stochastic volatility model, compound option model, displaced diffusion model, pure jump model, jump diffusion model and so on; but they failed without exception, because they do not touch the basic economic assumption.

Most stochastic models involve uncertainty arising mainly from lack of knowledge or from inherent vagueness. Traditionally those stochastic models are solved using probability theory and fuzzy set theory. There exist many practical situations where both types of uncertainties are present. For example, if the price of an option depends upon the nature of the volatility which changes randomly, then the volatility of the stock price movement which is estimated from the sample data is a random variable as well as a fuzzy number. Recently there has been a growing interest in using fuzzy numbers to deal with impreciseness (see [23] [24][25][19][8] for more details). Viewing the fuzzy numbers as random sets Dubois and Prade introduced the mean value as a closed interval bounded by the expectations calculated from its upper and lower distribution functions. In this section we provide a literature survey of some of the research done on fuzzy option pricing. Many authors have tried to deal fuzziness along with randomness in option pricing models. For example, recently, Cherubini determined the price of a corporate debt contract and provided a fuzzified version of the Black and Scholes model by means of a special class of fuzzy measures. On the other hand, Panayiotis [22] introduced artificial intelligence approach to price the options, using neural networks and fuzzy logic. They compare the result of artificial intelligence approach to that of Black-Scholes model, using stock indexes. Since the Black-Scholes option pricing formula is only approximate, which leads to considerable errors, Trenev [26] obtained a refined formula for pricing options. Due to the fluctuation of the financial markets from time to time, some of the input parameters in the Black-Scholes formula cannot always be expected in the precise sense. As a result, Wu applied fuzzy approach to the Black-Scholes formula. Zmeskal applied Black-Scholes methodology of appraising equity as a European call option. He used the input data in a form of fuzzy numbers in his approach to option pricing. Carlson and Fuller [3] use possibility theory to fuzzy real option valuation. Applications of fuzzy sets theory to volatility models have been studied by Thavaneswaran [24].

Guerra [19] analyzed the impact of LU parametric representation in the fuzzy option pricing. Guerra showed that the implementation of the Black-Scholes model involves the Greeks in the definition of the slopes in the parametrization of the fuzzy version of the underlying stock price, the volatility and the risk-free interest rate.However some of these models seemed to be soon unsatisfactory due to the incapability to capture the relevant stylized facts of real markets. Many attempts of fuzzy models have been recently proposed in the literature, but they either have the disadvantage of requiring a large amount of computations (e.g. constrained optimization problems) or they suffer a relative rigidity in representing and capturing the shapes of the fuzzy quantities (data and/or results). this model reveal their great flexibility and simplicity to model fuzzy financial quantities and information in option pricing (and other financial applications) and to perform quickly the required fuzzy calculus. An additional important advantage is that the overestimation effect, inherent to the use of the interval fuzzy arithmetic, is completely eliminated.

Some scholars more or less touched the fuzzy process and integral, but they did not view this problem from economics, corresponding to the consideration above, the paper gives a new thinking to use the fuzzy process to option pricing.

2 Basic Concepts in Fuzzy Theory

In this section, some concepts and results on fuzzy variables were recalled. Let $(\Theta, \mathcal{P}(\Theta), \mathrm{Cr})$ be a on credibility space, where Θ is a universe, $\mathcal{P}(\Theta)$ is the power set of Θ and Cr is a credibility measure defined on $\mathcal{P}(\Theta)$.

Definition 1. [13] A Fuzzy variable is defined as a function from a credibility space$(\Theta, \mathcal{P}(\Theta), \mathrm{Cr})$ to the set of real number.

Definition 2. [13] Let ξ be a fuzzy variable on the credibility space $(\Theta, \mathcal{P}(\Theta), \mathrm{Cr})$. Then its membership function is derived from the credibility measure by

$$\mu(x) = (2\mathrm{Cr}\{\xi = x\}) \wedge 1, \quad x \in \Re. \tag{1}$$

Definition 3. [13] A fuzzy variable ξ is said to be positive if and only if $\mathrm{Cr}\{\xi \leq 0\} = 0$.

Definition 4. [13] The fuzzy variables $\xi_1, \xi_2, \ldots, \xi_n$ are said to be independent if and only if

$$\mathrm{Cr}\{\xi_i \in \mathcal{B}_i, \ i = 1, 2, \ldots, n\} = \min_{i \geq 1} \mathrm{Cr}\{\xi_i \in \mathcal{B}_i\} \tag{2}$$

for any sets $\mathcal{B}_1, \mathcal{B}_2, \ldots, \mathcal{B}_n$ of $\Re$.

Definition 5. [13] The fuzzy variables $\xi_1, \xi_2, \ldots, \xi_m$ are said to be identically distributed if and only if

$$\mathrm{Cr}\{\xi_i \in \mathcal{B}\} = \mathrm{Cr}\{\xi_j \in \mathcal{B}\}, \quad i, j = 1, 2, \cdots, m \tag{3}$$

for any set $\mathcal{B}$ of $\Re$.

Definition 6. [15] Let ξ be a fuzzy variable on the credibility space $(\Theta, \mathcal{P}(\Theta), \mathrm{Cr})$. The expected value $E[\xi]$ is defined as

$$E[\xi] = \int_0^{+\infty} \mathrm{Cr}\{\xi \geq r\}\mathrm{d}r - \int_{-\infty}^0 \mathrm{Cr}\{\xi \leq r\}\mathrm{d}r \tag{4}$$

provided that at least one of the two integrals is finite. Especially, if ξ is a nonnegative fuzzy variable, then $E[\xi] = \int_0^{+\infty} \mathrm{Cr}\{\xi \geq r\}\mathrm{d}r$.

Definition 7. [13] Let ξ be a fuzzy variable with finite expected value e. Then the variance of ξ is defined by

$$V[\xi] = E[(\xi - e)^2]. \tag{5}$$

Proposition 1. *The distance between fuzzy variables is*

$$d_p(\xi, \eta) = (E[|\xi - \eta|^p])^{\frac{1}{p+1}}, \tag{6}$$

where $E[\cdot]$ is the expected value operator and $p \leq 0$.

Proposition 2. *Let ξ, η, ζ be fuzzy variables, and let $d_p(\cdot, \cdot)$ be the distance. Then we have*

(a) *(Nonnegativity)* $d_p(\xi, \eta) \geq 0$;
(b) *(Identification)* $d_p(\xi, \eta) = 0$ *if and only if* $\xi = \eta$;
(c) *(Symmetry)* $d_p(\xi, \eta) = d_p(\xi, \eta)$;
(d) *(Triangle Inequality)* $d_p(\xi, \eta) \leq d_p(\xi, \zeta) + d_p(\eta, \zeta)$.

Proposition 3. *Suppose that $\xi_1, \xi_2, \ldots, \xi_n$ are fuzzy variables with finite expected values defined on the credibility space $(\Theta, \mathcal{P}, \mathrm{Cr})$. We say that the sequence $\{\xi\}$ converges in mean to ξ_n if*

$$\lim_{i \to \infty} (E[|\xi_i - \xi|^p])^{\frac{1}{p+1}} = 0. \tag{7}$$

Definition 8. [16] Given an index set T and a credibility space $(\Theta, \mathcal{P}(\Theta), \mathrm{Cr})$, a fuzzy process is defined by [16] as a function from $T \times (\Theta, \mathcal{P}(\Theta), \mathrm{Cr})$ to the set of real numbers, That is, a fuzzy process $X(t, \theta)$ is a function of two variables such that function $X(t, \theta)$ is a fuzzy variable for each t.

For simplicity, we use the symbol X_t to replace $X(t, \theta)$ in the following section.

Definition 9. [16] A fuzzy process X_t is said to be have independent increments if

$$X_{t_1} - X_{t_0}, X_{t_2} - X_{t_1}, \ldots, X_{t_k} - X_{t_{k-1}}$$

are independent fuzzy variables for any times $t_0 \leq t_1 \leq \cdots \leq t_k$. A fuzzy process X_t is said to have stationary increments if, for any given $s \geq 0$, $X_{t+s} - X_t$ are identically distributed fuzzy variables for all t.

Especially, corresponding to Brownian motion in stochastic process, [16] introduced a special fuzzy process called C process.

Definition 10. [16]A fuzzy process is said to be a C process if

(i) $C_0 = 0$,
(ii) C_t has stationary and independent increments,
(iii) every increment $C_{s+t} - C_t$ is a normally distributed fuzzy variable with expected value et and variance $\sigma^2 t^2$ whose membership function is

$$\mu(x) = 2(1 + \exp(\frac{\pi|x - et|}{\sqrt{6}\sigma t}))^{-1}, -\infty < x < +\infty. \tag{8}$$

The C process is said to be standard if $e = 0$ and $\sigma = 1$.

Proposition 4. *Let X_t be a fuzzy process and C_t a standard C process. For any partition of closed interval $[a, b]$ with $a = t_1 \leq t_2 \leq \cdots < t_{n+1} = b$, the mesh is written as*

$$\Delta = \max_{1 \leq i \leq n} |t_{i+1} - t_i|.$$

Then the fuzzy integral of X_t with respect to C_t is

$$\int_a^b X_t dC_t = \lim_{\Delta \to 0} \sum_{i=1}^n X_{t_i}(C_{t_{i+1}} - C_{t_i}) \tag{9}$$

provided that the limit exists and is a fuzzy variable.

Proposition 5. *Let C_t be a standard Liu process, and $f(t, x), g(t, x)$ two continuous differentiable functions. Define $X_t = f(t, g(t, C_t))$ and denote $U_t = g(t, C_t)$. Then*

$$dX_t = \left(\frac{\partial f}{\partial t}(t, g(t, C_t)) + \frac{\partial f}{\partial U}(t, g(t, C_t))\frac{\partial U}{\partial t}(t, C_t)\right) dt + \frac{\partial f}{\partial U}(t, g(t, C_t))\frac{\partial U}{\partial x}(t, C_t)dC_t. \tag{10}$$

3 Fuzzy L Process

Let

$$dL_t = \xi\sqrt{dt}, \tag{11}$$

where ξ is standard normally distributed fuzzy variable.

Definition 11. A fuzzy process is said to be a standard L process if

(i) $L_0 = 0$,

(ii) L_t has stationary and independent increments,

(iii) every increment $L_{t_{i+1}} - L_{t_i}$ is a normally distributed fuzzy variable with expected value 0 and variance $t_{i+1} - t_i$.

$$E[L_{t_{i+1}} - L_{t_i}] = 0, V[L_{t_{i+1}} - L_{t_i}] = t_{i+1} - t_i,$$

$$E[(L_{t_{i+1}} - L_{t_i})^2] = t_{i+1} - t_i, V[(L_{t_{i+1}} - L_{t_i})^2] \approx 7.2384(t_{i+1} - t_i).$$

Proposition 6. *The distance between fuzzy variables is*

$$d_2(\xi, \eta) = (E[|\xi - \eta|^2])^{\frac{1}{3}} = (E[(\xi - \eta)^2])^{\frac{1}{3}}, \tag{12}$$

where $E[\cdot]$ is the expected value operator of a fuzzy variable.

Proposition 7. *Let L_t be a standard L process. For any partition of closed interval $[0, T]$ with $0 = t_1 \leq t_2 \leq \cdots < t_{n+1} = T$, the mesh is written as*

$$\Delta t = \max_{1 \leq i \leq n} |t_{i+1} - t_i|.$$

Let $\Delta L_i = L_{t_{i+1}} - L_{t_i}$ and $\Delta t_i = t_{i+1} - t_i$, then the fuzzy integral of L_t is

$$\int_0^T L_t dL_t = \frac{1}{2}(L_T^2 - L_0^2 - T). \tag{13}$$

Proof. *Since*

$$
\begin{aligned}
&\int_0^T L_t dL_t \\
&= \lim_{\Delta t \to 0} \sum_{i=1}^n L_{t_i} \Delta L_i \\
&= \lim_{\Delta t \to 0} \sum_{i=1}^n \frac{1}{2}((L_{t_i} + \Delta L_i)^2 - L_{t_i}^2 - \Delta L_i^2) \\
&= \lim_{\Delta t \to 0} \sum_{i=1}^n \frac{1}{2}(L_{t_{i+1}}^2 - L_{t_i}^2 - \Delta L_i^2) \\
&= \lim_{\Delta t \to 0} \frac{1}{2}(\sum_{i=1}^n L_{t_{i+1}}^2 - \sum_{i=1}^n L_{t_i}^2 - \sum_{i=1}^n \Delta L_i^2) \\
&= \lim_{\Delta t \to 0} \frac{1}{2}(L_T^2 - L_0^2 - \sum_{i=1}^n \Delta L_i^2).
\end{aligned} \tag{14}
$$

We want to show that

$$d_2(\sum_{i=1}^n \Delta L_i^2, T) = \lim_{\Delta t \to 0}(E[(\sum_{i=1}^n \Delta L_i^2 - T)^2])^{\frac{1}{3}} \to 0. \tag{15}$$

For $i \neq j$, the term ΔL_i and ΔL_j, Δt_i and Δt_j are independent, $\Delta L_i^2 - \Delta t_i$ and $\Delta L_j^2 - \Delta t_j$ are independent, so

$$E[\sum_{i=1}^{n}(\Delta L_i^2 - \Delta t_i)] = 0, \tag{16}$$

$$E[\sum_{i=1}^{n} \Delta L_i^4] = 8.4839 \Delta t_i^2, \tag{17}$$

$$V[\xi] = E[(\xi - E[\xi])^2] = E[\xi^2], \tag{18}$$

$$\begin{aligned}
\mathrm{Cr}\{\Delta L_i^2 \geq r\} &= \frac{1}{2}\left(1 + \sup_{\Delta L_i^2 \geq r} \mu(x) - \sup_{\Delta L_i^2 \leq r} \mu(x)\right) \\
&= (1 + \exp(\frac{\pi|x - et|}{\sqrt{6}\sigma t}))^{-1} \\
&= (1 + \exp(\frac{\pi|x|}{\sqrt{6}t}))^{-1}.
\end{aligned} \tag{19}$$

$$\begin{aligned}
E[\xi] &= \int_0^{+\infty} \mathrm{Cr}\{\xi \geq r\}dr \\
&= \int_0^{+\infty}(1 + \exp(\frac{\pi|x - et|}{\sqrt{6}\sigma t}))^{-1}dr \\
&= \int_0^{+\infty}(1 + \exp(\frac{\pi|x|}{\sqrt{6}t}))^{-1}dr,
\end{aligned} \tag{20}$$

$$\begin{aligned}
V[&\sum_{i=1}^{n}(\Delta L_i^2 - \Delta t_i)] \\
&= E[\sum_{i=1}^{n}(\Delta L_i^2 - \Delta t_i)^2] \\
&\leq \sum_{i=1}^{n} E[\Delta L_i^4 + \Delta t_i^2] \\
&= \sum_{i=1}^{n} \Delta t_i^2 + \Delta t_i^2 \\
&= \sum_{i=1}^{n} \Delta t_i^2.
\end{aligned} \tag{21}$$

As $\Delta t \to 0$, so

$$V[\sum_{i=1}^{n}(\Delta L_i^2 - \Delta t_i)] \leq \sum_{i=1}^{n} \Delta t_i^2 \leq \Delta t \sum_{i=1}^{n} \Delta t_i = T\Delta t \to 0, \tag{22}$$

$$\lim_{\Delta t \to 0}(E[(\sum_{i=1}^{n} \Delta L_i^2 - T)^2])^{\frac{1}{3}} \to 0. \tag{23}$$

The proof is complete.

$$d(\sum_{i=1}^{n} \Delta L_i, T) = \lim_{\Delta t \to 0} E[|\sum_{i=1}^{n}(\Delta L_i^2 - \Delta t_i)|] \to 0. \tag{24}$$

$$d(\sum_{i=1}^{n} \Delta L_i, T) = \lim_{\Delta t \to 0} E[|\sum_{i=1}^{n}(\Delta L_i^2 - \Delta t_i|] \le \sum_{i=1}^{n} E[|\Delta L_i^2 - \Delta t_i|] \ldots \to 0. \tag{25}$$

$$\int_0^T (dL_t)^2 = \int_0^T dt = T, \tag{26}$$

$$(dL_t)^2 = dL_t \times dL_t = dt. \tag{27}$$

Proposition 8. *Let L_t be a standard L process, and $f(t,x), g(t,x)$ two continuous differentiable functions. Define $X_t = f(t, U_t)$ and denote $U_t = g(t, L_t)$. Then*

$$dX_t = \frac{\partial f}{\partial t}dt + (\frac{\partial g}{\partial t})\frac{\partial f}{\partial U}dt + \frac{1}{2}(\frac{\partial g}{\partial L_t})^2 \frac{\partial^2 f}{\partial U^2}dU^2 + \frac{\partial g}{\partial L_t}\frac{\partial f}{\partial U}dU. \tag{28}$$

Proof. Assume that $\frac{\partial f}{\partial t}, \frac{\partial f}{\partial U}, \frac{\partial U}{\partial t}, \frac{\partial U}{\partial x}$ are bounded, since there exist continuous differential functions such that are bounded for each n and converge uniformly on compact subsets, respectively. Since $f(t,x)$ is a differentiable function, we immediately have

$$dU_t = \frac{\partial g}{\partial t}dt + \frac{\partial g}{\partial L_t}dL_t, \tag{29}$$

$$(dU_t)^2 = (\frac{\partial g}{\partial t})^2(dt)^2 + \frac{\partial g}{\partial t}\frac{\partial g}{\partial L_t}dL_tdt + (\frac{\partial g}{\partial L_t})^2(dL_t)^2, \tag{30}$$

$$(dL_t)^2 = dt, dL_t = \xi\sqrt{dt}, \tag{31}$$

$$dX_t = \frac{\partial f}{\partial t}dt + \frac{\partial f}{\partial U}dU + \frac{1}{2}\frac{\partial^2 f}{\partial U^2}dU^2 + \frac{\partial f}{\partial t\partial U}dtdU + \frac{1}{2}\frac{\partial^2 f}{\partial t^2}dt^2 + \text{higher order term.} \tag{32}$$

$$dX_t = \frac{\partial f}{\partial t}dt + (\frac{\partial g}{\partial t})\frac{\partial f}{\partial U}dt + \frac{1}{2}(\frac{\partial g}{\partial L_t})^2 \frac{\partial^2 f}{\partial U^2}dU^2 + \frac{\partial g}{\partial L_t}\frac{\partial f}{\partial U}dU. \tag{33}$$

4 Option Valuation Model for the Fuzzy L Process

In this section, considering two assets, a bond price B_t and a stock price S_t, where the bond price process B_t is riskless and the stock price process S_t is risky. Let r be the instantaneous interest rate, i.e., interest factor, on a bond. Let a bond price process B_t satisfy the ordinary differential equation

$$\mathrm{d}B_t = rB_t\mathrm{d}t. \tag{34}$$

The evolution stock price process $\{S_t, t \geq 0\}$ is assumed to be describe by the following equation

$$\mathrm{d}S_t = eS_t\mathrm{d}t + \sigma S_t\mathrm{d}L_t, \tag{35}$$

where $\{L_t, t \geq 0\}$ is a standard fuzzy L process, e is a constant expected return of the underlying asset, σ is standard variance of return on the underlying asset.

Let $V(t, S_t)$ denote the value of an option, S_t is the price of the stock at each time t. Then assume that $V(t, S_t)$ is a some function of variables S_t and t. Hence, By the Proposition 1, the following equation can be derived

$$\mathrm{d}V= \frac{\partial V}{\partial t}\mathrm{d}t + eS\frac{\partial V}{\partial S}\mathrm{d}t + \frac{1}{2}\sigma^2 S^2\frac{\partial^2 V}{\partial S^2}\mathrm{d}t + \sigma S\frac{\partial V}{\partial S}\mathrm{d}L_t. \tag{36}$$

Assuming numerical values

$$n = \text{number of shares of stock}$$

and setting up a appropriate portfolio consisting of n shares of stock, the value of the portfolio is

$$V(S_t, t) - nS_t = B_t \tag{37}$$

and its differential is

$$\mathrm{d}V(S_t, t) - n\mathrm{d}S_t = \mathrm{d}B_t. \tag{38}$$

Substituting equation(35) and(36) into equation 38 yields

$$\frac{\partial V}{\partial t}\mathrm{d}t + eS\frac{\partial V}{\partial S}\mathrm{d}t + \frac{1}{2}\sigma^2 S^2\frac{\partial^2 V}{\partial S^2}\mathrm{d}t + \sigma S\frac{\partial V}{\partial S}\mathrm{d}L_t - neS_t\mathrm{d}t - n\sigma S_t\mathrm{d}L_t = \mathrm{d}B_t. \tag{39}$$

Because the portfolio is riskless, that is to say the portfolio must instantaneously earn the same rate of return as the free risk bond, the $\mathrm{d}L_t$ term can be eliminated. Taking $n = \frac{\partial V}{\partial S}$, with this choice, not only do the $rmdL$ term cancel, but enS and $eS\frac{\partial V}{\partial S}$ cancel also. This leave

Substituting equation (34) into equation (39) yields

$$(\frac{\partial V}{\partial t} + \frac{1}{2}\sigma^2 S^2\frac{\partial^2 V}{\partial S^2})\mathrm{d}t = rB_t\mathrm{d}t. \tag{40}$$

From the equation (37), we can get

$$V(S_t, t) - nS_t = B_t, \tag{41}$$

Substitute equation (41) and $m = \frac{\partial V}{\partial S}$ into (40) to obtain

$$r(V - S_t\frac{\partial V}{\partial S})\mathrm{d}t = (\frac{\partial V}{\partial t} + \frac{1}{2}\sigma^2 S^2\frac{\partial^2 V}{\partial S^2})\mathrm{d}t. \tag{42}$$

We arrive at the equation

$$\frac{\partial V}{\partial t} - rV + rS_t\frac{\partial V}{\partial S} + \frac{1}{2}\sigma^2 S^2\frac{\partial^2 V}{\partial S^2} = 0. \tag{43}$$

This equation is called fuzzy partial differential equation. It has many solutions, corresponding to all the differential derivatives that can be defined with S as the underlying variable. The particular derivatives that is obtained when the equation is solved depend on the boundary conditions that are used.

If the option is of such a type that it can be exercised only on the expiration date itself, then it is called a European option. Let S_T be the price of the underlying asset at expiration time T. Then $g(\cdot)$, of a European style call option, at time T the key boundary condition is given by

$$g(S_{T-t}) = \max\{0, (S_T - K)\}, \tag{44}$$

From the equation (43) and (44), following equations can be obtained

$$\begin{cases} \frac{\partial V}{\partial t} - rV + rS_t\frac{\partial V}{\partial S} + \frac{1}{2}\sigma^2 S^2\frac{\partial^2 V}{\partial S^2} = 0, \\ V(T, S_T) = \max\{0, (S_T - K)\}. \end{cases} \tag{45}$$

In the case of a European put option, the key boundary condition is

$$h(S, T - t) = \max\{0, K - (S_T)\}, \tag{46}$$

By the equation (43) and (46), following equations can be obtained

$$\begin{cases} \frac{\partial V}{\partial t} - rV + rS_t\frac{\partial V}{\partial S} + \frac{1}{2}\sigma^2 S^2\frac{\partial^2 V}{\partial S^2} = 0, \\ V(T, S_T) = \max\{0, K - (S_T)\}. \end{cases} \tag{47}$$

5 Conclusion

This paper shows that options can be valued in fuzzy environment. Fuzzy L process is firstly is proposed,then we discuss that fuzzy L process can be successfully applied to the risk neutral option pricing model. This method is very general in the sense. Subsequently, some computational procedures are given to obtain the fuzzy options.

References

1. Bakshi, G., Cao, C., Chen, Z.: Empirical performance of alternative options pricing models. Journal of Finance 52(5), 2003–2049 (1997)
2. Black, F., Scholes, M.: The pricing of options and corporate liabilities. Journal of Political Economy 81, 637–654 (1973)
3. Carlson, C., Fuller, R.: A fuzzy approach to real option valuation. Fuzzy Sets and Systems 139, 297–312 (2003)
4. Cherubini, U.: Fuzzy measures and asset prices. Appl. Math. Finance 4, 135–149 (1997)
5. Cherubini, U., Della Lunga, G.: Fuzzy value-at-risk: accounting for market liquidity. Econom. Notes 30(2), 293–312 (2001)
6. Dubois, D., Prade, H.: Possibility Theory: An Approach to Computerized Processing of Uncertainty. Plenum, New York (1988)
7. Ghaziri, H., Elfakhani, S., Assi, J.: Neural networks approach to pricing options. Neural Network World 10, 271–277 (2000)
8. Han, L.Y., Zheng, C.L.: Fuzzy options with application to default risk analysis for municipal bonds in China. Nonlinear Analysis 63, 2353–2365 (2005)
9. Kaufman, A.: Introduction to the theory of fuzzy subsets. Academic Press, New York (1975)
10. Kazmerchuk, Y., Swishchuk, A., Wu, J.: The pricing of options for securities markets with delayed response. Mathematics and Computers in Simulation 75, 69–79 (2007)
11. Lee, C., Tzengb, G.: A new application of fuzzy set theory to the Black-Scholes option pricing model. Expert Systems with Applications 29, 330–342 (2005)
12. Liu, B.: A survey of credibility theory. Fuzzy Optimization and Decision Making 15, 387–408 (2006)
13. Liu, B.: Uncertainty Theory, 3rd edn., http://orsc.edu.cn/liu/ut.pdf
14. Liu, B.: Uncertainty theory: an introduction to its axiomatic foundations. Springer, Heidelberg (2004)
15. Liu, B., Liu, Y.: Expected value of fuzzy variable and fuzzy expected value models. IEEE Transactions on Fuzzy Systems 10, 445–450 (2002)
16. Liu, B.: Fuzzy process, hybrid process and uncertain process. Journal of Uncertain Systems 2(1) (2008)
17. Merton, C.: Theory of rational option pricing. Bell Journal of Economics Management, Science 4, 141–183 (1973)
18. Luciano, S., Laerte, S., Maria, L.G.: A parametrization of fuzzy numbers for fuzzy calculus and application to the fuzzy Black-Scholes option pricing. In: IEEE International Conference on Fuzzy Systems, pp. 16–21 (2006)
19. Maria, L.G., Laerte, S., Luciano, S.: Parametrized fuzzy numbers for option pricing, pp. 1–6. IEEE (2007)
20. Merton, C.: Option pricing when underlying stock returns are discontinuous. Journal of Financial Economics 3, 125–144 (1976)
21. Nahmias, S.: Fuzzy variables. Fuzzy Sets and Systems 1, 97–110 (1978)
22. Panayiotis, C.A., Chris, C., Spiros, H.M.: Pricing and trading european options by combining artificial neural networks and parametric models with implied parameters. European Journal of Operational Research 185, 1415–1433 (2008)
23. Thiagarajaha, K., Appadoob, S.S., Thavaneswaranc, A.: Option valuation model with adaptive fuzzy numbers. Computers and Mathematics with Applications 53, 831–841 (2007)

24. Thavaneswaran, A., Singh, J., Appadoo, S.S.: Option pricing for some stochastic volatility models. The Journal of Risk Finance 7, 425–445 (2006)
25. Thiagarajah, K., Thavaneswaran, A.: Fuzzy coefficient volatility models with financial applications. Journal of Risk Finance 7, 503–524 (2006)
26. Trenev, N.N.: A refinement of the Black-Scholes formula of pricing options. Cybernetics and Systems Analysis 37, 911–917 (2001)
27. Wu, H.: Pricing european options based on the fuzzy pattern of Black-Scholes formula. Computers & Operations Research 31, 1069–1081 (2004)
28. Wu, H.C.: European option pricing under fuzzy environments. International Journal of Intelligent Systems 20, 89–102 (2005)
29. Wu, H.: Using fuzzy sets theory and Black-Scholes formula to generate pricing boundaries of European options. Applied Mathematics and Computation 185, 136–146 (2007)
30. Yoshida, Y.: The valuation of european options in uncertain environment. European Journal of Operational Research 145, 221–229 (2003)
31. Yoshida, Y., Yasuda, M.: A new evaluation of mean value for fuzzy numbers and its application to American put option under uncertainty. Fuzzy Sets and Systems 157, 2614–2626 (2006)
32. Zadeh, L.: Fuzzy sets. Information and Control 8, 338–353 (1965)
33. Zadeh, L.A.: Fuzzy sets as a basis for a theory of possibility. Fuzzy Sets and Systems 1, 3–28 (1978)
34. Zhang, J., Du, H., Tang, W.: Pricing R&D option with combining randomness and fuzziness. In: International Conference on Intelligent Computing, vol. 2, pp. 798–808 (2006)
35. Zmeškal, Z.: Application of the fuzzy-stochastic methodology to appraising the firm value as a European call option. European Journal of Operation Research 135, 303–310 (2001)

Using Fuzzy Sentiment Computing and Inference Method to Study Consumer Online Reviews

Narisa Zhao and Yuan Li

Institute of Systems Engineering, Dalian University of Technology, Dalian, 116024, China
nmgnrs@dlut.edu.cn

Abstract. This paper considers the problem of online reviews sentiment mining based on the theory of consumer psychology and behavior. Given the fuzzy attribute nature of the online reviews, we have established fuzzy group bases of consumer psychology. Four fuzzy bases, including features, sense, mood and evaluation, are established. The consumer attitude elements are reflected by natural language reviews. A fuzzy sentiment computing algorithm of online reviews for consumer sentiment is developed, and a fuzzy rule base is also presented based on consumer decision-making process. Finally it shows by means of an experiment that the proposed approach is very well suited as an analysis tool for the online reviews sentiment mining problem.

Keywords: Online reviews, fuzzy group bases, fuzzy sentiment computing, fuzzy inference.

1 Introduction

Following the development of network technology, especially for the Web 2.0, the Internet is permeating almost every aspect of life [1]. One recent phenomenon is the popularity of online community. The attraction of the online community is mainly due to a new form of word-of-mouth (WOM) communication, comprising vast amounts of consumer information on opinions, attitudes, feelings, emotions and recommendations on products/services from experienced consumers [2]. Researchers often refer to this online review as electronic word-of-mouth (eWOM) [3-4]. Users tend to trust peer reviews more than advertising and other content created by marketing departments and advertising agencies [5], so people often make the buy/not buy decision on the basis of online reviews. Now the online reviews are regarded as the best to represent the interests of the potential consumers and reduce the inherent risks and anxiety in purchasing new products [6-7].

As the most convenient and abundant resources, the online reviews has become the important sources of experience information [8]. This sentimental information has a bright prospect in many fields, such as reputation analysis, public voice monitoring, opinion mining, product reviews, and personalized recommendation and so on [4, 9-10].

B.-Y. Cao and X.-J. Xie (Eds.): Fuzzy Engineering and Operations Research, AISC 147, pp. 297–305.
springerlink.com © Springer-Verlag Berlin Heidelberg 2012

Although the current text-based sentiment computing has made great progress, there is also much urgent improvement needed for the growing subjective information, particularly on the consumer sentiment analysis of online reviews which we are concerned with. The earlier sentiment computing did not focus much on the fuzzy attributes of natural language and also the consumer fuzzy sentiment and psychology.

In this paper, the research purpose are sentiment identification and behavioral inference of consumer online reviews based on the fuzzy sentiment group bases guided by cognitive linguistics and consumer psychology. We conduct the discussion from three levels, vocabulary level, review statement level and inference level. A fuzzy sentiment computing algorithm of online reviews for consumer sentiment is developed.

2 Fuzzification of Online Reviews

It is an important topic to measure or quantify the word meaning in complex system or decision-making process for a long time. In the traditional research of word meaning quantification, for example, Mosier's one-dimensional fixed-point or Osgood's multi-dimensional characterization [11], they all considered that meaning is accurate. For example, an orientation scale [-4, 4] is defined, then the word 'beautiful' can be assigned +2, 'old' can be assigned -1. The sentimental semantic quantification research also used this pattern mostly.

However, language is vague. Based on Zadeh's fuzzy theory, the meaning of a word corresponds to a fuzzy set instead of binary logic which cannot appropriately describe the fuzzy process of thinking. Natural language consists of basic words (atomic terms) and their composition (composite terms), which are defined as the elements and sets in the domain of natural language. Domain X is defined as an interpretation of the understanding or an expression of a word meaning. Suppose a special atomic term α in the domain of natural language, and a fuzzy set $\tilde{A}$ corresponding to its specific meaning in the interpretation domain X. The fuzzy set $\tilde{A}$ represents the mapping ambiguity between the atomic term α and its 'interpretation'. $\tilde{A}$ is characterized by a membership function $\mu_{\tilde{A}}(x)$ in interval [0, 1] which indicates the membership degree of the interpretation x of α in $\tilde{A}$. We call this "the natural language variable's 'value' can be defined by fuzzy sets $\mu_{\tilde{A}}(x)$".

Sentiment is a very broad concept, and has fuzzy attributes in nature. In the research of sentimental analysis of review text, it is necessary to make fuzzy processing to the sentimental words. The measurement of the meaning of sentimental words can be divided into five ranking separately on positive and negative category continuum, micro (A), small (B), neutral (C), large (D) and extreme (E). Each rank corresponds to a fuzzy membership function, namely,-E,-D,-C,-B,-A, +A, +B, +C, +D, +E. According to the subjective experience, that in a series of intensity, the possible psychological reaction distribution to a weak stimulate on the category continuum (weak-strong) is top-down, generally monotone decreasing, the peak of its curve is left-biased (weak side). To stimulus of moderate intensity, it seems like the normal distribution curve. For greater intensity, the peak is right-biased. To the strongest stimulus, the curve shows monotonically increasing, contrary with the weakest stimuli.

According to the principles of establishing membership functions, that should be convex fuzzy sets, symmetric, balanceable, and should conform to people's language sequence, avoid improper overlap, etc., the Gaussian function is chosen as a template to define the fuzzy membership functions for these 10 sentimental ranks in the domain [-4, 4]:

$$\mu_w(x) = gaussmf_w(x, \sigma_w, a_w) = \exp\left(\frac{-(x-a_w)^2}{2\sigma_w^2}\right) \quad . \tag{1}$$

Here, $w \in \{-E, -D, -C, -B, -A, +A, +B, +C, +D, +E\}$, σ_w, a_w are the expectation and the standard deviation of Gaussian membership function respectively corresponding to the sentimental rank w. By the reason of the intersection of the membership function is neither for very low values nor for very high values, choose $\sigma_w = 0.4$. For negative pole, $x \in [-4, 0]$, $a_{-E} = -4$, $a_{-D} = -3$, $a_{-C} = -2$, $a_{-B} = -1$, $a_{-A} = 0$. For positive pole, $x \in [0, 4]$, $a_{+A} = 0$, $a_{+B} = 1$, $a_{+C} = 2$, $a_{+D} = 3$, $a_{+E} = 4$. For example, the membership functions of the variable 'evaluation' are shown in Figure 1.

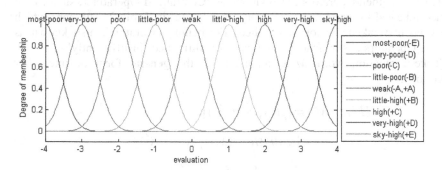

Fig. 1. Membership function

3   Sentence Fuzzy Sentiment Computing

3.1   *Fuzzy Operator of Qualifiers*

To achieve the fuzzy computing of mood and evaluation degree at the sentence level, we need to determine the language operators and characterize the semantic transfer caused by qualifiers.

Language operator indicates a class of prefix in language system, usually added in front of a phrase or word to adjust the meaning of it, such as the emphasized prefix or negative prefix. And this will infer to the sentiment transfer problem caused by Modified-Orientation. In order to resolve these two problems, we have defined two dictionaries: the Intensifier Dictionary and Privative Dictionary.

Intensifier operators. In our research, we use the word set in 'HowNet' which is created by Professor Dong, et al. [12]. After screening and refreshing, finally divide it into five ranks, namely, extreme/most, very, more, little and insufficiently. The following table lists some of the typical intensifier words (Table 1).

Table 1. Examples of some representative intensifier word ranks

Ranks	Extreme/Most	absolutely, amazingly, extremely, completely, exceedingly, beyond comparison, bitterly
	Very	considerably, especially, much, quite, particularly, too far, a lot, too much
	More	by far, comparatively, even more, further, furthermore, increasingly, relatively
	Little	a bit, a little, a little bit, fairly, more or less, passably, slightly, somewhat, some
	Insufficiently	a little less, just, less, merely, ultra, undue, unduly, surplus, to a fault

In the intensifier words, the ones which are used to strengthen the tone are called 'Strengthen Operators', also known as 'Centralized Operators', such as the ranks indicated in 'extreme/most', 'very' and 'more' in Table 2. The ones which are used to weaken the tone are called 'Freshening Operators', also known as 'Loose Operators', as indicated in the ranks of 'little' and 'insufficiently'.

Take evaluation word M as an example, the general form of intensifier operators is:

$$H_\lambda \mu_w(M) = [\mu(x, a_w \pm \lambda')]^\lambda$$
$$= [gaussmf_w(x, \sigma_w, a_w \pm \lambda')]^\lambda \ . \tag{2}$$
$$= \exp\left(\frac{-[x-(a_w \pm \lambda')]^2}{2\sigma_w^2} \cdot \lambda\right)$$

Here, $\sigma_w = 0.4$, w is the evaluation rank of the target word, H_λ is the Intensifier Operator, λ is a positive real number, when $\lambda > 1$, H_λ is a Centralized Operator, when $\lambda < 1$, H_λ is a Loose Operator. a_w is the desired value of the Gaussian function for evaluation rank w, for negative pole, it will shift $-\lambda'$ units, for positive pole, it will shift $+\lambda'$ units.

The five intensifier ranks should be determined according to the specific situations. We select the value of λ and λ' after experiment in the actual calculation as in Table 2.

Table 2. The value of variables λ and λ'

Intensifier Rank	Extreme/Most	Very	More	Little	Insufficiently
λ	4	2	1.5	0.5	0.25
λ'	+2	+1.5	+0.5	-1	-1.5

For each sentimental word with intensifier operator prefix, after transferring and width change of its membership function by means of Formula (2), we can achieve the corresponding change of sentiment degree.

3.2 Fuzzy Sentiment Computing Algorithm of Online Reviews

After the formulation of the modified rules, we can conduct fuzzy computing at the sentence-level.

Taking evaluation as an example, the Fuzzy Sentiment Computing Algorithm of Online Reviews (FSCA-OR) go as follows:

Step 1. First, conduct part-of-speech tagging and syntactic analysis (take use of the Language Technology Platform (LTP) for online presentation developed by Harbin Institute of Technology Laboratory of the language information retrieval [13], then select all the evaluation words Mi ($i=1$, 2, ... n) of the target sentence and the corresponding Intensifier and Privative qualifiers, and then determine the sequence relationship between its qualifiers.

Step 2. From the established fuzzy group bases of consumer psychology, determine the fuzzy function $\mu_w(M_i)$ of each word, $w \in \{-E, -D, -C, -B, -A, +A, +B, +C, +D, +E\}$, as well as its corresponding Intensifier ranks 'I' (5-extreme/most, 4-very, 3-more, 2-little, 1-insufficiently, 0-none). The Privatives 'P' are denoted by N (Negative) and 0(none) respectively. If 'P' and 'I' do not appear at the same time, then record 'I' first, 'P' second. For example 'not well' can be analyzed as '0N+B'.

Step 3. For the analysis results of all the evaluation words Mi (i =1, 2, ... n), Using the operator of fuzzy language qualifier in section 3.1, the Modified-Strength can be calculated for each affective tagging data items created in the given table.

Step 4. The Sentence Fuzzy membership function is

$$\text{Sentence-Function} = \bigcup_{1}^{n} \mu_w(M_i), (i=1,2,\ldots n).$$

Step 5. Defuzzification = Centroid (Sentence-Function), get the evaluation degree of the sentence employing Centroid method to defuzzy.

The fuzzy computing algorithm of mood is similar. Thereby the evaluation and mood degree of a sentence can be calculated by the compiled program in Matlab 7.0.

4   Construction of the Inference Rule Base

There are a great number of methods of knowledge expression. In these various methods, the most common way is to express knowledge by the rules of natural language form:

IF premise (antecedent), THEN conclusion (consequent).

This knowledge expression, as it expresses the human experience and heuristic knowledge with their own language, has a superficial knowledge characteristic which is particularly suited to express the relationship between contexts. Usually these restrictions are established by the fuzzy sets and fuzzy relations.

To realize the consumer recommendation to a degree, based on online reviews, according to consumer decision making process, we take fuzzy variables 'evaluation' and 'mood' as inference antecedent, 'recommendation' as inference consequent, to establish Fuzzy Inference System (FIS).

The input variable 'evaluation' consults to Section 2. For the input variable 'mood', we put it into ten ranks too, respectively, super-bad: -E, very-bad: -D, bad: -C, little-bad: -B, so-so: -A, +A, little-good: +B, good: +C, very-good: +D, super-good: +E. The inference consequent 'recommendation' is divided into seven levels in domain [-1, 1] with $\sigma_{\underline{w}} = 0.1$ taking Gaussian function as membership function style, respectively strongly-resist, resist, negative, neutral, positive, recommend, strongly-recommend. The principle to set a rule is 'evaluation' occupy a leading position, 'mood' subordinate.

Some of the inference rules are established to realistic this operation, which are listed as below:

Rule 1. If (evaluation is very-poor) and (emotion is super-bad) then (recommendation is strongly-resist).

Rule 2. If (evaluation is little-poor) and (emotion is very-bad) then (recommendation is resist).

Rule 3. If (evaluation is so-so) and (emotion is bad) then (recommendation is negative).

A total of 52 rules are established here, and different rules give different weights respectively. Mamdani-based inference method is used. Defuzzification employs Centroid method.

5   Experiment

Using page collection tools 'bget_share', this experiment downloaded more than 1200 online reviews of a certain brand of notebook computer from the related posts of Baidu Post Bar (http://tieba.baidu.com/). On the corpus style, we chose relatively standardized, rigorous reviews as much as possible. Generally, the choice of reviews emphasized on the ones which have rich sentimental expression. 549 sentences with views are selected after screening. The so-called sentence with views refers to the sentence contains at least one orientation word. Once more, after the second screening based on typicality, ultimately 100 representative reviews were identified as the final corpus.

For each one of these 100 reviews, the mood and evaluation degree are calculated through the algorithm FSCA-OR, and then according to the inference rule base, we can obtain the recommendation degree of each review ultimately. Figure 2 shows the recommendation degree sorting from low to high. In order to be unified into a table, the degree was normalized to range in [-1, 1], that is,

$$\text{degree} = \text{degree (evaluation or mood)} / 4 \qquad (3)$$

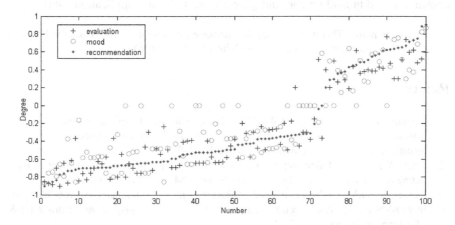

Fig. 2. Diagrammatic presentation of the results of a certain brand of laptop

From Figure 2, we can see that, for this part of reviews, recommendation degree that less than zero occupies 72%, that greater than zero only 28%. This is because many online reviewers tend to express their sentiment of dissatisfaction. A small number of reviews both have higher value of sentiment and recommendation which are inconsistent with the majority, that because, here, we do not rule out the role of soft advertising reviews.

In addition, the emotional tendency of reviewers and the recommendation degree remain consistent basically, and the evaluation degree keeps consistent changes with mood. Their combination decides the degree of the recommendation. In Figure 2, 17% of the sentence does not contain the mood words, and the mood values distribution is relatively scattered, which largely because it has something to do with the situation at that time. We can see that this laptop brand reputation in the online reviews is poor, and more people do not recommend others to purchase. The experiment also shows the validity of the algorithm.

6   Conclusion

This paper studied consumer online reviews using fuzzy sentiment computing and inference method through three levels. In vocabulary level, fuzzy sentiment modeling for consumer online review texts is discussed based on fuzzy mathematics, and fuzzy group bases of consumer psychology is established. In review statement level, after sentiment membership functions' shift and transformation, a fuzzy sentiment computing algorithm of review sentences is proposed. Finally in inference level, a series of fuzzy inference rules are made for consumer recommendation through sentiment mining. Our experiment showed well performance in Precise and Recall. Our results yield interesting and important insights for both academic researchers and practitioners. It has an important scientific significance on some of the basic theory of text sentimental analysis and consumer aid-decision-making. In mining customer perspective and understanding the market response of the products, establishing interactive relations between producers and consumers, and in guiding consuming behavior, it also has application value.

Acknowledgements. This work is partially supported by research grant from the Program of National Natural Science Foundation of China (No. 61072128).

References

1. Cheung, M.Y., Luo, C., Sia, C.L., Chen, H.: Credibility of electronic word-of-mouth: informational and normative determinants of on-line consumer recommendations. International Journal of Electronic Commerce 13, 9–38 (2009)
2. Chen, Y., Xie, J.: Online consumer review: word-of-mouth as a new element of marketing communication mix. Management Science 54, 477–491 (2008)
3. Lee, J., Lee, J.N.: Understanding the product information inference process in electronic word-of-mouth: An objectivity–subjectivity dichotomy perspective. Information & Management 46, 302–311 (2009)
4. Clemons, E.K., Gao, G., Hitt, L.M.: When online reviews meet hyperdifferentiation: a study of the craft beer industry. Journal of Management Information Systems 23, 149–171 (2006)
5. Park, C., Lee, T.M.: Antecedents of online reviews' usage and purchase influence: an empirical comparison of U.S. and Korean consumers. Journal of Interactive Marketing 23, 332–340 (2009)
6. Dellarocas, C., Zhang, X., Awad, N.F.: Exploring the value of online product reviews in forecasting sales: the case of motion pictures. Journal of Interactive Marketing 21, 23–45 (2007)
7. Godes, D., Mayzlin, D.: Using online conversation to study word of mouth communication. Marketing Science 23, 545–560 (2004)
8. Duan, W., Gu, B., Whinston, A.B.: Do online reviews matter? — an empirical investigation of panel data. Decision Support Systems 45, 1007–1016 (2008)
9. Yang, H.C., Lee, C.H.: Semantic matching and annotation of images by self-organizing maps. International Journal of Innovative Computing, Information and Control 5, 677–688 (2009)

10. Lee, J., Park, D.H., Han, I.: The effect of negative online consumer reviews on product attitude: an information processing view. Electronic Commerce Research and Applications 7, 341–352 (2008)
11. Zétényi, T.: Fuzzy sets in psychology. Elsevier scientific publishers, Amsterdam (1988)
12. Dong, Z.D., Dong, Q.: HowNet: Computer language information center of CAS, http://www.keenage.com/html/e_index.html
13. Liu, T.: Language Technology Platform (LTP). Harbin institute of technology information retrieval laboratory, http://ir.hit.edu.cn/demo/ltp/

A Quantitative Reservoir Evaluation Method Based on Fuzzy Comprehensive Appraisal and Analytic Hierarchy Process

A Case Study of Xujiahe Zu in Baojie Area

Yan He

The State key Laboratory of Oil and Gas Reservoir Geology and Exploitation,
Southwest Petroleum University, China
swpuheyan@163.com

Abstract. The classification and evaluation of reservoir is not only critical part of reservoir study, but also the fundamental of reservoir description. Based on the interpretation ambiguity and mutual contradictions of the evaluation methods in the past, this paper proposes a new quantitative reservoir evaluation method that combines fuzzy comprehensive appraisal and analytic hierarchy process to conduct a quantitative reservoir evaluation. The combination of those two technologies takes advantages of only using either of them and applied to the case study of Xjiahe Zu Formation. The reservoir parameters, such as reservoir porosity (Φ), permeability (K), reservoir effect pay (H) and the ratio of sand thickness and stratum thickness, as well as clay content, are used to determine the functions and membership grade. Analytic hierarchy process (AHP) was applied to determine the weight indexes based on above parameters to evaluate the reservoir. The results show the calculated weight of reservoir porosity (Φ), permeability (K), reservoir effect pay (H) and the ratio of sand thickness and stratum thickness, clay content are 0.1948, 0.4052, 0.2034, 0.0981, 0.0985, respectively. The results match very well with the actual field data of Xujiahe Zu Formation, Baojie Area.

Keywords: Reservoir evaluation, fuzzy comprehensive appraisal, analytic hierarchy process, BaoJie Area.

1 Introduction

The detailed reservoir researches study involves in not only anatomizing reservoir bed and ascertaining geological development characteristic, also evaluating and classifying reservoir quantitatively. There are many factors which affect the reservoir including porosity, permeability, net thickness, pore texture, clay content, heterogenity etc. In the past evaluation process, mono-factor was usually used, or evaluated with the expert marked combining Fuzzy math. There were contraries due to mono-factor evaluating, the more parameters the more contraries.

B.-Y. Cao and X.-J. Xie (Eds.): Fuzzy Engineering and Operations Research, AISC 147, pp. 307–314.
springerlink.com   © Springer-Verlag Berlin Heidelberg 2012

There are interpretation ambiguities due to different experts with different opinions when using expert marked combined with Fuzzy math. Considering the past methods with so many problems to evaluate reservoir, the author proposes a new quantitative reservoir evaluation method that combines fuzzy comprehensive appraisal and analytic hierarchy process, and applies to reservoir evaluation of XuJiahe Zu in the BaoJie area. The evaluation results are proved to match well with the actual production test data.

2 Fuzzy Comprehensive Appraisal

Fuzzy comprehensive appraisal, also named fuzzy comprehensive decision, is to make a comprehensive evaluation based on various interfering fuzzy factors.

Assumed domain $U = (u_1, u_2 ; \cdots , u_n)$ and $V = (v_1, v_2 ; \cdots , v_n)$, R is a fuzzy relation from U to V. If any fuzzy subset A of U and subset B of V, they are fit for

$$B = AR, \tag{1}$$

where

$$R = (r_{ij})_{n \times m} \quad 0 \le r_{ij} \le 1, \tag{2}$$

In the formula: r_{ij} *is the membershipgrade value for factor i to the j noto.*

3 The Method for Reservoir Comprehensive and Quantitative Evaluation

3.1 Determination of Parameter Classification and Criteria

According to the actual field data and the importance of the reservoir parameters at Xujiahe Zu in Baojie Area, mean porosity ($\overline{\phi}$), mean permeability ($\overline{K}$), net pay (H), sand-formation ratio ($R_{a/f}$) and shale content (SH) are selected in the model. Using the parameters as factors, we get factors set.

$$U = (\overline{\phi}, \overline{K}, H, R_{a/f}, SH). \tag{3}$$

The Xujiahe Zu Reservoir are classified into 6 categories ranging from excellent reservoir quality to non-reservoir, signed as Pure, I , II , III , IV , V respectively (shown in sheet 1). A appraisal set is written as

$$V = (\text{pure}, \text{ I }, \text{ II }, \text{ III }, \text{ IV }, \text{ V }). \tag{4}$$

According to the sand reservoir classification criteria made by China National Petroleum Company, most of the reservoirs in Xujiahe Zu are low porosity, low permeability or extremely low porosity, extremely low permeability. The classification in Sheet 1 is based on the reservoir quality of this case study.

Table 1. Cassification criterion xujiahe zu baojie area

Type of reservoir	$\bar{\phi}$, %	$\bar{K}$, $10^{-3}\mu m^2$	H, m	$R_{a/f}$	SH, %
pure I	>15	>20	>10	>0.9	<5
I	13~15	5~20	7~10	0.8~0.9	5~20
II	11~13	3~5	4~7	0.65~0.8	20~30
III	9~11	1~3	1~4	0.5~0.65	30~40
IV	7~9	0.2~1	0.2~1	0.4~0.5	40~50
V	<7	<0.2	<0.2	<0.4	>50

3.2   Membership Grade

The membershipgrade is the appraisal value for each single factor in whole factor set. There are many methods to calculate it, such as the fuzzy statistics testing, the two-element comparing sequencing, the expert remarking, and the piecewise function etc [7-9]. According to the reservoir characteristic of the study area, after many times pilot calculation and verification, we selected ridge-like serial position curve as membership function of mono-factor classification. The curves include small, big and medium 3 types [1].

Fixing any mono-factor, we can get its classification membershipgrade according to membership function (ridge-like serial position curve). Usually the type big and medium curves are used to get those mono-factor membership where the reservoir quality increases with the increases of the mean porosity, mean permeability, net thickness, sand-formation ratio. Otherwise using the type small and medium ridge-like function for the worse reservoir quality with the increase of clay content. For example, when the permeability $\bar{K}$ >20×10^{-3}μm^2, the chance of class I reservoir is 100%, $r_1 = 1$; The chance of being pure class I reservoir can be get by calculating ridge-like function $r_2 = \frac{1}{2} + \frac{1}{2}\sin[\frac{\pi}{40}(x-20)]$, meanwhile the chance of being class II, III, IV is 0, then $r_3 = r_4 = r_5 = r_6 = 0$; when $5\times10^{-3} < \bar{K} < 20\times10^{-3}$μm^2, the chance of being pure class is $r_1 = \frac{1}{2} + \frac{1}{2}\sin[\frac{\pi}{15}(x-\frac{5+20}{2})]$, the chance of being class I is $r_2 = 1$;the chance of being class II is $r_3 = \dfrac{1}{2} - \dfrac{1}{2}\sin[\dfrac{\pi}{15}(x - \dfrac{5+20}{2})]$; The

chance of being the other is zero, then $r_4 = r_5 = r_6 = 0$; and so on.. We can

get the $\overline{K}$ membership grade within other range. The same method can be
applied to obtain any mono-factor's membership grade.

Mono-factor appraisal has one bias, since every single factor's contribution
determined by its weight, the weight distribution directly affects the result of
comprehensive appraisal and classification, all parameters' weight make a weight
fuzzy matrix.

$$A = (a_1, \ a_2, \ ..., \ a_n). \tag{5}$$

3.3 The Method of Analytic Hierarchy Process to Determine Weight

Methods to determine weight were used to the principal component statistics and
the expert remarks. The author here uses the method of analytic hierarchy process
to determine weight [12-13]. There are many factors affecting the reservoir
comprehensive evaluation. Assumed there are 错误！未找到引用源。 factors,
making appraising matrix.

$$P = (p_{ij})_{nn}, \tag{6}$$

where p_{ij} is the value of importance degree which the factor i influences the
objective strata comparing with factor j .

If p_i and p_j are same important, then p_{ij} is 1, if p_i is slightly more
important than p_j , then p_{ij} is 3, if p_i is more important than p_j , then p_{ij} is

5, and so on, if between these, can use 2,4,6,8 etc. let $p_{ij} = \dfrac{1}{p_{ji}}$ get pairs of

comparable appraisal matrix

$$P = \begin{pmatrix} p_{11} & p_{12} & \cdots & p_{1n} \\ p_{21} & p_{22} & \cdots & p_{2n} \\ \vdots & \vdots & \ddots & \vdots \\ p_{n1} & p_{n2} & \cdots & p_{nn} \end{pmatrix}. \tag{7}$$

Calculate the importance sequence of the pairs of comparable appraisal matrix using common sum method. Normalizing matrix P to get $\overline{P}$:

$$\overline{P} = \begin{pmatrix} \overline{P}_{11} & \overline{P}_{12} & \cdots & \overline{P}_{1n} \\ \overline{P}_{21} & \overline{P}_{22} & \cdots & \overline{P}_{2n} \\ \vdots & \vdots & \ddots & \vdots \\ \overline{P}_{n1} & \overline{P}_{n2} & \cdots & \overline{P}_{nn} \end{pmatrix}, \tag{8}$$

where

$$\overline{P}_{ij} = \frac{P_{ij}}{\sum\limits_{i=1}^{n} P_{ij}},$$

Adding $\overline{P}$ in row to get $\overline{W}_i = \sum\limits_{j=1}^{n} \overline{P}_{ij}$, then normalizing $\overline{W} = (\overline{W}_1, \overline{W}_2, \cdots, \overline{W}_n)^{\mathrm{T}}$ to obtain

$$W_i = \frac{\overline{W}_i}{\sum\limits_{j=1}^{n} \overline{W}_j}, \tag{9}$$

The vector $W = (W_1, W_2, \cdots, W_n)$, named as feature root vector. Calculate and normalize the biggest feature root vector of the appraisal matrix to calculate the affecting factors weight matrix. Then apply the method of analytic hierarchy process to calculate the weight coefficients of mean porosity, mean permeability, net thickness, sand-stratum ratio and clay content in Xujiahe Zu of Baojie Area.

$$A = (0.1948, 0.4052, 0.2034, 0.0981, 0.0985). \tag{10}$$

We obtain the membership grade matrix, based on the mono-factor evaluation in equation (1), and analytic hierarchy process method.

4   Case Study

4.1   Geology Background

Baojie Area belongs to the Longnvsi structural group of Weiyuan in the middle of Sichuan basin, westwards to the Siliujing Group of the Huangjiachang Strusture and Longchang Structure; Northern is the Dazu-Anyue Syncline; Southeast joints with Xishan Structure [10]. It is a big syncline structure dipping towards northeast with many small nose-like structures developed, covering 1600km²[10]. The reservoir rock in Xujiahe Group are mainly grey to white median grain sands with mainly Quartz, mixed with feldspar and debris. A small amount fine or coarse sands are seen in the reservoir too. The reservoirs are low porosity and low permeability [11].

4.2 Evaluation on Classification

The data in well 24 in Baojie Area is used in the study. The mean porosity of this well in Xujiahe Zu is 10.44%, mean permeability is $66.16\times10^{-3}m^2$, net thickness is 9.25m, sand-formation ratio is 0.76, shale content is 30.57%. Applying the method above in this paper, the membership of its mean porosity, mean permeability, net thickness, sand-formation ratio, shale content by ridge- like function are calculated as followings:

Excellent　　good　-------------------------------------- non-reservoir

$$R = \begin{bmatrix} 0 & 0 & 0.61376 & 1 & 0.38624 & 0 \\ 1 & 0.28372 & 0 & 0 & 0 & 0 \\ 0.72134 & 1 & 0.27866 & 0 & 0 & 0 \\ 0 & 0.79834 & 1 & 0.30166 & 0 & 0 \\ 0 & 0 & 0.92357 & 1 & 0.07643 & 0 \end{bmatrix} \begin{matrix} \overline{\varphi} \\ \overline{K} \\ H \\ R_{a/f} \\ SH \end{matrix} \cdot \quad (11)$$

According to the weight fuzzy matrix (10)of the reservoir classification of Xujiahe Zu Baojie Area, using fore Quation(1) and fuzzy transformation, we can get the membership matrix of this grid data.

$$B = (0.18325, 0.48963, 0.20137, 0.07632, 0.04943, 0). \quad (12)$$

Using the maximum membership principle in fuzzy reasoning, we got class I reservoir.

Based on the above theory, we made fuzzy comprehensive appraisal program, classified the sand reservoir of Xujiahe Zu in Baojie Area. Using the software of Geomap, we made the reservoir types areal distribution map (Fig. 1) of T3X2 Formation based on referencing the sedimentary microfacies, porosity, permeability, net thickness in this area.

Comparing the results of classification evaluation with the actual production test in Table 2, we know that class I and class II reservoir correspond with high production reservoir, class III reservoir corresponds with lower production, class IV reservoir gets trace gas production or dry. The reservoirs with over 104 m3/d gas production mainly belong to class I and class II, class III gets lower gas production, and class IV only gets trace gas. The result of evaluation matched well with production test data.

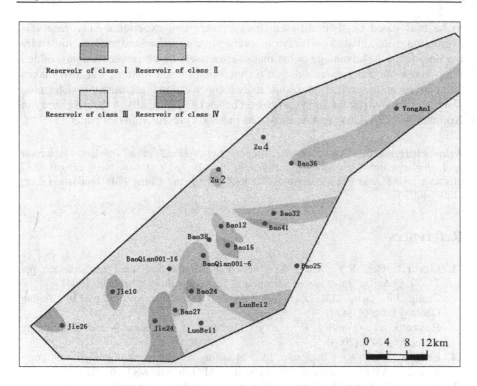

Fig. 1. Reservoir types map of T3X2 in Xujiahe Zu, Baojie Area

Table 2. Reservoir types of production test zone in Xujiahe Zu, Baojie Area

Well name	Production test	Interval (m)	result			Reservoir type
			oil/	gas/	water/(m³	
Bao24	Xu2	1756-		6.38		I
Bao27	Xu2	1684-	0.58	1.65	18.8	II
	Xu4	1564-		0.07		IV
Bao36	Xu2	2286-	0.2	1.05	12	II
	Xu4	2116-		trace		IV
Bao Qian001-16	Xu4	1697-		trace	12.6	IV
	Xu6	1584-		1.25		III
Zu2	Xu4			0.36		III

5 Conclusion

The traditional mono-factor reservoir evaluation methods usually result in the different evaluation to the same reservoir. The expert remark method often tends to be bias based on their different backgrounds and experience. The reservoir evaluation method based on fuzzy comprehensive appraisal and analytic hierarchy not only takes the advantage of the mathmatics itself which is a continuous-valued logic and more accurate calculation on the complicated matters, but also makes the classification evaluation result more objective and accurate by determine weight using analytic hierarchy process. The successful results in the case study at Xujiahe Zu, Baiji Area , further prove the methods presented in this paper.

Acknowledgements. Thanks to the support by National major project "Reservoir geological research for developing efficiently complicated and fault-block oil-gas field offshore in the Asian-Pacific region" (2008ZX05000) and China PhD foundation item (20060400307).

References

1. Liu, Y.J., Gen, X.Y., Xiao, C.Y.: Petroleum engineering fuzzy mathematics, pp. 71–72. ChengDu University of Science and Techology Press, ChengDu (1994)
2. Zhang, J.F., Deng, B.R., Zhu, Y.X.: Applied fuzzy mathematics, p. 114. Beijing Geology Press (1988)
3. Hambalek, N., González, R.: Fuzzy Logic applied to lithofacies and permeability forecasting. SPE (2003)
4. Elise, B., Mark, A.P., Benjamin, J.R.: Modeling secondary oil migration with core-scale data: Viking Formation, Alberta basin. AAPG Bulletin 86(1), 63–85 (2002)
5. Cheng, S.Y., Long, H.: The use of fuzzy integrated decision analysis in oil and gas resource Evaluation. Computing Techniques for Geophysical and Geochemical Exploration 24(4), 318–320 (2002)
6. Wu, Y.J., Cai, Z.Q.: Oilfield developmental geology, p. 36. Beijing Petroleum Industry Press, BeiJing (2000)
7. Li, J.Q.: Main factors controlling filling degree of lithologic reservoirs and its fuzzy comprehensive evaluation. Linnan Sag.Petroleum Geology and Recovery Ration 15(3), 36–38 (2008)
8. Xu, P.D., Dai, J.S., Lin, B.: Identifying spatial distribution of the interlayer with fuzzy evaluation and stochastic modeling—taking Guantao 6 Member in Zhongyi area of Gudao Oilfield as an example. Petroleum Geology and Recovery Ration 14(3), 61–63 (2007)
9. Xin, Z.G.: Evaluation of watered out behavior of reservoirs using fuzzy comprehensive assessment. Petroleum Geology and Recovery Ration 14(6), 88–90 (2007)
10. Zhu, S.J., Huang, J.X.: Stratigraphic Classification and Tracing of Xiangqi Group in central Sichuan-South Sichuan Transitional Belt. Journal of South west petroleum university 18(2), 1–7 (1996)
11. Yin, X.M., Peng, J., He, Y.: Reservoir characteristics of Xujiahe Formation of Upper Triassic in Baojie Area. Fault-Block Oil and Gas Field 15(5), 19–22 (2008)
12. Yan, K., Wu, Z., Jing, F., He, T.: The deficiency of AHP(Analytic Hierarchy Process) and its improved in multiple objective decisions, vol. 5, pp. 10–11 (2007)
13. Wen, W.B.: The theoretical basis for prediction. Petroleum Industry Press (1984)

A Flocculation and Deposit Model of Asphaltene in Porous Media for Damage Evaluation

Zhou-hua Wang, Ping Guo, and Jian-fen Du

State Key Laboratory of Oil and Gas Geology
and Exploitation, Southwest Petroleum University, Chengdu 610500, China
wangzhouhua@126.com

Abstract. The flocculation and deposit of asphaltene in porous media is concerned all the time by researchers home and overseas; because of its complexity, the study of testing method, flocculation and deposit theory is **a** hot topic for all researchers, especially for the problem of flocculation and deposit of asphaltene in porous media. Based on some related researches, this paper establishes one dimension gas-liquid-solid three phase flocculation and deposit model of asphaltene in porous media, simulates practical experiments with it, and contrasts the calculating results with experimental ones. The results suggest that the model show good agreement with the experiments; at the same flow speed, the lower porosity and permeability the core has, the greater effect is of porous media on the flocculation and deposit of asphaltene; the flow speed is inversely propotional to the amount of deposited asphaltene.

Keywords: Asphaltene, flocculation and deposit, porous media damage, model.

1   Introduction

The flocculation and deposit of organic solid phase in reservoir fluid has always been a serious problem in petroleum industry. It lies in the oil reservoir bed, exploitation facilities, pipeline transport and process equipment. In the process of gas injection for enhancing recovery, the asphaltene will flocculate from the crude oil, which will clog the wellbore and production equipment and affect the production of oil field seriously. Normally, if the oil field develops containing asphaltene with depletion developing, rich gas driving or CO2 injection, there will be a flocculation and deposit of asphaltene. Study has showed that flocculated asphaltene will strongly affect the porosity and permeability [1,2,3]. In regards to the flocculation of asphaltene, based on different phase theory, there are various models; there are seldom any researches on the influence of porous media on asphaltene flocculation, and theory research is not enough. This is the very first time to study the flocculation and deposit of asphaltene in porous media theoretically home. Thus, studying the mechanism of flocculation and deposit of asphaltene in porous media has an important theoretical value, and it is important for enhancing oil and gas recovery ratio of our country.

B.-Y. Cao and X.-J. Xie (Eds.): Fuzzy Engineering and Operations Research, AISC 147, pp. 315–324.
springerlink.com   © Springer-Verlag Berlin Heidelberg 2012

2 Damage Description of Flocculation and Deposit of Asphaltene

It normally is believed that the flocculation and deposit of asphaltene is due to the variation of temperature, pressure and component of reservoir fluid. When flocculating, asphaltene will be positively charged usually, and it tends to adhere to the negatively charged rock, which turns the water wetted surface into oil wetted surface. So, the reason why the flocculation and deposit of asphaltene causes damage is that the asphaltene will block the pore throat and reduce the relative permeability of crude oil. The variation of asphaltene particle size after flocculating is the main cause of formation damage. In the bituminous crude oil titration experiment, it is easy to find out particle diameter change before and after the experiment (Fig 1). When flocculation of asphaltene happens in the matrix rock, the bigger conglomeration clog the pore throat, the smaller ones will flow with the fluid and cause bridge plug at the pore throat which will reduce the relative permeability. So the main purpose of this research is how to evaluate the damage caused by flocculation and deposit of asphaltene.

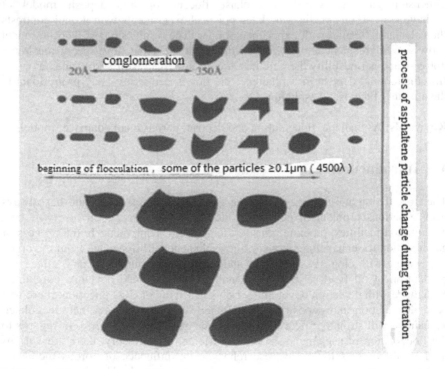

Fig. 1. Change of asphaltene grain in titration process

3 Flocculation Model of Aphaltene

There are two typical theories which can describe the state of asphaltene in crude oil, colloid liquid theory and molecular solution theory. The colloid liquid theory points out that asphaltic crude do not exist as solution but as stable dispersoid. The core of discrete phase in the system is aphaltene, the outer part of it is colloid, and the micelle of this sort is dispersed in the oil. In the liquid theory, the asphaltene is regarded as solid phase solute, crude oil without asphaltene and wax is solvent, asphaltene is dissolved in crude as molecule, and the solution is a real solution[5]. Technicians abroad tend to build flocculation model based on molecular solution theory, considering that it reflects the process of real flocculation and deposit of asphaltene. In 1984, Hirschberg [3] utilized molecular solution theory to simulate the flocculation and deposit of asphaltene for the first time and proposed the following model.

$$\phi_A = \exp\left[\frac{V_A}{V_L} - 1 - \frac{V_A}{RT}(\delta_A - \delta_L)^2\right].$$ (1)

4 Deposit Model of Asphaltene in Porous Media

This paper has established one dimension gas-liquid-solid three phase flocculation and deposit model of asphaltene in porous media, ignoring the gravitational effect during the experiment. The flocculation and deposit model of asphaltene is reflected by gas-liquid mass conservation equation, momentum conservation equation, asphaltene mass conservation equation, instantaneous porosity of porous media equation and permeability formula.

Gas phase mass conservation equation:

$$\frac{\partial}{\partial t}(\phi S_V \rho_V + \phi S_L \rho_L w_{GL}) + \frac{\partial}{\partial x}(\rho_V u_V + \rho_L u_L w_{GL}) = 0 .$$ (2)

Liquid phase mass conservation equation

$$\frac{\partial}{\partial t}(\phi S_L \rho_L w_{oL}) + \frac{\partial}{\partial x}(\rho_L u_l w_{oL}) = 0.$$ (3)

Asphaltene mass conservation equation:

$$\frac{\partial}{\partial t}(\phi C_A \rho_A + \phi \rho_L w_{AL}) + \frac{\partial}{\partial x}(\rho_L u_L w_{SAL} + \rho_L u_L w_{AL}) = -\rho_A \frac{\partial E_A}{\partial t}.$$ (4)

Momentum conservation equation

$$u_L = -\frac{kk_{RL}}{\mu_L}\frac{\partial P}{\partial x} ,$$ (5)

$$u_V = -\frac{kk_{RV}}{\mu_V}\frac{\partial P}{\partial x} .$$ (6)

Flocculation rate of aphaltene:

$$\frac{\partial E_A}{\partial t} = \alpha C_A \phi - \beta E_A (v_L - v_{cr,L}) + \gamma u_L C_A . \tag{7}$$

Saturability equation:

$$S_V + S_L + C_A = 1. \tag{8}$$

Instantaneous porosity of porous media equation and permeability formula:

$$\phi = \phi_0 - E_A , \tag{9}$$

$$k = f_p k_0 \left(\frac{\phi}{\phi_0} \right)^3 . \tag{10}$$

5   Model Solution

In order to get the numerical solution of the model, additional mathematical treatment needs to be made. First, combining Eqs. (3) and (4) and considering that $w_{OL} + w_{AL} = 1$, yields:

$$\frac{\partial \left[\phi (\rho_A C_A + \rho_L) \right]}{\partial t} + \rho_L u_L \frac{\partial C_A}{\partial x} = -\rho_A \frac{\partial E_A}{\partial t} . \tag{11}$$

Then, in order to ensure the closure of the equations, an equation of state is added:

$$Pv_L = nRT . \tag{12}$$

Finally, Eqs. (2),(5),(7),(11) and (12) constitute a set of close partial differential equations as follows:

$$\begin{cases} \dfrac{\partial (\phi w_{OL})}{\partial t} + u_L \dfrac{\partial w_{OL}}{\partial x} = 0, \\[2mm] \dfrac{\partial \left[\phi (\rho_A C_A + \rho_L) \right]}{\partial t} + \rho_L u_L \dfrac{\partial C_A}{\partial x} = -\rho_A \dfrac{\partial E_A}{\partial t}, \\[2mm] \dfrac{\partial E_A}{\partial t} = \alpha C_A \phi - \beta E_A (v_L - v_{cr,L}) + \gamma u_L C_A, \\[2mm] \dfrac{\partial P}{\partial x} = \dfrac{-u_L \mu_L}{f_p k_o (\phi/\phi_o)^3} = f(\phi), \\[2mm] C_A = \exp \left(\dfrac{v_A}{v_L} - 1 - \dfrac{v_A}{RT} (\delta_A - \delta_L)^2 \right), \\[2mm] Pv_L = nRT. \end{cases} \tag{13}$$

Considering the physical characteristics of convection transportation, upwind difference scheme is used to solve the model above. The difference scheme can be expressed by:

$$
\begin{cases}
\dfrac{P_i^n - P_{i-1}^n}{\Delta x} = f\left(\phi_i^n\right), \\[2mm]
\left(v_L\right)_i^n = \dfrac{nRT}{P_i^n}, \\[2mm]
\left(C_A\right)_i^n = \exp\left(\dfrac{v_A}{\left(v_L\right)_i^n} - 1 - \dfrac{v_A}{RT}\left(\delta_A - \delta_L\right)^2\right), \\[2mm]
\left(E_A\right)_i^{n+1} = \left(E_A\right)_i^n + \Delta t\left[\alpha(C_A)_i^n\,\phi_i^n - \beta(E_A)_i^n\left((v_L)_i^n - v_{cr,L}\right) + \gamma u_L(C_A)_i^n\right], \\[2mm]
\left[\rho_A(C_A)_i^n + \rho_L\right]\dfrac{\phi_i^{n+1} - \phi_i^n}{\Delta t} + \dfrac{(C_A)_i^n - (C_A)_{i-1}^n}{\Delta x} = -\rho_A\dfrac{(E_A)_i^{n+1} - (E_A)_i^n}{\Delta t}, \\[2mm]
\phi_i^n\dfrac{(w_{OL})_i^{n+1} - (w_{OL})_i^n}{\Delta t} + u_L\dfrac{(w_{OL})_i^n - (w_{OL})_{i-1}^n}{\Delta x} = 0.
\end{cases}
\tag{14}
$$

The model above can be solved by explicit forward difference method. By using Matlab programming language, the problem **is** solved with 12 grid nodes divided and 127 lines of program. In the solution procedure, we recommend pressure to be the first parameter to be calculated.

6   Example Calculation

The fluid data in this paper was quoted from literature [3], and experiment data quoted from literature [4]. Fluid composition is shown in list 1,with initial reservoir temperature of 50 °C, and bubble point pressure of 13.36 MPa. L.Minssieux (1997) used this sample of oil for asphaltene damage test with cores of different porosity and permeability. Tests were carried out by constant rate displacement, and the cumulative injection volume of oil was about 70 PV. Because of the limitation of experiment techniques, the amount of asphaltene deposit could not be measured accurately. But porosity and permeability of the core could be measured, so the asphaltene deposit could be known by measuring permeability of the core before and after the displacement. Basic information of the cores used in the test is shown in table 2. The results calculated by the model are shown in Figure 2 and 3.

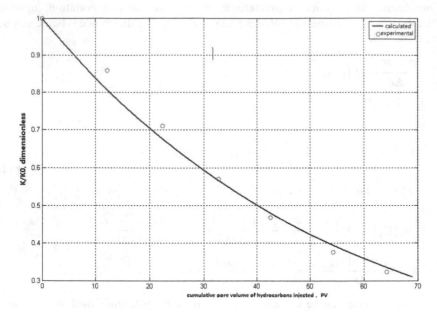

Fig. 2. 1[#] Comparison between calculation results and experiment data

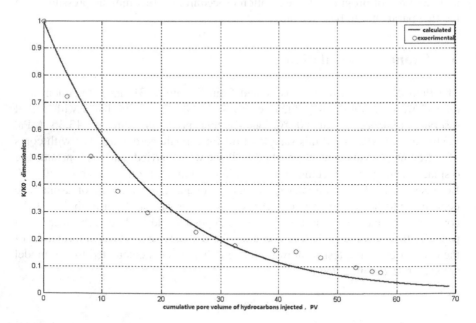

Fig. 3. 2# Comparison between calculation results and experiment data10. Li Baozhen, Li Xiangfang, Kamy Sepehrnoori, et al. : Optimization of the Injection and production schemes during CO_2

Table 1. Fluid composition

Component	mol, %	Component	mol, %
CO_2	0.4	iC_5	1.66
N_2	33.91	nC_5	1.66
C_1	6.22	C_6	2.45
C_2	5.07	C^+_7	45
C_3	1.05	C_7^+ molecular weight 224; Oil molecular weight 221.5; Oil gravity; 0.873	
iC_4	2.90		
nC_4	1.34		

Table 2. Basic data of the experiment

Data of cores (initial)			Experiment data		
Parameter	$1^\#$	$2^\#$	Parameter	$1^\#$	$2^\#$
k, md	77.4	29	Flow rate, cm^3/h	10	10
Φ, %	13.7	24.7	T, ℃	50	50
L, cm	6.0	6.0	Pressure, MPa	15	15
D, cm	2.3	2.3			

From Table 1 and 2, we can see that, as the volume of injection oil increases, the ratio(k / k_0) of the instantaneous permeability to initial permeability of the porous media gradually decreases, showing that some flocculated asphaltene deposit blocks the flow path in the porous media, so the permeability of porous media decreases. Since the porosity and permeability of the core $1^{\#}$ are higher than core $2^{\#}$, the flocculated asphaltene is easier to pass, so the asphaltene deposit is less. In the figures, the decrease tendency and relative decreasing amount of k / k_0 of core $1^{\#}$ are lower than that of core $2^{\#}$, permeability of core $1^{\#}$ decreases 52.3, by 67.5%; however, permeability of core $2^{\#}$ decreases 26.72, by 90.58%. From the data above we can see that, the damage of asphaltene deposit to cores with low porosity and low permeability is much bigger than damage to cores with high permeability. Such phenomenon is of significance for actual production process of oil and gas field. Damage of asphaltene deposit to reservoirs with low porosity and low permeability is more obvious than damage to reservoirs with high permeability. So in the actual production process, it is very important to predict, prevent and relieve the asphaltene deposit, to lower the damage to the reservoir.

7   Conclusion

1) The results of the model built in this paper is in accordance with the experiment data, the model can reflect the actual process of asphaltene deposit in experiment.

2) The results of experiment and model calculation show that, asphaltene deposit has great effect on porosity and permeability of cores; it is necessary to consider the asphaltene deposit in actual production.

3) As the volume of injection fluids increases, the decrease in porosity and permeability of the core is more obvious.

4) The model built in this paper still has differences with actual production process of reservoirs, and needs further improvement. In addition, quantitative test of asphaltene deposit in porous media requires further study.

References

1. Leontaritis, K.J., Mansoori, G.A.: Asphaltene Deposition: A survey of field experience and research approaches. Peetroil. Sci. and Eng. 11, 229–239 (1988)
2. Burke, N.E., Hobbs, R.E., Kashou, S.F.: Measurement and modeling of asphaltene precipitation. Petrol. Technol. 12, 1440–1456 (1990)
3. Hirschberg, A., DeJong, L.N.J., Schipper, B.A., Meijer, J. G.: Influence of temperature and pressure on asphaltene flocculation. SPEJ 24, 283–293 (1984)
4. Minssieux, L.: Core core damage from crude asphaltene deposition. In: SPE, vol. 37250, pp. 401–419 (1997)
5. Min, L., Shilun, L., Ping, G.: Vapor-liquid-solid three phase equilibrium calculation. Acta Petrolei Sinica 23(1), 98–101 (1997)

Nomenclature

ϕ_A --Volume percentage of asphaltene dissolved in oil

V_A --Molar volume of asphaltene(assumed to be constant)

V_L -- Molar volume of liquid phase

T -- Absolute temperature, K

R --Gas constant

δ_A --Solubility parameter of asphaltene, $\delta_A = 20.04*(1-hT)$

h --Specific constant

δ_L --Dissolubility parameter of liquid phase, $\delta_L = \left(\dfrac{\Delta U_{vaporization}}{V_L} \right)^{1/2}$

$\Delta U_{vaporization}$ --Change of evaporating internal energy of liquid phase per unit volume, $du = C_v dt$, $kJ / kg \cdot k$

ϕ --Porosity of porous media

ρ_V, ρ_L --Density of gas phase, liquid phase kg / m^3

u_V, u_L --Darcy velocity of gas phase, liquid phase, $m / \sec$

w_{GL} --Mass fraction of gas dissolved in liquid phase

w_{oL} -- Mass fraction of oil in liquid phase

C_A -- Volume fraction of deposited asphaltene suspended in liquid phase

ρ_A --Density of asphaltene, kg / m^3

w_{SAL} -- Mass fraction of deposited asphaltene suspended in liquid phase

w_{AL} -- Mass fraction of asphaltene dissolved in liquid phase

E_A --Porosity of deposited asphaltene

k --Instantaneous permeability, μm^2

μ_L --Viscosity of liquid phase, $mPa \cdot s$

P --Pressure, kPa

α --Ground deposition rate coefficient

β --Carrying rate coefficient

γ --Clogging rate coefficient

v_L --Liquid superficial speed, $v_L = u_L / \phi$, $m / \sec$

$v_{cr,L}$ --Liquid critical carrying rate (constant), $m / \sec$

S_V, S_L --Gas, liquid saturation

ϕ --Instantaneous porosity

ϕ_0 --Initial porosity

f_p --Correction coefficient of permeability

k_0 --Initial permeability, μm^2

k_{RV} , k_{RL} --Gas, liquid relative permeability

When $v_L > v_{cr,L}$, $\beta = \beta_i$; $v_L < v_{cr,L}$, $\beta = 0$

When $D_{pt} \leq D_{ptcr}$, $\gamma = \gamma_i (1 + \sigma E_A)$; $D_{pt} > D_{ptcr}$, $\gamma = 0$

γ_i --Deposition clogging rate coefficient

σ --Snowball effect coefficient

D_{pt} --Average diameter of pore throat, m

D_{ptcr} -- Critical diameter of pore throat of asphaltene deposit

The Study of Modular Software System Maintenance Cost Based on Markov Chain

Xiao-mei Zhu, Zhi-gang Guo, and Cheng Yuan

School of Computer Science, Southwest Petroleum University
School of Economics and Management, Southwest Petroleum University
School of Sciences, Southwest Petroleum University, Chengdu 610500, China
Zhuxiaomei_75@126.com

Abstract. How to evaluate software system maintenance cost is a very important problem for software producers, and this paper proposes a valuable means to predict module software system maintenance cost based on markov chain. And simulation shows that the way is of high scientific, easy to use and has good predicting result.

Keywords: Software, module, maintenance cost, markov chain.

1   Introduction

With the development of computer technology and diversify demand for computer software function, computer software scale is becoming more and more large, and then software maintenance costs are tended to more and more big. How to estimate software maintenance cost and how much the scale is are the most concerned problems for software developers. To improve the software quality, and to reduce maintenance cost simultaneously, modular development technology is an effective way and has been commonly used. Focusing on how to estimate the modular software system's maintenance cost, this paper proposes a forecast way for double markov chains. The forecast means is available for software producer's decision-making to analysis the cost benefit in the process of software development and maintenance.

2   Markov Theory

Markov forecast method is a kind of modern prediction way, it has the character of high scientific, accurate and adaptability, and holds very important position in modern prediction methods.

2.1   Markov Process Basic Principle

According to the development of system, time usually can be discrete into n = 0,1,2,3, ⋯, i, ⋯. Then every state of the system can be expressed as available

B.-Y. Cao and X.-J. Xie (Eds.): Fuzzy Engineering and Operations Research, AISC 147, pp. 325–330.
springerlink.com   © Springer-Verlag Berlin Heidelberg 2012

random variables, which with a certain probability correspondingly named as state probability. When the system transfer the state by one phase to another, in this state phase transfer process exists a transfer probability called transition probability.

If the transition probability is only relate to the change of adjacent states, which is to say that the system next state is only decided by the now state and has nothing to do with all the past states, then the process of transfer randomly of discrete states is called markov process.

Mathematical model of Markov process can be expressed as follows:

Suppose the system state space is S. Then in every system phrase there may be one possible state such as: S1, S2, …, Sn.

Suppose the system state is changed with the chance of time. Then according to the character of the markov chain, the system state in different phases can be expressed by state vector. Set the initial stage of the system state vector as $\pi(0)$, then the ith stage of the system can be expressed as$\pi(i)$. the probability to transfer the stage from now a time stage S_i to S_j is $p_{ij}(1 \leq i \leq n, \ 1 \leq j \leq n)$, a matrix is made up of p_{ij} is called system state transition matrix, noted by P.

$$P = \begin{bmatrix} p_{11} & p_{12} & \cdots & p_{1n} \\ p_{21} & p_{22} & \cdots & p_{2n} \\ \cdots & \cdots & \cdots & \cdots \\ p_{n1} & p_{n2} & \cdots & p_{nn} \end{bmatrix},$$

where, the ith row of P is the probability for system to transfer stages from current state Si to next state such as S1, S2, …, Sn,

$$\sum_{j=1}^{n} P_{ij} = 1, i = 1, 2, \cdots, n \cdot$$

And then, according to the character of markov each system stage vector in different stage is expressed as follows respectively:$\pi(1)=\pi(0)P$, $\pi(2)=\pi(1)P$, …, $\pi(i)=\pi(i-1)P$, i=1, 2, …, n, so stage vector satisfies the equation$\pi(k)=\pi(0)P^k$. When the system operation endlessly, the final equilibrium system state distribution vector π can be obtained according to the character of steady for k step transfer probability.

At this time the steady system state vector π satisfied the condition: $\pi P=\pi$, π is the invariant transfer vector for P.

Record $\pi=(x_1, x_2, …, x_n)$, then satisfied the condition

$$\sum_{i=1}^{n} x_i = 1 \cdot$$

2.2   Markov Chain

The integer of limited markov processes is called markov chain. The analysis of Markov chain is mainly focus on the states of limited markov process and mutual relationship, and then to forecast the chain's future condition and make some valuable decision correspondingly. According to the composition of the markov chain, the process has the following three features: the character of discrete; the character of randomness; the character of no following effects.

2.3   Markov Chain Application Steps

The system with the three characters all can be researched and can be forecasted by markov process. The application process for the markov chain can follow the steps below:

(1) Making system states and ascertaining the corresponding state probability;
(2) Calculating state transition matrix by the state transition;
(3 Calculating state vector for each state according to transition probability matrix;
(4) Having analysis, forecasting, and decision-making in stable condition.

3   Maintenance Cost of Module Software System

Software module refers to a series of interface process for accepting and saving data in order to realize one kind of functions system. A complex software system is usually composed of different modules together packaged as whole. And software maintenance cost is made up by each different modules maintenance costs. In the running processes, owing to the affection of lots of factors such as software environment, operation process and so on, module may be failure, therefore leading the software system to collapse. If without enough self-adjust function, then the failure module of the software system needs to be repaired to ensure the software system continue to work. The maintenance software costs in this paper refer to the failure module repair cost to ensure software system to keep good state. So in the module software system, maintenance cost mainly relates to the failure frequency of the module and repair costs.

In the running process of software system, suppose the system is divided into several limited independent function modules and modules call process obey a Markova process, that is to say, under the condition of knowing state of module at any time, the running process of system with character of "no aftereffects".

And also assuming that in the running process, each independent module final states include only normal and failure two kinds of states, and the final states change process also belong to Markova process, so the running of the module can be seen as a two state markov process. Therefore the software system running results can be looked at as a double markov process.

4   Modular Software System Maintenance Cost Forecasting Model

Modular software system maintenance costs refer to maintenance cost arose by module failure in unite time in the process of software system running. Because when software running to the stable condition, each function module calls frequency and time tend to stable, and the failure rate of each module tends to be stable too. So the basic thought to predict maintenance cost are: calculating each module unit time failure probability firstly, and then get each module average repair cost by statistic, and calculate the software system weighted maintenance cost by taking module failure probability as weight in the last.

Based on the ideas mentioned as above, software unit time maintenance cost forecast model can be depicted as below:

$$COST = c_1\lambda_1 \sum_{i=1}^{n} b_i p_{i1} + c_2\lambda_2 \sum_{i=1}^{n} b_i p_{i2} + \cdots + c_n\lambda_n \sum_{i=1}^{n} b_i p_{in},$$

where COST means software system expect maintenance cost in unit time, n means the number of modules in software system, k means the number of the states of each module, c1, c2, ..., c_n mean the average maintenance cost of each function module, bi means running probability of the ith module which can be got after the software system has been tested, p_{ij} refers to transfer probability from module i to module j in next phrase, λi refers to the steady failure probability of ith module which get after the module has been tested.

If define $\gamma_j = \sum_{i=1}^{n} b_i p_{ij}$, then γ_j means proportion of running time of module j in all the system running time.

Define

C=(c_1, c_2, ..., c_n), λ=(λ_1, λ_2, ..., λ_n), γ=(γ_1, γ_2, ..., γ_n).

Then : $COST = C'\lambda\gamma'$.

5   Data Simulation

A personnel management system which includes six function modules: input module, inquires module, report module, printing module, security module and data backup and restore module. After analysis, knowing that the process of system calls modules accords with Markova process. And each module state after the module has been run into end also accord with Markova process. Then get test data show as Table 1 after one week time software test:

Table 1. Origin test data

M	O	E	R	T
1	108	14	13	12
2	214	15	15	25
3	165	8	8	21
4	158	6	6	17
5	136	12	11	15
6	185	22	22	23
Total	966	77	75	113

where M means module identifier, O means experimental operation times, E means operation failure times, R means failure repair times, T means each module actual execution time (unit h) during system test. After test statistic, each module task time share in one week is $\pi(0)=(0.11,0.22,0.19,0.15,0.13,0.2)$, the system module state transfer probability matrix in one week is:

$$P = \begin{bmatrix} 0.2 & 0.15 & 0.15 & 0.35 & 0.1 & 0.05 \\ 0.15 & 0.3 & 0.05 & 0.1 & 0.2 & 0.2 \\ 0.05 & 0.15 & 0.3 & 0.1 & 0.1 & 0.3 \\ 0.05 & 0.25 & 0.2 & 0.05 & 0.3 & 0.15 \\ 0.1 & 0.15 & 0.25 & 0.1 & 0.1 & 0.3 \end{bmatrix}.$$

Then according to $\pi(k)=\pi(0)P^k$, by the means of MATLAB, when k=7, the module in this system task time tend to steady, namely satisfies the equation $\pi(7)P =\pi(7)$, and now $\gamma=\pi(7)=(0.1086,0.2233,0.1718,0.1705,0.1423,0.1871)$.

At the same time, according to the software system test results calculated each module state transition matrix Y shown as Table 2 below:

Table 2. Each module state transfer matrix

Module	Matrix factor			
	y_{11}	y_{12}	y_{21}	y_{22}
1	0.89	0.11	0.93	**0.07**
2	0.94	0.06	1	**0**
3	0.95	0.05	1	**0**
4	0.96	0.04	1	**0**
5	0.91	0.09	0.92	**0.08**
6	**0.93**	**0.07**	1	**0**

where the first state is normal state and the second state is failure state.

Before software system running, suppose the state of every module is normal state, then the initial state vector of t each module is $\pi(0)=(1,0)$. Then predict failure probability of each module by means of markov chain, the result shows in Table 3:

Table 3. Module failure probability forecast

Time (week)	Module failure probability forecast result					
	λ_1	λ_2	λ_3	λ_4	λ_5	λ_6
1	0.1100	0.0600	0.0500	0.0400	0.0900	0.0700
2	0.1076	0.0574	0.0485	0.0394	0.0865	0.0671
3	0.1063	0.0568	0.0476	0.0385	0.0862	0.0661
4	0.1058	0.0566	0.0476	0.0385	0.0857	0.0654
5	0.1058	0.0566	0.0476	0.0385	0.0857	0.0654
6	0.1058	0.0566	0.0476	0.0385	0.0857	0.0654

As shown in table 3, has been run for four weeks long, the module failure probability tends to steady. Then get

$\lambda= (0.1058,0.0566,0.0476,0.0385,0.0857,0.0654)$.

If the module average maintenance cost vector in this software system is $C=(1450,2500,1300,2500,1800,2650)(¥)$.

From the deduce process above, we can see when the software system tend to steady after seven weeks running. Then by using the means proposed in this paper we can get the software system expect maintenance cost after seven weeks running. And the forecast result is:

$$COST = C'\lambda\gamma' = 101.24 \ (¥).$$

So the software system annually maintenance predict cost is:

$$101.24\times48=4859.48(¥).$$

6   Conclusion

By using markov chain, this paper established a modular software system maintenance cost forecasting model, and this model can bring some application value for software developers to have cost benefit analysis.

References

1. Black, F., Scholes, M.: The valuation ofoptions and corporate liabities. Journal of Political Economy (81), 637–654 (1973)
2. Merton, R.C.: Option Pricing when Underlying Stock Return A Discontinuous. Journal of Financial Economics 3, 125–144 (1976)
3. Huang, Y.A., Lee, W.E.: A Cooperative Intrusion Detection Systemfor Ad Hoc Networks. In: Proceedings of the 1st ACM Workshop on Security of Ad Hoc and Sensor NetWorks (SASN), Fairfax, Virginia, October 31 (2003)
4. Mohamed, Y. A., Abdullah, A. B: Immune inspired Approach for Securing Wireless Ad Hoc Networks. International Journal of Computer Science and Network Security 9(7) (2009)
5. Marianne, A.A., Sherif, M.E., Magdy, S.E.: A survey on anomaly detection methods for ad hoc networks. Ubiquitous Computing and Communication Journal 2(3), 67–76 (2005)

A Research of Fuzzy Comprehensive Evaluation on Carbonate Reservoir

Tie-jun Li[1], Da-li Guo[1,2], Yu-jun Gong[1], Zhi-hao Tang[1], and Ke-quan Chen[1]

[1] School of Sciences, Southwest Petroleum University, Chengdu 610500,
  P.R. China
  ltj@swpu.edu.cn
[2] State Key Laboratory of Oil and Gas Reservoir Geology and Exploitation,
  Southwest Petroleum University, Chengdu 610500, P.R. China

Abstract. The carbonate reservoir has a sound heterogeneity and low-permeability, which brings certain difficulties to the prediction of the re-moulding of the reservoir. The effective comprehensive evaluation method of reservoir plays a significant role in the modification of the layers of well. Aiming at the characteristics of the reservoir geology, the paper builds up reservoir evaluation criteria based on the principles of geological evaluation and construct the membership functions and models of analytic hierarchy process, secondly, by the means of fuzzy analytic hierarchy process, the paper obtains the proportion of well logging, well testing, geology and geophysical prospecting to the importance of reservoir modification. Last, by the means of fuzzy comprehensive evaluation method, it gains the conclusion about whether the reservoir needs to be modification or not. The rebuilding results of reservoir shows that the above method can effectively resolve the consistency problem among various appraisal parameters, improve the accuracy of evaluation and the efficiency of the modification and reduce invalid production. This method has been successfully applied to the evaluation of low-permeability reservoirs in Tarim oilfield.

Keywords: Carbonate reservoir, the characteristics of the geology, analytic hierarchy process, fuzzy comprehensive evaluation, the remoulding of reservoir.

1 Introduction

The carbonate reservoir has a sound heterogeneity and low-permeability, which brings certain difficulties to the prediction of the remoulding of reservoir. However, a majority of the layers of well have no natural capacity [1,10], which needs to make some measures to reach industrial aerial currents. In

B.-Y. Cao and X.-J. Xie (Eds.): Fuzzy Engineering and Operations Research, AISC 147, pp. 331–337.
springerlink.com

order to improve remoulding measure effectively and reduce invalid production by the greatest extent, a crucial issue is to find the effective comprehensive evaluation method for geology. After years of trial and effort, based
on a great lot of data such as well-logging, well-measure, test, geology and
geophysical prospecting, The scholars utilized the principles of the comprehensive evaluation of reservoir to choose various appropriate reservoir evaluation parameters, and then established reservoir comprehensive evaluation
method [2,4,5,8,9]. Recently, the authors commonly utilized empirical method
or expert scoring method [6,11] to evaluate the reservoir, however, the consequences agree little with the actual results. For the purpose of overcoming the
human element of evaluation, this paper takes advantage of fuzzy theory and
analysis method [3,7] with the help of the first-hand data, deals with various
parameters comprehensively and builds up a system of evaluation method
of carbonate reservoir. This method provides the evidence for remoulding of
carbonate reservoir, which is also of effectively guiding significance for solving problems on the spot production. It has shown that the proposed method
is effective and practicable, and possesses good application prospect in the
evaluation of carbonate reservoir.

2 Comprehensive Evaluation of Reservoir

Fuzzy mathematics is a theory method of studying and dealing with fuzzy
system analysis. Utilize fuzzy mathematics to assess carbonate reservoir comprehensively, based on the principle of the greatest membership. It chooses
the i−th evaluation grades which is corresponding to the greatest membership as the result of the synthetical evaluation of reservoir. The key influence
that applied fuzzy mathematics exerts on the comprehensive evaluation is
how to construct the membership function and to establish proportion.

2.1 Establishing Evaluation Parameters and Evaluation Criteria

Basing on the choosing of the evaluating parameters, the synthetical evaluation of reservoir imposes a comprehensive assessment on the single impactors
in the reservoir, and then obtains a comprehensive evaluation index, according to which we can classify the reservoir.

Many factors influence the carbonate reservoir, so the accumulating spaces
change sophisticatedly. When choosing parameters, we need to analyze all factors affecting oil-gas well capacity globally. Combine the information gained
from the exploration, we can take the parameter mostly linked with carbonate reservoir capacity as an index for synthetical evaluation of reservoir.
Based on the analysis and research on the carbonate oil reservoirs in Tarim
field, we have chosen well-logging, well-measure, test, geology and geophysical prospecting which are able to reflect the macro and micro feature of

reservoir as the index for comprehensive evaluation of reservoir, composing an element set to have a correct recognition on the feature of carbonate oil reservoirs. Taking an example of the well-logging, the parameter which is related to production includes all hydrocarbon growth, natural gamma, bulk density, neutron porosity, acoustic time, deep lateral, shallow lateral, fracture porosity and best logging category. Excluding the natural gamma and bulk density running against the theory, influenced by the peak of the parameter, the three sensitive parameters are porosity, all hydrocarbon growth and acoustic time, because they are close to the measure. We carry out a synthetical evaluation from the three main factors of well-logging, and then classify the reservoir. Accordingly, we assess other parameters such as well-measure, test, geology and geophysical prospecting. As a result, we establish the criterion for comprehensive evaluation on reservoir. See Table 1:

Table 1. Criterion for comprehensive evaluation on carbonate reservoir

class	well-logging	well-measure	test	geology	geological prospecting
I	0.7	0.6	0.7	0.7	0.7
II	0.5	0.5	0.6	0.6	0.5
III	0.3	0.4	0.3	0.4	0.2
IV	0.1	0.2	0.1	0.2	0.1

2.2 Constructing Membership Functions

Let H be a universal set, the fuzzy subset F on H be a mapping from H to the unit interval $[0, 1]$, and the membership degree $F(x)$ represents the degree x belonging to F. Then the function F is called a membership function of H.

If we want to establish a criterion for evaluation of the choosing of parameter assessment of reservoir, we need to assess parameters individually, among which, the most crucial one is how to construct the membership function. How to establish the membership function's expression is an important mathematical issue. Currently, we mostly rely on experiment and expert assessment in application. In order to rise above the insufficiency of experience, in this research we add up assessment parameters in reservoir of wells. Using the theory of fuzzy mathematics, we analyze and pre-handle implement lower semi-trapezoidal method to construct a membership function basing on five parameters towards four assessment degree.

$$\eta_I = \begin{cases} 1, & x \geq x_I, \\ \dfrac{x - x_{II}}{x_I - x_{II}}, & x_{II} < x < x_I, \\ 0, & x \leq x_{II}, \end{cases} \tag{1}$$

$$\eta_{II} = \begin{cases} 0, & x \geq x_I \quad or \quad x \leq x_{III}, \\ \dfrac{x_I - x}{x_I - x_{II}}, & x_{II} \leq x < x_I, \\ \dfrac{x - x_{III}}{x_{II} - x_{III}}, & x_{III} < x < x_{II}, \end{cases} \tag{2}$$

$$\eta_{III} = \begin{cases} 0, & x \geq x_{II} \quad or \quad x \leq x_{IV}, \\ \dfrac{x_{II} - x}{x_{II} - x_{III}}, & x_{III} \leq x < x_{II}, \\ \dfrac{x - x_{IV}}{x_{III} - x_{IV}}, & x_{III} < x < x_{IV}, \end{cases} \tag{3}$$

$$\eta_{IV} = \begin{cases} 0, & x \geq x_{III}, \\ \dfrac{x_{III} - x}{x_{III} - x_{IV}}, & x_{IV} < x < x_{III}, \\ 1, & x \leq x_{IV}, \end{cases} \tag{4}$$

where $\eta_I(x)$, $\eta_{II}(x)$, $\eta_{III}(x)$ and $\eta_{IV}(x)$ be I, II, III, IV's membership function of reservoir evaluation relatively. x is a actual value of evaluation parameter. x_I, x_{II}, x_{III}, x_{IV} are I, II, III, IV's standard value. From which, we drew the membership degree of a membership function. The established membership function expression conforms to the reality after testing.

2.3 Proportion Establishment

The importance of a certain assessment factor accounting for general feature is proportion. Calculating different index proportion in comprehensive assessment is searching for the quantitative relation in the inside inter-relative of an substance. Because there are so many ways, we used to take expert assessment on proportion. In order to avoid subjective factors, this paper uses fuzzy analytic hierarchy process, we can count $A = (d(x_1), d(x_2), d(x_3), d(x_4), d(x_5))$.

According to the five assessment in carbonate reservoir, coupled with the reality, we establish a basic structural model for assessing layers factors. See Figure 1:

2.4 Fuzzy Composite Operator

The key point of whether we can modify fuzzy mathematics comprehensive evaluation lies in the selection of fuzzy composite operators. As different fuzzy composite operator has different perturbation which will attribute to discrepancy, this paper selects max-min composite operator $M(\wedge, \vee)$ whose perturbation is minimal. Let $M(\wedge, \vee) = A \circ R$, where, A denote an $m-$dimensional

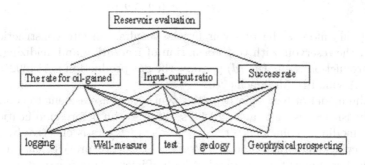

Fig. 1. Analytic hierarchy process

vector and R denote a $m \times n$ matrix. Let $A \circ R = B$. Then the specific form is in the following:

$$M(\wedge, \vee) : b_j = (a_1 \wedge r_{1j}) \vee (a_2 \wedge r_{2j}) \vee \cdots \vee (a_m \wedge r_{mj}), (j = 1, 2, \cdots, n). \quad (5)$$

3   An Example

Use the model of fuzzy mathematics synthetical assessment to conduct a comprehensive assessment among 18 test oil in Tarim before modification. Take interval $5780(m) - 5796(m)$ for example to further expound the calculation. We use well-logging, well-measure, test, geology and geophysical prospecting as a factor vector $A = (a_1, a_2, a_3, a_4, a_5)$ and meanwhile use must-modification suggested-modification, non-modification and need not-modification as a decision vector $B = (b_1, b_2, b_3, b_4)$.

Let the data measured from the spot production be $A = (0.61, 0.43, 0.52, 0.42, 0.66)$. Combining with the data from Table 1, put membership function into calculation, and a relation matrix for fuzzy synthetical evaluation is gained as follows:

$$R = \begin{pmatrix} 0.55 & 0.45 & 0.00 & 0.00 \\ 0.00 & 0.30 & 0.70 & 0.00 \\ 0.00 & 0.73 & 0.27 & 0.00 \\ 0.00 & 0.10 & 0.90 & 0.00 \\ 0.80 & 0.20 & 0.00 & 0.00 \end{pmatrix}.$$

Proportion is obtained from analytic hierarchy process, which is shown in the form of vector A':

$$A' = (0.15, 0.2, 0.1, 0.15, 0.4).$$

Synthetical evaluation of interval $5780.00(m) - 5796.00(m)$ is gained by max-min composite operator.

$$B = A' \circ R = (0.4, 0.2, 0.2, 0).$$

We should modify the reservoir by the results. On site construction we modified the reservoir with transformation of fracturing and acidizing, then the oil production is $46.1(m^3/d)$ per day, the gas production is $154692(m^3/d)$ per day. A sound result is achieved.

Use the model of fuzzy mathematics synthetical assessment to conduct a comprehensive assessment among 18 test oil in Tarim before modification. 15 layers is modified, coincidence rate is 83.3%, 12 layers is effective by modification, coincidence rate is 66.7%. Practical evidence proves that the fuzzy synthetical assessment is a method of high efficiency and pragmatic.

4 Conclusions and Advancement

1) It is the first time that we use parameters such as well-logging, well-measure, test, geology and geological prospecting to evaluate carbonate reservoir, reflecting the feature of reservoir of this kind, reasonably, effectively and objectively.
2) Making use of the method of fuzzy comprehensive evaluation towards carbonate reservoir conforms to the reality, which shows that the method is highly creditable, accurate and workable.
3) Fuzzy comprehensive evaluation is conducted unitless when using original data and the comprehensive usage of multi-parameter can improve accuracy highly.
4) It is necessary to enhance the accuracy of the parameter because of the high demands of fuzzy comprehensive evaluation towards parameters.

Acknowledgement. Thanks to the support by major projects of educational commission of Sichuan province (No.10ZA150) and the natural science foundation of southwest petroleum university (No.2010XJZ196).

References

1. Chen, X., Liu, H.L., Wang, H.Y., Zhao, Q.: Gas-seepage characteristics of coalbed reservoirs with different water saturation in the Qinshui basin. Acta Petrolei Sinica 3, 500–503 (2011)
2. Chen, G.B., Fan, J.G., Tao, Y.G.: The application of multiple information for predictive of carbonate karst reservoir and effect analysis. Geophysical Prospecting For Petrole 44, 34–36 (2005)
3. Li, Z.F., He, S.L.: Application of fuzzy mathematics to evaluating the carbonate reservoirs in Changqing asg field. Natural Gas Industry 3, 55–57 (2005)
4. Luo, T.T., Zhou, L.F., Liu, Z.W.: Application of NMR technology in ordovician reservoir evaluation in the Weihe Basin. Special Oil and Gas Reservoirs 18, 37–41 (2011)

5. Tang, H.F., Peng, S.B., Zhao, Y.C.: The classification and stochastic modeling of flow units in tight gas reservoir. Journal of Jilin University (Earth Science Edition) 3, 469–474 (2007)
6. Wang, S.: Analysis of rock pore structural characteristic by nuclear magnetic resonance. Xinjiang Petroleum Geology 6, 768–770 (2009)
7. Wei, X.D., Zhang, Y.Q., Cao, L.L., Wang, Y.N.: Applications of amplitude spectrum gradient for reservoir evaluation in WC project. Oil Geophysical Prospecting 46, 282–286 (2011)
8. Yang, Z.M., Zhang, Y.Z., Hao, M.Q., Liu, X.G., Shan, W.W.: Comprehensive evaluation of reservoir in low-permeability oilfields. Acta Petrolei Sinica 2, 64–67 (2006)
9. Zhang, P.B., Jiang, L.P., Ge, Y.X.: Evaluation of reservoir in Baimiao gas field. Fault-Block Oil and Gas Field 5, 40–43 (2008)
10. Zhou, Y., Tan, X.C., Liu, H., Yang, J.L., Yao, Y.B., Li, J.L., Zhong, H., Lin, J.P.: Evaluation of porous carbonate reservoir of Jia 2 Member in Moxi Structure of Sichuan basin. Acta Petrolei Sinica 3, 372–378 (2009)
11. Zou, L.Z., Liu, Q.H.: The logging interpretation model for reservoir parameters of nomber 1 interval of Liushagang group of fushan depression. World Well Logging Technology 1, 21–24 (2011)

Theoretical Analysis on Initial Pheromone Values for ACO

Quan-feng Qiu and Xiang-jun Xie

School of Science, Southwest Petroleum University, Chengdu, 610500, China
i7366@126.com

Abstract. Generally considering the update rules in ant system (AS) and ant colony system (ACS), the basic theory of setting initial pheromone values for ant colony optimization (ACO) algorithm and the conditions that initial pheromone values on edges have to satisfy are presented. This paper also proposes the evaluating method of the initial pheromone values, which is the function of $\Delta\tau$, ρ, M and T_2. At last, the theory is used to analyze the commonly used initial pheromone settings. The analysis of those cases indicates that it is highly recommended to make use of c^{nn} when setting the initial pheromone values.

Keywords: Initial pheromone value, ACO, c^{nn}.

1 Introduction

Ant Colony tries to communicate by a certain stigmergy named pheromone. Thus the ants behave in an organized society, and they work collaboratively and efficiently to do a certain job [7,8]. Based on the biological behaviors, Dorigo [1,5,6] and some other researchers presented a novel biomimetic optimization algorithm, which is ACO algorithm. The basic ant colony algorithm makes use of the pheromone variation on a graph to guide all the ants to a certain trail, so as to find the shortest rout from their nest to the food source. As a result, the algorithm finds the optimal solution. The applications in many different fields have been studied a lot, and many theoretical researches have been carried out too. The applications have been a great successful, yet the theories develop very slowly. And little of the discussion of setting initial pheromone values has come out. At present, there are mainly two setting methods: one simply sets the value to a constant (such as 1), the other makes use of the nearest neighbor method, which is the currently mostly used method. In this paper we try to understand what the influence of the initial pheromone value is in the process and how it works.

To simplify the discussing, it doesn't really take into account an ant's probability of visiting an edge. During the constructing process of an ACO algorithm, it is supposed that sufficient iterations ensures that almost all ants visit those edges that are of the optimal solution, and on the other hand, those edges that are not of the optimal solution are almost always not visited.

B.-Y. Cao and X.-J. Xie (Eds.): Fuzzy Engineering and Operations Research, AISC 147, pp. 339–349.
springerlink.com © Springer-Verlag Berlin Heidelberg 2012

2   Ant Colony Algorithm

Each ant colony algorithm at least includes three sub-processes: preparation for the constructing, processing construction, updating pheromone values. The processes can be depicted as:

1) Initialization: putting initial pheromone τ_0 on all edges of a graph G, putting m ants randomly on the graph;
2) All the ants move to the next nodes according to a certain state transition rule, until all edges are visited. When each of them finishes visiting all edges on the graph once, an iteration of this algorithm is fulfilled, which means all ants finish a cycle on the graph. Different ACO algorithms work by different state transition rules. Presented in literature [3], ACS worked by pseudo-random-proportional rule, whereas in literature [2] AS worked by random-proportional rule. No matter whatever the transition rules are in all ACO algorithms, they must be or approximately are

$$
p_{ij}^{k} = \begin{cases} \dfrac{[\tau(i,j)]^{\alpha}[\eta(i,j)]^{\beta}}{\displaystyle\sum_{(i,u)\notin tabu_k} [\tau(i,u)]^{\alpha}[\eta(i,u)]^{\beta}} & (i,u)\notin tabu_k, \\ 0, & otherwise. \end{cases} \tag{1}
$$

3) If termination condition is satisfied, the optimal solution s^{*} is returned. Otherwise, the pheromone values on all directed edges, so as to be ready for the next iteration.

Pheromone values can be updated locally or globally. How to update the pheromone values depends on how it is put in practice. Here in this paper only the following globally updating rules are taken into account:

$$
\tau_{ij}(t) = (1-\rho)\tau_{ij}(t-1) + \sum_{h=1}^{m}\Delta\tau_{ij}^{h}(t), \tag{2}
$$

or

$$
\tau_{ij}(t) = (1-\rho)\tau_{ij}(t-1) + \rho\sum_{h=1}^{m}\Delta\tau_{ij}^{h}(t), \tag{3}
$$

where ρ is evaporation probability, $\Delta\tau_{ij}^{h}(t)$ is the laid pheromone on the edge (i,j) at the t-th iteration of ant h. If ant h doesn't visit the edge, $\Delta\tau_{ij}^{h}(t)=0$.

The above processes imply that initial pheromone τ_0 on graph G effects on the basal of a construction of a solution.

3   Theoretical Analysis of Initial Pheromone Values

The graphs discussed here are always completely connected. If a graph is not completely connected, it will be translated into a completely connected one. Ant colony algorithm will discover a Hamilton cycle at the minimum cost. The minimum cost means minimum weights or shortest path. So, the algorithms discussed here are depicted as a minimization problem.

Let $(\bar{S}, f, \Omega)$ be the minimization problem, where $\bar{S}$ is state, f is the objective, and Ω is the constraint. $\tilde{S} \subseteq \bar{S}$ is the candidates set. The feasible region S is the states in $\tilde{S}$ that satisfy Ω. Let $G = (V, E)$ be a completely connected undirected graph, where V is the vertexes set, E is the edges set, $v_i \in V$, $(i, j) = (v_i, v_j) \in E$, $i \neq j$, $1 \leq i, j \leq n = |V|$. Then all possible states of V consists of $\bar{S}$. Let $s_i = (v_{i1}, v_{i2}, \cdots, v_{il})$, $1 \leq l \leq n$. Then $s_i \in \tilde{S} \subseteq \bar{S}$. If $s_i \in \tilde{S} \cap \Omega = S$, s_i is a feasible state. The optimal solution s^* for the minimization problem will satisfy $f(s^*) \leq f(s), \forall s \in S$.

For the above graph G, the weight of e_{ij} is denoted as w_{ij}, and if $i = j$, $w_{ij} = 0$. By the means, the minimization problem can be translated into finding the solution that has the summary minimum weight. The maximum and minimum weights on edges are w_{max} and w_{min} respectively. The maximum and minimum number of edges on feasible solutions are l_{max} and l_{min} respectively. s_{max} is a series consists of l_{max} edges which have weight w_{max}, and s_{min} is a series consists of l_{min} edges which have weight w_{min}. According to Doctor Liu [4], $\forall s \in S$, pheromone increment $\Delta \tau$ satisfies $g(s_{max}) \leq \Delta \tau \leq g(s_{min})$ (theory 2.1 in [4]), where the quality function $g(x)$ is non-increasing.

If pheromone is updated globally by (3), to ensure that the pheromone evaporation on edges of optimal solution is less than the pheromone increment, the following has to be satisfied.

$$\rho \bullet \tau_{min} < \rho \bullet \Delta \tau,$$

that is to say

$$\tau_{min} < \Delta \tau = g(\hat{s}) \leq g(s^*) \leq g(s_{min}),$$

where $\hat{s}$ is the present optimal solution, s^* is the global optimal solution. But ant colony algorithm can be misled to a local optimal solution s_{local}. So the latter text will use s_{best} to represent s^* and s_{local}.

If $\tau_{min} = g(s_{max}) - \gamma$, γ is a very small number. And because $g(s_{max})$ is a lower bound of pheromone increment, which is a conservative evaluation, τ_{min} can also be $g(s_{max})$. Commonly $g(x) = \dfrac{Q}{g'(x)}$, where $g'(x)$ is the summation of series weights.

If global pheromone updated according to (2),

$$\tau_{ij}(t) = (1-\rho)^t \tau_{ij}^0 + \sum_{r=1}^{t}(1-\rho)^{t-r}\Delta\tau_{ij}(r). \tag{4}$$

Let $t \to \infty$. Then

$$\tau_{ij} = \lim_{t\to\infty}\tau_{ij}(t) = \lim_{t\to\infty}\left[(1-\rho)^t \tau_{ij}^0 + \sum_{r=1}^{t}(1-\rho)^{t-r}\Delta\tau_{ij}(r)\right]$$
$$= \lim_{t\to\infty}\left[\sum_{r=1}^{t}(1-\rho)^{t-r}\Delta\tau_{ij}(r)\right]. \tag{5}$$

In (5), $\Delta\tau_{ij}(r) = \sum_{h=1}^{m}\Delta\tau_{ij}^h(r)$.

Because $\inf_r\{\Delta\tau_{ij}^h(r)\} = 0$, $\sup\{\Delta\tau_{ij}^h(r)\} = g(s^*)$. Let $\Delta\tau = mg(s_{best})$. Then $\Delta\tau_{ij}(r) \le \Delta\tau$, $\Delta\tau \in [0, mg(s*)]$, by (5),

$$\tau_{ij} \le \lim_{t\to\infty}\left[\sum_{r=1}^{t}(1-\rho)^{t-r}\Delta\tau\right]$$
$$= \Delta\tau\lim_{t\to\infty}\left[\sum_{r=1}^{t}(1-\rho)^{t-r}\right] \tag{6}$$
$$= \frac{\Delta\tau}{\rho}.$$

And, obviously $\tau_{ij} \ge 0$ in Inequality (6).

If global pheromone updated according to (3), in turn, similarly there'll be

$$\tau_{ij}(t) = (1-\rho)^t \tau_{ij}^0 + \rho\sum_{r=1}^{t}(1-\rho)^{t-r}\Delta\tau_{ij}(r), \tag{7}$$

$$\tau_{ij} = \lim_{t\to\infty}\tau_{ij}(t) = \lim_{t\to\infty}\left[\rho\sum_{r=1}^{t}(1-\rho)^{t-r}\Delta\tau_{ij}(r)\right], \tag{8}$$

$$\tau_{ij} \le \Delta\tau \tag{9}$$

From (5) to (9), it is obvious that all initial pheromone is evaporated, so long as the iteration times are enough, the left pheromone will be none. For example, when $\rho = 0.5$, $t = 50$, $(1-0.5)^{50} = 8.881784\times10^{-16}$; when $\rho = 0.1$, $t = 50$ or $t = 340$, respectively $(1-0.1)^{50} = 5.1538\times10^{-4}$ and $(1-0.1)^{340} = 2.7698\times10^{-16}$. It means that no matter how much ρ is, the initial pheromone will disappear rapidly. And of course the evaporated pheromone includes not only the initial one, but also the laid one at the last iteration, which is why (5) and (7) shows the pheromone increment coefficients decrease from the present iteration to the first iteration. (6) indicates that the pheromone values on some edges can reach the limit $\frac{\Delta\tau}{\rho}$, where $\Delta\tau \in [0, mg(s^*)]$. Obviously, these edges can only be on the optimal solution s_{best}. (6) suggests that there exists an iteration, after this iteration, the pheromone values on edges of s_{best} will be equal or nearly equal to $\frac{\Delta\tau}{\rho}$. But comparatively, pheromone values not on edges of s_{best} will be 0 or almost 0.

But the above analysis doesn't say that the initial pheromone values are not important and can be arbitrary. The following will prove that initial pheromone values are important to guide ants to a good solution, especially at the early stage of the algorithm. According to (1), if initial pheromone values are 0, the algorithm behavior is turned into a random search, and doesn't make use of the guidelines, which is contrary to original idea of the designed ant colony algorithm. Ant colony algorithm requires that $\tau_0 > 0$, and for all the edges on s_{best}, $\exists T_1 > 0$, if $t > T_1$, pheromone values on these edges are almost monotone non-decreasing, i.e.,

$$\tau_{ij}(t) \leq \tau_{ij}(t+1). \tag{10}$$

Calculated by (2) and (4), (11) is obtained.

$$(1-\rho)^t \tau_{ij}^0 + \sum_{r=1}^{t}(1-\rho)^{t-r}\Delta\tau_{ij}(r) \leq (1-\rho)^{t+1}\tau_{ij}^0 + \sum_{r=1}^{t+1}(1-\rho)^{t+1-r}\Delta\tau_{ij}(r). \tag{11}$$

So,

$$\tau_{ij}^0 \leq \frac{1}{\rho}\left[\sum_{r=1}^{t+1}(1-\rho)^{1-r}\Delta\tau_{ij}(r) - \sum_{r=1}^{t}(1-\rho)^{-r}\Delta\tau_{ij}(r)\right]. \tag{12}$$

Based on the algorithm behaviors, there exists $T_2 \geq T_1 > 0$, if $t > T_2$, almost all ants discover the optimal solution s_{best}. So, when $t+1 > t \geq r > T_2$, $\Delta\tau_{ij}(t+1)$ and $\Delta\tau_{ij}(r)$ are almost $\Delta\tau$. And because $\rho\sum_{r=1}^{T_2}(1-\rho)^{-r}\Delta\tau_{ij}(r)$ is limited, from (12), Inequality (13) is concluded.

$$0 \leq \tau_{ij}^0 \leq \frac{\Delta\tau}{\rho}\cdot(1-\rho)^{-T_2} - \frac{\Delta_0\tau}{\rho}. \tag{13}$$

In (13), $\Delta_0 \tau = \rho \sum_{r=1}^{T_2} (1-\rho)^{-r} \Delta \tau_{ij}(r)$, $\forall (i,j) \in s_{best}$. Let $t = 1$. The particular calculation is got from (12):

$$0 \le \rho(1-\rho)\tau_{ij}^0 \le \Delta \tau_{ij}(2) - \Delta \tau_{ij}(1).$$

It demonstrates that initial pheromone is decided by ρ and solution quality constructed at the first and second iteration. Generally, Inequality (13) suggests that initial pheromone values are function of T_2, ρ and the quality of s_{best}. T_2 and initial pheromone values affect each other, let ρ and the quality of s_{best} be constant. (13) also indicates that $\Delta_0 \tau$ is monotone non-decreasing with T_2. It is important to remember that in (13), so long as t satisfies that $t > T_2$, (13) will be true. Now, a way to evaluate T_2 is presented. Let $M = \max_{1 \le r \le T_2} \{\Delta \tau_{ij}(r)\}$, then from (13),

$$\rho \tau_{ij}^0 - M \le \Delta \tau \cdot (1-\rho)^{-T_2} - M(1-\rho)^{-T_2}. \tag{14}$$

Suppose that $\rho \tau_{ij}^0 - M < 0$, then $0 \le \tau_{ij}^0 < \dfrac{M}{\rho}$, $(1-\rho)^{-T_2} \ge \dfrac{\Delta \tau - M}{\rho \tau_{ij}^0 - M}$. So,

$(1-\rho)^{-T_2} \ge \dfrac{\Delta \tau - M}{\rho \tau_{ij}^0 - M} > 0$, $M > \max\{\rho \tau_{ij}^0, \Delta \tau\} = \Delta \tau$. That is to say, the first T_2

iterations have constructed a solution better than the optimal solution, which contradicts the designed algorithm behavior. Ant colony algorithm shows convergence at $t > T_2$, and when the algorithm converges, almost all solutions constructed are better than those at $t \le T_2$. So, $\rho \tau_{ij}^0 - M \ge 0$. If $\rho \tau_{ij}^0 - M = 0$,

$\tau_{ij}^0 = \dfrac{M}{\rho}$, for $\forall T_2 > 0$, (10) is satisfied. If $\rho \tau_{ij}^0 - M > 0$. Let equality holds in (14),

then

$$\tau_{ij}^0 = \frac{M}{\rho} + \frac{\Delta \tau - M}{\rho} \cdot (1-\rho)^{-T_2}. \tag{15}$$

(15) indicates that if $\Delta \tau$, M and ρ are constant, initial pheromone values are monotone increasing with T_2. It also says that the point that algorithm starts constructing a high quality solution and τ_{ij}^0 change at the same direction. So, if τ_{ij}^0 is relatively bigger, T_2 will be bigger, which results in a better exploration in a longer time. On the contrary, a smaller initial pheromone value means that it is easy to be trapped in a bad solution. So, a reasonable evaluation of $\Delta \tau$ and M plus

ρ and the purpose of T_2 decides a reasonable initial pheromone value. And if initial pheromone value is decided by T_2, it is possible to keep a good exploration in a longer time. From (15), it is easy to get a simple conclusion:

$$\tau_{ij}^0 > \frac{\Delta\tau}{\rho}. \tag{16}$$

If global pheromone updated by (3) and (10) is satisfied, by the same reasoning process, the following will be obtained.

$$0 \le \tau_{ij}^0 \le \Delta\tau \cdot (1-\rho)^{-T_2} - \Delta_0\tau, \tag{17}$$

$$\tau_{ij}^0 - M \le \Delta\tau \cdot (1-\rho)^{-T_2} - M(1-\rho)^{-T_2}. \tag{18}$$

Obviously, $\tau_{ij}^0 - M \ge 0$. So, either $\tau_{ij}^0 = M$, or when equality in (18) holds,

$$\tau_{ij}^0 = M + (\Delta\tau - M) \cdot (1-\rho)^{-T_2}. \tag{19}$$

A simple analysis can show that pheromone updating methods (2) and (3) leads to the same conclusion. However (19) and (15) suggest that (3) can serve the algorithm a better exploration. Whatever the updating method is, if τ_{ij}^0 is very small, T_2 will be very small too, which results in a rapid convergence to some edges, and thus the algorithm is easily trapped in a local optimal solution. Just like (16), inequality (20) can be got from (19),

$$\tau_{ij}^0 > \Delta\tau. \tag{20}$$

As is given, pheromone values on all edges of optimal solution s_{best} must be monotone non-decreasing. But for those that are not on optimal solution, $\exists T_3$, for $t > T_3$, pheromone values must be monotone non-increasing, except only a few edges at only a few iterations. And $\lim_{t \to \infty} \tau_{ij}(t) = 0$ is true for $\forall(i,j) \notin s_{best}$. So, for $\forall(i,j) \notin s_{best}$,

$$\tau_{ij}(t+1) < \tau_{ij}(t). \tag{21}$$

According to the state transition rule and positive feedback of ant colony algorithm, whatever the update method is, if $t > T_3$, for $\forall(i,j) \notin s_{best}$, almost no ants visit these edges. So, there's almost always $\Delta\tau_{ij}^h(r)=0$, $1\leq h\leq m$, $(1-\rho)^{t-r}\Delta\tau_{ij}(r)\to 0$,

and almost at any time and any move, $\Delta\tau_{ij}(r)=0$ and $\lim_{t\to\infty}\tau_{ij}(t)=0$. All of these

suggests that, for $t>T_3$, (21) can almost always be true. Although when $t\geq r>T_3$,
according to (1), the edge (i,j) might probably be visited by one or two ants at a
very small possibility. So, initial pheromone is not important for those edges that
can't be on s_{best}, but very important to those on s_{best}, so is to ant colony algorithm.

4   Case Analysis

According to lemma (2.1) in literature [4], pheromone values on all edges satisfy

$\tau_{ij}\leq\tau_{max}=g(s^*)\leq g(s_{min})$. So, the initial pheromone values τ_0 on all edges can

be set as $g(s_{min})$. But $g(s_{min})$ enlarges $g(s^*)$ a lot, such as finding a minimum
cost Hamilton cycle, because in this instance $g(s_{min})$ is l_{min} (mostly, $l_{min}=n$)
times of minimum weight. Assuming that s^* has n' (commonly, $n'=1$) edges of

minimum weight, if $n'\ll n$, $0<\dfrac{n'}{n}\ll 1$ (much less than 1).

Let $\tau_0=g(s_{min})$, $g(x)=\dfrac{Q}{g'(x)}$, by (13), there is

$$0<g(s_{min})=\frac{Q}{g'(s_{min})}=\frac{Q}{l_{min}\bullet w_{min}}\leq\frac{\Delta\tau}{\rho}\bullet(1-\rho)^{-T_4}-\frac{\Delta_0\tau}{\rho}\,,\tag{22}$$

where $\Delta_0\tau=\rho\sum_{r=1}^{T_4}(1-\rho)^{-r}\Delta\tau_{ij}(r)$, $\Delta\tau_{ij}(r)=\sum_{h=1}^{m}\Delta\tau_{ij}^h(r)$. Then

$$T_4\geq-\frac{\ln\left(\rho g(s_{min})-M\right)-\ln\left(\Delta\tau-M\right)}{\ln(1-\rho)}.\tag{23}$$

Generally, by (13), the initial pheromone values satisfy

$$T_4\geq-\frac{\ln\left(\rho\tau_{ij}^0-M\right)-\ln\left(\Delta\tau-M\right)}{\ln(1-\rho)}\tag{24}$$

$\Delta \tau_{ij}^{h}(r)$ is the pheromone increment at the r-th iteration of ant h. From (22), (23) can be obtained. As a result t at least must be

$$\left\lfloor -\frac{\ln\left(\rho g(s_{min})-M\right)-\ln\left(\Delta \tau - M\right)}{\ln(1-\rho)} \right\rfloor \quad (\lfloor \bullet \rfloor : \text{a rounding operation to get an integer}$$

not smaller than the given number). Generally, (22) is true, and so will be $g(s_{min}) > \Delta \tau / \rho$. Because $\Delta \tau$ can at most be weights sum of the n biggest weights, and usually $0 < \dfrac{n'}{n} \ll 1$. If $g(s_{min}) \gg \Delta \tau$ is true, and because $\Delta \tau$, M and ρ are relatively stable, T_2 is mainly decided by $g(s_{min})$, whose value is relatively big. It is seen that when n' is close to $n\rho$, T_2 can be reached almost at once. So, it can be concluded that $\tau_0 = g(s_{min})$ is generally adequate. But when n' is too close to $n\rho$ initial pheromone value should be enlarged some.

Now, the initial pheromone values got from nearest neighbor method will be studied. In [2], the settings are at Table 1. In Table 1, C^{nn} is the evaluation of solution constructed by nearest neighbor method.

Table 1. Initial pheromone values of ant colony from literature [2]

ACO algorithm	AS	EAS	AS$_{rank}$	MMAS	ACS
Initial pheromone values	m/C^{nn}	$(e+m)/(\rho C^{nn})$	$0.5\sigma(\sigma-1)/(\rho C^{nn})$	$1/(\rho C^{nn})$	$1/(nC^{nn})$

Some other settings can also be found. For example, Ichiro Iimura[10], etc. let $\tau_0 = 1/(mC^{nn})$.

Let (13) be written as

$$0 \le \tau_{ij}^0 \le \Delta \tau \cdot \frac{(1-\rho)^{-T_i}}{\rho} - \frac{\Delta_0 \tau}{\rho}.$$

Considering $\tau_0 = 1/C^{nn}$, it is obviously an approximate of $\Delta \tau$ at a local region. In Table 1, the different settings try to improve $\tau_0 = 1/C^{nn}$ so as to get closer to $\Delta \tau$. All these settings can be a function of $1/C^{nn}$ and can be written as $\tau_0 = f(1/C^{nn})$.

For τ_0 , there's always T , such that if $t > T$, (13) is satisfied. Let $g_{max} \leq \Delta \tau_{ij}(r) \leq g_{min}$, according to (13)

$$-\frac{\ln(\rho \tau_0 - g_{max}) - \ln(\Delta \tau - g_{max})}{\ln(1-\rho)} \leq T \leq -\frac{\ln(\rho \tau_0 - g_{min}) - \ln(\Delta \tau - g_{min})}{\ln(1-\rho)} \quad (25)$$

According to (25), at most after $\left\lfloor -\dfrac{\ln(\rho \tau_0 - g(s_{max})) - \ln(\Delta \tau - g(s_{max}))}{\ln(1-\rho)} \right\rfloor$

iterations, pheromone will converge to the solution s_{best} . For example, $\tau_0 = f(1/C^{nn})$, and g_{max} , g_{min} are the best and worst solution obtained by nearest

neighbor method, or $g_{max} = g(s_{max})$, $g_{min} = g(s_{min})$.

During the process of constructing solutions of ants, the main effect of pheromone is guidance, which is the key of ant colony algorithm. If let s_{nn} be a series appropriate to C^{nn} , s_{nn} is obtained by nearest neighbor method and may be close to the optimal solution, $\tau_0 = f(1/C^{nn})$ will work well to guide the algorithm to the solution s_{best} . Except for nearest neighbor method, genetic algorithm and simulated annealing method and other metaheuristic methods can also help decide the initial pheromone values, such as the initial value $\tau_S = \tau_c + \tau_G$ in [9].

But, by the discussing in this paper, the setting method in [11] is not an ideal method.

5   Conclusion

This paper presents the general theory of initial pheromone settings, proposes the conditions that initial pheromone values have to satisfy and the theory of iterations and initial pheromone values. And also, the evaluation method of initial pheromone values is presented. The theoretical analysis proves that $\tau_0 = f(1/C^{nn})$ is a good means to set initial pheromone values. And the paper presents that $\tau_0 = f(1/C^{nn})$ can be decided by $\Delta \tau$, M , ρ and the expected iteration times, shown in (15) and (19).

References

1. Dorigo, M.: Optimization, learning and natural algorithms, Ph.D. dissertation, DEI, Politecnico di Milano, Italy (1992)
2. Dorigo, M., Stützle, T.: Ant Colony Optimization. The MIT Press, Cambridge (2004)

3. Dorigo, M., GambardellaL, M.L.: Ant colony system: A sooperative learning approach to the traveling salesman problem. IEEE Transactions on Evolutionary Computation 1(1) (1997)
4. Liu, Y.P.: Research on Ant Colony Optimization and Its Application. Zhejiang University, Hanzhou (2007)
5. Colorni, A., Dorigo, M., Maniezzo, V.: Distributed optimization by ant colonies. In: Proceedings of European Conference on Artificial Life, Paris, France, pp. 134–142 (1991)
6. Dorigo, M., Maniezzo, V., Colorni, A.: Positive feedback as a search strategy. Italy: Technical Report 91-016, Dipartimento di Elettronica, Politecnico di Milano (1991)
7. Goss, S., Aron, S., Deneubourg, J.L., Pasteels, J.M.: Self-organized shortcuts in the argentine ant. Naturwissenschaften 76, 579–581 (1989)
8. Beckers, R., Deneubourg, J.L., Goss, S.: Trails and U-turns in the selection of the shortest path by the ant Lasius Niger. J. Theoretical Biology 159, 397–415 (1992)
9. Ding, J.L., Chen, Z.Q.: On the combination of genetic algorithm and ant algorithm. Journal of Computer Research and Development 40(9) (2003)
10. Chin, H.C., Chung, R.J., Su, Y.S., Sun, H.H.: Application of the ant colony system for open wye-open delta adjustment. IEEE (2004)
11. Ichiro, I., Toshiya, I., Shigeru, N.: A study of stimulative queen ant strategy in ant colony optimization method. In: Seventh International Conference on Parallel and Distributed Computing. Applications and Technologies (PDCAT 2006), pp. 180–184 (2006)

Separability of Products Based on Fuzzy Finite State Machines

Qian Chen, Dao-xing Tu, and Chun-lan Zhao

College of Sciences, Southwest Petroleum University,
Sichuan 610500, P.R. China
chenqian0204@163.com

Abstract. In this paper,we investigate separability of the products with infinitely many fuzzy finite state machines, then their weak covering of the products with infinitely many components are discussed.

Keywords: Fuzzy finite state machine, separability, weak covering.

1 Introduction

Inspired by the work of Zadeh [1,2], Wee introduced the notion of a fuzzy automata [3]. In [4,5], Malik, Mordeson and Sen began the algebraic study of fuzzy automata. Other results can be found in [6]. The concepts of transformation semigroup,covering, cascade product, and wreath product and other algebraic theories play a prominent role in the study of automata [7]. Thereafter, there were considerable authors, such as H.V.Kumbhojkar and S.R.Chaudhari [8], Youn-Hee Kim, Jae-Gyeom Kim and Sung-Jin Cho [9], Tatjana Petković [10] et al. [11-13] having contributed in this field. By defining several especial functions of states transition of ffsms,General direct infinite products, full direct infinite products, restricted direct infinite products , cascade infinite products, wreath infinite products of fuzzy finite state machines are introduced [13], the results in [4,8] are generalized.In order to overcome some of the difficulties which arise from the fuzzification of these concepts, in this paper, we study their separability and weak covering, and so on. These completes previous works on the same subject. It is important to study algebraic theory of fuzzy automata systematically. A fuzzy finite state machine (ffsm) M is a triple $M = (Q, X, \mu)$ where Q and X are nonempty sets and μ is a fuzzy subset of $Q \times X \times Q$, i.e., μ is a function of $Q \times X \times Q$ into the closed interval [0,1]. Q is called the set of states and X is called the set of input symbols. Let X^* denote the set of all words of finite length over X. Let λ denote the empty word in X^* and $|x|$ denote the length of x, $\forall x \in X^*, X^*$ is a semigroup with identity λ with respect to the binary operation concatenation of two words.

B.-Y. Cao and X.-J. Xie (Eds.): Fuzzy Engineering and Operations Research, AISC 147, pp. 351–361.
springerlink.com   © Springer-Verlag Berlin Heidelberg 2012

2 Preliminaries

Definition 1. [5] Let $M = (Q, X, \mu)$ be a fuzzy finite state machine (ffsm). Define $\mu^* : Q \times X^* \times Q \to [0, 1]$ by

$$\mu^*(q, \lambda, p) = \begin{cases} 1, \ q = p, \\ 0, \ \text{otherwise}, \end{cases}$$

and $\mu^*(q, ax, p) = \vee\{\mu(q, a, r) \wedge \mu^*(r, x, p) | r \in Q\}, \forall x \in X^*, a \in X$.

Lemma 1. [5] Let $M = (Q, X, \mu)$ be a ffsm, $\forall p, q \in Q, \forall x, y \in X^*$. Then $\mu^*(q, xy, p) = \vee\{\mu^*(q, x, r) \wedge \mu^*(r, y, p) | r \in Q\}$.

Theorem 1. [5] Let $M = (Q, X, \mu)$ be a ffsm. Define a relation $\equiv$ on X^* by $\forall x, y \in X^*, x \equiv y$ if and only if $\mu^*(q, x, p) = \mu^*(q, y, p), \forall p, q \in Q$. Then $\equiv$ is a congruence relation on X^*.

Let $x \in X^*, [x] = \{y \in X^* | x \equiv y\}$ and $E(M) = \{[x] | x \in X^*\}$.

Theorem 2. [5] Let $M = (Q, X, \mu)$ be a ffsm. Define a relation $*$ on $E(M)$ by $\forall [x], [y] \in E(M), [x] * [y] = [xy]$. Then $(E(M), *)$ is a finite semigroup with identity.

Definition 2. [5] A fuzzy transformation semigroup(fts) is a triple (Q, S, ρ), where Q is a finite nonempty set, S is a finite semigroup and ρ is a fuzzy subset of $Q \times X \times Q$ such that

   (i) $\rho(q, uv, p) = \vee\{\rho(q, u, r) \wedge \rho(r, v, p) | r \in Q\}, \forall u, v \in S, \forall q, p \in Q$;

   (ii) If S contains the identity e, then $\rho(q, e, p) = 1$, if q=p and $\rho(q, e, p) = 0$, if $q \neq p, \forall q, p \in Q$. If, in addition, the following property holds, then (Q, S, ρ) is called faithful.

   (iii) Let $u, v \in S$. If $\rho(q, u, p) = \rho(q, v, p), \forall q, p \in Q$, then u=v.

Theorem 3. [5] Let $M = (Q, X, \mu)$ be a ffsm. Let $E(M)$ be defined as before. Then $(Q, E(M), \rho)$ is a faithful fts where $\rho(q, [x], p) = \mu^*(q, x, p), \forall q, p \in Q, x \in X^*$.

Let $M = (Q, X, \mu)$ be a ffsm. Then by Theorem 3, $(Q, E(M), \rho)$ is a fuzzy transformation semigroup which we denote by FTS(M). We call FTS(M) the fuzzy transformation semigroup associated with M.

Definition 3. [4] Let $M_i = (Q_i, X_i, \mu_i), i = 1, 2$ be ffsms. Let $\overline{X}$ be a finite set and f a function from $\overline{X}$ into $X_1 \times X_2$. Let π_i be the projection map of $X_1 \times X_2$ onto X_i, i=1,2, $\forall (q_1, q_2), (p_1, p_2) \in Q_1 \times Q_2, \forall (a_1, a_2), (x_1, x_2) \in X_1 \times X_2, \forall a \in \overline{X}$, define:

   (1) The machine $M_1 \times M_2 = (Q_1 \times Q_2, X_1 \times X_2, \mu_1 \times \mu_2)$ is called the (full) direct product of M_1 and M_2, where $\mu_1 \times \mu_2 : (Q_1 \times Q_2) \times (X_1 \times X_2) \times$

$(Q_1 \times Q_2) \longrightarrow [0,1]$ is defined as follows: $\mu_1 \times \mu_2((q_1, q_2), (x_1, x_2), (p_1, p_2)) = \mu_1(q_1, x_1, p_1) \wedge \mu_2(q_2, x_2, p_2)$.

(2) The machine $M_1 * M_2 = (Q_1 \times Q_2, \overline{X}, \mu_f)$ is called the general direct product of M_1 and M_2, where $\mu_f : (Q_1 \times Q_2) \times \overline{X} \times (Q_1 \times Q_2) \longrightarrow [0,1]$ is defined as follows: $\mu_f((q_1, q_2), a, (p_1, p_2)) = \mu_1 \times \mu_2((q_1, q_2), (\pi_1(f(a)), \pi_2(f(a))), (p_1, p_2))$. Recall $\mu_1 \times \mu_2((q_1, q_2), (a_1, a_2), (p_1, p_2)) = \mu_1(q_1, a_1, p_1) \wedge \mu_2(q_2, a_2, p_2)$.

(3) Let $\omega : Q_2 \times X_2 \longrightarrow X_1$ be a mapping. Then the machine $M_1 \omega M_2 = (Q_1 \times Q_2, X_2, \mu^\omega)$ is called a cascade product of M_1 and M_2, where $\mu^\omega : (Q_1 \times Q_2) \times X_2 \times (Q_1 \times Q_2) \longrightarrow [0,1]$ is defined as follows: $\mu^\omega((q_1, q_2), x_2, (p_1, p_2)) = \mu_1(q_1, \omega(q_2, x_2), p_1) \wedge \mu_2(q_2, x_2, p_2)$.

(4) Let $X_1^{Q_2} = \{f | f : Q_2 \longrightarrow X_1\}$. Then the machine $M_1 \circ M_2 = (Q_1 \times Q_2, X_1^{Q_2} \times X_2, \mu^\circ)$ is called a wreath product of M_1 and M_2, where $\mu^\circ : (Q_1 \times Q_2) \times (X_1^{Q_2} \times X_2) \times (Q_1 \times Q_2) \longrightarrow [0,1]$ is defined as follows: $\mu^\circ((q_1, q_2), (f, x_2), (p_1, p_2)) = \mu_1(q_1, f(q_2), p_1) \wedge \mu_2(q_2, x_2, p_2)$.

Let $\Pi_{i=1}^n {}^* M_i = M_1 * M_2 * \ldots * M_n$, $\Pi_{i=1}^n {}^\circ M_i = M_1 \circ M_2 \circ \ldots \circ M_n$, $\Pi_{i=1}^n {}^\omega M_i = M_1 \omega M_2 \omega \ldots \omega M_n$.

3 Separability of Products of Ffsms

Let $M_t = (Q_t, X_t, \mu_t), t \in I$ be ffsms, I is an infinite set. Let $\overline{X}$ be a finite set and f a function from $\overline{X}$ into $\Pi_{t \in I} X_t$. Let π_t be the projection map of $\Pi_{t \in I} X_t$ onto $X_t, t \in I$, $\forall \{q_t\}_{t \in I}, \{p_t\}_{t \in I} \in \Pi_{t \in I} Q_t, \forall \{x_t\}_{t \in I} \in \Pi_{t \in I} X_t$. We generalize the concepts above as follows:

Definition 4. [13] The machine $\Pi_{t \in I} {}^* M_t = (\Pi_{t \in I} Q_t, \overline{X}, \mu_f)$ is called the general direct infinite product of $M_t, t \in I$, where $\mu_f : \Pi_{t \in I} Q_t \times \overline{X} \times \Pi_{t \in I} Q_t \longrightarrow [0,1]$ is defined as follows: $\mu_f(\{q_t\}_{t \in I}, a, \{p_t\}_{t \in I}) = \Pi_{t \in I} \mu_t(\{q_t\}_{t \in I}, \{\pi_t(f(a))\}_{t \in I}, \{p_t\}_{t \in I})$. Recall $\Pi_{t \in I} \mu_t(\{q_t\}_{t \in I}, \{a_t\}_{t \in I}, \{p_t\}_{t \in I}) = \bigwedge_{t \in I} \mu_t(q_t, a_t, p_t), \forall \{a_t\}_{t \in I} \in \Pi_{t \in I} X_t, a \in \overline{X}$.

Definition 5. [13] The machine $\Pi_{t \in I} M_t = (\Pi_{t \in I} Q_t, \Pi_{t \in I} X_t, \Pi_{t \in I} \mu_t)$ is called the (full) direct infinite product of $M_t, t \in I$, where $\Pi_{t \in I} \mu_t : \Pi_{t \in I} Q_t \times \Pi_{t \in I} X_t \times \Pi_{t \in I} Q_t \longrightarrow [0,1]$ is defined as follows:

$$\Pi_{t \in I} \mu_t(\{q_t\}_{t \in I}, \{x_t\}_{t \in I}, \{p_t\}_{t \in I}) = \bigwedge_{t \in I} \mu_t(q_t, x_t, p_t).$$

Proposition 1. Let $M_t = (Q_t, X, \mu_t), t \in I$ be ffsms. Define the relation $\equiv_I$ on X^* by $x \equiv_I y$ if and only if $x \equiv_t y, \forall t \in I$, where $\equiv_t$ is the congruence relation on M_t defined in Theorem 1, $t \in I$. Then $\equiv_I$ is a congruence relation on X^*.

Definition 6. [13]Let $M_t = (Q_t, X_t, \mu_t), t \in I$ be ffsms. Let ω be a function of $\Pi_{t \neq \alpha} Q_t \times \Pi_{t \neq \alpha} X_t \longrightarrow X_\alpha, t, \alpha \in I$. Let $Q = \Pi_{t \in I} Q_t$. Define: $\mu^\omega : Q \times \Pi_{t \neq \alpha} X_t \times Q \longrightarrow [0, 1]$ as follows: $\forall (\{q_t\}_{t \in I}, \{x_t\}_{t \neq \alpha}, \{p_t\}_{t \in I}) \in Q \times \Pi_{t \neq \alpha} X_t \times Q, \mu^\omega(\{q_t\}_{t \in I}, \{x_t\}_{t \neq \alpha}, \{p_t\}_{t \in I}) \quad = \quad \mu_\alpha(q_\alpha, \omega(\{q_t\}_{t \neq \alpha}, \{x_t\}_{t \neq \alpha}), p_\alpha) \wedge (\bigwedge_{t \neq \alpha} \mu_t(q_t, x_t, p_t))$. Then $M = (Q, \Pi_{t \neq \alpha} X_t, \mu^\omega)$ is a ffsm. M is called the cascade infinite product of $M_t, t \in I$ and we write $M = \Pi_{t \in I}{}^\omega M_t$.

μ^ω is called separable: if $\forall \{q_t\}_{t \in I}, \{p_t\}_{t \in I} \in Q, \forall \Pi_{i=1}^n(\{x_{ti}\}_{t \neq \alpha}) = \{x_{t1}\}_{t \neq \alpha} \{x_{t2}\}_{t \neq \alpha} \cdots \{x_{tn}\}_{t \neq \alpha} \in (\Pi_{t \neq \alpha} X_t)^*, \alpha \in I, x_t = x_{t1} x_{t2} \cdots x_{tn} \in X_t^*, \forall t \in I.$

$\mu^{\omega*}(\{q_t\}_{t \in I}, \Pi_{i=1}^n(\{x_{ti}\}_{t \neq \alpha}), \{p_t\}_{t \in I}) \quad = \quad \mu_\alpha^*(q_\alpha, \omega(\{q_t\}_{t \neq \alpha}, \{x_{t1}\}_{t \neq \alpha}) \omega(\{q_t^{(1)}\}_{t \neq \alpha}, \{x_{t2}\}_{t \neq \alpha}) \omega(\{q_t^{(2)}\}_{t \neq \alpha}, \{x_{t3}\}_{t \neq \alpha}) \cdots \omega(\{q_t^{(n-1)}\}_{t \neq \alpha}, \{x_{tn}\}_{t \neq \alpha}), p_\alpha) \wedge (\bigwedge_{t \neq \alpha} \mu_t^*(q_t, x_t, p_t))$, for some $\{q_t^{(i)}\}_{t \neq \alpha} \in \Pi_{t \neq \alpha} Q_t, i = 1, 2, \cdots n - 1.$

Theorem 4. Let $M_t = (Q_t, X_t, \mu_t), t \in I$ be ffsms. Let $M = \Pi_{t \in I}{}^\omega M_t$ for some ω, $\forall \{q_t\}_{t \in I}, \{p_t\}_{t \in I} \in Q, \forall \Pi_{i=1}^n(\{x_{ti}\}_{t \neq \alpha}) = \{x_{t1}\}_{t \neq \alpha} \{x_{t2}\}_{t \neq \alpha} \cdots \{x_{tn}\}_{t \neq \alpha} \in (\Pi_{t \neq \alpha} X_t)^*, \alpha \in I$. Then $\mu^{\omega*}(\{q_t\}_{t \in I}, \Pi_{i=1}^n(\{x_{ti}\}_{t \neq \alpha}), \{p_t\}_{t \in I}) = \vee\{\mu_\alpha^*(q_\alpha, \omega(\{q_t\}_{t \neq \alpha} \{x_{t1}\}_{t \neq \alpha}) \omega(\{q_t^{(1)}\}_{t \neq \alpha}, \{x_{t2}\}_{t \neq \alpha}) \omega(\{q_t^{(2)}\}_{t \neq \alpha}, \{x_{t3}\}_{t \neq \alpha}) \cdots \omega(\{q_t^{(n-1)}\}_{t \neq \alpha}, \{x_{tn}\}_{t \neq \alpha}), p_\alpha) \wedge (\bigwedge_{t \neq \alpha} \{\mu_t(q_t, x_{t1}, q_t^{(1)}) \wedge \mu_t(q_t^{(1)}, x_{t2}, q_t^{(2)}) \wedge \cdots \wedge \mu_t(q_t^{(n-1)}, x_{tn}, p_t)\}) \{q_t^{(i)}\}_{t \neq \alpha} \in \Pi_{t \neq \alpha} Q_t, i = 1, 2, \cdots n - 1\}.$

Proof. $\mu^{\omega*}(\{q_t\}_{t \in I}, \Pi_{i=1}^n(\{x_{ti}\}_{t \neq \alpha}), \{p_t\}_{t \in I}) = \vee\{\mu^\omega(\{q_t\}_{t \in I}, \{x_{t1}\}_{t \neq \alpha}, \{q_t^{(1)}\}_{t \in I}) \wedge \mu^\omega(\{q_t^{(1)}\}_{t \in I}, \{x_{t2}\}_{t \neq \alpha}, \{q_t^{(2)}\}_{t \in I}) \wedge \cdots \wedge \mu^\omega(\{q_t^{(n-1)}\}_{t \in I}, \{x_{tn}\}_{t \neq \alpha}, \{p_t\}_{t \in I}) | \{q_t^{(i)}\}_{t \in I} \in Q, i = 1, 2, \cdots n - 1\} = \vee\{[\mu_\alpha(q_\alpha, \omega(\{q_t\}_{t \neq \alpha}, \{x_{t1}\}_{t \neq \alpha}), q_\alpha^{(1)}) \wedge (\bigwedge_{t \neq \alpha} \mu_t(q_t, x_{t1}, q_t^{(1)}))] \wedge [\mu_\alpha(q_\alpha^{(1)}, \omega(\{q_t^{(1)}\}_{t \neq \alpha}, \{x_{t2}\}_{t \neq \alpha}), q_\alpha^{(2)}) \wedge (\bigwedge_{t \neq \alpha} \mu_t(q_t^{(1)}, x_{t2}, q_t^{(2)}))] \wedge \cdots \wedge [\mu_\alpha(q_\alpha^{(n-1)}, \omega(\{q_t^{(n-1)}\}_{t \neq \alpha}, \{x_{tn}\}_{t \neq \alpha}), p_\alpha) \wedge (\bigwedge_{t \neq \alpha} \mu_t(q_t^{(n-1)}, x_{tn}, p_t))] | \{q_t^{(i)}\}_{t \in I} \in Q, i = 1, 2, \cdots n - 1\} = \vee\{[\mu_\alpha(q_\alpha, \omega(\{q_t\}_{t \neq \alpha}, \{x_{t1}\}_{t \neq \alpha}), q_\alpha^{(1)}) \wedge \mu_\alpha(q_\alpha^{(1)}, \omega(\{q_t^{(1)}\}_{t \neq \alpha}, \{x_{t2}\}_{t \neq \alpha}), q_\alpha^{(2)}) \wedge \cdots \wedge \mu_\alpha(q_\alpha^{(n-1)}, \omega(\{q_t^{(n-1)}\}_{t \neq \alpha}, \{x_{tn}\}_{t \neq \alpha}), p_\alpha)] \wedge \{\bigwedge_{t \neq \alpha} [\mu_t(q_t, x_{t1}, q_t^{(1)}) \wedge \mu_t(q_t^{(1)}, x_{t2}, q_t^{(2)}) \wedge \cdots \wedge \mu_t(q_t^{(n-1)}, x_{tn}, p_t)]\} | \{q_t^{(i)}\}_{t \in I} \in Q, i = 1, 2, \cdots n - 1\} = \vee\{\mu_\alpha^*(q_\alpha, \omega(\{q_t\}_{t \neq \alpha}, \{x_{t1}\}_{t \neq \alpha}) \omega(\{q_t^{(1)}\}_{t \neq \alpha}, \{x_{t2}\}_{t \neq \alpha}) \omega(\{q_t^{(2)}\}_{t \neq \alpha}, \{x_{t3}\}_{t \neq \alpha}) \cdots \omega(\{q_t^{(n-1)}\}_{t \neq \alpha}, \{x_{tn}\}_{t \neq \alpha}), p_\alpha) \wedge (\bigwedge_{t \neq \alpha} \{\mu_t(q_t, x_{t1}, q_t^{(1)}) \wedge \mu_t(q_t^{(1)}, x_{t2}, q_t^{(2)}) \wedge \cdots \wedge \mu_t(q_t^{(n-1)}, x_{tn}, p_t)\}) | \{q_t^{(i)}\}_{t \neq \alpha} \in \Pi_{t \neq \alpha} Q_t, i = 1, 2, \cdots n - 1\}.$

Theorem 5. *Let $M_t = (Q_t, X_t, \mu_t), t \in I$ be ffsms. Let $Im(\mu)$ denote the image of μ. If $Im(\bigwedge_{t \neq \alpha} \mu_t) = \{0, 1\}$, then μ^ω is separable.*

Proof. $\forall \{q_t\}_{t \in I}, \{p_t\}_{t \in I} \in Q, \forall \ \Pi_{i=1}^n(\{x_{ti}\}_{t \neq \alpha}) = \{x_{t1}\}_{t \neq \alpha}\{x_{t2}\}_{t \neq \alpha}$
$\cdots \{x_{tn}\}_{t \neq \alpha} \in (\Pi_{t \neq \alpha} X_t)^*, \alpha \in I, x_t = x_{t1}x_{t2}\cdots x_{tn} \in X_t^*, \forall t \in I.$
By Theorem 4, $\mu^{\omega*}(\{q_t\}_{t \in I}, \Pi_{i=1}^n(\{x_{ti}\}_{t \neq \alpha}), \{p_t\}_{t \in I}) = \vee\{\mu_\alpha^*(q_\alpha, \omega(\{q_t\}_{t \neq \alpha},$
$\{x_{t1}\}_{t \neq \alpha})\omega(\{q_t^{(1)}\}_{t \neq \alpha}, \{x_{t2}\}_{t \neq \alpha}) \ \omega(\{q_t^{(2)}\}_{t \neq \alpha}, \{x_{t3}\}_{t \neq \alpha}) \cdots \omega(\{q_t^{(n-1)}\}_{t \neq \alpha},$
$\{x_{tn}\}_{t \neq \alpha}), p_\alpha) \wedge (\bigwedge_{t \neq \alpha} \{\mu_t(q_t, x_{t1}, q_t^{(1)}) \wedge \mu_t(q_t^{(1)}, x_{t2}, q_t^{(2)}) \wedge \cdots \wedge \mu_t(q_t^{(n-1)},$
$x_{tn}, p_t)\})|\{q_t^{(i)}\}_{t \neq \alpha} \in \Pi_{t \neq \alpha} Q_t, i = 1, 2, \cdots n - 1\}.$

If $Im(\bigwedge_{t \neq \alpha} \mu_t) = 0, \exists t_0 \in I, t_0 \neq \alpha, Im(\mu_{t_0}) = 0, Im(\mu_{t_0}^*) = 0.$ Thus
$\bigwedge_{t \neq \alpha} \mu_t^*(q_t, x_t, p_t) = 0.$ If $Im(\bigwedge_{t \neq \alpha} \mu_t) = 1, \forall t \in I, t \neq \alpha, Im(\mu_t) = 1, Im(\mu_t^*) = 1.$
Thus $\bigwedge_{t \neq \alpha} \mu_t^*(q_t, x_t, p_t) = 1.$ Then

Case 1: $\bigwedge_{t \neq \alpha} \mu_t^*(q_t, x_t, p_t) = 1.$ In fact, $\forall t \in I, t \neq \alpha, \vee\{\mu_t(q_t, x_{t1}, q_t^{(1)})$
$\wedge \mu_t(q_t^{(1)}, x_{t2}, q_t^{(2)}) \wedge \cdots \wedge \mu_t(q_t^{(n-1)}, x_{tn}, p_t)|q_t^{(i)} \in Q_t, i = 1, 2, \cdots, n - 1,$
$t \in I, t \neq \alpha\} = \mu_t^*(q_t, x_t, p_t) = 1.$ Hence $\forall t \in I, t \neq \alpha, \exists q_t^{(1)}, q_t^{(2)},$
$\cdots, q_t^{(n-1)} \in Q_t,$ such that $\mu_t(q_t, x_{t1}, q_t^{(1)}) \wedge \mu_t(q_t^{(1)}, x_{t2}, q_t^{(2)}) \wedge \cdots \wedge \mu_t(q_t^{(n-1)},$
$x_{tn}, p_t) = 1.$ Thus, $\mu^{\omega*}(\{q_t\}_{t \in I}, \Pi_{i=1}^n(\{x_{ti}\}_{t \neq \alpha}),$
$\{p_t\}_{t \in I}) = \vee\{\mu_\alpha^*(q_\alpha, \omega(\{q_t\}_{t \neq \alpha}, \{x_{t1}\}_{t \neq \alpha})\omega(\{q_t^{(1)}\}_{t \neq \alpha}, \{x_{t2}\}_{t \neq \alpha})\omega($
$\{q_t^{(2)}\}_{t \neq \alpha}, \{x_{t3}\}_{t \neq \alpha}) \cdots \omega(\{q_t^{(n-1)}\}_{t \neq \alpha}, \{x_{tn}\}_{t \neq \alpha}), p_\alpha)|\bigwedge_{t \neq \alpha} \{\mu_t(q_t, x_{t1}, q_t^{(1)}) \wedge$
$\mu_t(q_t^{(1)}, x_{t2}, q_t^{(2)}) \wedge \cdots \wedge \mu_t(q_t^{(n-1)}, x_{tn}, p_t)\} = 1, q_t^{(i)} \in Q_t, t \in I, t \neq \alpha, i =$
$1, 2, \cdots, n - 1\}.$ Since μ_α^* is finite valued, the supremum is attained at some
$q_t^{(1)}, q_t^{(2)} \cdots q_t^{(n-1)} \in Q_t, t \in I, t \neq \alpha.$ Hence $\exists q_t^{(1)}, q_t^{(2)}, \cdots, q_t^{(n-1)} \in$
$Q_t, t \in I, t \neq \alpha,$ such that $\mu^{\omega*}(\{q_t\}_{t \in I}, \Pi_{i=1}^n(\{x_{ti}\}_{t \neq \alpha}), \{p_t\}_{t \in I})$
$= \mu_\alpha^*(q_\alpha, \omega(\{q_t\}_{t \neq \alpha}, \{x_{t1}\}_{t \neq \alpha})\omega(\{q_t^{(1)}\}_{t \neq \alpha}, \{x_{t2}\}_{t \neq \alpha})\omega(\{q_t^{(2)}\}_{t \neq \alpha}, \{x_{t3}\}_{t \neq \alpha})$
$\cdots \omega(\{q_t^{(n-1)}\}_{t \neq \alpha}, \{x_{tn}\}_{t \neq \alpha}), p_\alpha) = \mu_\alpha^*(q_\alpha, \omega(\{q_t\}_{t \neq \alpha}, \{x_{t1}\}_{t \neq \alpha})\omega(\{q_t^{(1)}\}_{t \neq \alpha},$
$\{x_{t2}\}_{t \neq \alpha})\omega(\{q_t^{(2)}\}_{t \neq \alpha}, \{x_{t3}\}_{t \neq \alpha}) \cdots \omega(\{q_t^{(n-1)}\}_{t \neq \alpha}, \{x_{tn}\}_{t \neq \alpha}), p_\alpha) \wedge 1 =$
$\mu_\alpha^*(q_\alpha, \omega(\{q_t\}_{t \neq \alpha}, \{x_{t1}\}_{t \neq \alpha})\omega(\{q_t^{(1)}\}_{t \neq \alpha}, \{x_{t2}\}_{t \neq \alpha})\omega(\{q_t^{(2)}\}_{t \neq \alpha}, \{x_{t3}\}_{t \neq \alpha}) \cdots$
$\omega(\{q_t^{(n-1)}\}_{t \neq \alpha}, \{x_{tn}\}_{t \neq \alpha}), p_\alpha) \wedge (\bigwedge_{t \neq \alpha} \mu_t^*(q_t, x_t, p_t)).$

Case 2: $\bigwedge_{t \neq \alpha} \mu_t^*(q_t, x_t, p_t) = 0.$ Then $\exists t \in I, t \neq \alpha, \mu_t^*(q_t, x_t, p_t) = 0,$
$\forall q_t^{(1)}, q_t^{(2)}, \cdots, q_t^{(n-1)} \in Q_t, \mu_t(q_t, x_{t1}, q_t^{(1)}) \wedge \mu_t(q_t^{(1)}, x_{t2}, q_t^{(2)}) \wedge \cdots \wedge$
$\mu_t(q_t^{(n-1)}, x_{tn}, p_t) = 0.$ Thus $\mu^{\omega*}(\{q_t\}_{t \in I}, \Pi_{i=1}^n(\{x_{ti}\}_{t \neq \alpha}), \{p_t\}_{t \in I}) = 0$
$= \mu_\alpha^*(q_\alpha, \omega(\{q_t\}_{t \neq \alpha}, \{x_{t1}\}_{t \neq \alpha})\omega(\{q_t^{(1)}\}_{t \neq \alpha}, \{x_{t2}\}_{t \neq \alpha})\omega(\{q_t^{(2)}\}_{t \neq \alpha}, \{x_{t3}\}_{t \neq \alpha})$
$\cdots \omega(\{q_t^{(n-1)}\}_{t \neq \alpha}, \{x_{tn}\}_{t \neq \alpha}), p_\alpha) \wedge 0 = \mu_\alpha^*(q_\alpha, \omega(\{q_t\}_{t \neq \alpha}, \{x_{t1}\}_{t \neq \alpha})$

$$\omega(\{q_t^{(1)}\}_{t\neq\alpha}, \quad \{x_{t2}\}_{t\neq\alpha})\omega(\{q_t^{(2)}\}_{t\neq\alpha}, \quad \{x_{t3}\}_{t\neq\alpha})\cdots\omega(\{q_t^{(n-1)}\}_{t\neq\alpha},\{x_{tn}\}$$
$$_{t\neq\alpha}),p_\alpha)\wedge(\bigwedge_{t\neq\alpha}\mu_t^*(q_t,x_t,p_t)),\forall q_t^{(1)},q_t^{(2)}\cdots q_t^{(n-1)}\in Q_t,t\in I,t\neq\alpha.$$

Definition 7. [13] Let $M_t=(Q_t,X_t,\mu_t),t\in I$ be ffsms. Let f be a function from $\Pi_{t\neq\alpha}Q_t$ into X_α. Define $\mu^\circ:Q\times(X_\alpha^{\Pi_{t\neq\alpha}Q_t}\times\Pi_{t\neq\alpha}X_t)\times Q\longrightarrow[0,1]$ as follows: for all $(\{q_t\}_{t\in I},(f,\{x_t\}_{t\neq\alpha}),\{p_t\}_{t\in I})\in Q\times(X_\alpha^{\Pi_{t\neq\alpha}Q_t}\times\Pi_{t\neq\alpha}X_t)\times Q,\quad\mu^\circ(\{q_t\}_{t\in I},(f,\{x_t\}_{t\neq\alpha}),\{p_t\}_{t\in I})=\mu_\alpha(q_\alpha,f(\{q_t\}_{t\neq\alpha}),p_\alpha)\wedge(\bigwedge_{t\neq\alpha}\mu_t(q_t,x_t,p_t))$. Then $M=(Q,X_\alpha^{\Pi_{t\neq\alpha}Q_t}\times\Pi_{t\neq\alpha}X_t,\mu^\circ)$ is a ffsm. $M=\Pi_{t\in I}^\circ M_t$ is called the wreath infinite product of $M_t,t\in I$.

μ° is called separable, if $\forall\{q_t\}_{t\in I},\{p_t\}_{t\in I}\in Q,\forall(f_1,\{x_{t1}\}_{t\neq\alpha})(f_2,\{x_{t2}\}_{t\neq\alpha})\cdots(f_n,\{x_{tn}\}_{t\neq\alpha})\in(X_\alpha^{\Pi_{t\neq\alpha}Q_t}\times\Pi_{t\neq\alpha}X_t)^*,\mu^{\circ*}(\{q_t\}_{t\in I},(f_1,\{x_{t1}\}_{t\neq\alpha})(f_2,\{x_{t2}\}_{t\neq\alpha})\cdots(f_n,\{x_{tn}\}_{t\neq\alpha}),\{p_t\}_{t\in I})=\mu_\alpha^*(q_\alpha,f_1(\{q_t\}_{t\neq\alpha})f_2(\{q_t^{(1)}\}_{t\neq\alpha})\cdots f_n(\{q_t^{(n-1)}\}_{t\neq\alpha}),p_\alpha)\wedge(\bigwedge_{t\neq\alpha}\mu_t^*(q_t,x_{t1}x_{t2}\cdots x_{tn},p_t))$, for some $\{q_t^{(i)}\}_{t\neq\alpha}\in\Pi_{t\neq\alpha}Q_t,i=1,2,\cdots,n-1$.

Proposition 2. Let $M_t=(Q_t,X_t,\mu_t),t\in I$ be ffsms. Let $M=\Pi_{t\in I}^\circ M_t$ is the wreath infinite product of $M_t,t\in I$. Then for all $\{q_t\}_{t\in I},\{p_t\}_{t\in I}\in Q$, for all $(f_1,\{x_{t1}\}_{t\neq\alpha})(f_2,\{x_{t2}\}_{t\neq\alpha})\cdots(f_n,\{x_{tn}\}_{t\neq\alpha})\in(X_\alpha^{\Pi_{t\neq\alpha}Q_t}\times\Pi_{t\neq\alpha}X_t)^*,\mu^{\circ*}(\{q_t\}_{t\in I},\quad(f_1,\{x_{t1}\}_{t\neq\alpha})\quad(f_2,\{x_{t2}\}_{t\neq\alpha})\quad\cdots(f_n,\{x_{tn}\}_{t\neq\alpha}),\quad\{p_t\}_{t\in I})=\quad\vee\{\mu_\alpha^*(q_\alpha,f_1(\{q_t\}_{t\neq\alpha})\quad f_2(\{q_t^{(1)}\}_{t\neq\alpha})\cdots f_n(\{q_t^{(n-1)}\}_{t\neq\alpha}),\quad p_\alpha)\quad\wedge\quad(\bigwedge_{t\neq\alpha}[\mu_t(q_t,x_{t1},q_t^{(1)})\quad\wedge\quad\mu_t(q_t^{(1)},x_{t2},q_t^{(2)})\quad\cdots$
$\mu_t(q_t^{(n-1)},x_{tn},p_t)])|q_t^{(1)},q_t^{(2)}\cdots q_t^{(n-1)}\in Q_t,t\in I,t\neq\alpha\}$.

Proof. $\mu^{\circ*}(\{q_t\}_{t\in I},\quad(f_1,\{x_{t1}\}_{t\neq\alpha})\quad(f_2,\{x_{t2}\}_{t\neq\alpha})\quad\cdots(f_n,\{x_{tn}\}_{t\neq\alpha}),\{p_t\}_{t\in I})$
$=\vee\{\mu_\alpha(q_\alpha,f_1(\{q_t\}_{t\neq\alpha}),q_\alpha^{(1)})\wedge(\bigwedge_{t\neq\alpha}\mu_t(q_t,x_{t1},q_t^{(1)}))\wedge\mu_\alpha(q_\alpha^{(1)},f_2(\{q_t^{(1)}\}_{t\neq\alpha}),$
$q_\alpha^{(2)})\wedge(\bigwedge_{t\neq\alpha}\mu_t(q_t^{(1)},x_{t2},q_t^{(2)}))\wedge\cdots\wedge\mu_\alpha(q_\alpha^{(n-1)},f_n(\{q_t^{(n-1)}\}_{t\neq\alpha}),p_\alpha)\wedge$
$(\bigwedge_{t\neq\alpha}\mu_t(q_t^{(n-1)},x_{tn},p_t))\,|q_t^{(i)}\in Q_t,q_\alpha^{(i)}\in Q_\alpha,i=1,2,\cdots,n-1,t\neq$
$\alpha\}=\vee\{[\mu_\alpha(q_\alpha,f_1(\{q_t\}_{t\neq\alpha}),q_\alpha^{(1)})\wedge\mu_\alpha(q_\alpha^{(1)},f_2(\{q_t^{(1)}\}_{t\neq\alpha}),q_\alpha^{(2)})\wedge\cdots\wedge$
$\mu_\alpha(q_\alpha^{(n-1)},f_n(\{q_t^{(n-1)}\}_{t\neq\alpha}),p_\alpha)]\wedge[\bigwedge_{t\neq\alpha}(\mu_t(q_t,x_{t1},q_t^{(1)})\wedge\mu_t(q_t^{(1)},x_{t2},q_t^{(2)})\wedge$
$\cdots\wedge\mu_t(q_t^{(n-1)},x_{tn},p_t))]|q_t^{(i)}\in Q_t,q_\alpha^{(i)}\in Q_\alpha,i=1,2,\cdots,n-1,t\neq$
$\alpha\}=\vee\{\mu_\alpha^*(q_\alpha,f_1(\{q_t\}_{t\neq\alpha})f_2(\{q_t^{(1)}\}_{t\neq\alpha})\cdots f_n(\{q_t^{(n-1)}\}_{t\neq\alpha}),p_\alpha)\wedge(\bigwedge_{t\neq\alpha}[\mu_t(q_t,$
$x_{t1},q_t^{(1)})\wedge\mu_t(q_t^{(1)},x_{t2},q_t^{(2)})\wedge\cdots\wedge\mu_t(q_t^{(n-1)},x_{tn},p_t)])|q_t^{(1)},q_t^{(2)},\cdots,q_t^{(n-1)}\in$
$Q_t,t\in I,t\neq\alpha\}$.

Theorem 6. Let $M_t = (Q_t, X_t, \mu_t), t \in I$ be ffsms. If $Im(\bigwedge_{t \neq \alpha} \mu_t) = \{0, 1\}$, then μ° is separable.

Proof. $\forall \{q_t\}_{t \in I}, \{p_t\}_{t \in I} \in Q, \forall (f_1, \{x_{t1}\}_{t \neq \alpha})(f_2, \{x_{t2}\}_{t \neq \alpha}) \cdots (f_n, \{x_{tn}\}_{t \neq \alpha}) \in (X_\alpha^{\Pi_{t \neq \alpha} Q_t} \times \Pi_{t \neq \alpha} X_t)^*$. By Proposition 2, $\mu^{\circ *}(\{q_t\}_{t \in I}, (f_1, \{x_{t1}\}_{t \neq \alpha})(f_2, \{x_{t2}\}_{t \neq \alpha}) \cdots (f_n, \{x_{tn}\}_{t \neq \alpha}), \{p_t\}_{t \in I}) = \vee \{\mu_\alpha^*(q_\alpha, f_1(\{q_t\}_{t \neq \alpha}) f_2(\{q_t^{(1)}\}_{t \neq \alpha}) \cdots f_n(\{q_t^{(n-1)}\}_{t \neq \alpha}), p_\alpha) \wedge (\bigwedge_{t \neq \alpha} [\mu_t(q_t, x_{t1}, q_t^{(1)}) \wedge \mu_t(q_t^{(1)}, x_{t2}, q_t^{(2)}) \cdots \mu_t(q_t^{(n-1)}, x_{tn}, p_t)]) | q_t^{(1)}, q_t^{(2)} \cdots q_t^{(n-1)} \in Q_t, t \in I, t \neq \alpha\}$.

Case 1: $\bigwedge_{t \neq \alpha} \mu_t^*(q_t, x_{t1} x_{t2} \cdots x_{tn}, p_t) = 1$. In fact, $\forall t \in I, t \neq \alpha, \vee \{\mu_t(q_t, x_{t1}, q_t^{(1)}) \wedge \mu_t(q_t^{(1)}, x_{t2}, q_t^{(2)}) \wedge \cdots \wedge \mu_t(q_t^{(n-1)}, x_{tn}, p_t) | q_t^{(i)} \in Q_t, i = 1, 2, \cdots, n-1\} = \mu_t^*(q_t, x_{t1} x_{t2}, \cdots, x_{tn}, p_t) = 1$. Hence, $\forall t \in I, t \neq \alpha, \exists q_t^{(1)}, q_t^{(2)}, \cdots, q_t^{(n-1)} \in Q_t$, such that $\mu_t(q_t, x_{t1}, q_t^{(1)}) \wedge \mu_t(q_t^{(1)}, x_{t2}, q_t^{(2)}) \wedge \cdots \wedge \mu_t(q_t^{(n-1)}, x_{tn}, p_t) = 1$. Thus, $\mu^{\circ *}(\{q_t\}_{t \in I}, (f_1, \{x_{t1}\}_{t \neq \alpha})(f_2, \{x_{t2}\}_{t \neq \alpha}) \cdots (f_n, \{x_{tn}\}_{t \neq \alpha}), \{p_t\}_{t \in I}) = \vee \{\mu_\alpha^*(q_\alpha, f_1(\{q_t\}_{t \neq \alpha}) f_2(\{q_t^{(1)}\}_{t \neq \alpha}) \cdots f_n(\{q_t^{(n-1)}\}_{t \neq \alpha}), p_\alpha) | \bigwedge_{t \neq \alpha} \{\mu_t(q_t, x_{t1}, q_t^{(1)}) \wedge \mu_t(q_t^{(1)}, x_{t2}, q_t^{(2)}) \wedge \cdots \wedge \mu_t(q_t^{(n-1)}, x_{tn}, p_t)\} = 1, q_t^{(i)} \in Q_t, t \in I, t \neq \alpha, i = 1, 2, \cdots, n-1\}$. Since μ_α^* is finite valued, the supremum is attained at some $q_t^{(1)}, q_t^{(2)} \cdots q_t^{(n-1)} \in Q_t, t \in I, t \neq \alpha$. Hence $\exists q_t^{(1)}, q_t^{(2)}, \cdots, q_t^{(n-1)} \in Q_t, t \in I, t \neq \alpha$, such that $\mu^{\circ *}(\{q_t\}_{t \in I}, (f_1, \{x_{t1}\}_{t \neq \alpha})(f_2, \{x_{t2}\}_{t \neq \alpha}) \cdots (f_n, \{x_{tn}\}_{t \neq \alpha}), \{p_t\}_{t \in I}) = \mu_\alpha^*(q_\alpha, f_1(\{q_t\}_{t \neq \alpha}) f_2(\{q_t^{(1)}\}_{t \neq \alpha}) \cdots f_n(\{q_t^{(n-1)}\}_{t \neq \alpha}), p_\alpha) \wedge 1 = \mu_\alpha^*(q_\alpha, f_1(\{q_t\}_{t \neq \alpha}) f_2(\{q_t^{(1)}\}_{t \neq \alpha}) \cdots f_n(\{q_t^{(n-1)}\}_{t \neq \alpha}), p_\alpha) \wedge (\bigwedge_{t \neq \alpha} \mu_t^*(q_t, x_{t1} x_{t2} \cdots x_{tn}, p_t))$.

Case 2: $\bigwedge_{t \neq \alpha} \mu_t^*(q_t, x_{t1} x_{t2} \cdots x_{tn}, p_t) = 0$. Then there exists $t \in I, t \neq \alpha$, such that $\mu_t^*(q_t, x_{t1} x_{t2}, \cdots, x_{tn}, p_t) = 0$. For all $q_t^{(1)}, q_t^{(2)}, \cdots, q_t^{(n-1)} \in Q_t, \mu_t(q_t, x_{t1}, q_t^{(1)}) \wedge \mu_t(q_t^{(1)}, x_{t2}, q_t^{(2)}) \wedge \cdots \wedge \mu_t(q_t^{(n-1)}, x_{tn}, p_t) = 0$. Thus, $\mu^{\circ *}(\{q_t\}_{t \in I}, (f_1, \{x_{t1}\}_{t \neq \alpha})(f_2, \{x_{t2}\}_{t \neq \alpha}) \cdots (f_n, \{x_{tn}\}_{t \neq \alpha}), \{p_t\}_{t \in I}) = 0 = \mu_\alpha^*(q_\alpha, f_1(\{q_t\}_{t \neq \alpha}) f_2(\{q_t^{(1)}\}_{t \neq \alpha}) \cdots f_n(\{q_t^{(n-1)}\}_{t \neq \alpha}), p_\alpha) \wedge 0 = \mu_\alpha^*(q_\alpha, f_1(\{q_t\}_{t \neq \alpha}) f_2(\{q_t^{(1)}\}_{t \neq \alpha}) \cdots f_n(\{q_t^{(n-1)}\}_{t \neq \alpha}), p_\alpha) \wedge (\bigwedge_{t \neq \alpha} \mu_t^*(q_t, x_{t1} x_{t2} \cdots x_{tn}, p_t)), \forall q_t^{(1)}, q_t^{(2)}, \cdots, q_t^{(n-1)} \in Q_t, t \in I, t \neq \alpha$.

As usual, let $M_i = (Q_i, X_i, \mu_i), i = 1, 2$ be ffsms. Let η be a function of Q_2 onto Q_1 and let ξ be a function of X_1 onto X_2. Extend ξ to a function ξ^* of X_1^* into X_2^* by $\xi^*(\lambda) = \lambda$ and $\forall x \in X_1^*, \xi^*(x) = \xi^*(x_1) \xi^*(x_2) \cdots \xi^*(x_n)$, where $x = x_1 x_2 \cdots x_n$ and $x_i \in X_i, i = 1, 2, \ldots, n$. Then (η, ξ) is called a covering of M_1 by M_2 in [4], written $M_1 \leq M_2$, if and only if $\forall q_2 \in Q_2, q_1 \in Q_1$, and $x \in X_1^*$,

$$\mu_1^*(\eta(q_2), x, q_1) = \vee \{\mu_2^*(q_2, \xi^*(x), r_2) | \eta(r_2) = q_1, r_2 \in Q_2\}.$$

We are now interested in obtaining covering results involving the cascade infinite product and the wreath infinite product of fuzzy finite state machines [13]. It can be easily seen that these relations are not necessarily single-valued. Hence, we replace $E(M)$ with X^* in the following definition. Then we prove the weak covering relations of $FTS(\Pi^{\circ}_{t\in I} M_t)$ by $\Pi^{\circ}_{t\in I} FTS(M_t)$ and $FTS(\Pi^{\omega}_{t\in I} M_t)$ by $\Pi^{\circ}_{t\in I} FTS(M_t)$.

Definition 8. [4] Let $M = (Q, X, \mu)$ and $M_i = (Q_i, X_i, \mu_i), i = 1, 2$ be ffsms. Let η be a function of $Q_1 \times Q_2$ onto Q and ζ a function of X^* into $E(M_1)^{Q_2} \times E(M_2)$. Then (η, ζ) is said to be a *weak covering* of $FTS(M)$ by $FTS(M_1) \circ FTS(M_2)$ if $\rho(\eta(q_1, q_2), [x], \eta(p_1, p_2)) = \vee\{\hat{\mu}^{\circ}((q_1, q_2), \zeta(x), (r_1, r_2)) | \eta(r_1, r_2) = \eta(q_1, q_2), (p_1, p_2), (r_1, r_2) \in Q_1 \times Q_2\}, \forall x \in X^*$, and $(q_1, q_2), (p_1, p_2) \in Q_1 \times Q_2$.

Now we research the weak covering of the products with infinitely many fuzzy finite state machines.

Theorem 7. *Let* $M_t = (Q_t, X_t, \mu_t), t \in I$ *be ffsms. If* μ° *is separable, then*

$$FTS(\Pi^{\circ}_{t\in I} M_t) \leq_{weakly} \Pi^{\circ}_{t\in I} FTS(M_t).$$

Proof. Let $Q = \Pi_{t\in I} Q_t$, $\Pi^{\circ}_{t\in I} M_t = (Q, X_\alpha^{\Pi_{t\neq\alpha} Q_t} \times \Pi_{t\neq\alpha} X_t, \mu^{\circ})$, where $\mu^{\circ}(\{q_t\}_{t\in I}, (g, \{x_t\}_{t\neq\alpha}), \{p_t\}_{t\in I}) = \mu_\alpha \left(q_\alpha, g(\{q_t\}_{t\neq\alpha}) \, p_\alpha\right) \wedge \left(\bigwedge_{t\neq\alpha} \mu_t(q_t, x_t, p_t)\right)$, for all $(\{q_t\}_{t\in I}, (g, \{x_t\}_{t\neq\alpha}), \{p_t\}_{t\in I}) \in Q \times (X_\alpha^{\Pi_{t\neq\alpha} Q_t} \times \Pi_{t\neq\alpha} X_t) \times Q$. And, $FTS(\Pi^{\circ}_{t\in I} M_t) = (Q, E(\Pi^{\circ}_{t\in I} M_t), \rho^{\circ})$, where $\rho^{\circ}(\{q_t\}_{t\in I}, [(g_1, \{x_{t1}\}_{t\neq\alpha}) (g_2, \{x_{t2}\}_{t\neq\alpha}) \cdots (g_k, \{x_{tk}\}_{t\neq\alpha})], \{p_t\}_{t\in I}) = \mu^{\circ*}(\{q_t\}_{t\in I}, (g_1, \{x_{t1}\}_{t\neq\alpha}) (g_2, \{x_{t2}\}_{t\neq\alpha}) \cdots (g_k, \{x_{tk}\}_{t\neq\alpha}), \{p_t\}_{t\in I}), \forall (g_1, \{x_{t1}\}_{t\neq\alpha}) (g_2, \{x_{t2}\}_{t\neq\alpha}) \cdots (g_k, \{x_{tk}\}_{t\neq\alpha}) \in (X_\alpha^{\Pi_{t\neq\alpha} Q_t} \times \Pi_{t\neq\alpha} X_t)^*$. Also, $\Pi^{\circ}_{t\in I} FTS(M_t) = (Q, E(M_\alpha)^{\Pi_{t\neq\alpha} Q_t} \times \Pi_{t\neq\alpha} E(M_t), \hat{\mu}^{\circ}), \forall \{x_{t1}\}_{t\neq\alpha}\{x_{t2}\}_{t\neq\alpha} \cdots \{x_{tk}\}_{t\neq\alpha} \in (\Pi_{t\neq\alpha} X_t)^*$, where $\hat{\mu}^{\circ}(\{q_t\}_{t\in I}, (f, [\{x_{t1}\}_{t\neq\alpha}\{x_{t2}\}_{t\neq\alpha} \cdots \{x_{tk}\}_{t\neq\alpha}]), \{p_t\}_{t\in I}) = \rho_\alpha(q_\alpha, f(\{q_t\}_{t\neq\alpha}), p_\alpha) \wedge \left(\bigwedge_{t\neq\alpha} \rho_t(q_t, [x_{t1}x_{t2} \cdots x_{tk}], p_t)\right) = \mu_\alpha^*(q_\alpha, x_\alpha, p_\alpha) \wedge \left(\bigwedge_{t\neq\alpha} \mu_t^*(q_t, x_{t1}x_{t2} \cdots x_{tk}, p_t)\right)$, where $f(\{q_t\}_{t\neq\alpha}) = x_\alpha$, and $x_\alpha \in X_\alpha^*$ is selected below. Let η be the identity map of Q. Define $\xi : (X_\alpha^{\Pi_{t\neq\alpha} Q_t} \times \Pi_{t\neq\alpha} X_t)^* \longrightarrow E(M_\alpha)^{\Pi_{t\neq\alpha} Q_t} \times \Pi_{t\neq\alpha} E(M_t)$, by $\xi((g_1, \{x_{t1}\}_{t\neq\alpha})(g_2, \{x_{t2}\}_{t\neq\alpha}) \cdots (g_k, \{x_{tk}\}_{t\neq\alpha})) = (f, [\{x_{t1}\}_{t\neq\alpha}\{x_{t2}\}_{t\neq\alpha} \cdots \{x_{tk}\}_{t\neq\alpha}])$. Now $(g_1, \{x_{t1}\}_{t\neq\alpha}) (g_2, \{x_{t2}\}_{t\neq\alpha}) \cdots (g_k, \{x_{tk}\}_{t\neq\alpha}) = (h_1, \{y_{t1}\}_{t\neq\alpha})(h_2, \{y_{t2}\}_{t\neq\alpha}) \cdots (h_j, \{y_{tj}\}_{t\neq\alpha}) \Longleftrightarrow k = j$ and $(g_i, \{x_{ti}\}_{t\neq\alpha}) = (h_i, \{y_{ti}\}_{t\neq\alpha}), i = 1, 2, \cdots k. \Longleftrightarrow k = j, g_i = h_i$, and $\{x_{ti}\}_{t\neq\alpha} = \{y_{ti}\}_{t\neq\alpha}, i = 1, 2, \cdots k$. Thus ξ is single-valued. Then $\rho^{\circ}(\{q_t\}_{t\in I}, [(g_1, \{x_{t1}\}_{t\neq\alpha}) (g_2, \{x_{t2}\}_{t\neq\alpha}) \cdots (g_k, \{x_{tk}\}_{t\neq\alpha})], \{p_t\}_{t\in I}) = \mu^{\circ*}(\{q_t\}_{t\in I}, (g_1, \{x_{t1}\}_{t\neq\alpha}) (g_2, \{x_{t2}\}_{t\neq\alpha}) \cdots (g_k, \{x_{tk}\}_{t\neq\alpha}), \{p_t\}_{t\in I}) = \vee \{\mu_\alpha^*(q_\alpha, g_1(\{q_t\}_{t\neq\alpha}) \, g_2 (\{q_t^{(1)}\}_{t\neq\alpha}) \cdots g_k (\{q_t^{(k-1)}\}_{t\neq\alpha}), p_\alpha)$

$$\wedge \ (\ \bigwedge_{t\neq\alpha} (\mu_t(q_t, x_{t1}, q_t^{(1)}) \ \wedge \ \mu_t(q_t^{(1)}, x_{t2}, q_t^{(2)}) \ \cdots \ \mu_t(q_t^{(k-1)}, x_{tk}, p_t)))|q_t^{(1)}, q_t^{(2)}$$

$$\cdots q_t^{(k-1)} \ \in \ Q_t, t \ \in \ I, t \ \neq \ \alpha\}" \ = \ "\mu_\alpha^*(q_\alpha, x_\alpha, p_\alpha) \wedge (\bigwedge_{t\neq\alpha} \mu_t^*(q_t, x_{t1}x_{t2}$$

$$\cdots x_{tk}, p_t)) \ = \ \rho_\alpha(q_\alpha, \ f(\{q_t\}_{t\neq\alpha}), p_\alpha) \ \wedge \ (\bigwedge_{t\neq\alpha} \rho_t(q_t, [x_{t1}x_{t2}\cdots x_{tk}], p_t))$$

$= \hat{\mu}^\circ(\{q_t\}_{t\in I}, (f, [\{x_{t1}\}_{t\neq\alpha}\{x_{t2}\}_{t\neq\alpha}\cdots\{x_{tk}\}_{t\neq\alpha}]), \{p_t\}_{t\in I}) = \vee\{\hat{\mu}^\circ(\{q_t\}_{t\in I},$
$\xi((g_1, \{x \ _{t1}\}_{t\neq\alpha})(g_2, \{x_{t2}\}_{t\neq\alpha})\cdots(g_k, \{x_{tk}\}_{t\neq\alpha}), \{r_t\}_{t\in I})|\eta(\{r_t\}_{t\in I}) \ = $
$\{p_t\}_{t\in I}, \{r_t\}_{t\in I} \ \in \ \Pi_{t\in I}Q_t\}$. Since η is the identity map and μ° is separable, select $x_\alpha \ = \ g_1(\{q_t\}_{t\neq\alpha})g_2(\{q_t^{(1)}\}_{t\neq\alpha})\cdots \ g_k(\{q_t^{(k-1)}\}_{t\neq\alpha})$. So that $\{q_t^{(1)}\}_{t\neq\alpha}, \{q_t^{(2)}\}_{t\neq\alpha}, \cdots\{q_t^{(k-1)}\}_{t\neq\alpha}$ gives " $=$ " .Hence (η, ξ) is a weak covering of $FTS(\Pi_{t\in I}^\circ M_t)$ by $\Pi_{t\in I}^\circ FTS(M_t)$.

Theorem 8. *Let $M_t = (Q_t, X_t, \mu_t), t \in I$ be ffsms. If μ^ω is separable, then*

$$FTS(\Pi_{t\in I}^\omega M_t) \leq_{weakly} \Pi_{t\in I}^\circ FTS(M_t).$$

Proof. Let $Q = \Pi_{t\in I}Q_t$, $\Pi_{t\in I}^\omega M_t = (Q, \Pi_{t\neq\alpha}X_t, \mu^\omega)$, where $\omega : \Pi_{t\neq\alpha}Q_t \times \Pi_{t\neq\alpha}X_t \longrightarrow X_\alpha, t, \alpha \in I$, and $\mu^\omega(\{q_t\}_{t\in I}, \{x_t\}_{t\neq\alpha}, \{p_t\}_{t\in I}) = \mu_\alpha(q_\alpha, \omega(\{q_t\}_{t\neq\alpha}, \{x_t \ \}_{t\neq\alpha}), p_\alpha) \wedge (\bigwedge_{t\neq\alpha} \mu_t(q_t, x_t, p_t)), \forall(\{q_t\}_{t\in I}, \{x_t\}_{t\neq\alpha}, \{p_t\}_{t\in I}) \in Q \times \Pi_{t\neq\alpha}X_t \times Q$. $FTS(\Pi_{t\in I}^\omega M_t) = (Q, E(\Pi_{t\in I}^\omega M_t), \rho)$, where $\rho(\{q_t\}_{t\in I}, [\Pi_{i=1}^n(\{x_{ti}\}_{t\neq\alpha})], \ \{p_t\}_{t\in I}) \ = \ \mu^{\omega*}(\{q_t\}_{t\in I}, \Pi_{i=1}^n(\{x_{ti}\}_{t\neq\alpha}), \{p_t\}_{t\in I})$, for all $\{q_t\}_{t\in I}, \{p_t\}_{t\in I} \in Q$, for all $\Pi_{i=1}^n(\{x_{ti} \ \}_{t\neq\alpha}) \in (\Pi_{t\neq\alpha}X_t)^*$, let $x_t \ = \ x_{t1}x_{t2}\cdots x_{tn} \ \in \ X_t^*.\Pi_{t\in I}^\circ FTS(M_t) \ = \ (Q, E(M_\alpha)^{\Pi_{t\neq\alpha}Q_t} \times \Pi_{t\neq\alpha}E(M_t), \hat{\mu}^\circ), \forall\{x_t\}_{t\neq\alpha} = \{x_{t1}\}_{t\neq\alpha}\{x_{t2}\}_{t\neq\alpha}\cdots\{x_{tn}\}_{t\neq\alpha} \in (\Pi_{t\neq\alpha}X_t)^*$, where $\hat{\mu}^\circ(\{q_t\}_{t\in I}, (f, [\{x_{t1}\}_{t\neq\alpha}\{x_{t2}\}_{t\neq\alpha}\cdots\{x_{tn}\}_{t\neq\alpha}]), \{p_t\}_{t\in I}) = \rho_\alpha(q_\alpha, f(\{q_t\}_{t\neq\alpha}), p_\alpha)\wedge(\bigwedge_{t\neq\alpha} \rho_t(q_t, [x_t], p_t))$. Let η be the identity map of Q. Since μ^ω is separable, $\mu^{\omega*}(\{q_t\}_{t\in I}, \{x_{t1}\}_{t\neq\alpha}\{x_{t2}\}_{t\neq\alpha}\cdots\{x_{tn}\}_{t\neq\alpha}, \{p_t\}_{t\in I}) = \mu_\alpha^*(q_\alpha, \omega(\{q_t\}_{t\neq\alpha}, \{x_{t1}\}_{t\neq\alpha})\omega(\{q_t^{(1)}\}_{t\neq\alpha}, \{x_{t2}\}_{t\neq\alpha})\cdots\omega(\{q_t^{(n-1)}\}_{t\neq\alpha}, \{x_{tn}\}_{t\neq\alpha}), p_\alpha) \wedge (\bigwedge_{t\neq\alpha} \mu_t^*(q_t, x_t, p_t))$, for some $\{q_t^{(i)}\}_{t\neq\alpha} \in \Pi_{t\neq\alpha}Q_t, i = 1, 2, \cdots n-1$. Then we define $\xi : (\Pi_{t\neq\alpha}X_t)^* \longrightarrow E(M_\alpha)^{\Pi_{t\neq\alpha}Q_t} \times \Pi_{t\neq\alpha}E(M_t), \forall\Pi_{i=1}^n(\{x_{ti}\}_{t\neq\alpha}) \in (\Pi_{t\neq\alpha}X_t)^*$, let $a = \Pi_{i=1}^n(\{x_{ti}\}_{t\neq\alpha}), \xi(a) = (f_a, [a]), f_a(\{q_t\}_{t\neq\alpha}) = [x_\alpha], x_\alpha = x_{\alpha1}x_{\alpha2}\cdots x_{\alpha n}$, and $\omega(\{q_t\}_{t\neq\alpha}, \{x_{t1}\}_{t\neq\alpha}) = x_{\alpha1}, \omega(\{q_t^{(i-1)}\} \ _{t\neq\alpha}, \{x_{ti}\}_{t\neq\alpha}) = x_{\alpha i}, i = 1, 2, \cdots n$, for $\{q_t^{(i-1)}\}_{t\neq\alpha}, i = 1, 2, \cdots n$, so that $\vee\{\mu_\alpha^*(q_\alpha, \omega(\{q_t\}_{t\neq\alpha}, \{x_{t1}\}_{t\neq\alpha})\omega(\{q_t^{(1)}\}_{t\neq\alpha}, \{x_{t2}\}_{t\neq\alpha})\cdots\omega(\{q_t^{(n-1)}\}_{t\neq\alpha}, \{x_{tn}\}_{t\neq\alpha}), p_\alpha) \wedge (\bigwedge_{t\neq\alpha} (\mu_t \ (q_t, x_{t1}, q_t^{(1)})\wedge\mu_t(q_t^{(1)}, x_{t2}, q_t^{(2)})\wedge\cdots\wedge\mu_t(q_t^{(n-1)}, x_{tn}, p_t)))|\{q_t^{(i)}\}_{t\neq\alpha} \in \Pi_{t\neq\alpha}Q_t, i \ = \ 1, 2, \cdots n \ - \ 1\} \ = \ \mu_\alpha^*(q_\alpha, x_{\alpha1}x_{\alpha2}\cdots x_{\alpha n}, p_\alpha) \wedge (\bigwedge_{t\neq\alpha} \mu_t^*(q_t, x_{t1}x_{t2}\cdots x_{tn}, p_t))$. Now ξ is single-valued, since for all $\Pi_{i=1}^n(\{x_{ti}\}_{t\neq\alpha}) = \Pi_{j=1}^k(\{y_{tj}\}_{t\neq\alpha}) \Longleftrightarrow k = n$ and $\{x_{ti}\}_{t\neq\alpha} = \{y_{ti}\}_{t\neq\alpha}, \forall t \in I, t \ \neq \ \alpha, i \ = \ 1, 2, \cdots, n$. Then $\rho(\{q_t\}_{t\in I}, [\Pi_{i=1}^n(\{x_{ti}\}_{t\neq\alpha})], \{p_t\}_{t\in I}) = \mu^{\omega*}(\{q_t\}_{t\in I}, \Pi_{i=1}^n(\{x_{ti}\}_{t\neq\alpha}), \{p_t\}_{t\in I}) \ = \ \vee\{\mu_\alpha^*(q_\alpha, \omega(\{q_t\}_{t\neq\alpha}, \{x_{t1}\}_{t\neq\alpha})\omega (\{q_t^{(1)}\}_{t\neq\alpha}, \{x_{t2}\}_{t\neq\alpha})\omega(\{q_t^{(2)}\}_{t\neq\alpha}, \{x_{t3}\}_{t\neq\alpha})\cdots\omega(\{q_t^{(n-1)}\}_{t\neq\alpha}, \{x_{tn}\}_{t\neq\alpha}), p_\alpha)$

$$\wedge (\bigwedge_{t \neq \alpha} (\mu_t(q_t, x_{t1}, q_t^{(1)}) \wedge \mu_t(q_t^{(1)}, x_{t2}, q_t^{(2)}) \wedge \cdots \mu_t(q_t^{(n-1)}, x_{tn}, p_t)))|\{q_t^{(i)}\}_{t \neq \alpha} \in$$

$$\Pi_{t \neq \alpha} Q_t, i = 1, 2; \cdots n - 1\} = \mu_\alpha^*(q_\alpha, x_{\alpha 1} x_{\alpha 2} \cdots x_{\alpha n}, p_\alpha) \wedge (\bigwedge_{t \neq \alpha} \mu_t^*(q_t, x_{t1}, x_{t2}$$

$$\cdots x_{tn}, p_t)) = \rho_\alpha(q_\alpha, f(\{q_t\}_{t \neq \alpha}), p_\alpha) \wedge (\bigwedge_{t \neq \alpha} \rho_t(q_t, [x_t], p_t)) = \hat{\mu}^\circ(\{q_t\}_{t \in I},$$

$$(f, [\Pi_{i=1}^n(\{x_{ti}\}_{t \neq \alpha})]), \{p_t\}_{t \in I}) = \hat{\mu}^\circ(\{q_t\}_{t \in I}, \xi(\Pi_{i=1}^n(\{x_{ti}\}_{t \neq \alpha})), \{p_t\}_{t \in I}).$$

4 Conclusion

A sequential machine consists of two main structures, the transition structure and the output structure. The transition structure is an internal part of the machine while the output structure is the external part. Consequently, the output structure is of more interest for practical applications than the input structure, but the output structure is dependent on the transition structure. Hence the studying of the transition structure is quite important.

By defining several especial functions of states transition of ffsms, a set of infinite products (general direct, full direct, restricted direct and others) of fuzzy finite states machines are introduced. Then we generalize the products with two and finitely many components to the products with infinitely many components[13]. In this paper, some basic algebraic properties of these products are investigated of ffsms, these results generalize the previous conclusions in literature [4,8] and other papers. Therefore,we have established a fundamental framework of infinite products of ffsms. It is necessary to study the characterizations of the transition structure of ffsms systematically.

Acknowledgements. This work is supported by the Scientific Reserch Fund of SiChuan Provincial Education Department(12ZA199).

References

1. Zadeh, L.A.: Fuzzy sets. Inform. Control 8, 338–353 (1965)
2. Zadeh, L.: Fuzzy Sets and Systems. In: Proc. Symp. System Theory, Polytechnic Institute of Broodlyn, pp. 29–37 (1965)
3. Wee, W.G.: On generalizations of adaptive algorithm and application of the fuzzy sets concept to pattern classification. Ph.D.Thesis, Purdue University (June 1967)
4. Malik, D.S., Mordeson, J.M., Sen, M.K.: Products of fuzzy finite state machines. Fuzzy Sets and Systems 92, 95–102 (1997)
5. Malik, D.S., Mordeson, J.N., Sen, M.K.: Semigroups of fuzzy finite state machines. In: Wang, P.P. (ed.) Advances in Fuzzy Theory and Technology, Bookswright, Durham, North Carolina, vol. II (1994)
6. Kandel, A., Lee, S.C.: Fuzzy switching and automata: theory and applicaions. Crane Russak (1980)

7. Holcombe, W.M.L.: Algebraic Automata Theory. Cambridge Univ. Press, Cambridge (1982)
8. Kumbhojkar, H.V., Chaudhari, S.R.: On covering of products of fuzzy finite state machines. Fuzzy Sets and Systems 125, 215–222 (2002)
9. Kim, Y.H., Kim, J.G., Cho, S.J.: Products of T-generalized state machines and T-generalized transformation semigroups. Fuzzy Sets and Systems 93, 87–97 (1998)
10. Petković, T.: Congruences and homomorphisms of fuzzy automata. Fuzzy Sets and Systems 157, 444–458 (2006)
11. Mo, Z.W., Chen, Q.: Several Properties of Two Operators of Fuzzy Finite Aatomata. Fuzzy Systems and Mathematics 21(1), 75–81 (2007)
12. Chen, Q., Zhao, C.L., Mo, Z.W.: Characterizations of bifuzzy topology based on fuzzy finite automata. Chinese Journal of Engineering Mathematics 26(1), 17–22 (2009)
13. Chen, Q., Tu, D.X., Mo, Z.W.: Covering of infinite products of fuzzy finite automata. Fuzzy Systems and Mathematics 25(2) (2011)

Comprehensive Evaluation Model for Soil Heavy Metal Pollution

Chi-yuan Ren[1] and Xin Tian[2]

[1] College of Sciences, Southwest Petroleum University, Chengdu 610500, China
renchiyuan@163.com
[2] Apply Technology College, Southwest Petroleum University,
Chengdu 610500, China

Abstract. With the city's economic development and population growth, urban environmental quality became the focus of attention, how to apply these geochemical data to carry out evaluation of urban environmental quality has become a serious problem. Based on principal component analysis and the Borda score, the issue makes a two-step comprehensive evaluation method, and applies it to distinguish soil quality of five districts in a city. In the issue, first calculate the evaluate scores by principal component analysis and rank the data in order, then calculate the Borda score of each block, so the environmental quality order of districts is obtained. Evaluation results compared with the fuzzy comprehensive evaluation method indicate that the evaluation method is more effectively and fit of actual situation.

Keywords: Heavy metal pollution, comprehensive evaluation, principal component analysis, borda number.

1 Introduction

With the rapid economic development and population increasing, the impact to urban environmental quality by human activities is more and more evident. How to apply massive amounts of data to verify information obtained urban environmental quality assessment, under the influence of human activities. The evolution of geological environment model city has increasingly become the focus of attention. Divided by function, urban life can be divided into areas, industrial areas, mountains, trunk roads and parks, green areas and other areas, were recorded as Class 1 areas, Class 2 area, ..., Class 5 area, different areas of the environment by human the impact of different levels of activity, each area to determine the differences in the degree of pollution is an important issue.

Principal component analysis is a way to factor into the same system for multi-dimensional quantitative research and theory more complete multivariate statistical analysis methods, in solving many practical problems to achieve better results. In the computer hardware and software support, the improved application of principal component analysis of soil pollution in the city's comprehensive evaluation, in the comparable quantitative and qualitative analysis of the degree of

B.-Y. Cao and X.-J. Xie (Eds.): Fuzzy Engineering and Operations Research, AISC 147, pp. 363–369.
springerlink.com   © Springer-Verlag Berlin Heidelberg 2012

integration, such as selection index weights have better results. Borda method by C. de Borda proposed in 1784, the first vote in order to solve the problem. The basic idea is that by comparing the m-bit evaluators given the object being evaluated for the n relationship between the advantages and order to finalize the evaluation of n objects are sub-Borda.

Kongde Ying, Song Zhe uses the principal component analysis method is improved in order to decrease information loss during data standardization, and to make components coefficients implications more obvious when the coefficients value is similar [1]. Yao De, Sun Mei discusses the heavy metals as Cd, Cr, Cu, Ni, Pb and Zn. Study indicates that human activities lead to the increase of concentrations of Cd, Cu, Hg, Pb and Zn in top soil. Cr and Ni are mostly of geologic origin, but they are also affected by anthropogenic activity [2]. Fan MengJia apply fuzzy set theory to a pollution assessment system, a model of index of geo accumulation based on triangular fuzzy number is established. The concentration s of contaminants in the sediment and their values in the geo chemical background were used as triangular fuzzy numbers, while the pollution degrees of heavy metals were obtained by using cu t set technology and calculating the membership grade of interval numbers [3].

2 Preliminary

2.1 The Principal Component Analysis and Borda Number

Principal component analysis is the original number of variables into a few composite indicator of a statistical analysis method, from a mathematical point of view; this is a dimension reduction process technology. Assume there are n geographic samples, each sample a total of p variables described thus constitute a class of n × p matrix of geographic data:

$$X = \begin{bmatrix} x_{11} & x_{12} & \cdots & x_{1p} \\ x_{21} & x_{22} & \cdots & x_{2p} \\ \vdots & \vdots & & \vdots \\ x_{n1} & x_{n2} & \cdots & x_{np} \end{bmatrix}. \tag{1}$$

How so many variables from the data in the internal laws to seize the geography of things do? To solve this problem, the natural p-dimensional space to be examined, that is more troublesome. To overcome this difficulty, we need to reduce dimensionality, that is, with relatively few comprehensive index to replace the original variables more indicators, but also less comprehensive index both of these as much more indicators to reflect the original reflect the information, while between them is independent of each other. So, these composite indicators (that is, new variables) how to select it? Obviously, the simplest form is to take a linear combination of the original variable indicators, appropriate adjustments combination coefficients, so that the new indicator variables between the independent and representative best.

If you remember the original variable indicators $x_1, x_2, \cdots, x_p$, their composite indicator - the new variable indicators $z_1, z_2 \cdots, z_m$. Then

$$
\begin{cases}
z_1 = l_{11}x_1 + l_{12}x_2 + \cdots + l_{1p}x_p, \\
z_2 = l_{21}x_1 + l_{22}x_2 + \cdots + l_{2p}x_p, \\
\cdots\cdots\cdots\cdots\cdots\cdots\cdots\cdots\cdots \\
z_m = l_{m1}x_1 + l_{m2}x_2 + \cdots + l_{mp}x_p.
\end{cases}
\tag{2}
$$

This new decision variable index $z_1, z_2 \cdots, z_m$ are known as indicators of the original variables $x_1, x_2, \cdots, x_p$ of the first, second, ..., m-th principal component. Which, z_1 the total variance accounted for the largest proportion, $z_2, z_3, \cdots, z_m$ of variance followed by decline. In the actual analysis of the problem, often the largest selection of the first few principal components, so that not only reduces the number of variables, but also to seize the main contradiction, simplifying the relationship between variables.

Borda method is proposed first by Jena-Charles de Borda as a vote of the classic method, which is sort-style voting system. The approach is to let the voters vote to express not only what people most want elected, should also allow voters to those qualified candidates in mind to sort.

2.2 Calculation Process and the Steps

Referring by the principal component analysis method, this issue analyzes each area's heavy metal concentration data and obtains the sort of these sample points. Then by the Borda count method, the average Borda numbers of every area are got. The main steps are listed:

Step1: For clearing the impact of units of different various, at first these data must be standardized statistically; suppose there are n samples and k pollution indexes, and set them as $P_1, P_2, \cdots, P_m$. Set P_{ij} as the value of j th pollution in i th point. Do a change :

$$
Q_j = \frac{P_j - E(P_j)}{\sqrt{Var(P_j)}} \ (j = 1, 2, 3, ..., m).
\tag{3}
$$

Standardized matrix statistically

$$q_{ij} = \frac{p_{ij} - \overline{p}_j}{t_j}.$$

Here

$$\overline{p}_j = \frac{1}{n}\sum_{i=1}^{n} p_{ij}, \quad t_j^2 = \frac{1}{n-1}\sum_{i=1}^{n}(p_{ij} - \overline{p}_j)^2.$$

Step2: Based on the standardized matrix $Q = (q_{ij})_{m \times m}$, calculate correlation coefficient matrix of m types of pollution indexes $R = (r_{ij})_{m \times m}$, where

$$r_{ij} = \frac{\sum_{k=1}^{n}(p_{ki} - \overline{p}_i)(p_{kj} - \overline{p}_j)}{\sqrt{\sum_{k=1}^{n}(p_{ki} - \overline{p}_i)^2}\sqrt{\sum_{k=1}^{n}(p_{kj} - \overline{p}_j)^2}} \quad (i = 1,...,n; j = 1,...,m). \quad (4)$$

Step3: Calculate and sort the Eigen values of the matrix $\lambda_1 \geq \lambda_2 \geq ... \geq \lambda_m$ and regularization unit eigenvectors $l_i = (l_{1i}, l_{2i}, ... l_{mi})$, then the i th principal component may be wrote as:

$$Z_i = \sum_{k=1}^{m} l_{ki} P_k. \quad (5)$$

Step4: Determine the number of principal component by a control number α. Set

$$\frac{\sum_{i=1}^{u} \lambda_i}{\sum_{i=1}^{m} \lambda_i} \geq 1 - \alpha. \quad (6)$$

Then the minima of u is the number of principal component. In the issue α is set as 5%.

Step5: Calculate composite score : At first calculate the score of the k th principal component's score of i th sample point:

$$F_{ik} = \sum_{j=1}^{m} l_{jk} P_j. \quad (7)$$

Then set the Variance as power and calculate the composite score:

$$f_i = \sum_{k=1}^{u} F_{ik} \lambda_k (i = 1...n). \quad (8)$$

Step6: By the score, the pollution evaluation of these sample point may be sorted;
Step7: By the sort, the Borda number may be calculated and the average Borda number of every area may be calculated, then the final composite evaluation scores and the sort of the areas are got.

$$W_i = \frac{\sum_{k=1}^{e} f_k}{e} \quad , \tag{9}$$

where e is the sample number of the area.

3 Results of the Application

Use the method proposed in this issue, the main results are calculated as follow:

Table 1. Correlation table

$$
\begin{bmatrix}
1.0000 & 0.1330 & 0.5445 & 0.4249 & 0.3758 & 0.5969 & 0.2180 & 0.2458 \\
0.1330 & 1.0000 & 0.3468 & 0.6605 & 0.4857 & 0.2506 & 0.7713 & 0.7746 \\
0.5445 & 0.3468 & 1.0000 & 0.5758 & 0.3946 & 0.8277 & 0.4116 & 0.5349 \\
0.4249 & 0.6605 & 0.5758 & 1.0000 & 0.4492 & 0.4949 & 0.5981 & 0.6722 \\
0.3758 & 0.4857 & 0.3946 & 0.4492 & 1.0000 & 0.3255 & 0.5616 & 0.5486 \\
0.5969 & 0.2506 & 0.8277 & 0.4949 & 0.3255 & 1.0000 & 0.2659 & 0.4418 \\
0.2180 & 0.7713 & 0.4116 & 0.5981 & 0.5616 & 0.2659 & 1.0000 & 0.8264 \\
0.2458 & 0.7746 & 0.5349 & 0.6722 & 0.5486 & 0.4418 & 0.8264 & 1.0000
\end{bmatrix}
$$

Table 2. Contribution rate table of every principal component

Principal Component	Eigen value	Contribution Rate	Sum of Contribution Rate
1th	4.4954	56.19%	56.19%
2th	1.5301	19.13%	75.32%
3th	0.6662	8.33%	83.65%
4th	0.46618	5.83%	89.47%
5th	0.32959	4.12%	93.59%
6th	0.21592	2.70%	96.29%
7th	0.1694	2.12%	98.41%
8th	0.12718	1.59%	100.00%

Under the control parameter $\alpha = 5\%$, so the first six principal components may be used to analyze the sample point data, then the comprehensive evaluation of every area may be operated.

$$Z_1 = 0.259P_1 + 0.361P_2 + 0.360P_3 + 0.390P_4 + 0.324P_5 + 0.321P_6 + 0.378P_7 + 0.409P_8$$
$$Z_2 = -0.497P_1 + 0.401P_2 - 0.387P_3 + 0.027P_4 + 0.091P_5 - 0.496P_6 + 0.359P_7 + 0.251P_8$$
$$Z_3 = -0.418P_1 + 0.147P_2 + 0.266P_3 + 0.203P_4 - 0.778P_5 + 0.243P_6 - 0.037P_7 + 0.159P_8$$
$$Z_4 = -0.559P_1 - 0.169P_2 + 0.346P_3 - 0.538P_4 + 0.366P_5 + 0.297P_6 + 0.035P_7 + 0.153P_8$$
$$Z_5 = 0.409P_1 - 0.034P_2 - 0.048P_3 - 0.611P_4 - 0.379P_5 - 0.019P_6 + 0.497P_7 + 0.251P_8$$
$$Z_6 = -0.093P_1 - 0.738P_2 + 0.239P_3 + 0.338P_4 - 0.021P_5 - 0.344P_6 + 0.375P_7 + 0.119P_8$$

By the scores, the sort of these points is calculated and then the average Borda numbers of every area are got.

Table 3. Average Borda numbers of every area

Area	Border Number	Sort
Living	27.4	3
Industry	12.7	5
Mountain	45.7	1
Road	20.5	4
Green Park	36.1	2

The result show: the pollution of mountain area is least, and this of the industry area is most serious. The sort is: Industry > Road > Living > Green Park > Mountain, which meets the common sense.

4 Conclusion

Combined the principal component analysis method and Borda number method, the comprehensive evaluation of a city is calculated and obtain meaningful result. Compared by traditional methods, the result of the method has big advanced, which verifies that the method is fit for the evaluation of urban pollution. And this issue may supply meaningful information to control pollution.

References

1. Yao, D., Sun, M.: Environment geochemistry of heavy metals in urban soils of Qingdao city. China Geology 35, 539–550 (2008)
2. Li, J.W., Yang, L.H.: The application of improved principal component analysis method to the water quality appraisement in Baiyangdian Lake. Haihe Water Resource 3, 40–43 (2007)

3. Fan, M.J., Yuan, X.Z.: Assessment model for heavy metal pollution in river sediment based on triangular fuzzy numbers. Soils 43, 216–220 (2010)
4. Xu, H.J.: Soil heavy metal pollution assessment in the fuzzy integrated decision-making. Journal of Xingjiang University(Natural Science Edition) 11, 88–92 (1994)
5. Zhang, L., Song, F.B., Wang, X.B.: Heavy metal contamination of urban soils in China: status and countermeasures. Ecology and Environmental Sciences 13, 258–260 (2004)
6. Wang, H.Q., Liu, X.H., Li, G.: Soil Environmental Science. Higher Education Press, Beijing (2007)

Prediction of China's Energy Consumption Based on Combination Model

Xiao-yu Chen and Zhi-jie Lei

Southwest Petroleum University, SWPU, Chengdu 610500, China
chenxyu@163.com

Abstract. Considering the complexity and nonlinear characteristics of China's energy consumption system, neural networks and time series are used to establish individual forecasting models for China's energy consumption system, and each of the models was tested. The results showed that the models could be used as effective tools to predict China's future energy consumption. According to standard deviation method, suited weight was distributed to the prediction of each individual model, then a combination forecasting model was established. The combination model not only gets rid of defects of the former models, but it raised the accuracy of the prediction. Then the combination model was applied to predict China's energy consumption in the next six years. By 2015, China's energy consumption will be 4.19 billion tons of standard coal.

Keywords: Prediction, neural network, time series, combination Model.

1 Introduction

Energy is the important base of the economic development and society progress. And whether the energy supply can fulfill the sustainable development of our country's economy has become the topics of the most concern all around the world. Therefore, doing forecast analysis about the future energy consumption can provide a scientific reference to energy program and policy-making. In the meantime, it also has realistic effect on keeping national economy growing sustaining, steady and healthy. Many researchers and organizations have carried out broad scale research on energy demand, and they have come up with a great many ways of energy consumption predictions. Currently, the first one is the scenario analysis which means the energy consumption tendency is based on the assumption of the future economy development. Another one is the Trend Equation Method, such as Department analytic method, energy consumption elasticity coefficient method and Energy intensity method. The Trend Equation Method is on the basis of analysis about a long term energy requirement rule in the past. Since our country's energy consumption is closely interrelated with economical development, predictions of the future energy consumption are mainly in terms of relationship between energy

B.-Y. Cao and X.-J. Xie (Eds.): Fuzzy Engineering and Operations Research, AISC 147, pp. 371–378.
springerlink.com   © Springer-Verlag Berlin Heidelberg 2012

consumption and economical development. In the first place, making use of neural network and time series to establish monomial prediction model of our country's energy consumption system, then using standard balance method to conduct weight distribution of each model prediction in order to set up the combination forecast model of our country's future energy consumption.

2 Establishment of the Forecasting Model

2.1 Neural Network Forecasting Model

2.1.1 Brief Introduction to Model

The structure of BP network is showed as Fig. 1. BP network is a kind of neural network which consists of 3 or more than 3 levels neuron including input layer, interlayer (hidden layer) and output layer. Adjacent layers are fully connected, but neuron cells in every layer are not connected. After a pair of learning samples is provided to the network, activation numbers of neuron propagate from input layer and then various interlayer, finally reach output layer. As the result, every neuron cell of the output layer get BP network's input response. To reduce the error between target output and actual output, activation numbers of neuron propagate in the opposite way in order to correct weights of connection. This arithmetic is called "Error back propagation algorithm" also the BP method. The accuracy rate of response grows constantly because that the opposite propagation of error keeps revising.

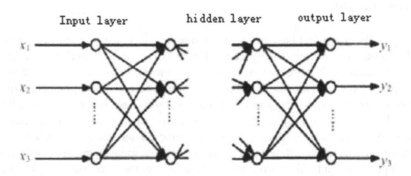

Fig. 1. Structure of BP neural network

BP network has an important principle that a continuous function in any closed interval can be approached with BP network of single hidden layer, thus a three-level network can accomplish mapping function from n to m dimensional space. Three-level BP network is widely used.

2.1.2 Modeling

Some rules to obey in selection of influencing factors in the process of modeling.

① The factor must be closely related to energy consumption relationship.

② The factor changes obviously, stationary and continuously, so that it won't produce big deviation.

GDP (on the basis of price in 1978), economy structure, population, and urbanization rate accord with rules above. The modeling expression is as follows:

$$Y = f(x_1, x_2, x_3, x_4) \tag{1}$$

In expression (1): Y gross of energy consumption, f neural network model, x_1: GDP, x_2: economy structure, x_3: population, x_4: urbanization rate.

2.1.3 Test on Results of Predictions

Use data between 1990 to 2003 to train the neural network, and take advantage of data between 2004 to 2009 to examine forecast accuracy. MATLAB set up the neural network model, and there are 36 neural cells in the hidden layer. All input or output variable should be normalized before being processed by network. The processed numbers are between 0 and 1. The results are shown in Picture 1.

Table 1. Chart 1 Testing results of neural network model/108 tons of standard coal

Year	True value	Predicted value	Relative error/%
2004	20.323	20.054	-1.324
2005	22.468	22.287	-0.806
2006	24.627	24.782	0.629
2007	26.558	26.896	1.273
2008	28.5	29.014	1.804
2009	31	32.658	5.348

Fitting average relative error between 1990 and 2003 is 2.17%, and average forecast error between 2004 and 2009 is 1.86%. Fitting accuracy and forecast accuracy of neural network model are both pretty high.

2.1.4 Sensitivity Test on Input Variable

The predictions need to estimate relevant indexes about society development, so it is necessary to test the sensitivity of every variable. The way to checkout a model is suitable or not is to discover the output changes accompanied by big changes of some variables. Take 2009 for example, the results are displayed in chart 2.

Table 2. Test results

Inspection item	Input values of inspection variable	Deviation between input values and true values	Deviation between model output results and primary results
GDP	71299.6	10%	6.94
	58336.0	-10%	-6.94
Economy structure	b1 = 10. 6, b2 = 49. 8,	∇ b1 = 0, ∇ b2 = 3,	7.9
	b3 = 39. 6	∇ b3 = - 3	
	b1 = 10. 6, b2 = 43. 8,	∇ b1 = 0, ∇ b2 = - 3,	-7.9
	b3 = 45. 6	∇ b3 = 3	
Population	136 474	3 ×103	5. 45
	136 474	- 3 ×103	- 5. 45
Urbanization rate	48. 6	2	5. 84
	44. 6	- 2	- 5. 84

Note: b_1, b_2, b_3 stands for the ratio of primary industry, secondary industry, tertiary industry, respectively.

2.2 Time Series Models

2.2.1 Brief Introduction

Utilize the time series models to imitate change trend of our country's future energy-consuming intensity. The expressions are below

$$Xt = f (t) + S + T + R \tag{2}$$

In expression (2): Xt -time series, f (t) -the long-term trend of certainty, S- season, T-period   random fluctuating factor

R comply with ARMA, the general expression of ARMA is below:

$$\Phi (B) ut =\Theta (B) at$$
$$\Phi (B) = 1 - \varphi1B - \varphi2B2 - \cdots - \varphi mBm ;$$
$$\Theta (B) = 1 - \varphi1B - \varphi2B2 - \cdots - \varphi nBn. \tag{3}$$

In expression (3): at white noise sequence, B backward shift operator ,times of B stands for retrusive number of periods.

2.2.2 Establishment and Test of Model

2.2.2.1 Deterministic Part

Approximately, per 10000 yuan of GDP (GDP is calculated on the basis of price in 1978) energy consumption of our country goes down exponentially since 1978, therefore it is necessary to start with logarithmic transformation. Suppose that the transformational energy consumption sequence is X_t, then choose X_t as dependent variable and year as independent variable, finally, carry out linear regression on X_t. Equation of linear regression is as [5, 6]:

$$X_t = 86.\ 816 - 0.\ 043\ t + u_t;$$

$$R2 = 0.\ 955D,\ W = 0.\ 174. \tag{4}$$

Goodness of fit of Equation is very high, and it can explain fluctuation of the majority of energy-consuming intensity.

2.2.2.2 Random Part

Suppose that residual sequence u_t is independent from each other, it means that u_t is white noise, the expression above is just a normal a yuan of linear regression model. However, D and W show that residual sequences u_t has close connections, and the residual sequence u_t is predictable. In the meantime, residual sequence u_t is stationary; hence, ARMA model based on can be built.

Observe pictures of self-correlation function and partial-correlation function, then compare optional models and build the certain AMRA model. (1, 1) Results are as follows:

$$u_t = 0.874u_{t-1} + 0.704a_{t-1} + a_t \tag{5}$$

$$R2 = 0.\ 891$$

Goodness of fit is quite rather high, and every parameter of model past the significance test. The auto carrel gram of residual sequence at proves that sample self-correlation function of residual and partial-correlation function are in the range of Bartlett's band width. Chart 3 shows that residual sequence is white noise sequence.

Table 3. Statistics test table

Delayed order	LB statistical energy	Value of P
6	5.953	0.428
12	13.558	0.330

2.2.3 Combines Model

Combine the certain part with the random part:

$$Xt = 81.196 - 0.04\ t + 0.884ut - 1 + 0.727at - 1 + at;$$
$$R2 = 0.996D, W = 1.570\ 6. \hspace{4cm} (6)$$

Goodness of fit is very high, and fitting average relative error is just 2.11%.

Every parameter of model has passed the significance test. On the condition that, a =0.01, check the examining critical value list, du = 1.42, du <D. W < 4 - du. It is defined that residual sequence tends to white noise, so the built model is appropriate.

2.2.4 Forecasting Results of Model

Table 4. Energy intensity prediction results of time sequence model

Year	Predicted value(SCE/ ten thousand)	Predicted value 95% confidence interval	
		UCL	LCL
2010	4.625	4.895	4.366
2011	4.400	4.897	3.942
2012	4.190	4.788	3.651
2013	3.994	4.644	3.416
2014	3.811	4.485	3.217
2015	3.639	4.431	3.042

3 Social Development Exceptions in the Coming 6 Years

According to our country's current situation and development target, assume that in the coming 6 years, the average rate of increase of GDP per year is 9%, the average rate of increase of primary industry's proportion reduces 0.3% per year and that of tertiary industry increases 0.8%, the total population raises 5‰ per year, urbanization increases 0.9% for one year.

4 Our Country's Energy Consumption Predictions in the Coming 6 Years

4.1 Forecast Results in Two Methods

Utilize data from 1990 to 2009 to train neural network (6 neural cells in hidden layer, fitting average relative error is 2.14%). Forecast results of our country's energy consumption in the future 6 years and time model are displayed in chart 5.

Table 5. Chart 5 forecasting results of total energy consumed

Year	Energy consumption	
	Neural network method	Time sequence method
2010	32.090	32.676
2011	34.090	33.884
2012	36.325	35.171
2013	38.819	36.543
2014	41.594	38.007
2015	44.675	39.558

It is clearly in chart 5, the forecast results of neural network model is higher than that of time sequence model, meanwhile; the difference between the two results becomes bigger as forecasting time grows longer. It is certain that the energy consumption per GDP reduces. Neural network model only reflects economy structure's influence on energy consumption per GDP reduction, the progress of science and enhancement of people's energy saving consciousness which have opposite impact have not been considered yet. Forecasting results of neural network model are a bit higher than actual results. Time sequence imitates energy consumption per GDP's trend to reduce, according to diminishing marginal effect rule; the reduction of energy consumption per GDP becomes slower. Our country's energy consumption per GDP reduces also more slowly, so the forecast results of time sequence are lower than actual results. Our country's future volume of energy consumption should be between the two forecast results.

4.2 All Predictions of Corresponding Weights

To make full use of single model's valid information, and conquer drawbacks, decrease randomness and increase forecasting precision, it is better to build combined model and use standard balance method to conform all corresponding weights.

Suppose the standard deviation of neural network model's and time sequence's forecast errors are σ_1 and σ_2 respectively, and $\sigma = \sum_{i=1}^{n} \sigma_i$, (n is the model number), metric $w_i = \dfrac{\sigma - \sigma_i}{\sigma(m-1)}$. Calculate each single model's weight vector w=

(0.464, 0.536) in order to get combination forecast model:

$$y= 0.464y_1 + 0.536y_2. \tag{7}$$

In Equation (7), y_1 means forecast results of neural network model; y_2 means forecast results of time sequence model.

4.3 Forecasting Results of Our Country's Energy Consumption from 2010 to 2015

In summary, all the composite models' predictions are showed in chart 6.

Table 6. Forecasting results/ 108 tons of standard coal

Year	2010	2011	2012
Energy consumption	32.404	33.980	35.706
Year	2013	2014	2015
Energy consumption	37.599	39.671	41.932

5 Conclusion

It is estimated that our country's energy consumption will exceed 4 billion tons of standard coal in 2015. The security situation of energy will be very severe, thus we should make efforts on three aspects:

Accelerate the research and spread of energy-efficient technology, and enhance people's consciousness of energy saving. Reduce energy consumption as much as possible.

Take all kinds of measures to ensure our country's energy supply, especially petroleum's supply, try to discover new supply channels. Exploit clear energy and alternate energy source to serve social and economical development.

References

1. The national statistics bureau of the PRC China statistical yearbook. China Statistics Press, Beijing (2008)
2. The national statistics bureau of the PRC National economy and social development statistics bulletin (2009)
3. Wang, Z.L., Hu, Y.H.: Applied time series analysis. Science Press, Beijing (2007)
4. Yu, C.H.: Pass and statistics analysis. Electronic Industry Press, Beijing (2007)
5. Liu, L.F., Yi, X.J.: Estimation of china energy demand and predict stimulate. Shanghai University of Finance and Economics Journal 10(4), 84–91 (2008)

Application Research on Greed Strategy-Based GA in Optimizing Curriculum Schedule Model in Universities

Yu-bin Zhong[1], Yan-qiang Li[1], and Yuan Duan[2]

[1] School of Mathematics and Information Sciences,
Guangzhou University, Guangzhou, China
zhong_yb@163.com
[2] Guangdong University of Science & Technology Guangdong, Dongguan 523083

Abstract. In this paper, we first analyze the relationship between curricula, teachers, classes, time and classrooms, which is a graph. Then on the basis of constraint conditions in Scheduling Curriculum practically in universities, we presents a Curriculum Schedule optimization model, in which the greed strategy is used to optimize Genetic Algorithm (GA), and a new GA encoding scheme is employed to design fitness function and punishment function for curriculum schedule problems. This model effectively improved the running performance, which provides a better implementation approach for improvements of the existing curriculum schedule systems. The experimental results show that fitness values of the improved GA algorithm are of obvious evolutional tendency, the chromosome encoding scheme and the fitness function can meet curriculum schedule requirements preferably, and the more adequate computation resources, the greater possibilities of no restoration for the obtained optimal individual.

Keywords: Curriculum schedule in university, greed strategy, genetic algorithm, optimization model, application research.

1 Introduction

In the last decade, researches on curriculum schedule problem have been very active both at home and abroad [1,2]. More and more algorithms are proposed to solve this problem, the main of which are graph coloring technique [3], Tabu Search (TS)[4], Simulated Annealing (SA)[5-6], Genetic Algorithm[7-9], etc. In this paper, by analyzing the relationship between curricula, teachers, classes, time and classrooms, we establish a relation model. Then model curriculum schedule problem in detail, and we propose a scheme that uses greed strategy to optimize GA, which is applied in solving curriculum schedule model.

B.-Y. Cao and X.-J. Xie (Eds.): Fuzzy Engineering and Operations Research, AISC 147, pp. 379–389.
springerlink.com

2 Constraints of Curriculum Schedule

In curriculum schedule problem, the main problem to be solved is the arrangement of curricula, teachers, classes, time and classrooms. In scheduling curricula, some constraints need to be met. And these constraints can fall into the following 2 categories: soft constraint conditions and hard constraint conditions [10-13].

2.1 Hard Constraint Conditions

Hard constraint conditions are those that must be met in scheduling curricula, otherwise, teaching jobs will not carry out normally and smoothly.

(1) The same teacher can't be scheduled two or more than two different curricula at the same time.

(2) The same classroom can't be scheduled two or more than two different curricula at the same time.

(3) The same class can't have two or more than two different curricula at the same time.

(4) Classrooms must be big enough, so that they can hold all of the students attending this class.

(5) Some curricula should be scheduled at specific time period or specific classroom.

2.2 Soft Constraint Conditions

Soft constraint conditions are those that may be met during scheduling curricula, which are criterions measuring curriculum schedule schemes.

(1) Curricula that have more class hour a week should be staggered reasonably.

(2) The distance between two classrooms for two continues curricula of one class should be as close as possible.

(3) The distance between two classrooms for two continues curricula of one teacher should be as close as possible.

(4) Curricula of a class in one week should distribute uniformly.

(5) Curricula of a teacher in one week should concentrate, the less the better.

(6) Each curriculum should be scheduled at the right time period.

(7) Meet teachers' lecture schedule requirements to the greatest extent.

(8) The number of a class should be as close as possible to that of the classroom can hold.

(9) The classroom meets demands of the curriculum.

3 Optimization Model for Curriculum Schedule

3.1 Basic Model

In curriculum schedule problem, the main problem to be solved is the arrangement of curricula, teachers, classes, time and classrooms. Treat curricula, teachers, classes, time and classrooms as vertices separately, join them with edges. We can establish the following relation graph (shown in Fig. 1) considering the fact that there are no strict requires between classes and teachers, classes and time, teachers and classrooms, classrooms and time.

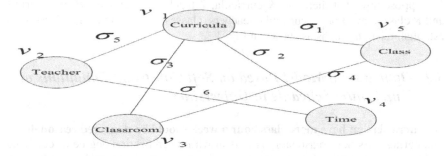

Fig. 1. The relation graph between curricula, teachers, classes, time and classrooms

This is an edge-sparse graph. v_i $(i=1,2,3,4,5)$ denote nodes of curricula, teachers, classes, time and classrooms, separately. σ_i $(i=1,\cdots,6)$ are weights of existing edges in Fig. 1,where

σ_1 indicates whether this class has this curriculum or not. If has, then $\sigma_1=1$, otherwise, $\sigma_1=0$.

σ_2 shows the effect of this curriculum scheduled at a certain time period. If $\sigma_2=0$, then it means this curriculum can't scheduled at this time period.

σ_3 indicates whether this classroom can meet requirements of a certain curriculum. σ_3 is a product of the following two values:

    (1) Whether equipments in this classroom can be met or not(denoted by e). If met, then e=1, otherwise, e=0.
    (2) The similarity degree between the type of a classroom and that of a curriculum is denoted by t. If a curriculum can't be scheduled in this classroom, then t=0.

Thus , σ_3 =e·t.

σ_4 shows the relation between the number of a class between that of a classroom can hold, given by the following formula.

$$\sigma_4 = \begin{cases} \dfrac{p}{q}, & p \leq q , \\ 0, & p > q \end{cases} \qquad (1)$$

where p is the number of a class, q is the number of a classroom can hold.

σ_5 indicates whether or not this teacher can take charge of a certain curriculum. If it can, then $\sigma_5 = 1$, otherwise, $\sigma_5 = 0$.

σ_6 expresses a teacher's favor to class hour. If $\sigma_6 = 0$, then it means this teacher can't teach at this time period.

Suppose now that there are K curricula, T teachers, C classes, P time periods and R classrooms. For all curricula, teachers, classes, time periods and classrooms, establish relation graphs as shown in Fig. 1.

3.2 Optimization Model Based on Soft Constraint Conditions of Curriculum Schedule in Universities

(1) Curricula that have more class hour a week should be staggered reasonably

Staggering class hour reasonably is to deal with discretization degree of curricula, which is a quantification criterion to assess a class' curricula combination schemes. If a class has a curriculum twice a week, then which two days are better for this curriculum? This is a problem that we must face during scheduling curricula. If twice a week (once per day),then the combinatorial number for two days in five days is : $C_5^2 = 10$; if three times a week, then the number becomes $C_5^3 = 10$. We can use discretization degree to measure each combination. For example, the discretization degree of the combination of Monday and Wednesday is 0.8, of Tuesday and Friday is 1.0, of Wednesday and Thursday is 0, of Monday, Tuesday and Wednesday is 0, and of Monday, Wednesday and Friday is 1.0, etc. For convenience, we assume the discretization degree is always 1.0 in the case of once a week.

We use the mean value of discretization degrees for all curricula to measure curriculum combination scheme of each class.

$$f_1' = \frac{\sum_{i=1}^{e_{l1}} s_{li}^{(1)} + \sum_{i=1}^{e_{l2}} s_{li}^{(2)} + \sum_{i=1}^{e_{l3}} s_{li}^{(3)}}{N_l}. \tag{2}$$

For f_1', the greater the better. So, the optimization model for this constraint is as follows.

$$Min \quad f_1 = 1 - \frac{1}{C}\sum_{l=1}^{C} f_1', \tag{3}$$

where $s_{li}^{(1)}, s_{li}^{(2)}$ and $s_{li}^{(3)}$ denote corresponding discretization degrees for class l in cases of once, twice and three times a week separately. N_l is the total number of curricula for class 1. e_{l1}, e_{l2} and e_{l3} separately denote total curricula number in cases of once, twice and three times a week for class 1, and $N_l = e_{l1} + e_{l2} + e_{l3}$.

(2) The distance between two classrooms for two continues curricula of one class should be as close as possible
Measuring distances between classrooms one by one is relatively troublesome, so here we employ estimate method.

Distance between classrooms in a same building can be regarded as 0, while this distance in different buildings can be got by estimating distance between two buildings. Divide the estimated longest distance equally according to the distribution of all estimated distances, then gradate them with level 1,2,3,etc from the near to the distant .Smaller intervals may needed so as to use levels to obviously represent the length of distances. Thus, levels between classrooms are 0, 1,2,3, etc. The greater the level is, the longer the distance becomes.

Two continues curricula are those scheduled in the same forenoon (or afternoon). For discontinues curricula in the same forenoon (or afternoon) , the distance level between classrooms can be set to 0. The mean distance level is used to measure status of changing classrooms for each class. The less the distance is, the better the curriculum schedule scheme is.

$$f_2' = \frac{1}{2d} \sum_{i=1}^{2d} \frac{s_{li}}{s_{max}} . \tag{4}$$

Therefore, the optimization model under this constraint is given by

$$Min \quad f_2 = \frac{1}{C} \sum_{l=1}^{C} f_2' \quad , \tag{5}$$

where s_{max} is the maximum distance level, s_{li} denotes distance level between classrooms in the same forenoon (or afternoon) for class 1. d is the number of working days.

(3) The distance between two classrooms for two continues curricula of one teacher should be as close as possible.
Similar to (2), we can get the following optimization model given by

$$Min \quad f_3 = \frac{1}{T} \sum_{h=1}^{T} \frac{\sum_{i=1}^{2d} s_{hi}}{2d \bullet s_{max}}, \tag{6}$$

where s_{hi} denotes distance level between classrooms in the same forenoon (or afternoon) for teacher h, s_{max} is the maximum distance level, and d is the number of working days.

(4) Curricula of a class in one week should distribute uniformly to the greatest extent

In a timetable, it is unscientific if one class has curricula through the day or has no curricula at all. In these cases, students either are tired to accepted lectures or have nothing to do, which has no good for students at all. Based on the analysis above, we should schedule students' curricula for one week as uniformly as possible, thus cases mentioned at the beginning can be avoided. Hence, we use the following formula to assess uniformity of curriculum schedule for one class.

$$f_4' = \sum_{k=1}^{d} \left(\frac{n_{lk} - \frac{1}{d}\sum_{j=1}^{d} n_{lj}}{\max\left\{ n_{l,\max} - \frac{1}{d}\sum_{j=1}^{d} n_{lj}, \frac{1}{d}\sum_{j=1}^{d} n_{lj} \right\}} \right)^2 . \tag{7}$$

For f_4' ,the less the better. The optimization model for curriculum schedule scheme under this constraint is as follows.

$$Min \quad f_4 = \frac{1}{C}\sum_{l=1}^{C} f_4' , \tag{8}$$

where n_{lk} is the number of curricula that class l has in the kth working day. $n_{l,\max}$ is the maximum number of curricula that class l has in all working days, d denotes the total number of working days within one week.

The less value of f_4 means more uniformity of class hours scheduled for a class per day, which avoids the case where curricula are concentrated in some continues days. It extremely improves the use rate of classrooms and eases relatively tense situation with the limited number of classrooms.

(5) Curricula of a teacher in one week should concentrate, the less the better

Since houses of teachers are usually far from universities, a teacher's curricula should be concentrated, so that it will save time in transportation. The mean curriculum concentration degree of all teachers is given by

$$f_5' = \frac{1}{T}\sum_{h=1}^{T} \frac{m_h}{d} , \tag{9}$$

where m_h is the number of working days for teacher h, d is the total number of working days within one week. The less value of f_5' means that teachers' curricula are more concentrated.

The fewer curricula for each teacher are the better. However, in general, curricula for a teacher would be no less than half of all curricula of which he/she can take charge. The following formula is used to measure this

$$f_5'' = \frac{1}{T} \sum_{h=1}^{T} \frac{c_h}{a_h} , \tag{10}$$

where c_h is the number of curricula for teacher h, a_h is the number of all curricula of which teacher h can take charge of. The less value of f_5'' means that curricula for each teacher tend to optimization.

By the above two formulas, we can establish the optimization model under this constraint as below.

$$M\,in \quad f_5 = \frac{f_5' + f_5''}{2} . \tag{11}$$

(6) Each curriculum should be scheduled at preferable time slice
In Fig. 2, if a certain curriculum is scheduled at some time slice, then it means that the edge between this curriculum and that time slice is selected. The greater the weight of edges is the better. So the optimization model for curriculum schedule under this constraint is given by

$$M\,in \quad f_6 = 1 - \frac{\displaystyle\sum_{i=1}^{K} \sum_{j=1}^{P} \left(w_2^{(ij)} \bullet \sigma_2^{(ij)} \right)}{\displaystyle\sum_{i=1}^{K} \sum_{j=1}^{P} w_2^{(ij)}} , \tag{12}$$

where $w_2^{(ij)}$ indicates whether or not curriculum i is scheduled in time slice j (if so, then $w_2^{(ij)} = 1$, otherwise, $w_2^{(ij)} = 0$). $\sigma_2^{(ij)}$ shows the effect on which curriculum i has in time slice j.

(7) Meet teachers' lecture schedule requirements to the greatest extent
Similar to (6), we get the following optimization model.

$$M\,in \quad f_7 = 1 - \frac{\displaystyle\sum_{i=1}^{T} \sum_{j=1}^{P} \left(w_6^{(ij)} \bullet \sigma_6^{(ij)} \right)}{\displaystyle\sum_{i=1}^{T} \sum_{j=1}^{P} w_6^{(ij)}} , \tag{13}$$

where $w_6^{(ij)}$ indicates whether or not teacher i is scheduled in time slice j (if so ,then $w_6^{(ij)} = 1$, otherwise, $w_6^{(ij)} = 0$). $\sigma_6^{(ij)}$ denotes favor of teacher i for time slice j.

(8) The number of a class should be as close as possible to that of the class-room can hold

Similar to (6), the optimization model for curriculum schedule under this constraint is given by

$$
Min \ = 1 - \frac{\sum_{i=1}^{C} \sum_{j=1}^{R} \left(w_4^{(ij)} \bullet \sigma_4^{(ij)} \right)}{\sum_{i=1}^{C} \sum_{j=1}^{R} w_4^{(ij)}} \ , \tag{14}
$$

where $w_4^{(ij)}$ indicates whether or not class i is scheduled in classroom j (if so ,then $w_4^{(ij)}$ =1,otherwise, $w_4^{(ij)}$ =0). $\sigma_4^{(ij)}$ denotes the relation between the number of class i and that of class j can hold.

(9) The classroom meets demands of the curriculum

We have the following model

$$
Min \ \ f_9 = 1 - \frac{\sum_{i=1}^{K} \sum_{j=1}^{R} \left(w_3^{(ij)} \bullet \sigma_3^{(ij)} \right)}{\sum_{i=1}^{K} \sum_{j=1}^{R} w_3^{(ij)}} \ , \tag{15}
$$

where $w_3^{(ij)}$ indicates whether or not curriculum i is scheduled in classroom j (if so, then $w_3^{(ij)}$ =1, otherwise, $w_3^{(ij)}$ =0). $\sigma_3^{(ij)}$ denotes the relation between curriculum i and classroom j.

To sum up, the optimization model for curriculum schedule problem under all soft constraint conditions is given by

$$
Min \ \ f = W_1 f_1 + W_2 f_2 + W_3 f_3 + W_4 f_4 + W_5 f_5 + W_6 f_6 + W_7 f_7 + W_8 f_8 + W_9 f_9, \tag{16}
$$

where W_i is weight which is on interval [0,1], for all i=1,2,3,...,9, and

$\sum_{i=1}^{9} W_i = 1$. Values of weight W_i can be selected according to the importance of each soft constraint condition.

4 Design of GA

GA mainly deals with hard constraint conditions that can be treated as conflicts, and also trains soft constraint conditions. For hard constraint conditions (4) and (5), conflicts can be avoided by properly handling initialization and mutation of population. As for hard constraint conditions (1), (2) and (5), they would be dealt with by using punishment functions when training, and finally by restoring the obtained optimal individual to avoid conflicts.

● **Encoding scheme**

Integer encoding scheme is employed in this paper to deal with curriculum schedule problem, and certain transform is made while encoding. For convenience in calculating, first, successively number classes, curricula, teachers, time slices and classrooms starting from 1, respectively.

● **Initializing population**

The following random method is employed when initializing population so as to make sure that the scheduled teachers, time slices and classrooms meet corresponding requirements. Steps for generating individual chromosome are as follows.

(1) Select the first gene position.

(2) Randomly select a teacher from those who can take charge of this curriculum according to the combination of classes and curricula to which the current gene position corresponds. Record the selected teacher's serial number.

(3)Randomly select a time slice that is available to both the current curriculum and the current teacher according to the combination of classes and curricula to which the current gene position corresponds , and the selected teacher. Record the selected time slice's serial number.

(4) Randomly select an available classroom for current curriculum according to the combination of classes and curricula to which the current gene position corresponds. Record the serial number of the selected classroom.

(5) Calculate serial number of combination of the teacher, the time slice and the classroom by using serial numbers of the selected teacher, time slice and classroom. The serial number is set to be the value in corresponding gene position.

(6)Select the next gene position, and go back to step (2) till all gene positions are scheduled well.

● **Design of punishment function**

In this design of GA, after decoding chromosome, hard constraint conditions (1),(2) and (3) are transformed into that the same teacher can't appear twice or more than twice in the chromosome at the same time, and that the same classroom (and class) can't appear twice or more than twice at the same time. The same teacher and the same time slice, and the same classroom and the same time slice may appear twice or more than twice in a chromosome in the case of a combined class, which is not regarded as a conflict. Conditions for combining classes are that teachers, time slices, classrooms and curricula are the same, and that the classroom can hold all students in the combinational class.

Suppose that conflicts number with respect to soft constraint conditions (1),(2) and (3) are l_1, l_2, and l_3, respectively. Let M be the length of chromosomes, then

$$g = \frac{l_1 + l_2 + l_3}{M}. \tag{17}$$

- **Design of fitness function**

The fitness function is designed according to the optimization model for curriculum schedule problem. By optimization model (16) and punishment function (2), we have the following fitness function.

$$Fitness(x) = \frac{f + g}{2} \cdot \tag{18}$$

- **Selection operator**

The selection operator is a method which combines elitism strategy and equal random.

- **Crossover operator**

Use straggled crossover operator: generate a random binary vector whose length equals to that of chromosomes. If the value of a position is 1, then select corresponding gene in this position from the first parent, otherwise, select gene from the second parent. In this way, we can form a child chromosome.

- **Mutation operator**

The mutation operator is performed under certain probabilities. If a certain gene position will mutate, the mutation process is similar to initialization steps (2),(3),(4) and (5) as mentioned in subsection 4.2. Only by doing so, can hand constraint conditions (4) and (5) be met.

5 Conclusion

Experiment parameters in this paper are set as follows. Population size is 50; maximum iteration is 200; crossover probability is 0.8 and mutation probability 0.01. Elite number is set to be 2. Weights $W = \{0.12 , 0.12 , 0.12 , 0.12, 0.12, 0.1, 0.1, 0.1, 0.1\}$ in optimization model (16). Experiment data comes from the data of the second semester from year 2009 to 2010 of Guangzhou University. The result shows that fitness values of the improved GA algorithm are of obvious evolutional tendency, the chromosome encoding scheme and the fitness function can meet curriculum schedule requirements preferably, and the more adequate computation resources, the greater possibilities of no restoration for the obtained optimal individual. Those results are fit for actual situations.

Acknowledgements. Thanks to the support by keynote Teaching Research Project of Guangzhou University.

References

1. Zhang, W.X., Liang, Y.: The mathematical basis of the genetic algorithm. Xi'an Jiaotong University Press (2000)
2. Radcliffe, N.J.: The Alegbra Of Genetic Algorithms. Annals of Math., & AI 10, 339–384 (1994)
3. Hu, S.R., Deng, Y., Wang, Z.: Graph Theory Research Based on College Time-Table System. Computer Engineering and Applications 10, 221–223 (2002)

4. Wu, J.R.: A Branch and Bound Algorithm for Time Table Problem. Operations Research and Management Science (2002)
5. Liu, J.Q., Chen, C.B.: Study on the Course-arranging Problem on the Base of Applying Simulated Annealing Algorithm. Journal of Wuhan Institute of Shipbuilding Technology 6, 45–46 (2003)
6. Li, Z.Z., Wang, Y.L., Chen, J.: Hybrid Simulated Annealing Algorithm for School Timetabling Problem. Journal of Xi'an Jiaotong University 37, 18–21 (2003)
7. Zhang, C.M., Xing, F.: Adaptive Genetic Algorithms for Solving University Timetable Problem. Acta Scientiarum Naturalium Universitatis Neimongol. 33, 15–16 (2002)
8. Wang, Y.: The Application of Parallel Genetic Algorithm in University Timetable System. Science and Technology Innovation Herald 7, 201 (2009)
9. Xu, K.S., Zhang, S.F.: Designing of Automatic Curriculum Scheduling Based on Genetic Algorithm. Network & Computer Security 7, 9–12 (2007)
10. Zhao, H., Qin, W.J.: A course dispatching method based on resources matching. Journal of Shenyang Polytechnic University 23, 112–119 (2001)
11. Yu, B.: The Approximate Solution to Problem of Schedule-Making. Journal of Yangzhou Polytechnie College 5, 30–34 (2001)
12. Sun, J.P., Mei, X.Y., Xiao, Z.H., Shi, Z.Z.: The Application of Association Rules in Intelligent University Timetable System. Computer Applications 22, 37–39 (2002)
13. Zheng, Y.F., Huang, D.C., Liu, D.Y.: Research on the application of genetic algorithm in solving TTP. Journal of Zhejiang University of Technology 34, 162–165 (2006)

Prediction of Coagulant Dosage in the Sewage Treatment Based on Fuzzy C-Means Clustering

Ya-xun Yang[1], Ci-yuan Xiao[1], Ming Ni[1], and Jing Chen[2]

[1] College of Science, Southwest Petroleum University, Chengdu 610500, China
[2] College of Science, Guizhou University, Guiyang 550025, China
yangyaxun2010@163.com

Abstract. This paper deals with the problem of prediction in the coagulation process of the sewage treatment where data sets are multivariable and multiple data. We propose a new predictor based on fuzzy c-means clustering algorithm. First, we will use fuzzy c-means clustering algorithm to cluster large sample. To reduce the size of sample, we replace all the sample data with cluster centre, then use these cluster centre as new sample set. We present three predictors-two arising from the multiple regression and one stemming from support vector regression(SVR).At last, properties of these predictors are discussed when they are used to predict values of response variable.

Keywords: Fuzzy c-means clusting, cluster centre, prediction.

1 Introduction

In industrial sewage and life sewage treatment, there is a kind of very important physico-chemical treatment method: coagulation method. Because this has high removal rate of pollution index, this water treatment method is used widely. In the process of coagulation method, we need to add appropriate coagulant into the water so that the colloidal material can separate from water and then form precipitation. Because the factors that influence flocculation effect is very complicated, we choose three of the most important factors: PH value, water temperature and turbidity. These three factors together influence coagulant dosage.

Now we have some data in the coagulation process of the sewage treatment. These 905 data consist of four-dimensional vector, the first three vectors are independent variables. They are PH value (x_1), water temperature (x_2), turbidity (x_3) and the last vector is response variable that denotes coagulant dosage (y).Everyday these data were recorded from 2008 to 2010.Let us assume that 723 data of 2008-2009 as the training set build a model, then regard the data of first half of 2010 as a test set and evaluate the property of model. Finally the model is applied to forecast coagulant dosage in future.

B.-Y. Cao and X.-J. Xie (Eds.): Fuzzy Engineering and Operations Research, AISC 147, pp. 391–398.
springerlink.com   © Springer-Verlag Berlin Heidelberg 2012

2 Fuzzy C-Means Clustering Algorithm

The Fuzzy C-means clustering(FCM) algorithm proposed by Bezdek in 1981 is one of most frequently used methods in pattern recognition. The FCM algorithm attempts to partition data set that have no clear boundaries between data. In this paper we have 905 data of the coagulation process. The difference of each dimension of the 905 data are not obvious. At the same time, the large sample training set of 723 data cause great difficulty in modeling. Based on the above two points, we choose to classify the large sample by using fuzzy c-means clustering algorithm before prediction. FCM is based on minimization of the objective function (1) to achieve good classifications.

$$J(U,V) = \sum_{k=1}^{n} \sum_{i=1}^{c} u_{ik}{}^{m} \left\| x_k - v_i \right\|^2 \tag{1}$$

Consider clustering the data set

$$X = \begin{pmatrix} x_1 \\ x_2 \\ \cdots \\ x_{723} \end{pmatrix} = \begin{pmatrix} x_{11} & x_{12} & x_{13} & y_1 \\ x_{21} & x_{22} & x_{23} & y_2 \\ \cdots & \cdots & \cdots & \cdots \\ x_{723\times1} & x_{723\times2} & x_{723\times3} & y_{723} \end{pmatrix},$$

X consist of four variables of 723 data, where n is the number of data points, $n = 723$. $V = \{v_1, v_2, \cdots, v_c\}$ is the set of corresponding cluster centre in data set X, where c is the number of clusters. In this paper, we assume that $c = 20$. u_{ik} is the subjection degree of data x_k to the cluster centre v_i. Meanwhile, u_{ik} has to satisfy the following conditions:

$$0 \le u_{ik} \le 1, \tag{2}$$

$$\sum_{i=1}^{20} u_{ik} = 1, \tag{3}$$

$U = (u_{ik})_{20\times723}$ is a fuzzy partition matrix. $\left\| x_k - v_i \right\|$ denotes the Euclidean distance between x_k and v_i. Parameter m is called the "fuzziness index", it is used to control the fuzziness of membership of each datum. The value of m should be within the range $m \in [1, \infty]$. There is no theoretical basis for the optimal selection of m . In this paper, we choose $m = c = 20$. The FCM algorithm can be performed by the following steps.

Step 1. Initialize fuzzy partition matrix U with a random value such that it satisfies conditions (2) and (3).

Step 2. Calculate the fuzzy centre $V^{(l)}$ using

$$V_i^{(l)} = \frac{\sum_{k=1}^{723} (u_{ik}^{(l-1)})^{20} x_k}{\sum_{k=1}^{723} (u_{ik}^{(l-1)})^{20}} \quad i = 1, 2, \cdots, 20 \tag{4}$$

Step 3. Update the fuzzy partition matrix U with

$$u_{ik}^{(l)} = \frac{1}{\sum_{j=1}^{20} (d_{ik}^{(l)} / d_{jk}^{(l)})^{\frac{2}{m-1}}}, \quad i = 1, 2, \cdots, 20; k = 1, 2, \cdots, 723, \tag{5}$$

where $d_{ik}^{(l)} = \left\| x_k - v_i^{(l)} \right\|$.

Step 4. Repeat Step (2) to (3) until $\max \left\{ \left| u_{ik}^{(l)} - u_{ik}^{(l-1)} \right| \right\} < \varepsilon_u$, where $\varepsilon_u > 0$ is a given termination condition.

During the clustering process, each data point is assigned to the cluster for which the subjection degree of the fuzzy partition matrix is maximal. And 20 cluster centre is considered to be representative of each class data, as the training sample.

3 Prediction Model

In section one, training set of 723 data have been concentrated into 20 cluster centre using by fuzzy c-means clustering algorithm. Then, we build model using the 20 cluster centre as training sample. At last, the data of first half of 2010 are used to build model to forecast coagulant dosage.

This is a quantitative prediction problem. From the relationship between independent variables (x_1, x_2, x_3) with response variable y , we should build a multivariate regression model. Multivariate regression model include multiple linear regression and multiple nonlinear regression. Below we will build respectively multiple linear regression model and multiple nonlinear regression model. At the same time, we also use a popular prediction algorithm, Support

Vector Regression (SVR) to forecast. The data of coagulant dosage of first half of 2010 will be predicted by using those three methods. At last, the prediction results of those three methods will be compared.

3.1 Multiple Linear Regression Model

The general form of multivariate regression model as follows:

$$y_i = b_0 + b_1 * x_{1i} + b_2 * x_{2i} + b_3 * x_{3i} \quad i = 1, 2, \cdots, n, \tag{6}$$

where b_0, b_1, b_2, b_3 are regression coefficients. We need solve the best estimation of regression coefficient by using the 20 cluster centre so as to minimize squares sum of the error between actual value and estimate value.

3.2 Multiple Nonlinear Regression Model

We draw respectively the scatter plot charts of x_1 with y, x_2 with y, x_3 with y. From the figures, we can analyze the general trend changes between independent variables (x_1, x_2, x_3) and response variable y. According to those, we build some models and choose the best model.

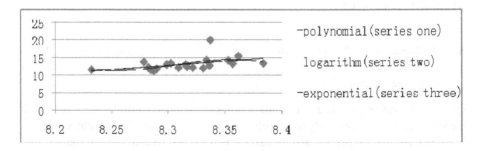

Fig. 1. The scatter plot chart of x_1 with y

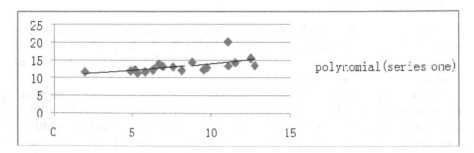

Fig. 2. The scatter plot chart of x_2 with y

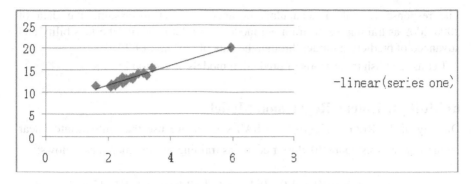

Fig. 3. The scatter plot chart of x_3 with y

In the first figure, it is seen that the relationship between x_1 with y can use not only linear ,quadratic and cubic polynomial to simulate the trend line of 20 cluster centre, but also logarithm function and exponential function to simulate the trend line. It is seen from the second figure that the relationship between x_2 with y can use quadratic to simulate the trend line. It is seen from the third figure that the relationship between x_3 with y can use linear to simulate the trend line. We also need to consider the cross correlation among independent variables x_1, x_2, x_3. According to the above information, we can build $C_5^1 * C_3^2 = 15$ kinds of multivariate nonlinear regression models.

3.3 Support Vector Regression (SVR)

The Support Vector Machine(SVM) introduced by Vapnik for classification and function estimation is an important methodology in the area of neural networks and nonlinear modeling. Typically, it is used to solve two types of classification problem, because of its generalization ability, we can also put classified thinking to establish support vector regression model. Support vector regression mainly solve nonlinear model in the low-dimensional space through improving dimension. Due to in high dimension space, we can construct linear decision function to realize linear regression. In order to adapt to the nonlinear of the training sample set, support vector regression algorithm uses kernel function instead of linear items of linear equation .It makes the original linear algorithm "non-linearization", which realized the nonlinear regression.

4 Experimental Results and Analysis

Now we have some data in the coagulation process of sewage treatment from 2008 to 2010. The data consist of three independent variables and a response variable. Independent variables are PH value(x_1), water temperature (x_2) , turbidity(x_3).

The response variable is coagulant dosage(y). Let us assume the data of 2008-2009 as training set to train out model. Then let us verify the feasibility and advanced of prediction model through the data of first half of 2010.

Let us establish three kinds of prediction models as section 2:

a) Multiple Linear Regression Model

Directly in the REG process of the SAS system, we use the multivariate linear regression analysis about 20 cluster centre as training set, the model as follows:

$$y = -32.26532 + 4.96481x_1 + 0.11349x_2 + 1.17233x_3 \quad (7)$$

The 183 data of first half of 2010 as a test set, we calculate sum of squares of deviations between predicted values and true values. Based on minimizing sum of squares of deviations, we judge performance of models. Results are showed in Table 1.

b) Multiple Nonlinear Regression Model

According to scatter plot charts, we can draw the trend line. Then we can build $C_5^1 * C_3^2 = 15$ kinds of multivariate nonlinear regression models to estimate regression coefficients. Directly in the SAS system, we can solve regression coefficients of 15 models through 20 cluster centre as training set. We use the same method as (1) to judge the performance of models. In this paper, we choose 6 models that have better performance in 15 models. Results are showed in Table 1.

c) Support Vector Regression Model

Importing libsvm to MATLAB toolbox, we assume the first three variables of 20 cluster centre as independent variables training set, the forth variable as the response variable train set. At the same time, the first three variables of 183 data of 2010 are viewed as independent variables test set, the forth variable as the response variable test set. At last, we call SVR program to predict data in MATLAB toolbox. It do not have a ascertain model in SVM. Results are showed in table 1.

Table 1. All models from the sum of squares of deviations in contrast

	Model	Sum of squares of deviations
Linear	$y = -32.26532 + 4.96481x_1 + 0.11349x_2 + 1.17233x_3$	472.7542
Non-linear	$y_1 = 24.7596 - 0.0307x_1^3 + 0.0102x_2^2 + 1.9843x_3$	161.7506
	$y_2 = 33.7982 - 0.3864x_1^2 + 0.0102x_2^2 + 1.9844x_3$	162.1734
	$y_3 = 61.7732 - 6.4978x_1 + 0.00449x_2^2 + 1.5346x_3 + 0.0423x_2x_3$	230.7009
	$y_4 = 123 - 54.4052\ln(x_1) + 0.00452x_2^2 + 1.5349x_3 + 0.0423x_2x_3$	230.9223
	$y_5 = 13.8281 - 0.00148\exp(x_1) + 0.00419x_2^2 + 1.5336x_3 + 0.0424x_2x_3$	232.5645
	$y_6 = 13.1335 - 0.00148\exp(x_1) + 0.00997x_2^2 + 1.9842x_3$	161.4508
SVM		446.2032

The results in Table 1 show:

1)The sixth of multiple nonlinear model is the best model because sum of squares of deviations is the minimum. At the same time, the results of prediction are the nearest to true values.

2) Compared with other methods, Multiple nonlinear regression models are better than multiple linear regression model and SVR model. The performance of the SVR model is not satisfactory. Maybe SVR model is not suitable for forecasting a large sample data.

5 Conclusion

The prediction of coagulant dosage is of great significance in the production and practice. This paper proposes a new method based on fuzzy c-means clustering methods to obtain 20 cluster centre to replace large sample data as new training set. On that basis, we draw respectively the scatter plot charts of independent variables with response variable using the 20 cluster centre. Analyze the trend line, we build 7 kinds of prediction models that contains one multiple linear regression model and 6 multiple nonlinear nonlinear regression model. We also use support vector regression to predict coagulant dosage. The experimental results show that the sixth multiple nonlinear model derives minimum sum of squares of deviations and captures the satisfying solution to predicting coagulant dosage. Compared with these models, we obtain the best prediction model:

$$y = 13.1335 - 0.00148\exp(x_1) + 0.00997x_2^{\,2} + 1.9842x_3 \qquad (8)$$

Acknowledgements. Thanks to post-doctoral Kejiang Zhang from Calgary University for their aid in sample collection.

References

1. Vapnik, V.N.: Statistical learning theory. Wiley, New York (1998)
2. Tan, P., Xing, Y.: Based on SVM speaker and FCM identification. Science Technology and Engineering 10(24), 1671–1815 (2010)
3. Ren, B., Zhao, M., Liu, Z., Wang, J.: The support vector machine forecasting of the dynamic index of oilfield development. Petroleum Planning and Engineering 19(6), 12–15 (2008)
4. Wang, X.: Shallow talk coagulation of water treatment method and coagulants. Heilong Ji ang Technical Information 20 (2010)
5. Meng, Y., Zhou, T.: The discussed about coagulation effect factors and improvement measures. The Water Supply Drains a Technique Dynamic State 2 (2000)
6. Xiang, C., Zhou, Z.: Chaos time series support vector machines forecast. Statistics and Decision-Making 1 (2010)
7. Xiao, C.: Engineering fuzzy system. Science press, Beijing
8. Cristianini, N., Shawe-Taylor, J.: An Introduction to Support Vector Machines and Other Kernel-based Learning Methods. Electronic Industry press.
9. Xie, Z.: MATLAB statistical analysis and applications: 40 case analysis. Beijing Aerospace University Press, Beijing (2010)
10. Ruan, G.: SAS statistical analysis of practical daqo. Tsinghua University Press, Beijing

Similarity Analysis of Oilfield Development Indices by Fuzzy C-Means Clustering

Yan Yang[1], Qing-you Liu[2], and Ying He[1]

[1] College of Sciences, Southwest Petroleum University,Sichuan 610500, China
[2] State Key Laboratory of Oil and Gas Reservoir Geology and Exploitation,
   Southwest Petroleum University,Sichuan 610500, China
   e-mail: christinyanyyh@hotmail.com

Abstract. Since the oilfield water development system is complex and there are numerous characterization indices, it is quite necessary to simplify the index system. This paper deals with the similarity analysis problem of oilfield development indices. At first, from the perspective of time series, 11 numerical characteristics of historical data of oilfield development indices are found, including statistical features, trend features, autocorrelation and partial autocorrelation coefficient, nonlinear features and spectral characteristics. Then based on these numerical characteristics, a new algorithm is proposed to make similarity analysis for oilfield development indices by fuzzy c-means clustering. In the end, a practical example is given, successfully solved, and the computational results are presented.

Keywords: Oilfield development indices, similarity analysis, fuzzy c-means clustering, numerical characteristics, time series.

1 Introduction

Since most oilfields in eastern China are in the very high water cut development stage, it is difficult to grasp their dynamic law. Accurate forecast of oilfield development indices (i. e., oil production and liquid production) is important for exploiting the oilfields efficiently, which can describe the state change of oilfield development and provide scientific basis for carrying on oilfield development evaluation, making or adjusting oilfield development plans and decision of oilfield construction investment. Various aspects of the problem of predicting development indices have been addressed in the open literature [1, 2, 4, 8, 9]. References [10, 11, 13] carried out system analysis and considered the oilfield water flooding development as a complex system since the relationship between oilfield performance (development indices) and its influencing factors is generally nonlinear. Thus, they established some nonlinear models to forecast development indices.

In fact, the selection of influencing factors plays a very important role in establishing the predicting model. It is required to select indices covering most of

B.-Y. Cao and X.-J. Xie (Eds.): Fuzzy Engineering and Operations Research, AISC 147, pp. 399–407.
springerlink.com

attributes and make the correlation degree of selected indices relatively low. In reference [10], two types of comprehensive index systems that describe the oilfield development system are obtained: the index system of geological status that represents natural reservoir quality of oilfield (8 indices,) and the dynamic control index system that represents development level subject to human control (10 indices). In book [14], by system analysis, the influencing factors of some important evelopment indices (such as oil production, liquid production and recovery efficiency) are got. From the professional perspective of petroleum, it is relatively brief to describe the gaint system of oilfield development with these indices. However, from the perspective of mathematical, these indices constitute high dimensional space of system characterization and it is very difficult to carry out theoretical analysis and geometric representation. Therefore, it is quite necessary to make similarity analysis to further simplify the index system.

This paper deals with the similarity analysis problem of oilfield development indices. At first, from the perspective of time series, 11 profiles of historical data of oilfield development indexes are found. Then a new algorithm by Fuzzy c-means (FCM) is used to make similarity analysis of oilfield development indices based on the numerical characteristics of their historical data.

2   The Numerical Characteristics of Historical Data

During oilfield development, a great amount of historical data is accumulated and there is usually much information behind it. The historic data sequence of a development index can be viewed as a time series, since a time series is a sequence of observations taken sequentially in time. In this section, from the perspective of time series, 11 numerical characteristics of historical data of oilfield development indices will be obtained, including statistical features, trend features, autocorrelation and partial autocorrelation coefficient, nonlinear features and spectral characteristics. Let $X = \{x(t)\}, t = 1, 2, \cdots, n$ be a time series. For example, it can represent the oil production of successive n months.

2.1   The Statistical Features of Historical Data

Statistical features have been applied widely in time series analysis[7], especially in additive model and auto-regressive integrated moving average model. Therefore, some statistics are extracted from historical data to reflect the position feature, dispersion and distribution shape of data.

Given $X = \{x(t)\}, t = 1, 2, \cdots, n$, then $x'(1) \leq x'(2) \leq \cdots \leq x'(n)$ is a new sequence constructed by rearrangement X. We now introduce some preliminary concepts as following[6]:

Mean

$$\overline{X} = \frac{1}{n} \sum_{t=1}^{n} x(t). \tag{1}$$

Median value

$$M = \begin{cases} x'(\frac{n+1}{2}), & n \text{ is a odd number;} \\ \frac{1}{2}[x'(\frac{n}{2}) + x'(\frac{n}{2}+1)], & \text{otherwise.} \end{cases} \tag{2}$$

Fractile

$$M_p = \begin{cases} x'[int(np)+1], & n \text{ is not a integer;} \\ \frac{1}{2}(x'[int(np)] + x'[int(np)+1]), & \text{otherwise.} \end{cases} \tag{3}$$

3-mean

$$\overline{M} = \frac{1}{4}Q_1 + \frac{1}{2}M + \frac{1}{4}Q_3, \tag{4}$$

where, $0 \le p < 1, Q_1 = M_{0.25}, Q_3 = M_{0.75}$, and $int()$ is the integral function.

$\overline{X}, M, M_p$ are from different emphases to reflect the position features of data, however, $3 - mean$ fully developed their advantages and good qualities is robust enough. Therefore, we extract the 3-mean $\overline{M}$ as a numerical characteristic for the valued position of data.

Variance

$$S^2 = \frac{1}{n-1} \sum_{t=1}^{n} [x(t) - \overline{X}]^2. \tag{5}$$

Standard deviation

$$S = \sqrt{S^2} = \sqrt{\frac{1}{n-1} \sum_{t=1}^{n} [x(t) - \overline{X}]^2}. \tag{6}$$

Coefficient of variation

$$CV = \frac{S}{\overline{X}}. \tag{7}$$

Quartiles

$$R_1 = Q_3 - Q_1. \tag{8}$$

Since CV is dimensionless and R_1 has good immunity for abnormal data, they can be used as numerical characteristics for the valued dispersion of data.

Coefficient of skewness

$$SKEW = \frac{n}{(n-1)(n-2)} \sum_{t=1}^{n} (\frac{x(t) - \overline{X}}{S})^3. \tag{9}$$

Coefficient of kurtosis

$$KURT = \frac{n(n+1)}{(n-1)(n-2)(n-3)} \sum_{t=1}^{n} (\frac{x(t) - \overline{X}}{S})^4 - \frac{3(n-1)^2}{(n-2)(n-3)}. \tag{10}$$

Therefore, $\overline{M}, CV, R_1, SKEM, KURT$ are obtained to depict the statistical features of historical data.

2.2   The Trend Character of Historical Data

As a whole, there is a trend downward or upward for historical data $X = \{x(t)\}$, $t = 1, 2, \cdots, n$ as a time series. In this subsection, the trend character of $\{x(t)\}$ is obtained by linear regression analysis. In linear regression, data are modeled using linear functions, and unknown model parameters are estimated from the data. We use ordinary least squares (the simplest and thus most common estimator, OLS)[12] to estimate the parameter b which is the slope of straight line after linear regression, where,

$$b = \frac{1}{n \sum_{t=1}^{n} t^2 - (\sum_{t=1}^{n} t)^2} [-(\sum_{t=1}^{n} t)(\sum_{t=1}^{n} x(t)) + n \sum_{t=1}^{n} t \cdot x(t)]. \quad (11)$$

2.3   Autocorrelation Coefficient and Partial Autocorrelation Coefficient of Historical Data

The autocorrelation coefficient is the cross-correlation of a signal with itself. Informally, it is the similarity between observations as a function of the time separation between them. The partial correlation coefficient describes the data in the partial regression residual plot. It measures the degree of association between two random variables, and is possible to remove the intervening correlations between $x(t)$ and $x(t - k)$.

For a time series $X = \{x(t)\}, t = 1, 2, \cdots, n$, there have:

Autocorrelation function

$$\rho_k = \frac{\sum_{t=1}^{n-k} [x(t) - \overline{x}][x(t+k) - \overline{X}]}{\sum_{t=1}^{n} [x(t) - \overline{X}]^2}. \quad (12)$$

Autocorrelation coefficient

$$\rho = \sum_{k=1}^{h} \rho_k = \frac{\sum_{k=1}^{h} \sum_{t=1}^{n-k} [x(t) - \overline{X}][x(t+k) - \overline{X}]}{\sum_{t=1}^{n} [x(t) - \overline{X}]^2}. \quad (13)$$

Partial autocorrelation function

$$\varphi_{k+1,k+1} = \frac{\rho_{k+1} - \sum_{j=1}^{k} \varphi_{kj} \rho_{k+1-j}}{1 - \sum_{j=1}^{k} \varphi_{kj} \rho_j}. \quad (14)$$

Partial autocorrelation coefficient

$$\varphi = \sum_{k=1}^{h} \varphi_{kk}, \quad (15)$$

where, $h = int[\frac{n}{4}], \varphi_{11} = \rho_1, \varphi_{k+1,j} = \varphi_{kj} - \varphi_{k+1,k+1}\varphi_{k,k+1-j}$.

Thus, ρ and φ can be used to depict the structure characteristics of historical data.

2.4   The Nonlinear Characteristics of Historical Data

There are various detecting nonlinear methods for time series, such as Lyapunov exponent, BDS test, Kolmogorov-Smirnov test, Bispectral test, RESET test, F test and so on. In this paper, the maximal Lyapunov exponent λ is used to depict the nonlinear characteristics of historical data, and Hilborn's algorithm[3]is used to estimate λ. The algorithm takes the following steps:

Step 1: For a time series $X = \{x(t)\}, t = 1, 2, \cdots, n$, construct a series of embedded vectors

$$X_t^h = \{x(t), x(t+1), \cdots, x(t+h-1)\}, \qquad t = 1, 2, \cdots, n-h+1. \quad (16)$$

Step 2: For any $t, t' \in \{1, 2, \cdots, n-h+1\}$, find the vector $X_{\hat{t}}^h$ such that

$$\|X_t^h - x_{\hat{t}}^h\| = min_{t' \neq t}\|X_t^h - X_{t'}^h\|, \quad (17)$$

where $\|X_t^h - X_{t'}^h\|$ is the Euler distance of X_t^h and $X_{t'}^h$.
Step 3: For any $t \in \{1, 2, \cdots, n-h+1\}$, let

$$d_t(i) = \|X_{t=i}^h - x_{\hat{t}+i}^h\|, \quad i = 0, 1, \cdots, min(n-h+1-t, n-h+1-\hat{t}). \quad (18)$$

Step 4: For any $i \in \{0, 1, \cdots, min(n-h+1-t, n-h+1-\hat{t})\}$, calculate

$$y(i) = \frac{1}{q\Delta t}\sum_{j=1}^{q} lnd_j(i), \quad (19)$$

where q is the number of nonzero $d_t(i)$, and Δt is the period of the time series X (for historical data, we let $\Delta t = n$).
Step 5: For $(0, y(0)), \cdots, (i, y(i)), \cdots$ obtained by step 4, calculate the slope of regressive line by means of least square method. The slop value is just the maximal Lyapunov exponent λ.

2.5   The Spectrum Characteristics of Historical Data

A important way of analyzing a time series is based on the assumption that it is made up of sine and cosine waves with different frequencies. Sample spectrum is an appropriate tool for analyzing time series that uses this idea. It can be used to detect and estimate the amplitude of a sine component, of known frequency, buried in noise. Fourier analysis [7] is one of the important method for sample spectrum. By discrete fourier transform (DFT), the time series $X = \{x(t)\}, t = 1, 2, \cdots, n$ is transformed into the sequence of $DFS[X] = \{\chi(k)\}, k = 1, 2, \cdots, n$ according to the formula:

$$\chi(k) = \sum_{t=1}^{n} x(t)e^{-j(\frac{2\pi}{n})kt}. \quad (20)$$

Let $k = 1$, then two Fourier coefficients can be given by

$$a_1 = \frac{1}{n} \sum_{t=1}^{n} x(t) cos(\frac{2\pi}{n}),$$ (21)

$$b_1 = \frac{1}{n} \sum_{t=1}^{n} x(t) sin(\frac{2\pi}{n}).$$ (22)

Thus, a_1 and b_1 can be used to depict the spectrum characteristics of historical data.

Therefore, for a time series $X = \{x(t)\}, t = 1, 2, \cdots, n$, 11 numerical characteristics of it can be obtained, which are denoted by a 11-dimensional vector $\widetilde{X} = \{\overline{M}, CV, R_1, SKEM, KURT, b, \rho, \varphi, \lambda, a_1, b_1\}$. Using $\widetilde{X}$ instead of X in the similarity analysis of oilfield development indices is more effective and has smaller operation.

3   The Similarity Analysis of Oilfield Development Indices

In this section, the similarity analysis of oilfield development indices will be investigated by fuzzy c-means (FCM) algorithm based on the numerical characteristics of their historical data.

The FCM algorithm [5] is often used to perform adaptive classifications, and it is an effective data clustering algorithm employing the least square principle. It is based on minimization of the following objective function:

$$J_m = \sum_{i=1}^{N} \sum_{j=1}^{C} u_{ij}^m \|x_i - c_j\|^2, \quad 1 \le m < \infty,$$ (23)

where m is any real number greater than 1, u_{ij} is the degree of membership of x_i in the cluster j, x_i is the ith of d-dimensional measured data, c_j is the d-dimension center of the cluster.

Next, a FCM algorithm based on the numerical characteristics of historical data is proposed for similarity analysis of oilfield development indices.

Algorithm 3.1 Given $X_i, i = 1, 2, \cdots, N$ are n oilfield development indices we collect, and $\{x_i(t)\}, t = 1, 2, \cdots, n$ is the corresponding time sequence of X_i.

Step1 : Non-dimensionalized X_i:

$$x_i'(t) = \frac{\vee_{j=1}^{n} x_i(j) - x_i(t)}{\vee_{j=1}^{n} x_i(j) - \wedge_{j=1}^{n} x_i(j)}$$ (24)

with $\vee$ and $\wedge$ denoting the usual maximum and minimum, respectively.

Step2 : Calculate the numerical characteristics of $\{x_i'(t)\}, t = 1, 2, \cdots, n$ by (4),(7)-(11),(13),(15),(19)-(21), and denoted them by a 11-dimensional vector $x_i = \{x_{i1}, x_{i2}, \cdots, x_{i11}\}$.

Step3 : Given the clustering number C and the termination criterion $\varepsilon \in [0, 1]$.
Step4 : Random initialize $U^{(0)} = [u_{ij}]$ matrix such that

$$\sum_{j=1}^{C} u_{ij}^0 = 1, \quad 0 \le u_{ij}^0 \le 1. \tag{25}$$

Step5 : Calculate the centers vectors $C^{(k)} = [c_j]$ with $U^{(k)}$ at k-step, where

$$c_j = \frac{\sum_{i=1}^{N} u_{ij}^k \cdot x_i}{\sum_{i=1}^{N} u_{ij}^k}. \tag{26}$$

Step6 : Update $U^{(k)}$ to obtain $U^{(k+1)}$, where

$$u_{ij}^{k+1} = \frac{1}{\sum_{k=1}^{C} \left(\frac{\|x_i - c_j\|}{\|x_i - c_k\|} \right)^{\frac{2}{m-1}}}. \tag{27}$$

Step7 : If $\|U^{(k+1)} - U^{(k)}\| < \varepsilon$, then go to the next step; otherwise return to Step 5.

Step8 : Calculate $H = U^{(k+1)} \cdot \begin{pmatrix} 1 \\ 2 \\ \vdots \\ C \end{pmatrix} = \begin{pmatrix} h_1 \\ h_2 \\ \vdots \\ h_C \end{pmatrix}$. For any $c, j \in \{1, 2, \cdots, C\}$,

if $c - 0.5 < H_j \le c + 0.5$ then the oilfield development index X_j oilfield development index belongs to the c-th class.

Given the clustering number C according to actual situation, the oilfield development indices can be divided into C categories. Select one oilfield development index from each class, then C oilfield development indices can be got.

4 A Practical Example

By qualitative Analysis, the factors influencing the monthly oil production for an oilfield in China include the number of active oil wells(N_O), the number of water injection well(N_W), the composite water cut(C_W), the monthly injection-production ratio(R_I), the production pressure difference(D_P), the oil production index(I_O), the total geologic reserve(R_T), the recoverable reserves(R_C), the remaining geologic reservesR_G), the injected water volume (v_W),the number of total stimulation treatments for old wells(N_T), the number of effective stimulation treatments for old wells(N_E). For these 12 indices, we collect their monthly data from January 1998 to January 2010. Table 1 shows their 11 numerical characteristics.

Let $\varepsilon = 10^{-4}$ and $C = 5$, then the clustering results based on the numerical characteristic given in Table 1 can be obtained by Algorithm 3.1 as follows:

Table 1. Numerical characteristics of oilfield development indices

	$\overline{M}$	CV	R_1	SKEM	KURT	b	ρ	φ	λ	a_1	b_1
N_O	0.9725	0.2669	0.3747	-0.8662	3.3826	0.0105	9.7759	4.9789	-0.0001	-0.1764	0.6274
N_W	1.0251	0.1764	0.1025	-4.7329	25.5709	0.0048	4.5543	2.1532	0.0002	-0.0852	0.6499
C_W	1.0237	0.1626	0.0201	-6.1117	36.4907	0.0024	2.7128	3.4149	0.0006	-0.0465	0.6505
R_I	1.0081	0.2279	0.3001	-2.9259	17.6986	0.0085	8.7413	4.4779	0.0001	-0.1463	0.6387
D_P	0.7602	1.1022	1.4174	2.3289	7.9288	0.0401	8.3513	4.3531	0.0032	-0.6305	0.5222
I_O	1.0500	0.3641	0.4821	-1.3066	2.0379	0.0142	9.3712	4.3446	-0.0005	-0.2348	0.6510
R_T	1.0891	0.2351	0.9732	-0.6902	-1.9991	0.0013	12.9657	6.0644	-0.0022	-0.1637	0.6422
R_C	1.0829	0.6652	1.4454	-0.8661	-1.2775	0.0236	11.2913	5.1776	-0.0041	-0.3955	0.7107
R_G	1.1911	0.8121	1.5961	-0.4229	-1.8289	0.0310	12.3152	5.7191	-0.0003	-0.5238	0.6720
V_W	0.7658	1.1903	1.3996	3.433	18.4648	0.0404	7.2479	4.0734	0.0003	-0.6322	0.5178
N_T	0.9652	0.5912	1.0275	0.1803	-1.1859	0.0257	12.9549	6.1439	0.0004	-0.4239	0.6164
N_E	1.0139	0.6063	1.0719	0.2056	-1.1493	0.0263	12.9144	6.2926	0.0001	-0.4337	0.6146

First class: the number of active oil wells, the oil production index.

Sencond class: the number of water injection well, the composite water cut,the monthly injection-production ratio.

Third class: the injected water volume, the production pressure difference.

Forth class: the number of total stimulation treatments for old wells, the number of effective stimulation treatments for old wells.

Fifth class: the total geologic reserve, the recoverable reserves, the remaining geologic reserves.

Therefore, we can select the number of active oil wells, the number of water injection well, the injected water volume, the number of effective stimulation treatments for old wells, the remaining geologic reserves as the influencing factors of the monthly oil production. If we denote these five indices and the monthly oil production by $x_1, x_2, x_3, x_4, x_5, y$, respectively, then based on their historical data, some nonlinear model $y = f(x_1, x_2, x_3, x_4, x_5)$ used to forecast the monthly oil production can be established by some nonlinear modeling and simulation methods.

5   Conclusion

In this paper, the oilfield water flooding development is considered as a gaint system. In order to simplify the index system, the similarity analysis problem of indices is investigated. A new algorithm based on fuzzy c-means cluster algorithm is presented to make similarity analysis. This method uses 11 numerical characteristics of historical data of oilfield development indexes instead of them to cluster. The result of test shows it is reasonable and credible.

Acknowledgements. This work is supported by the Scientific Reserch Fund of SiChuan Provincial Education Department(12ZA199, 11ZA024).

References

1. Chen, Y.Q.: Derivation and Application of the Generalized Weng Model. Natural Gas Industry Journal 16(2), 22–26 (1996) (in Chinese)
2. Feng, L.Y., Li, J.C., Pang, X.Q., et al.: Peak Oil Models Forecast China's Oil Supply and Demand. Oil & Gas Journal 14(2), 43–47 (2008)
3. Hilborn, R.C.: Chaos and nonlinear dynamics an introduction for scientists and engineers. Oxford University Press, New York (1994)
4. Hu, J.G., Chen, J.C., Zhang, S.Z.: A New Model for Predicting Production and Reserves of Oil and Gas Fields. Acta Petrolei Sinica 6(1), 70–86 (1995)
5. James, C.B., Robert, E., William, F.: FCM: the Fuzzy c-mean Clustering Algorithm. Computer & Geosciences 10(2), 191–203 (1984)
6. Michael, S.L.: Data analysis: an introduction. SAGE (1995)
7. Robert, H.S., David, S.S.: Time Series Analysis and Its Applications. Springer, Heidelberg (2006)
8. Shi, C.F., Xiao, W., Wang, F.L.: Prediction Model for Polymer Flooding Development Index. Acta Petrolei Sinica 26(5), 78–80, 84 (2005)
9. Zhao, Q.F., Chen, Y.Q.: Forecast and Analysis of China's Oil Consumption and Crude Oil Supply. Oil Forum 21(2), 344–348 (2010)
10. Liu, C.L., Xiao, W.: Index System of the Water Flooding Development of Oil Fields and its Structural Analysis. Petroleum Exploration and Development 37(3), 26–28 (2006)
11. Xiao, W.: Research on Systematic Evaluation of Water Flooding Effect in La-Sa-Xing Oilfirld. Daqing Oilfield Company, Daqing (2004) (in Chinese)
12. Wolberg, J.: Data Analysis Using the Method of Least Squares: Extracting the Most Information from Experiments. Springer (2005)
13. Zhong, Y.H., Zhao, L., Liu, Z.B., et al.: Using a Support Vector Machine Method to Predict the Development Indices of Very High Water Cut Oilfields. Petroleum Science 7(3), 379–384 (2010)
14. Ren, B.S., Liu, Z.B., Zhao, M.: The prediction, Early-warning and development planning of oil field during the middle and later stage. Petroleum Industry Press, Beijing (2005)

Parallel WNAD Method and Its Wave Field Simulations

Ya-li Chen[1], Xiang-jun Xie[1], and Guo-jie Song[1,2]

[1] College of Sciences, Southwest Petroleum University, Chengdu 610500, China
[2] Department of Computer Science and Technology,
   Tsinghua University, Beijing 100084, China
   cyl0326@126.com

Abstract. A fast and effective forward modeling algorithm for the acoustic wave equation is very important to the seismic. Many researches show that the weighted nearly-analytical discrete (WNAD) method is a new numerical technology which can effectively suppress the numerical dispersion when the coarse spatial grid is used. Recently, the multi-core CPU is very popular in the PC-market. It's a hot point to increase the computational efficiency by using the multi-core hardware, but the present computational programs almost all are single thread programs which make them cannot use the multi-core effectively. OpenMP is approved by many hardware industries and becomes one of the most popular parallel tools on the shared-memory platforms. In this paper, we combine the OpenMP and WNAD method to simulate the acoustic wave field. The numerical results show that OpenMP +WNAD method can suppress effectively the numerical dispersion, simulate accurately the wave field of acoustic and reduce effectively the simulation time to the 1/3 of that of the traditional program. OpenMP+WNAD method has a profound meaning to improve the seismic exploration.

Keywords: Shared-memory, openMP, parallel computation, weighted nearly analytic discrete method.

1 Introduction

Seismic prospecting is a geophysical exploration method to analyze and judge the discontinue layer, geological structure in stratum by studying the reflected and refracted wave. It is used in the oil, gas, coal exploration. The acoustic wave equations are the most widely used governing equation in the seismic exploration for its simple and easy to solve, so the research about how to solve the acoustic equation is very important.

In the large-scale seismic wave field simulation, especially for 3D large-scale seismic wave field simulation, the computational time is usually very large. It's important to accelerate the wave field simulation. In this paper we will study how to realize the goal of shorten the acoustic wave field simulation time.

B.-Y. Cao and X.-J. Xie (Eds.): Fuzzy Engineering and Operations Research, AISC 147, pp. 409–415.
springerlink.com   © Springer-Verlag Berlin Heidelberg 2012

As far as the software, choosing a good computational algorithm is important to improve the calculation efficiency. The Nearly Analytical Discretization (NAD) [1-5] method, presented by Yang et al, is an excellent numerical method for solving the acoustic and elastic wave equation. The NAD type methods not only use the displacement at each grid but also use the displacement gradient to reconstruct the high order spatial partial derivative at each grid. Researches show that NAD-type methods can keep more wave field information in the time marching because the wave tendency depends on the gradient of displacement. So the NAD type methods can suppress the numerical dispersion effectively even if the coarse grid is used. This means that we can improve the wave field simulation efficiency, save the storage space by using NAD type algorithm. That is why we choose the weighted nearly analytical discrete (WNAD) method as a wave field simulation tool in this paper.

At the same time, a fast, high efficiency computational program must consider how to use the hardware to speed up the simulation capacity. Recently, the multi core CPU is very popular in the PC-market, more and more CPU manufacturers develop their own multi core processor. Up to now, Dual-core, four-core, even the eight-core CPU has become the standard configuration in today's computer market. But the wave field simulation program designed by the most researchers is still used the single threaded to run their program. When this kind of program is running, many computational cores are idle and many computational resources are wasted. So it's important to recompile or redesign our traditional sequential execution code and make the program more suitable in the multi-core system. Parallel computation has become a fundamental skill for the researcher in this multi core era.

In this paper, we use the WNAD method to simulate wave propagation in the elastic medium on an ordinary public computer which can be bought easily in the PC market. In order to accelerate the simulation efficiency, we use the OpenMP technology to realize parallel computation.

2 The Weighted Nearly Analytical Discrete Method

At first, we will introduce the WNAD algorithm.

In the 2D case, the acoustic equation can be written as

$$\rho \frac{\partial^2 u}{\partial t^2} = \mu \left(\frac{\partial^2 u}{\partial x^2} + \frac{\partial^2 u}{\partial z^2} \right) + f \tag{1}$$

Here ρ is medium density, μ is the lame coefficient, u is displacement, f is the source term.

For simplicity, we introduce a vector symbol $U = \{u, \frac{\partial u}{\partial x}, \frac{\partial u}{\partial z}\}^T$ which is made by displacement u and its gradient. Symbol $W = \frac{\partial U}{\partial t}$ and $P = \frac{\partial W}{\partial t} = \frac{\partial^2 U}{\partial t^2}$ is the "particle speed" and "particle acceleration", respectively.

Following the idea of Yang[1], we can get the expression of displacement (or particle speed) in the (n+1)-th temporal layer represented by the displacement (or particle speed) and displacement (or particle speed) gradients in the n-th temporal layer,

$$W_{i,j}^{n+1} = W_{i,j}^n + \Delta t (W)_{i,j}^n + \frac{\Delta t^2}{2}(W)_{i,j}^n + \frac{\Delta t^3}{6}(W)_{i,j}^n$$

$$= W_{i,j}^n + \Delta t (P)_{i,j}^n + \frac{\Delta t^2}{2}(W)_{i,j}^n + \frac{\Delta t^3}{6}(P)_{i,j}^n \qquad (2)$$

and

$$U_{i,j}^{n+1} = U_{i,j}^n + \Delta t W_{i,j}^n + \frac{\Delta t^2}{2}(U)_{i,j}^n + \frac{\Delta t^3}{6}(W)_{i,j}^n + \frac{\Delta t^4}{24}(P)_{i,j}^n \qquad (3)$$

For the formula (3), considering that the latest particle velocity $W_{i,j}^{n+1}$ has been obtained by formula (2), just like the ideas of the Gauss-Seidel method [8] which is used to solve the linear equations, we can replace $W_{i,j}^n$ in formula (3) by the linear combination $W_{i,j}^{n+1}$ and $W_{i,j}^n$ to get a new numerical method. Thus, formula (3) is replaced by formula (4) as follow,

$$U_{i,j}^{n+1} = U_{i,j}^n + \Delta t \left[\omega W_{i,j}^n + (1 - \omega) W_{i,j}^{n+1} \right] + \frac{\Delta t^2}{2}(U)_{i,j}^n + \frac{\Delta t^3}{6}(W)_{i,j}^n + \frac{\Delta t^4}{24}(P)_{i,j}^n \qquad (4)$$

Here, weighted parameter ω is chose between 0 and 1.

The high order temporal partial derivatives of displacement U and particle velocity W should be calculated before we use the formula (2) and formula (4) to get a marching in the temporal layer. But the storage is very big if we discretize formula (2) and formula (4) directly. Consider that the high temporal derivatives in formula (2) and (4) are all even order. In the governing equation, the left hand is two order temporal partial derivatives and the right hand is two order spatial partial derivatives. So the governing equation exactly is a time-space transform relationship. Following the idea of Dablain[9], we can transfer the temporal derivatives into the spatial derivatives. At last, we use the high order partial derivative formula presented by Yang [1] to calculate these partial spatial derivatives in formula (2) and (4). This is so called the WNAD algorithm.

3 Parallel WNADM with OpenMP

In this section, we will study how to parallel the WNAD algorithm with the OpenMP.

3.1 The Selection of Evaluation Methods

Programing on a shared-memory parallel platform is much easier than that on a message passing platform. As a common standard of shared-memory platform,

OpenMP provide a group of powerful tool to support the traditional programing language such as Fortran, C and C++ by defining compile directive, library syntax and so on. Many examples have proof that OpenMP can fulfill the need of user in a large range.

Here, we need remember that only the compile directive syntax is given in the OpenMP. This means OpenMP lacks the programing debugging tool. The users have to ensure the code is right by themselves.

Next, we will show how to parallel a program by OpenMP. First, we will give a glance over the WNAD method. After a simply analysis, we find that the major calculation of WNAD algorithm is concentrated in 3 loop. Two of them are involved the spatial loop index and the other is about a temporal loop index. For the loop in temporal, we cannot split this loop into many segments to compute for each segments in temporal layer has causal relationship with each other. For the spatial loop, the loop index I and J which is spatial variable in x- and z-direction are independent, so we can split them into many segments and calculate parallel at the same time. Now, the only work we need do is adding the compile guidance syntax into the source code of the original WNAD algorithm. Add the compile guidance syntax "!$OMP Parallel" and "!$OMP DO" before the spatial loop statement and add the compile guidance syntax "!$OMP END DO" and "!$OMP END Parallel" after where the spatial loop is over, then we finish the parallel programing. As you see, it's very easy to parallel our program with OpenMP.

4 Numerical Experiments

At present, the acoustic wave equation is one of the most widely used governing equations in the seismic exploration. Acoustic wave propagation in the isotropic medium is chose as our first numerical example. In this example, we will study the parallel efficiency of WNAD algorithm. Assume that the simulation domain is a rectangle which width and depth are all 20km. The seismic source is located at the center of the computational domain. A force term vibrating with time is

$$f(t) = \sin{(2\pi f_0 t)} \exp{(-\frac{1}{4}\pi^2 f_0^2 t^2)}$$

Here, the main frequency $f_0 = 40$Hz.

The wave velocity is chose as $a = 4$km/s. Then the spatial step is $\Delta x = \Delta z = 50$m if we choose the sampling rate N=2 following the sampling formula,

$$N = \frac{v_{min}}{f\Delta x} \text{ or } \Delta x = \frac{v_{min}}{Nf}.$$

The temporal step is 1ms.

Figure1 show the computational resource configuration with (a) serial WNAD method is running, (b) parallel WNAD method with OpenMP is running. From the serial case (Figure 1(a)), it shows that the utilization ratio of CPU is 100%. But in fact, the true utilization ration is only 25% because we have 4 cores in our compute. From the parallel case (Figure 1(b)), we can see the utilization ratio of CPU is up to 326%. The true ratio is 81.9% which means all of the four cores are participate in the simulation project in the second program. Just because of this, the second program finishes in 173s while the first one needs 521s. Almost 3 time's computational time is saved by using the OpenMP.

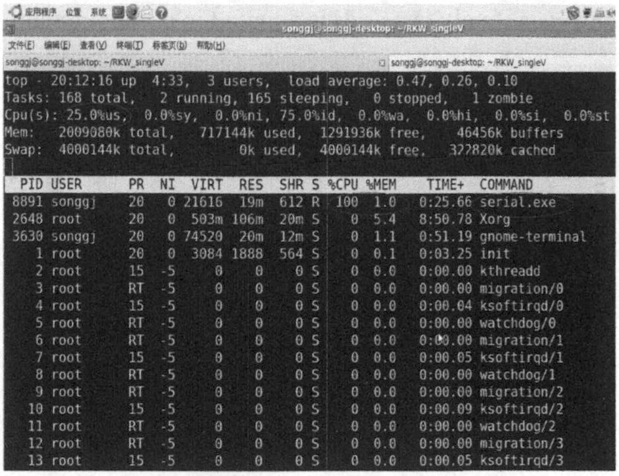

Fig. 1. (a)

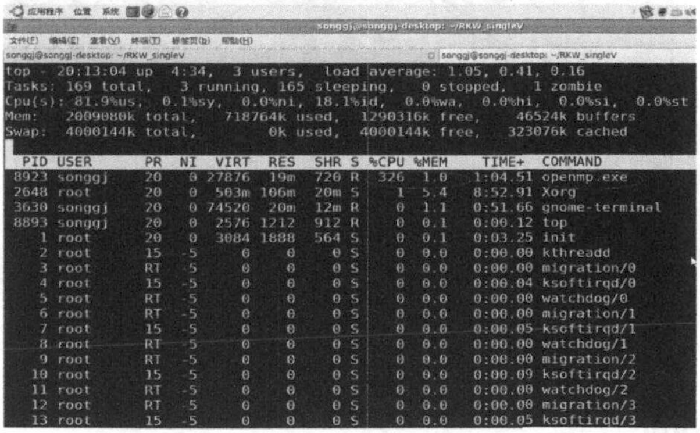

Fig. 1. (b)

The well-known SEG/Salt model is chose as our second model. The data of SEG/Salt model is obtained in the Mexico Gulf. As shown in Figure 2, there is a high speed domain (Salt dome) in the SEG/Salt model. The maximum wave velocity (about 4.4 km per second) of this model is in the salt dome while the velocity of the background is relatively small. The lowest velocity is 1.4 km per second. Numerical dispersion will appear for the huge velocity contrast in this model by using traditional numerical finite difference method if the coarse grid is used.

Figure 3 shows the surface seismic record simulated by WNAD algorithm. The spatial step is chose 50m, temporal step is 1ms. Only 2 grids are adopted in each minimum wave length. From Figure 3, we can see the seismic record is very clear without any numerical dispersion. This prove that the WNAD algorithm can suppress the numerical dispersion effectively and get a very clear wave field in the complex medium even if only two grids is adopted in each minimum wave field.

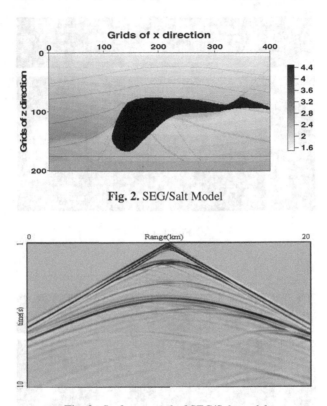

Fig. 2. SEG/Salt Model

Fig. 3. Surface record of SEG/Salt model

5 Conclusion

In this paper, we present a parallel WNAD algorithm by using the OpenMP on a personal computer. The parallel efficiency is quite satisfactory. The parallel program is about 3 time's fast than the original program on our 4 cores computer. The parallel efficacy is about 80%.

By using the parallel WNAD algorithm, we study the acoustic wave propagation in the isotropic model and SEG/Salt model. The numerical results show that the WNAD algorithm can suppress the numerical dispersion effectively even if only two grids are adopted in each minimum wave length. This means that we can use the large spatial step to simulate the wave propagation in elastic medium if we choose the WNAD method as our forwarding modeling algorithm in the seismic exploration. Furthermore, large spatial step also means much less storages and less computational time.

Numerical algorithm parallelization is very important in the multi core multi CPU era. The OpenMP, which is designed for the shard memory platform, is easier to learn for the customers who know little computer science and the ordinary computer in the PC-market are all shard memory machine. As shown in this paper, after a little modify with the ordinary code, we can get a more exciting speedup by using the OpenMP. It's advisable to parallel the traditional serial program with OpenMP.

Acknowledgements. Thanks to the support by Science Foundation of Southwest Petroleum University (No.192).

References

1. Yang, D.H., Wang, S.Q., Zhang, Z.J., Teng, J.W.: N-times absorbing boundary conditions for compact finite-difference modeling of acoustic and elastic wave propagation in the 2D TI medium. Bull. Seis. Soc. Am. 93(6), 2389–2401 (2003)
2. Yang, D.H., Song, G.J., Lu, M.: Optimally accurate nearly analytic discrete scheme for wave-field simulation in 3D anisotropic media. Bull. Seis. Soc. Am. 97(5), 1557–1569 (2007)
3. Yang, D.H., Hua, B.L., Song, G.J., Calandra, H.: Simulations of acoustic wave-Fields in heterogeneous media: a robust method with automatically suppressing numerical dispersion for large grid steps. Geophysics (2010)
4. Wang, L., Yang, D.H., Deng, X.Y.: A WNAD method for seismic stress-field modeling in heterogeneous media. Chinese J. Geophys. 52(6), 1526–1535 (2009)
5. Song, G.J., Yang, D.H., Chen, Y.L., Ma, X.: Non-uniform grid algorithm based on the WNAD method and elastic wave-field simulations. Chinese J. Geophys. 53(8), 1985–1992 (2010)
6. Gropp, W., Lusk, E., Skjellum, A.: Using MPI: portable parallel programming with the message-passing interface. The MIT press (1999)
7. Grama, A., Gupta, A., Karypis, G., Kumar, V.: Introduction to parallel computing. China Machine Press (2003)
8. Demmel, J.: Applied numerical linear algebra. SIAM (1997)
9. Dablain, M.A.: The application of high-order differencing to the scale wave equation Laser. Geophysics 1(51), 54–66 (1986)

Quantitative Analysis about 2010 World Expo on Tourism Investment in Shanghai

Dong-Hong Tian[1], Yong Wang[2,1], Rong Xiao[1], and Jian-Ying Xiao[1]

[1] School of sciences, Southwest Petroleum University, Chengdu, 610500, China
[2] State Key Laboratory of Oil and Gas Reservoir Geology and Exploitation,
Southwest Petroleum University, Chengdu, Sichuan 610500, China
tiandonghong@163.com

Abstract. Each World Expo can have many positive effects for its host country, especially on the tourism investment. This paper establishes a linear model of time series for the tourism investment in Shanghai from 1980 to 2002.With this linear model, the paper predicts the tourism investment values in Shanghai from 2003 to 2009.These values aren't affected by the right to host 2010 World Expo and are compared with the actual values which are affected by the right. Finally, the paper makes the one-way ANOVA test for the difference between the two groups of values.

Keywords: World Expo, tourism investment, one-way ANOVA, time series analysis.

1 Introduction

2010 World Expo was held in Shanghai In May 1, 2010, which was the first right to host World Expo in China. The total investment reached 450 billons more than the Olympic Games in Beijing and its scale was the largest in the history of World Expo.

Each World Pageant can have many positive effects for its host country. Although the Olympic Games in Beijing were further and further, the "Post Olympic effects" is increasingly showing its energy. Similarly, 2010 World Expo in Shanghai will also show its positive effects. World Expo is regarded as the embodiment of national comprehensive strength and the "economic Olympics" which reveals the latest results of economic, technology and culture. But in politics, economy, culture, science and technology, environment and other fields, the influence of successful bid for World Expo is the most obvious in economy and tourism .World Expo' influence on the tourism is also the most direct. Therefore, this paper chooses to quantitatively analysis the influence of 2010 World Expo on tourism investment in Shanghai.

Assume Shanghai did not obtain the right to host 2010 World Expo, the tourism investment in Shanghai from 2003 to 2009 should be another set of values. Firstly, we establish the suitable model based on the presented data from 1980 to 2002 about the tourism investment in Shanghai. Secondly, we predict the values about

B.-Y. Cao and X.-J. Xie (Eds.): Fuzzy Engineering and Operations Research, AISC 147, pp. 417–422.
springerlink.com

the tourism investment in Shanghai from 2003 to 2009 which are predicted without the influence of the right to host 2010 World Expo. Thirdly, we compare these values with the actual values from 2003 to 2009 which are collected in the references. Finally, we use the one-way ANOVA test to judge the influence of the right to host 2010 World Expo is remarkable or not.

2 The Descriptive Analysis of Tourism Investment in Shanghai from 1980 to 2009

Tourism investment includes direct investment and indirect investment. Direct investment is made up by permanent facilities of tourism and operating funds. Indirect investment is made up by public utilities investment (public transport, water, gas and other facilities investment), investment in power construction, municipal construction investment, tourism facilities investment (travel agency investment, the hotel industry investment) and so on. During 30 years of reform and opening up, with the rapid development of the tourism, Shanghai had been increasing the tourism investment. Especially in 2002, Shanghai got the successful bid to host 2010 World Expo, the tourism investment was far beyond previous investment levels. From 1980 to 2009, the tourism investment in Shanghai increased from 0.4 billons to 131.5 billons. In the yeas before 2002, the tourism investment in Shanghai was relatively less. The tourism investment reached 24.5 billons in 2003, and maintained the high level of investment after 2003.

Table 1. Tourism investment in Shanghai from 1980 to 2009

Year	X: the total investment (Billons)	Year	X: the total investment (Billons)
1980	0.4	1995	10.1
1981	0.4	1996	17.1
1982	2.5	1997	25.2
1983	0.6	1998	12.9
1984	0.6	1999	16.5
1985	1	2000	14.2
1986	1.3	2001	18.6
1987	1.8	2002	18.6
1988	2	2003	24.5
1989	1.6	2004	34.8
1990	1.6	2005	44.7
1991	2.5	2006	63.6
1992	3.3	2007	89.9
1993	7.6	2008	110.2
1994	12.7	2009	131.5

(Source: Shanghai statistical yearbook from 1979 to 2009).

This Table 1 shows that the tourism investment in Shanghai has been maintaining the high growth after 2002. The annual average increase on investment was 0.83 billons before 2002, but it was 16.12 billons after 2002. Only in 2002, the investment was more 0.4 billons than the total investment from 1980 to 2002. By 2009, the tourism investment in Shanghai amounted to 131.5 billons, which accounted 19.6% of the total investment from 1980 to 2009.

3 Building Time Series Model

To analyze the influence of World Expo on the tourism investment in Shanghai, it is necessary to predict the tourism investment without the right to host 2010 World Expo from 2003 to 2009. Then the paper will use the predicted data and the known actual data to do comparative analysis. So the authors need to predict the tourism investment in Shanghai from 2003 to 2009 by a fitting model according to the data about tourism investment in Shanghai from 1980 to 2002.

Firstly, it is necessary to observe the scatter of the data about tourism investment in Shanghai from 1980 to 2002, as Figure1.

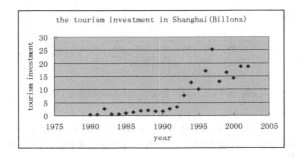

Fig. 1. The tourism investment in Shanghai from 1980 to 2002

From the figure1, the data points are relatively dispersed after 1995. So there is no consideration of directly establishing the fitting model with the curve fitting method. Because that the data is a set of time series data and is passed the stationary test after a first-order difference. We use the method of establishing stable linear model of time series to build a model.

The main steps are as follows:

1. We do a stationary test with the data of the tourism investment in Shanghai from 1980 to 2002. The horizontal fold line of sequence X is shown in Figure 2:

It is clear that X is unstable. So we do a first-order difference with the sequence X and get DX sequence. Secondly, the stationary test is done with DX, it can be thought stationary basically. Then we do a two-order difference to get DX2, the horizontal fold line of DX2 is basically similar to the line of DX. Therefore, we only need to do the first-order difference with the sequence.

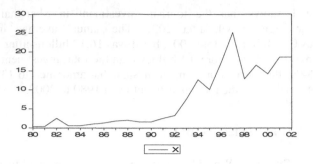

Fig. 2. Horizontal fold line of sequence X

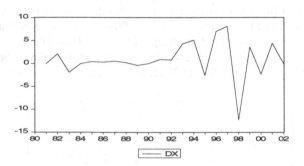

Fig. 3. Horizontal fold line of sequence DX

2. According to the tailing of the auto-correction coefficient and the truncation of the partial correlation coefficient of DX sequence, we can establish the MA (1) model, the model is:

$$DX_t = U_t + cU_{t-1} \ (U_t \text{ is the white noise sequence}).$$

Then, we use the command of equation estimate in the eviews3.1 software to estimate the parameter in the model. So the model is:

$$DX_t = U_t - 0.3097U_{t-1}.$$

The lag polynomial roots of the model are located in the unit circle, and the model satisfies the basic requirements of the smooth process. The adjusted- R^2 is relatively large, the model passed the test.

4   Predicting the Values of the Tourism Investment in Shanghai from 2003 to 2009

We use the model to predict the tourism investment in Shanghai from 2003 to 2009. Because E-Views software is suitable for short-term prediction, it is obviously unreasonable to predict 7 data by the 23 data. So the method used is: based on the data from 1980 to 2002, only predict one step and use the predicted data and the above data to reestablish the model to predict the data of the next year. Repeat 7 times, we can predict all the values about tourism investment in Shanghai from 2003 to 2009. Put these predicted data (if Shanghai had not obtained the right to host 2010 World Expo) and the actual data (Shanghai has won the right to host 2010 World Expo) to make a table. It is shown in Table2.

Table 2 Tourism investment in Shanghai from 2003 TO 2009

year	the tourism investment in Shanghai without the influence of the right to host 2010 World Expo	the tourism investment in Shanghai with the influence of the right to host 2010 World Expo	the difference between the two sets of values
2003	18.92	24.5	5.58
2004	20.33	34.8	14.47
2005	23.45	44.7	21.25
2006	27.67	63.6	35.93
2007	34.04	89.9	55.86
2008	39.36	110.2	70.84
2009	44.01	131.5	87.49

From the data in Table2, it can be seen that the tourism investment under the influence of the right to host 2010 World Expo in each year is higher than that tourism investment without the influence of the right. Each difference is obviously bigger than zero. But the problem is that the "bigger than" reflected from the data has statistical significance or not. Can the "bigger than" would be passed the statistical test? Can we say that the influence of the right to host 2010 World Expo in Shanghai is significant? We use the data of the tourism investment in Shanghai which is under and not under the influence of right to host 2010 World Expo to do the one-way ANOVA test. The conclusions are as follows:

Table 3. Analysis of variance

Variance	SS	df	MS	F	P-value	F crit
Between groups	6066.115	1	6066.115	7.042252	0.0210 34	4.7472 25
Within group	10336.66	12	861.3885			
Total	16402.78	13				

From the tested results, it can be seen that the two sets of data are from two different distributions. Therefore, we can think that the influence of the right to host World Expo is significant.

With the increasing tourism investment in Shanghai, it will promote tourism infrastructure in Shanghai and other hardware facilities and the high quality development of relative tourist departments in terms of software. It also can improve the travel agency with a higher level of management and the perfect system and institution, providing the international level of service. Give full play to their functions, so as to stimulate the development of tourism industry and form the sustainable benign loop with the investment.

Acknowledgements. This work was supported by the Sichuan Youth Science and Technology Fund (No. 2011JQ0044), the National Science Fund for Distinguished Young Scholars of China (Grant No. 51125019), the National Program on Key Basic Research Project (973 Program, Grant No. 2011CB201005) and the Scientific Research Fund (No. 10ZB113) of Sichuan Provincial Educational Department.

References

1. Xiao, H.Y.: Practical mathematical modeling and software application. Northwestern Polytechnical University Press, Xian (2008)
2. Hong, X.G.: The prediction and countermeasures of the visit number in Yangtze River during 2010 World Expo in Shanghai. Shanghai Economic Research, Shanghai (2004)
3. Gou, X.D.: The establishment of the background trend line of entry tourism in Shanghai and its application. Journal of Shanxi Normal University(Natural Science Edition) (2000)
4. Gao, R.X.: Comment on knowledge concentrated industry in Shanghai—Analysis based on principal component. Shanghai Economic Research, Shanghai (2007)
5. Shanghai Statistics: (September 10, 2010), http://www.stats-sh.gov.cn/
6. The Shanghai gazetteer office network (September 10, 2010), http://www.shtong.gov.cn/
7. China Shanghai net (September 10, 2010), http://www.shanghai.gov.cn/
8. The Shanghai Expo official website (September 10, 2010), http://www.expo2010.cn/

A Note on Wavelet and Fourier Methods for Solving the Sideways Heat Equation

Yi-long Wang[1] and Yong-ming Li[2]

[1] School of Sciences, Southwest Petroleum University, Chengdu, 610500, China
[2] State Key Laboratory of Oil and Gas Reservoir Geology and Exploitation,
  Southwest Petroleum University, Chengdu, 610500, China
  wangelongelone@163.com

Abstract. We consider an inverse heat conduction problem (IHCP), the sideways heat equation. The problem is ill-posed: a small perturbation in the data may cause dramatically large error in the solution. In reference [1] theory analysis and numerical implementation are not consistent. We solve this inconsistence by proposing a new regularization method. The method not only has the same accurate and computing cost but also keeps parts of high-frequency components of the data compared to the results in [1].

Keywords: inverse problem, ill-posed problem, Fourier analysis, heat conduction.

1 Introduction

The inverse heat conduction problem (IHCP) is a very important problem in many areas of heat transfer. In a one dimension setting, we consider the following ill-posed problem for the heat equation in the quarter plane: Find the temperature $u(x,t)$, $0 \le x \le 1$ from temperature measurements $g(t) = u(1,t)$, when $u(x,t)$ satisfies

$$\begin{cases} u_{xx} = u_t, 0 \le x, 0 \le t, \\ u(x,0) = 0, x \ge 0, \\ u(1,t) = g(t), t \ge 0, \qquad u\mid_{x \to \infty} bounded. \end{cases} \qquad (1.1)$$

Since $g(t)$ is measured, there will be measurement errors, and we would have as data some function $g_m \in L^2$, for which

$$\|g_m - g\| = \|g_m - u(1,t)\| \le \varepsilon , \qquad (1.2)$$

where the constant $\varepsilon > 0$ represents a bound on the measurement error. We know that for $x \ge 1$ one can solve a well-posed quarter plane problem using g_m as data.

B.-Y. Cao and X.-J. Xie (Eds.): Fuzzy Engineering and Operations Research, AISC 147, pp. 423–431.
springerlink.com   © Springer-Verlag Berlin Heidelberg 2012

For $0 \le x < 1$ we have the sideways heat equation.

$$\begin{cases} u_{xx} = u_t, 0 \le x, 0 \le t \\ u(x,0) = 0, x \ge 0 \\ u(1,t) = g_m(t), t \ge 0. \qquad u\mid_{x\to\infty} \ bounded. \end{cases} \tag{1.3}$$

This problem is known to be extremely ill-posed: a small perturbation in the data may cause a dramatically large error in the solution. For more details we refer to [1,2]. In reference [1], there is a gap between theory analysis and numerical implementation. The error estimate (Method I ,see the following Section 3) is not consistent with numerical implementation (Algorithm II ,see the following Section 4). In this paper we propose a regularization to match numerical implementation in reference [1]. Meanwhile we give the numerical implementation to match the error estimate in reference [1]. At last we give the numerical experiments.

2 Ill-Posed and Stabilization

The problem of solving the sideways heat equation is ill-posed. The ill-posedness can be seen by solving the problem in the Fourier domain .Let

$$\hat{g}(\xi) = \frac{1}{\sqrt{2\pi}} \int\limits_{-\infty}^{\infty} g(t)e^{-i\xi t}dt \text{ be the Fourier transform of the exact data function.}$$

The problem (1.1) can be formulated, in frequency space, as follows:

$$\begin{cases} \hat{u}_{xx}(x,\xi) = i\xi\hat{u}(x,\xi), x > 0 \ and \ \xi \in R, \\ \hat{u}(1,\xi) = \hat{g}(\xi), \xi \in R, \\ \hat{u}(x,\xi)\mid_{x\to\infty} \ bounded. \end{cases} \tag{2.1}$$

The solution to this problem, in frequency space, is given by

$$\hat{u}(x,\xi) = e^{\sqrt{i\xi}(1-x)}\hat{g}(\xi), \tag{2.2}$$

where $\sqrt{i\xi}$ denotes the principal value of the square root,

$$\sqrt{i\xi} = \begin{cases} (1+i)\sqrt{|\xi|/2}, \xi \ge 0, \\ (1-i)\sqrt{|\xi|/2}, \xi < 0. \end{cases} \tag{2.3}$$

In order to obtain this solution, we have used the bound on the solution at infinity. Since the real part of $\sqrt{i\xi}$ is positive and our solution $\hat{u}(x,\xi)$ is assumed to be in $L^2(R)$, we see that the exact data function, $\hat{g}(\xi)$, must decay rapidly as $\xi \to \infty$.

Now, we assume that the measured data function satisfies $g_m(t) = g(t) + \delta(t)$, where $\delta(t) \in L^2(R)$ is a small measurement error. If we try to solve the problem using g_m as data we get the solution

$$\hat{v}(x,\xi) = e^{\sqrt{i\xi}(1-x)}(\hat{g}(\xi) + \hat{\delta}(\xi)) = \hat{u}(x,\xi) + e^{\sqrt{i\xi}(1-x)}\hat{\delta}(\xi). \qquad (2.4)$$

Since we cannot expect the error $\hat{\delta}(\xi)$ to have the same decay in frequency as the exact data $\hat{g}(\xi)$ the solution $\hat{v}(x,\xi)$ will not, in general, be in $L^2(R)$. Thus, if we try to solve the problem (1.3) numerically, high-frequency components in the error δ, are magnified and can destroy the solution. We impose an a priori bound on the solution at $x = 0$. We consider the problem

$$\begin{cases} u_{xx} = u_t, 0 \le x, 0 \le t, \\ u(x,0) = 0, x \ge 0, \\ \|u(1,t) - g_m(t)\| \le \varepsilon, \\ \|u(0,t)\| \le M, \end{cases} \qquad (2.5)$$

where M is positive constant.

3 A Fourier Method

We know that the solution to (1.1) is showed to be

$$u(x,t) = \frac{1}{\sqrt{2\pi}} \int_{-\infty}^{\infty} e^{\sqrt{i\xi}(1-x)} \hat{g}(\xi) e^{i\xi t} d\xi. \qquad (3.1)$$

3.1 Method I

In reference [1], the way to stabilize the problem is to eliminate all high frequencies from the solution and instead only for $|\xi| \le \xi_{max}$. A regularized solution is

$$v(x,t) = \frac{1}{\sqrt{2\pi}} \int_{-\infty}^{\infty} e^{\sqrt{i\xi}(1-x)} \hat{g}_m(\xi) \chi_{max} e^{i\xi t} d\xi = \frac{1}{\sqrt{2\pi}} \int_{-\xi_{max}}^{\xi_{max}} e^{\sqrt{i\xi}(1-x)} \hat{g}_m(\xi) e^{i\xi t} d\xi, \qquad (3.2)$$

where χ_{max} is a characteristic function : $\chi_{max} = \begin{cases} 1, |\xi| \le \xi_{max} \\ 0, otherwise \end{cases}$ .The error esti-

mate is also given in Reference [1].

Theorem 3.1[1]. *Suppose that $u(x,t)$ is given by (3.1) with the exact data $g(t)$ and that $V(x,t)$ is given by (3.2) with measured data $g_m(t)$ .If we have a bound $\|u(0,t)\| \le M$ and the measured function satisfies $\|g(t) - g_m(t)\| \le \varepsilon$, and we choose $\xi_{max} = 2(\ln(\frac{M}{\varepsilon}))^2$, then we get the error bound*

$$\|u(x,t) - v(x,t)\| \le 2M^{1-x}\varepsilon^x . \tag{3.3}$$

3.2 Method II

In fact, this method eliminates all high frequencies components of $g_m(t)$, meanwhile it does the high-frequency components of $g(t)$. We propose a method to get the regularized solution

$$V(x,t) = \frac{1}{\sqrt{2\pi}} \int_{-\infty}^{\infty} e^{\sqrt{i\xi}\chi_{max}(1-x)} \hat{g}_m(\xi) e^{i\xi t} d\xi, \tag{3.4}$$

where χ_{max} is defined in (3.2).The regularization of (3.4) can keep parts of high-frequency components of data $g(t)$.

We will derive an error estimate for the approximate solution (3.4) and a bound on the difference between the solution (3.1) and (3.4). We assume that we have an a priori bound on the solution $\|u(0,t)\| \le M$.

Lemma 3.2 *Suppose that we have two regularized solutions V_1 and V_2 defined by (3.4) with data g_1 and g_2 , satisfying $\|g_1 - g_2\| \le \varepsilon$ .If we select $\xi_{max} = 2(\ln(\frac{M}{\varepsilon}))^2$,then we get the error bound*

$$\|v_1(x,t) - v_2(x,t)\| \le M^{1-x}\varepsilon^x \tag{3.5}$$

Proof. From the Parseval relation we get

$$\|v_1(x,t) - v_2(x,t)\|^2 = \|\hat{v}_1(x,\xi) - \hat{v}_2(x,\xi)\|^2$$

$$= \int_{-\infty}^{\infty} \left| e^{\sqrt{i\xi}\chi_{max}(1-x)} (\hat{g}_1 - \hat{g}_2) \right|^2 d\xi$$

$$= \int_{|\xi| > \xi_{max}} \left| \hat{g}_1 - \hat{g}_2 \right|^2 d\xi + \int_{|\xi| \le \xi_{max}} \left| e^{\sqrt{i\xi}(1-x)} (\hat{g}_1 - \hat{g}_2) \right|^2 d\xi$$

$$\leq \int_{|\xi|>\xi_{max}} |\hat{g}_1 - \hat{g}_2|^2 \, d\xi + e^{2\sqrt{\frac{\xi_{max}}{2}}(1-x)} \int_{|\xi|\leq\xi_{max}} |\hat{g}_1 - \hat{g}_2|^2 \, d\xi$$

$$\leq e^{2\sqrt{\frac{\xi_{max}}{2}}(1-x)} \int_{-\infty}^{\infty} |\hat{g}_1 - \hat{g}_2|^2 \, d\xi = e^{2\sqrt{\frac{\xi_{max}}{2}}(1-x)} \varepsilon^2 = M^{2(1-x)}\varepsilon^{2x}.$$

We obtain $\left\|v_1(x,t)-v_2(x,t)\right\| \leq M^{1-x}\varepsilon^x$.

Lemma 3.3. *Let $u(x,t)$ and $v(x,t)$ be the solution (3.1) and (3.4) with the same exact data $g(t)$, and let $\xi_{max} = 2(\ln(\frac{M}{\varepsilon}))^2$. Suppose that* $\left\|u(0,t)\right\| \leq M$. *Then*

$$\left\|u(x,t)-v(x,t)\right\| \leq M^{1-x}\varepsilon^x \tag{3.6}$$

Proof. As in Lemma 3.1 we start with the Parseval relation, we get

$$\left\|u(x,t)-v(x,t)\right\|^2 = \left\|\hat{u}(x,\xi)-\hat{v}(x,\xi)\right\|^2$$

$$= \int_{-\infty}^{\infty} \left|(e^{\sqrt{i\xi}(1-x)} - e^{\sqrt{i\xi}\chi_{max}(1-x)})\hat{g}\right|^2 \, d\xi$$

$$= \int_{|\xi|>\xi_{max}} \left|e^{\sqrt{i\xi}(1-x)} - 1\right|^2 |\hat{g}|^2 \, d\xi = \int_{|\xi|>\xi_{max}} \left|e^{-x\sqrt{i\xi}} - e^{-\sqrt{i\xi}}\right|^2 \left|e^{\sqrt{i\xi}}\hat{g}\right|^2 \, d\xi$$

$$= \int_{|\xi|>\xi_{max}} \left|e^{-2x\sqrt{i\xi}}\right| \left|1-e^{(x-1)\sqrt{i\xi}}\right|^2 \left|e^{\sqrt{i\xi}}\hat{g}\right|^2 \, d\xi$$

$$\leq e^{-2x\sqrt{\frac{\xi_{max}}{2}}} \int_{|\xi|>\xi_{max}} \left|1-e^{(x-1)\sqrt{i\xi}}\right|^2 \left|e^{\sqrt{i\xi}}\hat{g}\right|^2 \, d\xi.$$

Since

$$\left|1-e^{(x-1)\sqrt{i\xi}}\right|^2 = (1-e^{(x-1)\sqrt{\frac{|\xi|}{2}}}\cos((x-1)\sqrt{\frac{|\xi|}{2}}))^2 + (e^{(x-1)\sqrt{\frac{|\xi|}{2}}}\sin((x-1)\sqrt{\frac{|\xi|}{2}}))^2.$$

Suppose $\lambda = (x-1)\sqrt{\frac{|\xi|}{2}}$, we obtain

$$\left|1-e^{(x-1)\sqrt{i\xi}}\right|^2 = (1-e^{\lambda}\cos\lambda)^2 + (e^{\lambda}\sin\lambda)^2 = 1 - 2e^{\lambda}\cos\lambda + e^{2\lambda}.$$

Since $\lambda < 0$, so $0 < e^{\lambda} < 1$, $\left|1-e^{(x-1)\sqrt{i\xi}}\right|^2 \leq 1$,

So $\left\|u(x,t)-v(x,t)\right\|^2 \leq e^{-2x\sqrt{\frac{\xi_{max}}{2}}} \int_{|\xi|>\xi_{max}} \left|e^{\sqrt{i\xi}}\hat{g}\right|^2 \, d\xi$

$$= e^{-2x\sqrt{\frac{\xi_{max}}{2}}} M^2 = M^{2(1-x)}\varepsilon^{2x}.$$

We obtain
$$\|u(x,t) - v(x,t)\| \le M^{1-x}\varepsilon^x.$$

Theorem 3.4. *Suppose that* $u(x,t)$ *is given by (3.1) with exact data* $g(t)$ *and* $v(x,t)$ *is given by (3.4) with measured data* $g_m(t)$. *If we have a bound* $\|u(0,t)\| \le M$, *and the measured function* $g_m(t)$ *satisfies* $\|g(t) - g_m(t)\| \le \varepsilon$, *and if we choose* $\xi_{max} = 2(\ln(\frac{M}{\varepsilon}))^2$, *then we get the error bound*

$$\|u(x,t) - v(x,t)\| \le 2M^{1-x}\varepsilon^x. \tag{3.7}$$

Proof. Let $v_1(x,t)$ be the solution defined by (3.4) with exact data $g(t)$. Then by using the triangle inequality and the two previous lemmas we get

$$\|u(x,t) - v(x,t)\| \le \|u(x,t) - v_1(x,t)\| + \|v_1(x,t) - v(x,t)\| \le 2M^{1-x}\varepsilon^x.$$

From Theorem 3.4 we find that (3.4) is an approximation of the exact solution, $u(x,t)$. The approximation error depends continuously on the measurement error.

4 Numerical Implementation

Now we give the algorithms to implement (3.2) and (3.4) numerically.

4.1 Algorithm I

The numerical implementation of the regularized solution (3.2) will be given. The regularized solution (3.2) is approximated as follows:

$$V(0,:) = F^H \Lambda F G_m, \tag{4.1}$$

where G_m is a vector containing samples from g_m on an equidistant grid $\{t_k\}_{k=0}^{n-1}$, F is the Fourier Matrix [6] and Λ is a diagonal matrix. The diagonal elements of Λ are

$$(\Lambda)_{k,k} = \begin{cases} \exp(\sqrt{i\xi_k}), |\xi_k| < \xi_{max} \\ 0, |\xi_k| \ge \xi_{max} \end{cases}, \tag{4.2}$$

where the ξ_k are defined as follows .The unique trigonometric polynomial inter-

polating g_m on the grid $\{t_k\}_{k=0}^{n-1}$ can be written

$$\tilde{g}(x,t) = \frac{1}{\sqrt{2\pi}} \sum_{k=-n/2}^{n/2-1} \hat{g}_k e^{i\xi_k t}, \xi_k = 2\pi k.$$ The product of F and a vector can be

computed using FFT . When using FFT algorithm we implicitly assume that the vector G_m represents a periodic function. For more details we refer to [1].

4.2 Algorithm II

The reference [1] has given the numerical implementation and has discussed how to compute the regularized solution (3.4) numerically. The solution operator (3.4) is approximated as follows:

$$V(0,:) = F^H \exp(\sqrt{\tilde{\Lambda}})FG_m , \qquad (4.3)$$

where G_m and F have the same meaning in (4.1) and $\tilde{\Lambda}$ is a diagonal matrix. The diagonal elements of $\tilde{\Lambda}$ are

$$(\tilde{\Lambda})_{k,k} = \begin{cases} i\xi_k, |\xi_k| < \xi_{\max} \\ 0, |\xi_k| \geq \xi_{\max} \end{cases}, \qquad (4.4)$$

where the ξ_k have the same meaning in (4.2).

So we see that the error estimate and computing cost using these two methods to regularize are same. Algorithm I and II match the Method I and II.

5 Numerical Experiments

The tests are performed in the following way: we select a solution $u(0,t) = f(t), 0 \leq t \leq 1$ and compute data functions $u(1,t) = g(t)$ by solving a well-posed quarter plane problem for the heat equation using a finite difference scheme. Then we add a normally distributed perturbation of variance 10^{-3} to each data function , giving vector g_m. Our error estimates using the signal-to-noise

rate M / ε and there we compute $\dfrac{M}{\varepsilon} \approx \dfrac{\|f\|}{\|g - g_m\|}$ [1]. We solve the problem

(1.1) using a discontinuous function and continuous function as the exact solution. We use (4.1) to do numerical tests in Figure 1 and use (4.3) to do numerical tests in Figure 2. The error comparisons are given in Table 1.

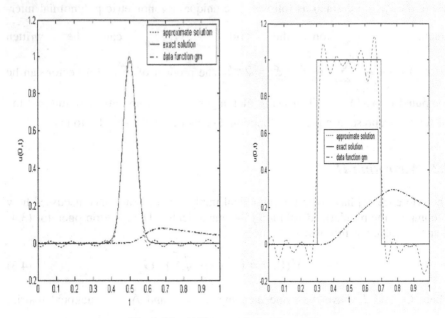

Fig. 1. Use (4.1) to regularize the problem

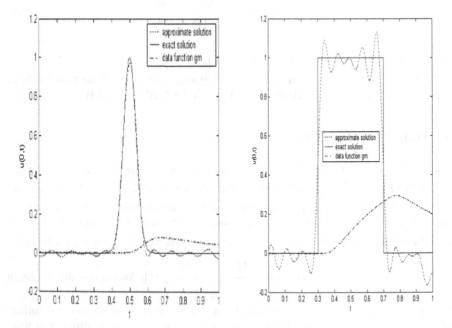

Fig. 2. Use (4.3) to regularize problem

Table 1. Error comparisons

	Discontinuous function		Continuous function	
	Use (4.1)	Use (4.3)	Use (4.1)	Use (4.3)
$\|v(0,t)-u(0,t)\|_2$	4.548613	4.547841	0.651051	0.651960

6 Conclusion

We consider an inverse heat conduction problem, the sideways heat equation. We give the theory analysis and numerical implementation to match the theory analysis andnumerical implementation. Compare these two methods. These two methods have the same error estimate, stability and computing complexity. The difference between them is (4.3) keeps parts of high-frequency components of the data. These two methods are available to this problem (IHCP).

References

1. Elden, L., Berntsson, F., Reginska, T.: Wavelet and Fourier method for solving the sideways heat equation. J. Sci. Comput. 21(6), 2187–2205 (2000)
2. Reginska, T., Elden, L.: Solving the sideways heat equation by a Wavelet-Galerkin method. Inverse Problems 13(5), 1093–1106 (1997)
3. Cannon, J.R.: The One Dimensional Heat Equation. Addision-Wesley, Reading (1984)
4. Hao, D.N., Reinhardt: On a sideways parabolic equation. Inverse Problems 13(4), 297–309 (1997)
5. Elden, L.: Numerical solution of the sideways heat equation by difference approximation in time. Inverse Problem 11, 913–923 (1995)
6. Van Loan, C.F.: Computational Frameworks for the fast Fourier Transform. SIAM, Philadelphia (1992)

On the Technology of Fuzzy Neural Network Identifying Complex Reservoir

Da-li Guo[1,2], Xiao-hui Zeng[3], Ruo-huai Lei[2],
Jin Chang[2], and Xi-jun Ke[2]

[1] State Key Laboratory of Oil and Gas Reservoir Geology and Exploitation,
  Southwest Petroleum University, Chengdu 610500, P.R. China
  guodali@sina.com
[2] School of Sciences, Southwest Petroleum University,
  Chengdu 610500, P.R. China
[3] School of Petroleum Engineering, Southwest Petroleum University,
  Chengdu 610500, P.R. China

Abstract. Reservoir identification plays an important role in the oil and gas exploration, especially in the complex reservoir analysis. The complex geological environment gives rise to the complexity of the reservoir, which is taken shape as a result of multiple geological factors, thus making the underground oil and gas reservoirs possess a strong nonlinearity and fuzzy features. This paper is based on the fuzzy neural network theoretically, and takes advantage of the fuzzy neural network technology to identify complex reservoir. In accordance with different complex reservoir and their features, this technology makes fully use of well-logging, well-drilling, core stone and test to choose different parameter for different complex reservoir as input parameter. In the light of different input parameter and the number of middle layer neurons, this skill uses genetic algorithm to optimize initial parameter, and takes BP algorithm to improve convergence speed. This technology applies to carbonate reservoirs in Tarim, carbonate reservoirs in Long Gang, Sichuan, Karamay volcanic reservoir and water-sensitive reservoir, its coincidence rate reaches 83.62%, achieving fruitful applied outcomes.

Keywords: Reservoir identification, complex reservoir, fuzzy neural network, model, genetic algorithm.

1 Introduction

Currently, the methods to identify the hydrocarbon reservoir include: recognition method based on the multivariate statistical [2], fuzzy comprehensive evaluation based on fuzzy mathematics and identification method of degree of approximation [8,10], gray multi-parameter weighted identification method

B.-Y. Cao and X.-J. Xie (Eds.): Fuzzy Engineering and Operations Research, AISC 147, pp. 433–440.
springerlink.com   © Springer-Verlag Berlin Heidelberg 2012

based on gray system theory [11], artificial intelligence-based expert system identification method [6,9], and so on. Nevertheless, the above-mentioned methods have shortcomings in some senses. For instance, most identification methods are linear, but the Fluid properties in complex reservoir and the difficult nonlinearity relation among the parameters needed explaining can not be fully reflected by the simple linear model, these methods cannot establish explanation model without real data. Neither can they build an explanation model based on the parameters chosen arbitrarily from the real information. Low-level of systemization is also the big problem of these methods in application. The fuzzy neural network [1,3,4,5,7] is a nonlinear mapping in possession of capacities of error-tolerance, self-adaptation, self-study and parallel-type. Utilizing the fuzzy neural network technology can conduct a comprehensive evaluation for the purpose of identifying complex reservoir more accurately.

2   The Technology of Fuzzy Neural Network Identifying Complex Reservoir

2.1   Fuzzy Neural Network Model

The fuzzy neural network constructed by this paper takes five networks and back-propagation training method multiple input can be fuzzifier in different conditions. And takes genetic algorithm when training to adjust proportion in every unit of hide layer. The network is shown in the following Figure 1.

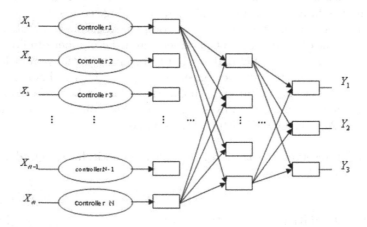

Fig. 1. Hybrid fuzzy neural network

Input layer (layer 1): every unit directly links with input component x_i, meanwhile, inputs variable $X = [x_1, x_2, \cdots, x_n]^T$ as an input of the next layer, among which T is a transposition, n is the number of the input variable, and $O_i^s = u_i^s, (i = 1, 2, \cdots, n)$, where s is a layer of the network, u_i^s is the i−th unit input of s−th layer, and O_i^s is the i−th unit output of s−th layer.

Condition for layer (layer 2): every unit represents a language variable. This layer is tailored for calculating membership function of all language variable values fuzzy set.

$$Q_{ij}^2 = e^{\frac{-u_{ij}^2 - c_{ij}^2}{\sigma_{ij}^2}} \quad (i = 1, 2, \cdots, n; j = 1, 2, \cdots, m), \tag{1}$$

where, m_i is a fuzzy partition number of x_i, c_{ij} and σ_{ij} is a core and width of the membership function. k−th moment's input is $u_{ij}^2(k) = O_i^1(k) + O_{ij}^2(k-1) \cdot \theta_{ij}$, θ_{ij} equals to linking proportion of recursive unit. In this layer $O_{ij}^2(k-1)$ records k-1 moment's information, thus being dynamic mapping.

Rule layer (layer 3): every unit represents a condition for fuzzy rule, which is aimed to perform utilization of every rule of fuzzy and operative accumulation.

$$O_i^3 = \prod_j u_{ij}^3 \quad (i = 1, 2, \cdots, n; j = 1, 2, \cdots, m). \tag{2}$$

Output layer (layer 4): output.

$$y_j = O_j^5 = \sum u_i^5 \omega_{ij}^5 \quad (j = 1, 2, \cdots, r), \tag{3}$$

where, r is the number of output variable, ω_{ij}^5 is the proportion of layer 5.

2.2 Study Algorithm

This paper combines genetic algorithm with BP algorithm to adjust every proportion and other parameters, because of the low searching speed of the former. So the combination of BP algorithm can settle this problem. Generally speaking, use the former first to select optimal resolution for initial parameters, then use latter to speed up convergence. Define the objective function as follows,

$$J = \frac{1}{2} \sum_i^r (y_{di} - y_i)^2. \tag{4}$$

A detailed training process:

1) encoding and decoding. Use genetic algorithm to conduct parameter training of neural network, encoding first, here it is a compact method, that is, every parameter gained from network is putted together by a certain sequence, the length of every parameter is binary number in L. The method is: the changes domain of α is the interval $[\alpha_{min}, \alpha_{max}]$, b represents binary number, the relation between the two parameters is

$$\alpha = \alpha_{min} + \frac{b}{2^L}(\alpha_{max} - \alpha_{min}).$$

2) fitness function is inverse of the objective function. i.e., $F = 1/J$.

3) genetics. In order to evaluate every individual among the group, the probability to F_i is P_i, $P_i = \frac{F_i}{\sum_{j=1}^{m} F_j}$ the method in training is select the fittest individual to inhere to the next generation.

4) cross. Conduct unit cross to network using probability P_c. First we need to pair up two individuals among the group arbitrarily. If one group is M, then half M pairs. Then to one in the pair, we set random the location afterward a locus as a intersection. According to the set cross probability, exchange two individual's part chromosome in intersection a new individual one come into being as a result.

5) mutation. Change individual's chromosome randomly by probability P_m. A new individual takes shape. Because of the influence P_c and P_m exerting on genetic algorithm, P_c is used to be $0.3 \sim 0.8$, and P_m $0.004 \sim 0.101$. However for accurate accumulation, a specified numerics needs to be ascertained after trial and error.

When the average value of adjustment of optimized individual and the adjustment of the group didn't change dramatically, use BP algorithm to speed up the network convergence. Let

$$\delta_i^5 = -\frac{\partial E}{\partial O_i^5} = y_{di} - y_i \tag{5}$$

be handled by proportion and membership function. Then we obtained the algorithm as follows,

$$\omega_{ij}(k+1) = \omega_{ij}(k) - \beta \frac{\partial E}{\partial \omega_{ij}} (i = 1, 2, \cdots, r; j = 1, 2, \cdots, m),$$

$$\begin{cases} c_{ij}(k+1) = c_{ij}(k) - \beta \dfrac{\partial E}{\partial \omega_{ij}}, \\[2em] \sigma_{ij}(k+1) = \sigma_{ij}(k) - \beta \dfrac{\partial E}{\partial \sigma_{ij}}, (i = 1, 2, \cdots, r; j = 1, 2, \cdots, m), \\[2em] \theta_{ij}(k+1) = \theta_{ij}(k) - \beta \dfrac{\partial E}{\partial \theta_{ij}}, \end{cases} \tag{6}$$

where, β is the rate of study, $\beta > 0$.

2.3 Total Structure

The fuzzy neural network technology is used to comprehensively evaluate complex reservoir, and the fuzzy neural network evaluation is classified into the following three parts: the construction of fuzzy neural network, the training of fuzzy neural network and comprehensively evaluate complex reservoir of fuzzy neural network. The flowchart is shown in Figure 2.

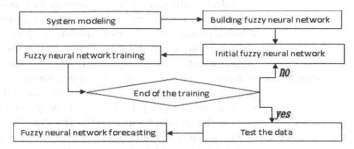

Fig. 2. The total structure of comprehensive evaluation on fuzzy neural network

where, building fuzzy neural network is to establish the input and output number of network, membership function and the number of parameter, fuzzy neural network training use data range in criteria of evaluation to train the network, and fuzzy neural network evaluation utilize well-trained fuzzy neural network to assess complex reservoir.

3   Emulated Experiment

According to the materials such as well-logging, well-drilling, core stone and test, the related parameters are used to recognize oil and gas reservoir. In the key project, since there is a correlation between sidewall parameter and core stone parameter, these parameters can be merged. That is, when the core stone parameter exists, we take the core stone parameter, otherwise we take sidewall parameter instead of it. In this way we could overcome the disadvantage of insufficient core stone information, and play the advantages of more accurate core stone parameters. In the following the parameters shown in Table 1 is chosen as candidate parameters of input layer in constructing a fuzzy neural network model.

In the above-mentioned candidate parameters, taking account of the feature of different complex reservoir and in accordance with different complex reservoir, we can choose different input parameters, as for carbonate reservoir, we chiefly select the quantity contained in the dolomite as the input parameter, as for volcanic reservoir, we chiefly select fracture condition as the input parameter, as for water-sensitive reservoir, we chiefly choose the absolute quantity contained of the clunch, the relative amount of illitesmectite formation as the input parameter. One parameter corresponds to an input layer neurons. As for the determination of the number of the neuron in

Table 1. Candidate parameters of input layer in a fuzzy neural network model

Class	the optimum parameter
General Parameters	horizon, lithology, core stone, total pore volume, effective drainage porosity, secondary porosity, depth, absolute permeability, effective permeability, relative permeability, terrane thickness, effective thickness, formation temperature gradient, pressure coefficient, formation fluid viscosity, reservoir water salinity.
Well-testing parameters	natural potential, natural gamma, formation true resistivity, shallow two-way resistivity, interval transit time, neutron, depth of the resistivity difference, the quantity contained in the clay
Well-logging parameters	fluorescent color, fluorescence level, percentage of total lithic sandstone, oil quantity accounting for the percentage of sandstone
Gas measured parameters	the value of all alkyl, all hydrocarbon growth, c1,c2,c3,c4,c5
Other parameters	capacity factor, water saturation, storage coefficient, Reservoir permeability factor, movable water index, media type factor, effective free water saturation, methane, oil length, the total water saturation, degree of oil long, length of traces of oil, fracture condition, length of the oil patch, depth of the resistivity ratio, undisturbed formation resistivity, the quantity contained in the dolomite, the absolute quantity in the clunch, the relative amount of illitesmectite formation

the middle layer of the fuzzy neural network, usually we select the neurons input layer of the system.

Under normal circumstance, if the neurons in middle layer are in short, the network will emphasize more of the general character than of individual sample attribute, what is worse is that the network cannot study at all. If there are too many neurons, the network will remember every sample instead of concluding the rules of all the samples. Therefore, the optimum neurons can be ascertained resulting from repeatedly testing the neurons in middle layer.

Conduct comprehensive evaluation to complex reservoir by using fuzzy neural network. Based on the feature of input and output data, firstly we construct neural network structure. Since input data is N items and output data are oil production, gas production and water production per day, we set input nodes as N and output nodes as 3, respectively. Secondly, we predict to fuzzy neural network with existing data, conduct training of fuzzy neural network by comparing the predicted values and expectations, and conduct training 20000 times repeatedly. Last, using trained network to predict on trained data, we present network training process shown in Figure 3.

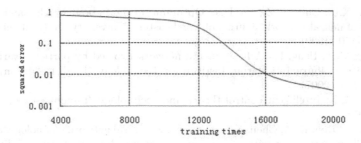

Fig. 3. Network training process

The trained fuzzy neural network is used to comprehensively evaluate complex reservoir. This technology applies to carbonate reservoirs in Tarim, carbonate reservoirs in Long Gang, Sichuan, Karamay volcanic reservoir and water-sensitive reservoir, its coincidence rate reaches 83.62%, achieving fruitful applied outcomes.

4 Conclusions and Identification

1) Based on the recognition method of fuzzy neural network, a predictive Model has been constructed by summarizing inherent regularity contained in known trained data, which could express complex nonlinear rule.
2) Fuzzy neural network can select the suitable parameter to build up forecasting model at a basis of actual information. Therefore, it boasts a sound flexibility and adjustment overcoming shortage of the neural network of a fixed structure.
3) The technology that using fuzzy neural network to conduct comprehensive evaluation of complex reservoir perfects well and confirms to the reality, which is an effective method of identifying complex reservoir.

Acknowledgements. Thanks to the support by major projects of national science and technology demonstration projects(No.2011ZX05062).

References

1. Deng, H.W.: Comprehensive evaluation research on fuzzy neural network. Southwest Normal University (2005)
2. Fan, X.Y., Huang, Y., Liu, Y.H., et al.: New method of reservoir fluid identification based on correspondence analysis theory. Drilling and Production Technology 1, 39–42 (2009)
3. Han, M., Sun, Y.N.: Multi-input fuzzy neural network and its application. Systems Engineering and Electronics 10, 1249–1251 (2003)

4. Hou, X.L., Chen, C.Z., Yu, H.J.: Optimum method about weights and thresholds of neural network. Journal of Northeastern University (Natural Science) 4, 447–450 (1999)
5. Peng, Z.P., Peng, H.: A fuzzy neural network evolved by particle swarm optimization. Journal of Harbin Institute of Technology (English Edition) 14(3), 316–321 (2007)
6. Sun, Z.Q.: Intelligent control theory and technology. Tsinghua Press, Beijing (1997)
7. Tan, T., Ren, K.C., Chen, X.L., et al.: Fuzzy neural network technology. Journal of Chongqing University of Arts and Sciences (Natural Science Edition) 2, 71–74 (2011)
8. Takagi, T., Sugeno, M.: Fuzzy identification of systems and its application to modeling and control. IEEE Trans. on System, Man, and Cybernetworkics 15(1), 21–24 (1985)
9. Wu, Q.: Performance evaluation of enterprises based on Bbck propagation network model. Journal of China University of Geosciences (Social Sciences Edition) 2, 91–95 (2008)
10. Wang, J.T., Ma, K.D.: Identifying hydrocarbon and water zones using fuzzy comprehensive evaluation method. Well Logging Technology 2, 23–25 (2006)
11. Wang, X.H., Pan, B.Z.: The application of genetic algorithm in mafic volcanic reservoir evaluation logging. World Well Logging Technology 2, 38–40 (2010)

High-Order and Unconditional Stability Difference Schemes for One-Dimensional Diffusion Equation

Jian-Ying Xiao[1], Xiao-Hua Liu[1], Yong-Tao Li[2], and Dong-Hong Tian[1]

[1] School of Sciences, Southwest Petroleum University, Chengdu, Sichuan 610500, China
[2] School of Chemistry and Chemical Engineering, Southwest Petroleum University, Chengdu, Sichuan 610500, China
xiaojianying1980@sina.com

Abstract. At first, a difference scheme is obtained for the diffusion equation without source. The difference scheme with high-order and unconditional stability is got by the analysis of the truncation error. Then, the scheme is also achieved for the diffusion equation with source. Finally, the effectiveness of the scheme is verified by using the numerical experiments.

Keywords: Difference Schemes, diffusion equation, unconditional stability, truncation errors.

1 Introduction

In the problem of predicting the actual problems accurately, the traditional schemes maybe have low-order or conditional stability, so it is necessary to analyze the new difference scheme with high-order and unconditional stability [1-6]. Further, the research on the new difference scheme of the diffusion equation can be extended to the construction of the difference schemes for other partial differential equations [4-7], because the diffusion equation is the foundation of many partial differential equations such as convection diffusion equation which can be transformed to the diffusion equation by using certain methods. Therefore, a new difference scheme which is more helpful to more accurate prediction is established in the paper.

2 The Construction of the New Difference Scheme and Stability

2.1 Analysis of Difference Schemes on Diffusion Equation without Source

Give a one-dimensional diffusion equation model:

B.-Y. Cao and X.-J. Xie (Eds.): Fuzzy Engineering and Operations Research, AISC 147, pp. 441–447.
springerlink.com   © Springer-Verlag Berlin Heidelberg 2012

$$\frac{\partial u}{\partial t} = a\frac{\partial^2 u}{\partial x^2} + f(x,t)(a>0, -\infty < x < +\infty, t > 0). \tag{1}$$

It is a diffusion equation with source when $f(x,t) \neq 0$, and without source when $f(x,t) = 0$ [1-4]. Assume the time step to be τ and the spatial step to be h and $t_n = n\tau$, $x_j = jh,(j = 0, \pm 1, \pm 2, \pm 3, \cdots, n = 0, 1, 2, \cdots)$. So, let $u(x_j, t_n)$ express the exact value of $u(x,t)$ at the node (x_j, t_n) and the difference solution u_j^n express the approximate value of $u(x_j, t_n)$. That is:

$$u_j^n \approx u(x_j, t_n) = u(jh, n\tau). \tag{2}$$

The diffusion equation without source is discussed about its difference scheme at first in the paper:

$$\frac{\partial u}{\partial t} = a\frac{\partial^2 u}{\partial x^2} (a>0, -\infty < x < +\infty, t > 0) \ . \tag{3}$$

2.1.1 Typical Difference Schemes

First, give the Central difference scheme [5]:

$$\frac{u_j^{n+1} - u_j^n}{\tau} = a\frac{u_{j+1}^n - 2u_j^n + u_{j-1}^n}{h^2} \tag{4}$$

Its precision   is $O(h^2)+O(\tau)$,and it has certain stablity in a small network parameters.

Second, give the Crank—Nicolson difference scheme [5]:

$$\frac{u_j^{n+1} - u_j^n}{\tau} = a\frac{u_{j+1}^n - 2u_j^n + u_{j-1}^n}{2h^2} + a\frac{u_{j+1}^{n+1} - 2u_j^{n+1} + u_{j-1}^{n+1}}{2h^2} \ . \tag{5}$$

Its precision is $O(h^2)+O(\tau^2)$,and it has unconditional stability.

2.1.2 New Difference Scheme

The new difference scheme   is established as follows:

$$A\frac{u_{j+1}^{n+1} - u_{j+1}^n}{\tau} + B\frac{u_j^{n+1} - u_j^n}{\tau} + C\frac{u_{j-1}^{n+1} - u_{j-1}^n}{\tau} = a(D\delta_x^2 u_j^{n+1} + E\delta_x^2 u_j^n), \tag{6}$$

$$\begin{cases} A+B+C=1 \\ D+E=1 \end{cases}, \tag{7}$$

where A, B, C, D, E satisfy (7), and $B > A + C$ since the matrix must be iteratively diagonal dominant.

2.1.3 Truncation Error Analysis

The expanded form of the function at the node $(x_j, t_{n+\frac{1}{2}})$ can be got [1-11] by Taylor formula:

$$\frac{u_{j+1}^{n+1} - u_{j+1}^{n}}{\tau} = [\frac{\partial u}{\partial t}(x_j, t_{n+\frac{1}{2}}) + h\frac{\partial^2 u}{\partial t \partial x}(x_j, t_{n+\frac{1}{2}}) + \frac{h^2}{2!}\frac{\partial^3 u}{\partial t \partial x^2}(x_j, t_{n+\frac{1}{2}}) + \frac{h^3}{3!}\frac{\partial^4 u}{\partial t \partial x^3}(x_j, t_{n+\frac{1}{2}}) + O(h^4)]$$

$$+\frac{\tau}{2}[\frac{\partial^2 u}{\partial t^2}(x_j, t_{n+\frac{1}{2}}) + h\frac{\partial^3 u}{\partial t \partial x^2}(x_j, t_{n+\frac{1}{2}}) + \frac{h^2}{2!}\frac{\partial^4 u}{\partial t \partial x^3}(x_j, t_{n+\frac{1}{2}})$$

$$+\frac{h^3}{3!}\frac{\partial^5 u}{\partial t \partial x^4}(x_j, t_{n+\frac{1}{2}}) + O(h^4)] + O(\tau^2) \tag{8}$$

$$\frac{u_j^{n+1} - u_j^{n}}{\tau} = \frac{\partial u}{\partial t}(x_j, t_{n+\frac{1}{2}}) + \frac{\tau}{2}\frac{\partial^2 u}{\partial t^2}(x_j, t_{n+\frac{1}{2}}) + O(\tau^2) \tag{9}$$

$$\frac{u_{j-1}^{n+1} - u_{j-1}^{n}}{\tau} = [\frac{\partial u}{\partial t}(x_j, t_{n+\frac{1}{2}}) - h\frac{\partial^2 u}{\partial t \partial x}(x_j, t_{n+\frac{1}{2}}) + \frac{h^2}{2!}\frac{\partial^3 u}{\partial t \partial x^2}(x_j, t_{n+\frac{1}{2}}) - \frac{h^3}{3!}\frac{\partial^4 u}{\partial t \partial x^3}(x_j, t_{n+\frac{1}{2}}) + O(h^4)]$$

$$+\frac{\tau}{2}[\frac{\partial^2 u}{\partial t^2}(x_j, t_{n+\frac{1}{2}}) - h\frac{\partial^3 u}{\partial t \partial x^2}(x_j, t_{n+\frac{1}{2}}) + \frac{h^2}{2!}\frac{\partial^4 u}{\partial t \partial x^3}(x_j, t_{n+\frac{1}{2}}) - \frac{h^3}{3!}\frac{\partial^5 u}{\partial t \partial x^4}(x_j, t_{n+\frac{1}{2}}) + O(h^4)] + O(\tau^2) \tag{10}$$

$$\frac{u_{j+1}^{n+1} - 2u_j^{n+1} + u_{j-1}^{n+1}}{h^2} = \frac{\partial^2 u}{\partial x^2}(x_j, t_{n+\frac{1}{2}}) + \frac{h^2}{12}\frac{\partial^4 u}{\partial x^4}(x_j, t_{n+\frac{1}{2}}) + \tau[\frac{\partial^3 u}{\partial x^2 \partial t}(x_j, t_{n+\frac{1}{2}}) + \frac{h^2}{12}\frac{\partial^5 u}{\partial x^4 \partial t}(x_j, t_{n+\frac{1}{2}})] + O(\tau^2) + O(h^4) \tag{11}$$

$$\frac{u_{j+1}^{n} - 2u_j^{n} + u_{j-1}^{n}}{h^2} = \frac{\partial^2 u}{\partial x^2}(x_j, t_{n+\frac{1}{2}}) + \frac{h^2}{12}\frac{\partial^4 u}{\partial x^4}(x_j, t_{n+\frac{1}{2}}) + O(\tau^2) + O(h^4) \ . \tag{12}$$

The truncation error can be got by substituting the equations from (8) and (12) into the Equation (6):

$$R_j^{n+\frac{1}{2}} = (A-C)ha\frac{\partial^3 u}{\partial x^3}(x_j, t_{n+\frac{1}{2}}) + [A+C-\frac{1}{6}(D+E)]\frac{h^2}{2}a\frac{\partial^4 u}{\partial x^4}(x_j, t_{n+\frac{1}{2}})$$

$$+(A-C)\frac{h^3}{6}a\frac{\partial^5 u}{\partial x^5}(x_j, t_{n+\frac{1}{2}}) + \frac{\tau}{2}a^2(A+B+C)\frac{\partial^4 u}{\partial x^4}(x_j, t_{n+\frac{1}{2}})$$

$$+\frac{\tau}{2}(A-C)ha\frac{\partial^5 u}{\partial x^5}(x_j, t_{n+\frac{1}{2}}) + (A+C-\frac{1}{3}D)\frac{1}{4}a^2 h^2 \tau\frac{\partial^6 u}{\partial x^6}(x_j, t_{n+\frac{1}{2}})$$

$$+(A-C)\frac{h^3 \tau}{12}a\frac{\partial^7 u}{\partial x^7}(x_j, t_{n+\frac{1}{2}}) + O(\tau^2) + O(h^4).$$

Obviously,if the following equations hold:

$$\begin{cases} A - C = 0, \\ A + C = \dfrac{1}{6}(D + E), \\ A + B + C = 2D, \\ A + C = \dfrac{1}{3}D. \end{cases} \tag{13}$$

That is, $A = \dfrac{1}{12}, B = \dfrac{5}{6}, C = \dfrac{1}{12}, D = \dfrac{1}{2}, E = \dfrac{1}{2}$, then the precision of the format is $O(h^4) + O(\tau^2)$. Hence the difference Scheme becomes:

$$A_{j-1} u_{j-1}^{n+1} + A_j u_j^{n+1} + A_{j+1} u_{j+1}^{n+1} = B_{j-1} u_{j-1}^n + B_j u_j^n + B_{j+1} u_{j+1}^n, \tag{14}$$

where

$$r = \frac{a\tau}{h^2}$$

$A_{j-1} = 1 - 6r, A_j = 10 + 12r, A_{j+1} = 1 - 6r, B_{j-1} = 1 + 6r, B_j = 10 - 12r, B_{j+1} = 1 + 6r.$

If $D = E = 1/2, B = 1, A = C = 0$, then (14)is the traditional Crank—Nicolson difference scheme whose precision is $O(h^2) + O(\tau^2)$.

Different difference schemes can be got if different A, B, C, D, E in (7) are given according to format (6) whose difference equation coefficient matrix is strictly diagonally dominant.Thus,higher precision can be obtained by adjusting the value of A, B, C, D, E.

2.1.4   The Stability Analysis of New Difference Scheme

Transition factor: let $r = a\dfrac{\tau}{h^2}$, $G = \dfrac{\lambda^{n+1}}{\lambda^n} = \dfrac{10 + 2\cos\sigma h - 12r(1 - \cos\sigma h)}{10 + 2\cos\sigma h + 12r(1 - \cos\sigma h)}$, because $1 - \cos\sigma h \geq 0$, $|G| \leq 1$. Obviously,$|G(k, \tau)| \leq 1$ no matter what values k, h are assigned. The difference scheme (14) is absolutely stable by the Von Neumann conditions [1-4].

2.1.5   Comparison with the Typical Schemes

The stability and precision are different among the traditional center difference scheme, Crank-Nicolson difference scheme and the new difference scheme mentioned in this paper, so the contrast list is obviously obtained as follows:

Table 1. Contrast list of different difference schemes

item \ type	Centerial difference scheme	Crank-Nicolson difference scheme	The new difference scheme
stability	$r \leq 1/2$	absolutely stable	absolutely stable
precision	$O(h^2) + O(\tau)$	$O(h^2) + O(\tau^2)$	$O(h^4) + O(\tau^2)$

From the table aboved, it can be conclued: crank-Nicolson difference scheme and the new difference scheme are absolutey stable schemes, but their precisions are different. The precision of the new difference scheme which has about 4 order accuracy is the highest, while crank-Nicolson scheme has 2 order accuracy. Besides, the center difference scheme and the crank-Nicolson scheme which have the same precision are different from their precisons. In a word, the new difference scheme has the highest precison and absolutely stability.

2.2 Analysis of Difference Schemes on Diffusion Equation with Source

Consider a model including source diffusion equation:

$$\frac{\partial u}{\partial t} = a\frac{\partial^2 u}{\partial x^2} + f(x,t)(a>0,-\infty<x<+\infty,t>0) \tag{15}$$

2.2.1 The New Difference Scheme

The scheme is given as follows:

$$A\frac{u_{j+1}^{n+1}-u_{j+1}^{n}}{\tau} + B\frac{u_{j}^{n+1}-u_{j}^{n}}{\tau} + C\frac{u_{j-1}^{n+1}-u_{j-1}^{n}}{\tau} = a(D\delta_x^2 u_j^{n+1} + E\delta_x^2 u_j^n)$$

$$+F(f_j^n + f_j^{n+1}) + G(f_{j-1}^{n+1} + f_{j+1}^{n+1}) + H(f_{j-1}^n + f_{j+1}^n), \tag{16}$$

where A,B,C,D,E,F,G,H are parameters and satisfy:

$$\begin{cases} A+B+C=1 \\ E+D=1 \\ 2F+2G+2H=1 \end{cases}, \tag{17}$$

where A,B,C,D,E,F,G,H satisfy (17) ,and $B>A+C$ since the matrix must be iteratively diagonal dominant.

2.2.2 Truncation Error Analysis

The function can be expanded at the node$^{(x_j,t_{n+\frac{1}{2}})}$ by using Taylor formula:

$$f_j^n = f\left(x_j,t_{n+\frac{1}{2}}\right) - \frac{\tau}{2}\frac{\partial f}{\partial t}\left(x_j,t_{n+\frac{1}{2}}\right) + O(\tau^2) \tag{18}$$

$$f_j^{n+1} = f\left(x_j,t_{n+\frac{1}{2}}\right) + \frac{\tau}{2}\frac{\partial f}{\partial t}\left(x_j,t_{n+\frac{1}{2}}\right) + O(\tau^2) \tag{19}$$

$$f_{j-1}^{n+1} = f\left(x_{j-1},t_{n+\frac{1}{2}}\right) + \frac{\tau}{2}\frac{\partial f}{\partial t}\left(x_{j-1},t_{n+\frac{1}{2}}\right) + O(\tau^2) \tag{20}$$

$$f_{j-1}^{n+1} = f\left(x_{j-1},t_{n+\frac{1}{2}}\right) + \frac{\tau}{2}\frac{\partial f}{\partial t}\left(x_{j-1},t_{n+\frac{1}{2}}\right) + O(\tau^2) \tag{21}$$

$$f_{j+1}^{n+1} = f\left(x_{j+1},t_{n+\frac{1}{2}}\right) + \frac{\tau}{2}\frac{\partial f}{\partial t}\left(x_{j+1},t_{n+\frac{1}{2}}\right) + O(\tau^2) \tag{22}$$

$$f_{j-1}^n = f\left(x_{j-1},t_{n+\frac{1}{2}}\right) - \frac{\tau}{2}\frac{\partial f}{\partial t}\left(x_{j-1},t_{n+\frac{1}{2}}\right) + O(\tau^2), \tag{23}$$

$$f_{j+1}^n = f\left(x_{j+1}, t_{n+\frac{1}{2}}\right) - \frac{\tau}{2}\frac{\partial f}{\partial t}\left(x_{j-1}, t_{n+\frac{1}{2}}\right) + O(\tau^2) \ . \tag{24}$$

Similarly, (18-24)are taken into(16),can be got:

$$A_{j-1}u_{j-1}^{n+1} + A_j u_j^{n+1} + A_{j+1}u_{j+1}^{n+1} = B_{j-1}u_{j-1}^n + B_j u_j^n + B_{j+1}u_{j+1}^n + F_j^{n+1} + F_j^n , \tag{25}$$

where

$$A = \frac{1}{12}, B = \frac{5}{6}, C = \frac{1}{12}, D = \frac{1}{2}, E = \frac{1}{2}, F = \frac{5}{12}, G = H = \frac{1}{24}$$

$$A_{j-1} = 1 - 6r, A_j = 10 + 12r, A_{j+1} = 1 - 6r, B_{j-1} = 1 + 6r, B_j = 10 - 12r, B_{j+1} = 1 + 6r$$

$$F_j^{n+1} = \frac{\tau}{2}(10 f_j^{n+1} + f_{j+1}^{n+1} + f_{j-1}^{n+1}), F_j^n = \frac{\tau}{2}(10 f_j^n + f_{j+1}^n + f_{j-1}^n)$$

Because the low order f_j^n and f_j^{n+1} do not affect the stability of the difference format (12), we only consider the relevant passive stability in the analysis of stability on the equation with source. Using the same method,it is easily to see that the format (25) is absolutely stable [2].

3   Numerical Example

Consider the pure diffusion equation

$$\begin{cases} \dfrac{\partial u}{\partial t} = \dfrac{\partial^2 u}{\partial x^2}, 0 \le x \le \pi \\ u(x,0) = \sin x \\ u(0,t) = u(\pi,t) = 0, t > 0 \end{cases} \tag{26}$$

The analytical solution of the problem is $u = e^{-t}\sin x$ .Figure 1 shows theabsolute value of the error and its configuration   between analytical solution and numerical solution and numerical solution and exact solutions using MATLAB programming when $h = 0.1, \tau = 0.01$.

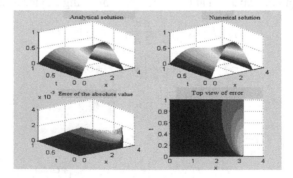

Fig. 1. Analytical solution, numerical solution, numerical solution and exact solution of the error of the absolute value of its top view when $h = 0.1, \tau = 0.01$.

4 Conclusion

A new difference scheme of one-dimensional diffusion equation is constructed in the paper.

1. The construction of the new difference scheme for diffusion equation based on numerical solution precision consideration and in contrast to the traditional typical difference scheme, can provide a powerful mathematical theory support for solving actual problem more accurately.

2. Different difference schemes can be got if different A, B, C, D, E in (7) are given according to format (6) whose difference equation coefficient matrix is strictly diagonally dominant. Thus, higher precision can be obtained by adjusting the value of A, B, C, D, E.

3. The format precision can be improved to $O(h^4) + O(\tau^2)$ by adjusting the value A, B, C, D.

4. The new difference scheme not only can be generalized to two-dimensional diffusion equations or other diffusion equations with higher-order but also can be used to difference schemes of other types of partial differential equations.

References

1. Sun, N.Z.: Groundwater pollution mathematical model and numerical method. Geological Press, Beijing (1989)
2. Li, R.H., Guo, B.Y.: Numerical solution of differential equation. Higher Education Press, Beijing (1996)
3. Guo, B.Y.: Partial differential equations of the finite difference method. Peking University Press, Beijing (1988)
4. Chen, G.Q.: A perturbational fourth-order upwind finite difference scheme for the convection-diffusion equation. Journal of Hydrodynamics 5(1), 82–97 (1993)
5. Yang, Z.F., Wang, X.: Research progress of High accuracy finite difference method. Progress in Natural Science 9(9), 769–779 (1999)
6. Strikwerda, J.C.: High-order-accurate schemes for incompressible viscous flow. Int. J. Numer Methods Fluids 24, 715–734 (1997)
7. Tian, Z.F., Zhang, Y.P.: Diffusion equation of high precision weighted difference scheme. Journal of University of Science and Technology of China (2), 237–241 (1999)
8. Ramos, J.I.: A: Piecewise-analytical method for singular perturbed parabolic problems. Applied Mathematics and Computation 161(2), 501–512 (2005)
9. Chen, H., Zhong, E.J.: High-order and unconditional stability difference scheme for diffusion equation. Journal of University of Electronic Science and Technoloav of China 4, 429–431 (2007)

Natural Gas Pipeline Corrosion Rate Prediction Model Based on BP Neural Network

Chi-yuan Ren[1], Wei Qiao[2], and Xin Tian[3]

[1] College of Sciences, Southwest Petroleum University, Chengdu 610500, China
renchiyuan@163.com
[2] Gas Transportation Department, Petrol China Southwest Oil & Gas Field Company,
Nanchong 637000, China
[3] Apply Technology College, Southwest Petroleum University, Chengdu 610500, China

Abstract. The transfer medium of oil and natural gas pipelines are flammable, explosive material, and contain various impurities to corrosion, so that internal and external corrosion of pipelines in the conditions are very complex, and pipe's flaws make the problem more serious. In case of explosion, leakage, shutdown and other accidents, it will lead to serious consequences. In recent years, pipeline spills have occurred, which cause great damage, and are harmful to the environment, so predicting the corrosion rate is very important and meaningful. In this issue, BP neural network is applied to predict the corrosion rate of long distance pipeline. Natural gas pipeline mileage, elevation difference, pipe inclination, pressure, Reynolds number, used as an input parameter and the maximum average corrosion rate of the pipeline used as an output parameter, a prediction model about a natural gas pipeline internal corrosion rate is established. The results show that: BP neural network is of better fitting precision and forecasting result, and prediction of corrosion rate is more reliable based on the model. The results compared with other methods show that, BP neural network algorithm is of fast convergence, with better prediction accuracy, which can predict effectively the corrosion rate of natural gas pipelines and meet the requirements of practical application accurately.

Keywords: BP neural network, natural gas pipeline, internal corrosion, corrosion rate, prediction model.

1   Introduction

Until 1996, China has built 395 long distance oil-gas pipelines, the total length of which is over 172,313,700 km, formed a Northeast, North, Central and Northwest regions of the vast network of underground pipelines. At present, because a considerable part of the long pipeline has entered the aging period, which results in frequent pipeline leak. Because the general length of the long pipelines are two to three hundred kilometers or more, after the accident it is difficult to find or determine the location immediately, and it may cause more accidents, so predicting the corrosion rate to natural gas pipeline is very important.

B.-Y. Cao and X.-J. Xie (Eds.): Fuzzy Engineering and Operations Research, AISC 147, pp. 449–455.
springerlink.com   © Springer-Verlag Berlin Heidelberg 2012

Artificial neural networks are very popular recently developed cross-disciplinary, which has a very wide range of applications. As the artificial neural network has a large-scale parallel information processing, adaptive and self-good and strong, and many other features, so the use of neural network in solving the complex problem of predicting nonlinear dynamic systems have a certain advantage. Prediction method based on neural network research in today's forecast is an important topic, with high theoretical and practical value. The design of BP neural network technology for gas gathering and transportation pipeline corrosion factors of complexity and uncertainty and the predicted remaining life of the pipeline corrosion problems, the establishment of corrosion prediction model.

Kongde Ying, Song Zhe use artificial neural network to find out the carbon steel and low alloy steel corrosion law in the sea environment, and establishes a forecast model of steel corrosion rate in different waters model [1]. However, they use the marine environment parameter data only include Qingdao, Zhoushan, Xiamen and Yulin area of the annual average, which does not fully reflect the different marine environments the corrosion rate. In addition, Kuo Chih-arc, Wang Yonghong, etc. will be applied to artificial neural network study of soil erosion, soil corrosion of carbon steel with seasonal variation[2]; Li Xiaogang, etc. in high-temperature sulfur corrosion, naphthenic acid corrosion in the application of artificial neural networks[3]; Cai Jian-ping , Ke-wei, the application of corrosion in the atmosphere of artificial neural networks, they have been good results, indicating the neural network in corrosion research has broad application prospects[4].

2   Preliminary

2.1   BP Neural Network Model of Pipeline Corrosion

Definition of neural network by T. Koholen: ``Artificial neural network is adaptive to a wide range of simple modules in parallel interconnection network, which can simulate the organization of biological nervous systems made of real world objects, interactive response." The brain may be looked as a neural network with more than 100 billion brain neurons.

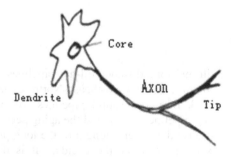

Fig. 1. Anatomy of neurons

Neuronal information transmission and processing is an electrochemical activity. As the role of Dendrites electrochemical stimulation received from outside; through cell axons in vivo activity reflected potential, when the axon potential reaches a certain value of the formation of nerve impulse or action potential; then passed to the other through the axons of peripheral neurons. From the control theory point of view; this process can be viewed as a multi-input single-output nonlinear system dynamic process.

Artificial neural network, the human brain as an abstract and simple simulation, is that people imitate the cerebral nervous system information processing capabilities of an intelligent system. Its emergence as people learns more about the mysteries of the human brain thinking a powerful tool. Although it is not a perfect model of the brain, but its unique non-linear adaptive information-processing capabilities, you can learn to get the knowledge and stored in the external network can solve computer problems is not easy to handle, especially in voice and image recognition, understanding, knowledge processing, portfolio optimization and intelligent control and a series of essentially non-counting problem, so that experts in the nervous system, pattern recognition, intelligent control, optimization, forecasting and other fields has been successfully applied. Artificial neural network combined with other traditional methods, artificial intelligence and will promote the continuous development of information processing technology.

Neural network is used bottom-up approach, from the brain to study the nervous system structure to brain function, study of a large number of simple neural information processing capacity of the group and its dynamic behavior. Neural Networks for thirty years now has been troubling signs deal with computer science and some of the problems can be more satisfactory solution, especially for those time and space to search for information storage and self-organization associated storage space data system described self-organization as well as from a number of interrelated activities, such as automatic access to the general problem solving knowledge, but also demonstrated its unique capabilities, which caused a widespread concern in intelligence researchers, and is generally considered suitable for low-level neural network mode processing.

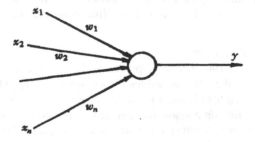

Fig. 2. Mathematical model of neural networks

In the figure, $x = (x_1, x_2, \cdots, x_m)$ is input vector, y is output, w_i is power parameter, the relationship between is:

$$y = f(\sum_{i=1}^{m} w_i x_i - \theta) \quad , \tag{1}$$

θ is threshold, $f(x)$ is stimulate function, which may be linear or non-linear function.

$$z = \sum_{i=1}^{m} w_i x_i - \theta \quad . \tag{2}$$

It may be given as

$$\text{sgn}(x) = \begin{cases} 1, & x > 0, \\ 0, & x \le 0. \end{cases} \tag{3}$$

or

$$y = f(z) = \begin{cases} 1, & \sum_{i=1}^{m} w_i x_i > \theta, \\ 0, & \sum_{i=1}^{m} w_i x_i \le \theta. \end{cases} \tag{4}$$

2.2   Calculation Process and the Steps

Artificial neural network is to imitate the structure and function of brain cells, brain structure and brain function such as thinking, deal with the problem of new information processing systems. As the artificial neural network with complex dynamic characteristics, parallel processing mechanism, learning, association and memory functions, as well as its highly self-organizing, adaptive capacity and flexibility of natural areas by the extensive attention of scholars.

BP neural network model to predict pipeline corrosion rate is calculated as follows:

1) Train the neural network

Affect the corrosion rate of each input variable initial values, the role of the hidden layer weight of the obtained values of the middle layer, and then by the role of the output layer weights obtained predicted output value. The predicted values and measured values for comparison, if the error is not within the required accuracy, the results need to iteratively adjust the hidden layer and output layer of weights, up until the accuracy to meet the requirements.

2) Predict corrosion rate

The use of trained neural networks can predict the factors on pipeline corrosion rate. BP neural network algorithm can be summarized as follows:

Set the input vector of a training to X_k, output vector to Y_k,

$X_K = (x_{K1}, x_{K2}, ..., x_{Kn})^T$, $Y_K = (y_{k1}, y_{k2}, ..., y_{km})^T$, the expected output of X_k is

$Y_k = (yc_{k1}, yc_{k2}, ..., yc_{kn})^2$, the error of which is

$$E_k = \frac{1}{2} \sum_{j=1}^{m} \left(yc_{kj} - y_{kj} \right)^2 . \tag{5}$$

The BP neural network prediction model of corrosion rate, based on three single forecasting model simulated values of corrosion rate as a network of training samples, i.e., input layer has three neurons; the corrosion rate measured as the network desired output, that output only one layer of neurons; the number of hidden layer neurons is still no theoretical guidance to determine, usually based on experience or the use of a spreadsheet formula to determine where, and are the hidden layer, input layer and output layer neuron number.

Neural network specific steps are as follows:

Step1: The network connection weights and thresholds of each randomly assigned to the interval [0, 1] value;

Step2: The input data is normalized to the interval [- 1, 1];

Step3: The establishment of BP neural network, which set up the network structure, selection, training functions;

Step4: Set the least squares objective function upper bound of the deviation, maximum number of iterations, learning efficiency and the potential rate;

Step5: With different hidden nodes of the network training, choose a good network generalization ability to survive;

Step6: Using the trained network to predict.

Once the network weights and biases are initialized, the network can begin to train. We can train the network to do function approximation (nonlinear back), model combination, or pattern classification. Training process requires a proper example of network operations - network inputs p and target outputs t. Network during the training period and the weighted deviation of the network performance continuously to minimize the function.

Back-propagation learning algorithm is the simplest application of the performance function along the direction of steepest increase - the negative gradient direction of the weight and bias update. This recursive algorithm can be written as:

$$x_{k+1} = x_k - a_k g_k . \tag{6}$$

Where x_k the current weight and bias vectors is, g_k is the current gradient, a_k is the learning rate. There are two different approaches to achieve gradient descent algorithm: increase mode and batch mode. Increasing mode, enter the network every once submitted, a gradient calculation and update the weights. In batch mode, when all inputs are submitted to the network was only after the update. The following two sections will discuss the increase and batch mode.

3   Results of the Application

For a long distance gas pipeline corrosion factors: distance, elevation difference, pipe inclination, pressure, Reynolds number, liquid holdup case, the use of orthogonal design table design of the 10 group experiments. To five factors as the network input, the natural gas pipeline in the soil maximum corrosion rate for the network output. After repeated spreadsheet to determine the hidden layer neurons for the two predictions when the second best. In table 1, the results can be seen. Using BP neural network prediction model, the predicted value and the real pipeline corrosion rate values is highly consistent, and the average relative error is only 1.33%, which meets actual production needs. The training Performance and the predicted result are considerable good.

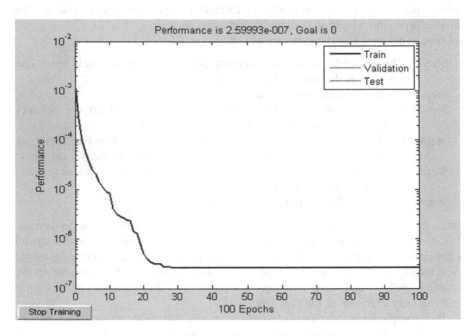

Fig. 3. Training Performance

Table 1. Station No1- 9 line test results compared with the predicted (slug)

Mlies,m	Angle	P,Pa	Re	Water Ratio,%	Real rate, mm/a	Calc result	Error,%
400.1	3.03	4836	22607	0.01155	0.0711	0.0712	0.0014
429.4	3.03	4835	22608	0.01154	0.0622	0.0622	0.0064
916.8	15.5	4835	22607	0.0147	0.0889	0.0889	0

Table 1. (*continued*)

1110	6.5	4835	22607	0.01249	0.0533	0.0533	0
1456	1.2	4845	22601	0.01104	0.0267	0.0267	0.0599
1463	1.2	4844	22602	0.01104	0.0756	0.0756	0.0317
1521	1.2	4844	22602	0.01103	0.0667	0.0667	0.0299

4   Conclusion

Artificial neural network method does not require consideration of various factors on the corrosion of the influence of the form, simply target sample, you can use the trained artificial neural network to predict the corrosion rate of the pipeline, if the increase in network training samples, can be more accurate predict the service life of pipelines. The neural network prediction method with high precision, especially for nonlinear data processing results more reasonable for gas gathering and transportation pipeline corrosion predict complex situations, is more popular methods at home and abroad, has been widely used in all aspects of engineering technology . Visible to mileage, elevation difference, pipe inclination, pressure, Reynolds number, liquid holdup as a model input variables, BP neural network model can be used to describe the natural gas pipeline in different corrosive environments the maximum corrosion rate.

References

1. Zhou, H., Wang, X.G.: BP neural networks combinatorial forecasting method for gas pipeline's corrosion rate. Corrosion Science and Protection Technology 22, 162–165 (2010)
2. Hu, Q.S., S.X.: BP neural network-based prediction model for internal corrosion rate of oil pipelines. Oil & Gas Storage and Transportation 29, 448–451 (2010)
3. Kong, X.D., Han, J.W., Jiang, H.J., Wang, C.: Study on the model-predicting of carbon dioxide corrosion. Inner Mongolia Petrochemical Industry 1, 11–12 (2011)
4. Yuan, G., Wang, S.G., Huang, Y.: Application of markov chain in prediction of corrosion conditions for buried gas steel pipeline. Journal of Harbin Institute of Technology 42, 1328–1331 (2010)
5. Qu, L.S., Li, X.G., Du, C.W., et al.: Corrosion rate prediction model of carbon steel in regional soil based on BP artificial neural network. Journal of University of Science and Technology Beijing 31, 1569–1575 (2009)
6. Li, G.M., He, R.Y., Cheng, H.W., et al.: Fluid field simulation-based prediction and inspection method of pipeline inner-corrosion. Pipeline Technique and Equipment 5, 23–26 (2009)

Energy Consumption Analysis of Crude Oil Gathering System without Heating

Xiao-yu Chen, Chuan Zeng, and Juan She

Southwest Petroleum University, SWPU, Chengdu 610500, China
chenxyu@163.com

Abstract. In this article, we use multivariate analysis methods to analyze the influencing factors of energy consumption of the gathering system without heating. According to orthogonal array, we arrange experiments, and use range method to analyze to the results obtained. Finally we get the impact of various factors in energy consumption. So we can provide a theoretical guidance to the crude oil gathering system without heating on saving energy.

Keywords: Crude oil gathering system, energy consumption analysis, orthogonal experiment, range analysis.

1 Introduction

With the deepening of oil field development, the integrated water content of crude oil is rising, and the content of many wells have over 90%. The viscosity of Crude oil is reducing, and the mobility is enhancing. But the majority of crude oil gathering process is still transported by way of mixed water. This way requires a lot of energy, so it limits oil production costs. Therefore, this article analyzes the energy consumption of crude oil gathering system without heating to provide theoretical guidance on energy saving [1]-[5].

2 Energy Consumption Factors of Gathering System

2.1 Choice of Energy Consumption Factors

In this paper, the crude oil gathering system's pipe network structure has been determined. It means the water mixing tube diameter, set pipe diameter, pipe insulation structure has been determined. In addition to water temperature in collecting stations, water pressure, wellhead produced fluid temperature, moisture, liquids production will be as well the production process such as dynamic change, and the other factors are stable. Therefore, selecting the dynamic factors as water temperature to analysis is more meaningful than selecting the static analysis as

B.-Y. Cao and X.-J. Xie (Eds.): Fuzzy Engineering and Operations Research, AISC 147, pp. 457–464.

diameter. At the same time, the distance is different from the wellhead to metering station according to wellhead location. So, it must consider the average distance of well station when we choose the factors.

2.2 Determine of the Energy Consumption Factors Levels

According to the actual production situation, the wellhead back pressure generally between 0.1 ~ 0.3MPa; the distance from wellhead to metering station is 200m or so, and in some individual remote wells, it is 1000m; the wellhead liquid temperature between 40 and 60 after into stable production cycle, the moisture content between 40% to 90%; the liquid production of single heavy well is lower, generally between the 5 to 15t /d; in the collecting stations, the water temperature between 60°C to 75°C; the water production is generally considered as 1 times as the oil production. In summary, the factors and the factor levels are selected in Table 1.

Table 1. Factors and factor levels selected

Factor's Name	Liquid Temperature	Liquid Producing Capacity	Water Content	Wellhead Back Pressure	Water Production	Water Temperature	Average Distance of Well Station
Factor's Code	A	B	C	D	E	F	G
Units	$°C$	t/d	Dimensionless	MPa	t/d	$°C$	m
Level 1	40	5	0.4	0.15	$0.5G_{oil}$	55	150
Level 2	50	10	0.6	0.25	G_{oil}	65	200
Level 3	60	15	0.8	0.35	$1.5G_{oil}$	75	250
Note	Factors A, B, C's value is the average sense; G_{oil} is the single well oil production.						

3   The Selection of Energy Analytical Targets

Changes in the factors will not only cause changes in energy consumption, but also lead to changes in pressure and temperature distribution. When we use multivariate analysis, it is necessary to examine the impact of the factors on system energy consumption and the factors on the incoming temperature and pressure of the collecting stations; it means that the analytical targets include system energy consumption and the incoming temperature and pressure of the collecting stations. For comparison purposes, we choose the minimum of the incoming temperature and pressure of the collecting stations as analytical targets (hereinafter called "incoming temperature" and "incoming pressure" means the minimum of the incoming temperature and pressure of the collecting stations.)

4   Experimental Design of Energy Consumption Factors

Orthogonal experimental design is a design method to study multi-factor and multi-level, which is based on orthogonal selected some representative points from the full-scale testing to test. These representative points are evenly distributed, and neat comparable. Orthogonal design is a high efficiency, fast and economical method of experimental design.

According to the number of factors and factor levels, we select the L_{18} (37) orthogonal experimental table to arrange experiments, so for a total of 18 simulation experiments. According to the results we draw the effect curve of each factor on the experimental targets, as shown in Table 2.

Table 2. Factor analysis schedule

Factors	A	B	C	D	E	F	G
Experiment 1	40	5	0.4	0.2	$0.5G_{oil}$	55	150
Experiment 2	40	10	0.6	0.3	$1G_{oil}$	65	200
Experiment 3	40	15	0.8	0.4	$1.5\,G_{oil}$	75	250
Experiment 4	50	5	0.4	0.3	$1G_{oil}$	75	250
Experiment 5	50	10	0.6	0.4	$1.5\,G_{oil}$	55	150
Experiment 6	50	15	0.8	0.2	$0.5G_{oil}$	65	200
Experiment 7	60	5	0.6	0.2	$1.5\,G_{oil}$	65	250
Experiment 8	60	10	0.8	0.3	$0.5\,G_{oil}$	75	150
Experiment 9	60	15	0.4	0.4	$1\,G_{oil}$	55	200
Experiment 10	40	5	0.8	0.4	$1\,G_{oil}$	65	150
Experiment 11	40	10	0.4	0.2	$1.5\,G_{oil}$	75	200
Experiment 12	40	15	0.6	0.3	$0.5\,G_{oil}$	55	250
Experiment 13	50	5	0.6	0.4	$0.5\,G_{oil}$	75	200
Experiment 14	50	10	0.8	0.2	$1\,G_{oil}$	55	250
Experiment 15	50	15	0.4	0.3	$1.5\,G_{oil}$	65	150
Experiment 16	60	5	0.8	0.3	$1.5\,G_{oil}$	55	200
Experiment 17	60	10	0.4	0.4	$0.5\,G_{oil}$	65	250
Experiment 18	60	15	0.6	0.2	$1\,G_{oil}$	75	150

5   The Analysis of Energy Consumption Factors

According to Table 2, we arranged to simulate the experiments, and the results in Table 3.

Table 3. Experiment results

Factors	A	B	C	D	E	F	G	Experiment Results		
								Energy Consumption	Incoming Temperature	Incoming Pressure
Experiment 1	1	1	1	1	1	1	1	3.806	27.978	-1.643
Experiment 2	1	2	2	2	2	2	2	6.167	36.863	0.27
Experiment 3	1	3	3	3	3	3	3	14.409	41.876	0.333
Experiment 4	2	1	1	2	2	3	3	9.963	40.106	0.282
Experiment 5	2	2	2	3	3	1	1	4.836	44.205	0.371
Experiment 6	2	3	3	1	1	2	2	3.521	42.591	0.165
Experiment 7	3	1	2	1	3	2	3	13.992	46.604	0.181
Experiment 8	3	2	3	2	1	3	1	6.11	49.02	0.279
Experiment 9	3	3	1	3	2	1	2	1.084	49.311	-0.902
Experiment10	1	1	3	3	2	2	1	24.669	36.826	0.382
Experiment11	1	2	1	1	3	3	2	6.897	43.917	0.17
Experiment12	1	3	2	2	1	1	3	1.643	33.122	0.257
Experiment13	2	1	2	3	1	3	2	8.347	34.417	0.385
Experiment14	2	2	3	1	2	1	3	7.571	40.553	0.171
Experiment15	2	3	1	2	3	2	1	2.982	47.898	-1.468

Table 3. (*continued*)

Experiment16	3	1	3	2	3	1	2	18.743	44.828	0.281
Experiment17	3	2	1	3	1	2	3	1.711	43.561	-0.662
Experiment18	3	3	2	1	2	3	1	3.589	53.772	0.146
Note	Incoming Pressure is negative means that under the given conditions, the mixture of oil and water can not be transported to the flow stations.									

We use range analysis method to analyze the simulation results. The range of factors on the test indicators in Table 4.

Table 4. The range analysis of the results

Indicators		Factor A	Factor B	Factor C	Factor D	Factor E	Factor F	Factor G
Energy Consumption	Mean 1	9.599	13.253	4.407	6.563	4.190	6.280	7.665
	Mean 2	6.203	5.549	6.429	7.601	8.840	8.840	7.460
	Mean 3	7.538	4.538	12.504	9.176	10.310	8.219	8.215
	Range	3.396	8.715	8.097	2.613	6.120	2.560	0.755
Incoming Temperature	Mean 1	36.764	38.460	42.129	42.569	38.448	39.999	43.283
	Mean 2	41.628	43.020	41.497	41.973	42.905	42.391	41.988
	Mean 3	47.849	44.762	42.616	41.699	44.888	43.851	40.970
	Range	11.085	6.302	1.119	0.870	6.440	3.852	2.313
Incoming Pressure	Mean 1	-0.038	-0.022	-0.704	-0.135	-0.203	-0.244	-0.322
	Mean 2	-0.016	0.100	0.268	-0.016	0.058	-0.189	0.062
	Mean 3	-0.113	-0.245	0.269	-0.016	-0.022	0.266	0.094
	Range	0.097	0.345	0.973	0.119	0.261	0.510	0.416

In order to visually illustrate the problem, we mapped the response curves of factors on the corresponding experimental Indicators according to table 4, as shown in Figure 1 to Figure 3.

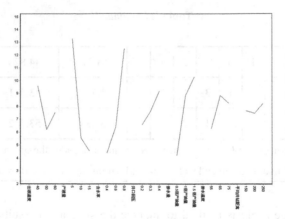

Fig. 1. The effect curve of each factor on the system energy consumption

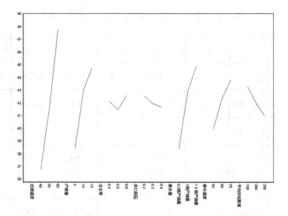

Fig. 2. The effect curve of each factor on incoming temperature

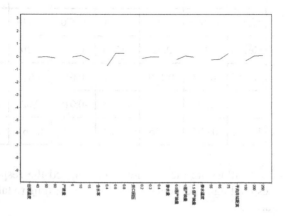

Fig. 3. The effect curve of each factor on incoming pressure

Comprehensive range analysis and response curve, according to the significance of impact of various factors on the Indicators, we can sort the results in Table 5.

Table 5. Sequencing of significance of factors

Target	The Sequencing of Significance of Factors
System energy consumption	B>C>E>A>D>F>G
Incoming temperature	A>E>B>F>G>C>D
Incoming pressure	C>F>G>B>E>D>A

6   Conclusion

Energy Consumption: Form Table 5, we can see the influence of liquids production capacity (factor B) and moisture content (factor C) is the top two; the influence of Average Distance of Well Station (factor G) is weak. The influence of factors B and C on energy consumption is the essence of the influence of oil production on energy consumption. According to energy consumption model, the greater the fluid production and the lower water content, so the oil production is larger and the energy consumption is smaller. In other words, the trend of B and C on the energy consumption is the opposite. Water Production (Factor E) also have significant effect on energy consumption. This is because when the water production increased, moisture content corresponding increased in transmission, so the system energy consumption has increased significantly.

Incoming Temperature: Form Table 5, we can see the influence of wellhead fluid temperature (factor A) is most significant, and wellhead back pressure (factor D) is lowest. In the case of the other things being equal, the higher the temperature of the wellhead fluid, the higher the energy of input system, which means fluid temperature and incoming temperature are positively correlated. The stress factor in the temperature Model is not involved so the effect of wellhead back pressure on the incoming temperature is lowest.

Incoming Pressure: Form Table 5, we can see the influence of water content (factor C) is the top. It is the essence that the impact on the type of emulsion of oil and water. When the moisture content of oil-water mixture is higher than the moisture content in the turning point, its viscosity and the flow resistance is also reduced, so the incoming pressure is higher; the contrary, the incoming pressure is reduced.

In addition, from Table 3 and table 4 we can attain the trends of the experimental targets under the combinations of different factors levels through the range analysis. In this article of the factors and factors levels: the combination of factors $(A_1B_1C_3D_3E_3F_2G_3)$ corresponding to the highest level of energy consumption, which indicates that the well which has low temperature, low liquid production, high moisture content, high wellhead back pressure and long distance, if it uses a large middle-temperature water capacity to mix, its energy consumption is very high; and the combination of factors $(A_2B_3C_1D_1E_1F_1G_2)$ corresponding to the lowest level of energy consumption, which means that the well which has medium liquid temperature, high liquid production, low moisture content, low wellhead back pressure and short distance, if it uses small low-temperature water capacity to mix, that may get better energy efficiency. Therefore, to achieve crude oil gathering system without heating energy consumption, we should adjust water capacity and temperature according to different well.

Acknowledgements. Thanks to the support by crude oil gathering and transportation without heating technology research project of Nanyang oilfield.

References

1. Li, X., Yu, S., et al.: Study on Prediction Method of Energy Consumption in Crude Oil Pipelines. Oil & Gas Storage and Transportation 28(6), 2–3 (2009)
2. Liu, W.: Elementary Introduction to affect primary factor on the Energy Consumption in the Hot Oil Pipeline. Pipeline Technique and Equipment (3), 1–3 (2001)
3. Hu, S., Zhou, G., Miao, Q.: Energy Consumption Analysis on Eastern Oil Pipeline Network. Oil & Gas Storage and Transportation 29(5), 1–2 (2010)
4. Lou, Q.: Energy Consumption Analysis and Energy Saving of Oil Pump Station. Oil & Gas Storage and Transportation 23(2), 1–2 (2004)
5. Liu, C.: The Energy Consumption Analysis and Calculation for Proposed Linyi-Ji' Nan Oil Pipeline. Oil & Gas Storage and Transportation 24(1), 2–6 (2005)

The Analysis of Oilfield Development Indexes and Impact Factors Based on Weighted FCM

Zhi-bin Liu[1,2], Shao-xian Bing[1,3], and Juan Hu[1]

[1] Department of Graduate, Southwest Petroleum University
[2] State Key Laboratory of Oil and Gas Reservoir Geology and Exploitation,
  Southwest Petroleum University
[3] Geology Scientific Research Institute, Shengli Oilfield, Sinopec Corp.

Abstract. One of the most important tasks in oilfield development systemic engineering is to identify the development indexes and impact factors. In this paper, both the indexes and factors are regarded as a sample space, the method of weighted FCM and grey related analysis were applied to classify the sample space, and then the system of development indexes and impact factors were established. It makes the oilfield development program become more effective and reasonable.

Keywords: Weighted FCM, grey relation analysis, development indexes, impact factors.

1   Introduction

Oilfield development is a systematic project, it is significant to master dynamical rules of oilfield accurately and frame scientific development plan by having accurate comprehension of dynamic changes of oilfield development. One of the most important tasks in formulating development program is how to classify the development indexes and impact factors from a sample space [1-3]. The development indexes and impact factors, who can reflect dynamic characters of oilfield or development unit, have varied action but close relation.

In this paper, proceed from raw data which was provided by oilfield, regard the development indexes and impact factors as a sample space by making use of fuzzy classification quantitative analysis method, adopt weighted FCM clustering method to classify the factors, and then obtain the fuzzy clustering hierarchical diagram to classify the development and impact factors dynamic. Then, calculate the grey relation grades of development indexes about each impact factor. Ascertain the typical impact factors of each development index according to the value of relation grades and set the systems of oilfield development index and impact factor.

2   The Method of Weighted FCM

With the development of data mining technology, clustering analysis is considered as an effective tool to solve the problem of data analysis and data visualized.

B.-Y. Cao and X.-J. Xie (Eds.): Fuzzy Engineering and Operations Research, AISC 147, pp. 465–473.
springerlink.com   © Springer-Verlag Berlin Heidelberg 2012

Fuzzy clustering was extensive used because it can describe medium of the sample and reflect the real world objectively. At present, FCM is one of the most popular and operative clustering method. FCM is a kind of local search optimization technology which gets optimization solution by minimizing the objective function in essence. The objective function may have large extreme points, the clustering process based on the objective function is the process of searching extreme point, so algorithm is easy to fall in local extreme point if initialization inappropriate and the clustering results prone to errors. The shortcoming is most acute in clustering sample [4-5]. Therefore, there still have some problems for the traditional FCM algorithm to be settled urgently. In this paper, we adopt weighted Euclidean fuzzy clustering method to fix reasonable clustering number and center, and then apply the traditional FCM algorithm. It solved the problem of how to apply FCM algorithm to clustering in information incomplete and reduced time consumption. The Simulation Experiments proved the feasibility, actuality and effectiveness of the algorithm.

2.1   *Weighted Euclidean Distance*

Let data objects $X = \{x_1, x_2, \cdots, x_n\} \subset R^d$ is a set of limited data object, element x_i is k - dimensional vector. x_{it} is the value of t th element of i th sample. Classify n samples, take Euclidean distance as classification statistics [6], where

$$d(i, j) = \sqrt{\left|x_{i1} - x_{j1}\right|^2 + \cdots + \left|x_{ik} - x_{jk}\right|^2} \ . \tag{1}$$

Give a weight to each variable according to the importance, the weighted Euclidean distance is

$$d(i, j) = \sqrt{w_1 \left|x_{i1} - x_{j1}\right|^2 + \cdots + w_k \left|x_{ik} - x_{jk}\right|^2} \ . \tag{2}$$

Rational use of weight can reflect different actions of variable in data, and improve the clustering result. The determination of the weight ask the person to possess the specialty knowledge to fit in with the needs, however, it is difficult to come true in practical [5]. So, in this paper, we set the weight of each index in accordance with density degree of different indexes to clustering influence. To assign a value to each variable by applying Weighted Euclidean distance FCM method to incarnate different effect on various in clustering procedure. So, it both incarnates different influence of each variable and takes full advantage of characteristic of data itself to make the clustering result accurately.

2.2   *FCM Algorithm*

In 1977, J.C.Bezdek put forward fuzzy classification by utilizing the concept of fuzzy set on the basis of traditional clustering. They believed the clustering object affiliated a certain class with certain membership degree. The membership degree was used to fix each data point belong to a certain class. By repetitious alter clustering center and classified matrix to realize dynamic iteration clustering. And then search a fuzzy classified matrix $\left(u_{ki}\right)_{c\times n}$ and c clustering center as $V=\left\{v_1,v_2,\cdots,v_c\right\}$. It makes the objects in the same class have maximum similarity, and minimum similarity among different classes [6-8].

Let domain $U=\left\{x_1,x_2,\cdots,x_n\right\}$ be classified objects, each object has p factors to show the characters, that is to say $x_i=\left\{x_{i1},x_{i2},\cdots,x_{ip}\right\}$ $(i=1,2,\cdots,n)$.

$J\left(U,V\right)=\min\sum_{i=1}^{n}\sum_{k=1}^{c}\left(u_{ki}\right)^{m}d^2\left(x_i,v_k\right)$ is the objective function of fuzzy clustering, where $U=\left[u_{ki}\right]$ is a fuzzy classified matrix, and $0\leq u_{ki}\leq 1$. v_i $(i=1,2,\cdots,c)$ is the i-th clustering center, where $V\in\left[v_i\right].m\in\left[1,+\infty\right)$ is a fuzzy weighted index. The membership degree u_{ki} represents the level of vector x_i to center v_k. $d\left(x_i,v_k\right)$ is the distance from target data x_i to v_k. $J\left(U,V\right)$ represents the weighted distance square. The smaller $J\left(U,V\right)$, the better the clustering result. Weight is the sample m th power to i class' membership degree[6-8].

2.3　The Step of Weighted FCM Algorithm [6-8]

Step 1. Input the clustering number c which obtained from the weighted Euclidean distance method and fuzzy exponent m and stop parameter.

Step 2. Let $l=0$. And input the initial clustering center vector which obtained from the weighted Euclidean distance clustering method.

Step 3. Calculating the distance of each vector to the clustering center, according to

$$d_{ik}=\left|x_k-v_i\right|=\left[\sum_{j=1}^{d}\left(x_{kj}-v_{ij}\right)^2\right]^{1/2}. \tag{3}$$

Step 4. Calculating membership matrix $U^{(l)}$, according to

$$u_{ij}^{(l)}=1\Big/\sum_{j=1}^{c}\left(d_{ik}/d_{jk}\right)^{2/(m-l)},1\leq k\leq n. \tag{4}$$

Step 5. Let $l = l + 1$, according to $v_i^{(l)} = \sum_{k=1}^{n} (u_{ik})^m x_k \Big/ \sum_{k=1}^{n} (u_{ik})^m$,

where $1 \leq i \leq c$, to calculate new clustering center.

Step 6. Calculating objective functions $J_m^{(l)}$ according to

$$J_m = \min \sum_{i=1}^{n} \sum_{k=1}^{c} (u_{ki})^m d^2 (x_i, v_k). \qquad (5)$$

Judge convergence condition, if $\left| J_m^{(l)} - J_m^{(l-1)} \right| > \xi$,continue to iteration, or else, break the loop, and obtain the optimal clustering result.

3   Grey Correlation Analysis

Correlation analysis is to study the interrelation, nature and compactness among the variable. That is to say, the task of correlation analysis is to quantitatively describe the correlation. Coefficient is a critical parameter to measure the linear correlation degree among the variable [9]. The overall coefficient was calculated by all the data, and the sample coefficient was calculated by sample data.

Grey correlation analysis is a kind of data theory and technology. It gives a new method, in the condition of lower requirement of sample, to analyze the factor of the system. In the same time, it effectively recovers the defect of traditional method such as regression analysis, variance analysis and principal component analysis. Meanwhile, grey correlation analysis is fit to time series data, which fully consider the character of data object changing with time [9-10]. And grey correlation analysis reflects the changes rapidly and exactly under the condition that time series data changes with time.

In correlation analysis, ascertain the factors which need to be enquired firstly, and then analyze and study the correlation analysis among the factors by extracting the sample value in time series of factors' curve. In order to seek the reality law of the series, we propose the notion of correlation degree to describe both strong or weak relation and great or small relation between the development indexes and impact factors in the middle and later periods of oilfield development. The change trend (direction, value, speed) of the sample data will be consistent basically, and then the correlation degree is big, otherwise, it is small.

3.1   Computational Method of Correlation Degree [10-11]

Step 1. Set series analysis.
According to real problem to fix a factor as dependent variable and others are independent variables. Regard time series of dependent variable as reference series $x_0^{'}$,

the other series of independent variables compose the comparatives series $x_0'(i=1,2,3\cdots,n)$ and $n+1$ data series compose the matrix marks as following:

$$
(x_0',x_1',\cdots,x_n') =
\begin{bmatrix}
x_0'(1) & x_1'(1) & \cdots & x_n'(1) \\
x_0'(2) & x_1'(2) & \cdots & x_n'(2) \\
\cdots & \cdots & \cdots & \cdots \\
x_0'(N) & x_1'(N) & \cdots & x_n'(N)
\end{bmatrix}_{N\times(n+1)}
, \tag{6}
$$

where $x_i' = (x_i'(1),x_i'(2),\cdots,x_i'(N))^T, i=0,1,2,\cdots,n$ N is the length of dependent variable series.

Step 2. Do the non-dimensional indicators of variable series.

Since the primitive indexes time series have different dimensional or magnitude order, it is necessary to dimensionless the time series to protect reliability of the analysis result. The conventional methods are established in the following.

Average method

$$
x_i(k) = \frac{x_i'(k)}{\dfrac{1}{N}\displaystyle\sum_{k=1}^{N} x_i'(k)} \quad \begin{pmatrix} i=0,1,2,\cdots,n \\ k=1,2,\cdots,N \end{pmatrix}, \tag{7}
$$

and initial method

$$
x_i(k) = \frac{x_i'(k)}{x_i'(1)} \quad \begin{pmatrix} i=0,1,2,\cdots,n \\ k=1,2,\cdots,N \end{pmatrix}. \tag{8}
$$

So, the non-dimensional matrix of the indexes is given as follows,

$$
(x_0,x_1,\cdots,x_n) =
\begin{bmatrix}
x_0(1) & x_1(1) & \cdots & x_n(1) \\
x_0(2) & x_1(2) & \cdots & x_n(2) \\
\cdots & \cdots & \cdots & \cdots \\
x_0(N) & x_1(N) & \cdots & x_n(N)
\end{bmatrix}. \tag{9}
$$

Step 3. Calculate time sequence difference, maximum and minimum differences.

Calculate absolute value between reference series and comparatives series to compose the absolute value matrix as follows,

$$
\begin{bmatrix}
\Delta_{01}(1) & \Delta_{02}(1) & \cdots & \Delta_{0n}(1) \\
\Delta_{01}(2) & \Delta_{02}(2) & \cdots & \Delta_{0n}(2) \\
\cdots & \cdots & \cdots & \cdots \\
\Delta_{01}(N) & \Delta_{02}(N) & \cdots & \Delta_{0n}(N)
\end{bmatrix} ,
\tag{10}
$$

where, $\Delta_{0i}(k) = |x_0(k) - x_i(k)|, i = 1, 2, \cdots, n; k = 1, 2, \cdots, N$.

The maximum and minimum in the matrix is just maximum and minimum difference.

$$
\Delta(\max) = \max_{\substack{1 \le i \le n \\ 1 \le k \le N}} \left\{ \Delta_{0i}(k) \right\} ,
\tag{11}
$$

$$
\Delta(\min) = \min_{\substack{1 \le i \le n \\ 1 \le k \le N}} \left\{ \Delta_{0i}(k) \right\}.
\tag{12}
$$

Step 4. Calculate correlative coefficient.

Do the transformation to the data in absolute difference matrix and obtain the correlative coefficient matrix as follows,

$$
\xi_{0i}(k) = \frac{\Delta(\min) + \rho\Delta(\max)}{\Delta_{0i}(k) + \rho\Delta(\max)},
\tag{13}
$$

$$
\begin{bmatrix}
\xi_{01}(1) & \xi_{02}(1) & \cdots & \xi_{0n}(1) \\
\xi_{01}(2) & \xi_{02}(2) & \cdots & \xi_{0n}(2) \\
\cdots & \cdots & \cdots & \cdots \\
\xi_{01}(N) & \xi_{02}(N) & \cdots & \xi_{0n}(N)
\end{bmatrix}_{N \times n} ,
\tag{14}
$$

where, ρ is discrimination coefficient, the influence of $\Delta(\max)$ to data conversion was controlled by ρ. It improves remarkable of the difference among the correlative coefficient when ρ is small, so $\rho \in (0,1)$, ordinary circumstances $\rho \in (0.1, 0.5)$. $\xi_{0i}(k) \in (0,1)$, the smaller the $\Delta_{0i}(k)$, the bigger the $\xi_{0i}(k)$, and it reflects the correlative degree of the comparatives series x_i of i-th index to reference series x_0 in the k-th time sequence.

Step 5. Calculate correlative degree among the indicators,

$$
\gamma_{0i} = \frac{1}{N} \sum_{k=1}^{N} \xi_{0i}(k) \qquad (i = 1, 2, 3, \cdots, n),
\tag{15}
$$

γ_{0i} represents the correlative degree of reference series x_0 and comparative series x_i $(i = 0, 1, 2, \cdots, n)$.

Step 6. Calculate the average value of the variables' coefficient degree.

Step 7. The correlative degrees are listed in its value order, and ascertain correlative order and strong or weak relation. It provides quantitative scientific basis for analyzing correlative relation among the index curves.

4 Simulation Analysis

4.1 Data Acquisition of Oilfield Development Indexes

In this paper, it proposed a new correlation analysis method to study the indexes and factors, and in order to test and verify the method's high accuracy and efficiency; we make use of the data which were got from the oilfield.

In initial data there are 21 indicators such as original oil in place, remaining original oil in place, the quantity of production well etc. Preliminary screening of indicators, get rid of similar and inapplicable indicators(oil yield, cumulative oil production, cumulative oil production, fluid yield production, cumulative fluid production, oil yield of new well, increased oil of old well in measured year, water injection yield), the remained indicators are original oil in place, remaining original oil in place, the quantity of production well and new well, the quantity of well startup, oil-producing monthly, fluid-producing monthly, composite water, the quantity of new well startup in production, old wells measures the total wells times, old wells measures the total operative wells times, the quantity of water injection well, water injection startup well number injection-production ration in monthly.

Input the initial data, take advantage of Matlab to calculate weighted Euclid distance and obtain initial clustering number, and then simulate FCM clustering process of indexes and factors, obtain the dynamic clustering figure.

The sample space was divided into two classes.

The first class : original oil in place (OOP), remaining original oil in place(ROOP), the quantity of production well(QPW), the quantity of well startup (QWS), composite water (CW), injection-production ration monthly (IPRM), oil-producing monthly (OPM), fluid-producing monthly (FPM), the quantity of water injection well (QWJW), water injection startup well number (WISWN).

The second class : the quantity of new well (QNW), the quantity of new well startup in production (NWSP), old wells measures the total wells times (OWMTWT),old wells measures the total operative wells times (OWMTOWT).

From the clustering result, the factors and indexes cannot be separated if the clustering number is two, because the development indexes were selected in the first class. The elements in the second class are few, and the relationship between the indexes and factors is not strong. So, it is necessary to continue to cluster the first class. Repeat the above-mentioned process to get new clustering result until the result is satisfied.

Take advantage of grey correlation analysis to study the variable and get the correlation symmetry matrix via MATLAB.

$$R = \begin{pmatrix} 1 & 0.81 & 0.62 & 0.53 & 0.83 & 0.37 & 0.37 & 0.34 & 0.33 & 0.61 & 0.6 & 0.72 \\ & 1 & 0.55 & 0.58 & 0.88 & 0.37 & 0.38 & 0.35 & 0.32 & 0.68 & 0.66 & 0.78 \\ & & 1 & 0.44 & 0.58 & 0.33 & 0.33 & 0.32 & 0.32 & 0.48 & 0.47 & 0.56 \\ & & & 1 & 0.57 & 0.37 & 0.37 & 0.32 & 0.3 & 0.75 & 0.78 & 0.59 \\ & & & & 1 & 0.38 & 0.39 & 0.35 & 0.31 & 0.68 & 0.66 & 0.79 \\ & & & & & 1 & 0.79 & 0.5 & 0.47 & 0.38 & 0.37 & 0.37 \\ & & & & & & 1 & 0.5 & 0.45 & 0.38 & 0.38 & 0.38 \\ & & & & & & & 1 & 0.62 & 0.32 & 0.32 & 0.31 \\ & & & & & & & & 1 & 0.31 & 0.31 & 0.3 \\ & & & & & & & & & 1 & 0.91 & 0.71 \\ & & & & & & & & & & 1 & 0.7 \\ & & & & & & & & & & & 1 \end{pmatrix}.$$

From the clustering result and the calculation method of correlation degree, calculate the correlation degree of indexes and each factor, the result is shown in table1.

Table 1. Class and correlative degree of indexes and factor

Class	Factor	Oil/month	Fluid/month
1	OOP	0.9828	0.9826
	ROOP	0.9839	0.9816
2	QPW	0.9804	0.9771
3	QWS	0.9786	0.9786
	CW	0.9817	0.9849
4	IPRM	0.9782	0.9877
5	QWJW	0.9787	0.9835
	WISWN	0.9796	0.9858
6	QNW	0.8469	0.8624
	NWSP	0.8600	0.8759
7	OWMTWT	0.8396	0.8586
	OWMTOWT	0.8337	0.8523

The result shows that the development has two classes: oil-producing monthly, and fluid-producing monthly, respectively. The indicators in the same class have high correlative relation and the influence of them to development indexes is similar. Based on the value of correlative degree, selected the variables which have high correlative with the indexes as class representative.

5 Conclusion

The method of weighted FCM algorithm reduced the dependence to initializing of traditional clustering method, settled the problem of convergence in local minimum, and decreased the complexity. In addition, it provide a foundation to set a reasonable and effective development program for oilfield, by applying the weighted FCM algorithm and grey correlative analysis to classify the indexes and factors of oilfield.

Acknowledgements. Thanks to the support by National Natural Science Foundation of China (No. 70572099).

References

1. Beliakova, N.: Hydrocarbon field planning tool for medium to long term production forecasting form oil and gas fields using integrated subsurface-surface models. SPE 65610 (2000)
2. Wackowski, R.K., et al.: Applying rigorous decision analysis methodology to optimization of a tertiaary recovery project: rangely weber sand unit, colorado. SPE 24234 (1992)
3. Yan, P., Roland, N.: Horne. Improved methods for multivariate optimization of field development scheduling and well placement design. SPE 49055 (1988)
4. Li, S.J., Gu, L.B.: Evaluation of region economy with clustering analysis method. China Rural Survey 1, 2–8 (2001)
5. Gao, X.B.: Fuzzy cluster analysis and its applications. Shanxi Xidian University Press (2004)
6. Casermeriro, E.M., Perez, J.M.: MREM.An associative autonomous recurrent Network. Journal of Intelligent and Fuzzy Systems 4, 163–173 (2002)
7. Ozbay, Y., Ceylan, R., Karlik, B.: A fuzzy clustering neural network architecture for classification of ECG arrhythmias. Computers in Biology and Medicine 36, 376–388 (2006)
8. Kong, X.Y., Zou, K.Q.: Evaluation method of fuzzy integral for the complex development of the society. Journal of Dalian University 27, 41–44 (2006)
9. Guo, R., Guo, D.: Random fuzzy variable foundation for grey differential equation modeling. Soft Computing-A Fusion of Foundations, Methodologies and Applications 13, 185–201 (2009)
10. Kung, C.Y., Lee, C.K., Wang, C.M.: Evaluation of entry, mode of overseas investment using grey relation method. Journal of Grey System 10, 96–104 (2007)
11. Xia, X.T., Wang, Z.Y., Chen, X.Y.: Evaluation for optimum technical plan of rolling bearing vibration using grey system theory. Journal of Grey System 1, 9–14 (2006)

An Intelligent Prediction Model for Oilfield Production Based on Fuzzy Expert System

Yi-hua Zhong[1], Ming-xia Zhu[1], and Zhi-yin Zhang[1,2]

[1] School of Science, Southwest Petroleum University, Chengdu, Sichuan 610500, China
[2] Shanxi Zoom lion Earth Working Machinery Ltd., Co., Weinan, Shan xi 714000, China
Zhongyh_65@126.com

Abstract. An accurate prediction of oilfield production can benefit the oilfield enterprise in formulating a best development planning. Many prediction methods of oilfield production have been proposed in recent years, but these methods are independent relatively, and their applicable development phases and their model parameters are different. Because the oil-gas field development system is a nonlinear, time-varying, and dynamic complex system, it is hard for predictor to choose an appropriate prediction method according to the changing oilfield information. After analyzing the applicability of the existed production prediction methods and the expert experience of choosing the prediction method, this paper proposed an intelligent predicting method for the prediction of oilfield production based on the theory and method of fuzzy expert system and decision support system, whose key technology is fuzzy reasoning technology. The results of simulation experiment show that the intelligent prediction model established in this paper is correct, which can automatically select the optimum prediction model by the computer according to the oilfield information, solves the prediction problem of complex dynamic system and opens a new study way for the prediction theory and practices of oilfield development indices.

Keywords: Intelligent prediction, oilfield production, fuzzy expert system, fuzzy rule, fuzzy reasoning.

1 Introduction

At present, our country advocates the construction of digital oilfield energetically. The prediction of development indexes is an important task in the oil-gas field development programming. The reliability of the prediction results decides the oilfield development level and economic benefits. Oilfield production is the most important index in the oilfield development indexes. Thus domestic and foreign oil workers made a lot of research work to present many prediction methods, but each of they has itself suitable conditions and application scopes, that is different prediction method suits to different type reservoirs and development stages [1]. In prediction practice, it is very difficult to choose an optimal from so many prediction methods and models, for there is no a uniform criterion how to select it. In order to accelerate the construction of oilfield information, improve quality and

B.-Y. Cao and X.-J. Xie (Eds.): Fuzzy Engineering and Operations Research, AISC 147, pp. 475–484.
springerlink.com   © Springer-Verlag Berlin Heidelberg 2012

efficiency of prediction, it is very important and significant to study the intelligent prediction of oilfield production. Intelligent prediction is to find out the best prediction method corresponding to the actual situation of forecast objects on the basis of the traditional prediction methods, which can make people input relevant parameters directly and choose the optimal prediction method intelligently by the computer to solve practical prediction problems. So far there have been some researches in the theory of intelligent prediction and its application [2-5], but nobody has studied the automatic forecast problem of oil output, whose prediction model is suitable for different reservoirs type and development phase may be adaptively selected by computer. There is an urgent need to design an intelligent forecasting system based on a thoroughly research of the existing prediction methods; realize the intelligent prediction of oilfield output prediction. As we know, the oilfield development system is a nonlinear, time-varying, fuzzy and complex dynamic system; the development indexes which describe the development status are time-varying and fuzzy too, furthermore, development indexes have some relevance relation, so it is difficult to select an appropriate prediction method of development index. This paper plans to study the adaptive prediction method for oilfield production by using the theory and methods of fuzzy expert system and intelligent decision support system; this intelligent forecasting method can select the most appropriate method automatically from many existing prediction methods by computer to predict adaptively output of oilfield or block for the different reservoirs and development stages.

This paper is organized as follows: review about prediction method of oilfield production in Section 2, the intelligent prediction model of oilfield production in Section 3, simulation experiment in Section 4, and conclusions in Section 5.

2   Review about Prediction Method of Oilfield Production

The existing prediction methods of oilfield production for water drive reservoir mainly includes analogy method, Water dynamics estimation method, water drive characteristic curve method, decreasing law method, prediction model method, united solutions, neural network method, differential simulation method, function simulation method based on time-varying and support vector machine (SVM) method etc. Above prediction methods of oil production almost aim at specific reservoir types and development phases. Therefore, in order to study the prediction methods of oilfield production which can automatically select the optimal prediction methods suitable for different reservoirs and development stages by the computer, the existing methods can be divided into three categories [1] by their applicability: one may apply to the low water cut development stage, it includes analogy method, water dynamics estimation method, differential simulation method; one may apply to middle and high water cut development stages, it includes water drive characteristic curve method, decreasing law method, prediction model method, united solutions, neural network method, differential simulation method, function simulation method based on time-varying, support vector machine (SVM) method; another may apply to very high water cut development stage, it includes function simulation method based on time-varying, support vector machine method.

3 The Intelligent Prediction Model of Oilfield Production

The following studies a new model of oilfield output by using the theory and structure of fuzzy expert system [6].

3.1 The Fuzzy Database of Intelligent Selecting Prediction Model

Fuzzy database is used to store the known initial information, basic input information, basic definitions, fuzzy intermediate conclusions and final conclusions from the system reasoning.

1. The Fuzzification of Input
According to the purpose of research, the main factors influencing selecting oilfield output prediction model and the knowledge and experience of experts selecting the method of output prediction, the reservoirs are divided into two types: thin oil ($\mu \leq 50$ $mpa \cdot s$) and heavy oil ($\mu > 50$ $mpa \cdot s$) by oil viscosity; the development stage are divided into three types: low water cut stage ($f_w \leq 20\%$), medium to high water cut stage (20%< $f_w \leq 90\%$) and very high water cut stage ($f_w \geq 90\%$) by water cut in this paper. Thus reservoir type and development stage are chosen as the antecedent of knowledge rule in intelligent forecast model; the prediction method are chosen as the consequent of reasoning rule. The fuzzification [7] is to convert definite value of the input (crude viscosity, water cut) into a fuzzy value by their membership functions.

2. The Membership Functions of Fuzzy Set
There are two fuzzy subsets of reservoir types — thin oil reservoir and heavy oil reservoir and three fuzzy subsets of development stage-low water stage, medium to high water cut stage and very high water cut stage. The membership functions of two fuzzy sets determined according to expert experience of the oilfield production prediction are described in formula (1) - (5) and Fig. 1.
1) The membership functions of thin oil reservoir

$$\tilde{Y}_1(\mu) = \begin{cases} 1, & 0 \leq \mu \leq 30, \\ \dfrac{1}{40}(70 - \mu), & 30 < \mu \leq 70. \end{cases} \quad (1)$$

2) The membership functions of heavy oil reservoir

$$\tilde{Y}_2(\mu) = \begin{cases} 1, & \mu \geq 70, \\ \dfrac{1}{40}(\mu - 30), & 30 \leq \mu < 70. \end{cases} \quad (2)$$

3) The membership functions of low water cut stage

$$\tilde{D}_1 = \begin{cases} 1, & 0.00 \leq f_w \leq 0.10, \\ 5(0.3 - f_w), & 0.10 < f_w \leq 0.30. \end{cases} \quad (3)$$

4)   The membership functions of medium to high water cut stage

$$\tilde{D}_2 = \begin{cases} 1, & 0.30 \leq f_w \leq 0.85, \\ \dfrac{1}{5}(f_w - 0.1), & 0.10 < f_w < 0.30, \\ \dfrac{1}{10}(0.95 - f_w), & 0.85 < f_w \leq 0.90. \end{cases} \tag{4}$$

5)   The membership functions of very high water cut stage

$$\tilde{D}_3 = \begin{cases} 1, & 0.95 \leq f_w \leq 1.0, \\ \dfrac{1}{10}(f_w - 0.85), & 0.85 < f_w < 0.95. \end{cases} \tag{5}$$

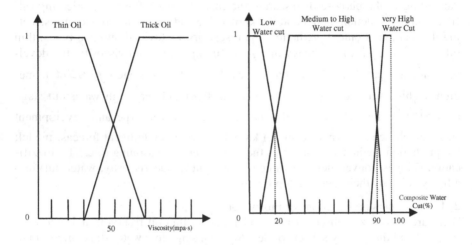

Fig. 1. Membership functions of reservoir types and development stages

3.2   *The Fuzzy Knowledge Base of Intelligent Selecting Prediction Model*

Fuzzy knowledge base is a key component in the intelligent forecast model of oil production. It stores the knowledge and rules of relevant facts, which affect production and select prediction method of production, which are concluded by oilfield prediction experts, knowledge engineers and book knowledge. The intelligent prediction model of oil production uses the semantic model of production rules, relational database to build the fuzzy knowledge base. The way of knowledge acquisition adopts the non-automatic access method.

1. The Prediction Method and Model Base
In order to realize automatical prediction of oilfield production for different reservoir types and different development stages, it is key to construct prediction

method and model base. Establishing method and model base will first classify the methods in section 2 according to their applicability. Then these methods are stored into a database by using the form of method base to be called by fuzzy reasoning machine. Because ten different prediction methods of oil production need different parameters, for example water drive characteristic curve prediction method needs the relevant historical data information that are cumulative oil output, cumulative water output, cumulative fluid output and water-oil ratio; functions simulation method based on time-varying needs more parameters information that are the remaining reserves, the total number of oil production wells, monthly injection-production ratio, water cut, the number of active water injection wells, the number of new production wells, and the number of effective stimulation treatments for old wells. etc. , a model base will be constructed by using the method of model base to store the required parameters of each method model and the interface parameters of the external software to implement specific method in the intelligent forecast system in order to the reasoning machine to select an optimal forecast model.

2. Fuzzy Rule Base

The fuzzy rules are an important foundation of the intelligent prediction for oil production. In this paper, fuzzy rules of prediction are established by using fuzzy production rules. The semantic model of fuzzy production rules [8] are described as follows:

$$\text{IF } (p1, p2 \ldots pn) \text{ THEN } q \text{ WITH CF } (R)$$

p1, p2, ..., pn is the condition of fuzzy rule, q is the conclusion of fuzzy rule, CF (R) is the credibility of the rule.

In order to establish fuzzy rule base in fuzzy expert system of oil output prediction, the following information are required according to the relationship of input and output in intelligent prediction model for oil production:

1) Methods of oil production prediction
   P11 : Analogy method
   P12 : Water dynamics estimation method
   P13 : Water drive characteristic curve method
   P14 : Decreasing law method
   P15 : Prediction model method
   P16 : United solutions
   P21 : Neural network method
   P22 : Differential simulation method
   P23 : Function simulation method based on time-varying
   P24 : Support vector machine (SVM) method
2) Development stages
   D1 : Low water cut ($f_w \leq 20\%$)

   D2 : Medium to high water cut ($20\% < f_w \leq 90\%$)

   D3 : Very high water cut ($f_w > 90\%$)

3) Reservoir types

    Y1 : Thin oil   Y2 : Heavy oil

Theoretically, there are two input variables (the reservoir type and the development stage), so there are 60 rules. But there are only 26 effective rules; the rule base of intelligent forecasting system based on fuzzy expert system can expressed by the structure of "IF... THEN" as follows:

R1 : IF Y1, D1 THEN P11 WITH CF(R1)

R2 : IF Y1, D1 THEN P12 WITH CF(R2)

R3 : IF Y1, D1 THEN P22 WITH CF(R3)

R4 : IF Y2, D1 THEN P11 WITH CF(R4)

R5 : IF Y2, D1 THEN P12 WITH CF(R5)

R6 : IF Y2, D1 THEN P22 WITH CF(R6)

R7 : IF Y1, D2 THEN P13 WITH CF(R7)

R8 : IF Y1, D2 THEN P14 WITH CF(R8)

R9 : IF Y1, D2 THEN P15 WITH CF(R9)

R10 : IF Y1, D2 THEN P16 WITH CF(R10)

R11 : IF Y1, D2 THEN P21 WITH CF(R11)

R12 : IF Y1, D2 THEN P22 WITH CF(R12)

R13 : IF Y1, D2 THEN P23 WITH CF(R13)

R14 : IF Y1, D2 THEN P24 WITH CF(R14)

R15 : IF Y2, D2 THEN P13 WITH CF(R15)

R16 : IF Y2, D2 THEN P14 WITH CF(R16)

R17 : IF Y2, D2 THEN P15 WITH CF(R17)

R18 : IF Y2, D2 THEN P16 WITH CF(R18)

R19 : IF Y2, D2 THEN P21 WITH CF(R19)

R20 : IF Y2, D2 THEN P22 WITH CF(R20)

R21 : IF Y2, D2 THEN P23 WITH CF(R21)

R22 : IF Y2, D2 THEN P24 WITH CF(R22)

R23 : IF Y1, D3 THEN P23 WITH CF(R23)

R24 : IF Y1, D3 THEN P24 WITH CF(R24)

R25 : IF Y2, D3 THEN P23 WITH CF(R25)

R26 : IF Y2, D3 THEN P24 WITH CF(R26)

3.3 Fuzzy Reasoning Machine of Intelligent Selecting Prediction Model

Fuzzy reasoning machine is another key component of fuzzy expert system. It judges from the known knowledge based on fuzzy rules and infers new knowledge. This paper uses "Mamdani" synthesis reasoning method and forward reasoning strategy.

1. Mamdani reasoning method

The membership functions of fuzzy implication relation of Mamdani [9] are as follows:

$$\tilde{A} \rightarrow \tilde{B} = \tilde{A} \times \tilde{B} ,$$

$$\tilde{R}_{\tilde{A} \rightarrow \tilde{B}} (x , y) = [\tilde{A} (x) \wedge \tilde{B} (x)] \cdot$$

Its reasoning statement of fuzzy condition is if $\tilde{A}$ and $\tilde{B}$ then $\tilde{C}$. And its fuzzy relation is $\tilde{R} = \tilde{A} \times \tilde{B} \times \tilde{C} = (\tilde{A} \times \tilde{B})^T \circ \tilde{C}$, where the operation T expresses that the relation matrix $(\tilde{A} \times \tilde{B})$ was written into $m \times n$ dimension columns vector.

The conclusion of corresponding fuzzy reasoning is $C_1 = [(A_1 \times B_1)^T]^L \circ R$, where $[(A_1 \times B_1)^T]^L$ expresses that the matrix $(\tilde{A}_1 \times \tilde{B}_1)$ of m rows and n columns is written into mn dimension columns vector.

2. The Strategy of Forward Reasoning

After fuzzification for input information of the reservoir type and development stage by membership functions, the reasoning machine begins fuzzy matching. If it successfully finds matching rule, it will let this value to the antecedent of the matched rule; otherwise it will continue to match until matching successfully. Therefore, after matching successfully, rule will be activated, corresponding prediction method model will be invoked, and the results of reasoning will be gained and stored in the fuzzy database. If there are lots of methods for prediction of oilfield production, the prediction model corresponding to high credibility of the rule will be chosen.

3.4 Fuzzy Explanation Module of Intelligent Selecting Prediction Model

Fuzzy explanation module is throughout the entire reasoning process of the intelligent prediction system in predicting oilfield production. It is used to explain the fuzzy reasoning results, tell the user that the conclusion obtained is resulted from which inferring rule, and provide users with the detailed process of prediction.

Because the conclusion of the fuzzy reasoning machine is a fuzzy quantity, it needs to be transformed into definite data values. There are several common defuzzification methods such as maximum membership, the median number, and gravity center method (weighted average method) [10], this paper adopts the maximum membership method to the defuzzy of output.

4 Simulation Experiment

The intelligent prediction model proposed in Section 3 is used to predict oil production of integrated conventional heavy oil reservoir A3 in China in its very high water cut stage by using Matlab software for simulation experiment, adopting Microsoft Access 2003 database system to store related parameters information of oilfield.

4.1 The Optimal Prediction Model of Intelligent Selection

1. The Man-machine Interface
According to prediction requirements, input oil viscosity (μ) and water cut (f_w) of the integrated oilfield A3; and store the data of variables and other parameters involved in the suitable prediction model for its development stage into database.

2. Fuzzy Database and Knowledge Base
Fuzzy database stores the fuzzy value of reservoir type and development stage as well its membership functions; Fuzzy knowledge base stores knowledge of selecting prediction methods which collected from oilfield prediction experts and knowledge engineers, and the new facts knowledge from reasoning machine reasoning.

3. Fuzzy Reasoning

In the command window of Matlab, type the command "Fuzzy", open a fuzzy logic window to input oil viscosity and water cut, and output production prediction method.

The membership functions of the oil reservoir type and development stage are edited like Fig. 2 and the membership function of output as Fig. 2.

After establishing the fuzzy reasoning system, rename the FIS file as "PREDICTION" and save to the current workspace, then type the some command in command window of Matlab as follows:

```
yuce=readfis ('PREDICTION.fis')
yuce =
        name: 'PREDICTION'
        type: 'mamdani'
   andMethod: 'min'
    orMethod: 'max'
 defuzzMethod: 'som'
   impMethod: 'min'
   aggMethod: 'max'
        input: [1x2 struct]
       output: [1x1 struct]
         rule: [1x26 struct]
evalfis([70 93], yuce)
ans=95
```

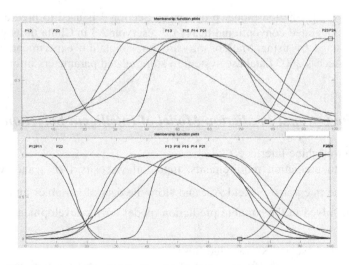

Fig. 2. Membership function of output

4. Defuzzification

Because oil viscosity of A3 oilfield is 70 $mpa \cdot s$, and water cut is 93 % (>90%), the reasoning machine can invoke the function simulation method based on time-varying (P23) and the support vector machine (SVM) method (P24) from fuzzy knowledge base according to matching rule R25 and rule R26 of in 3.2 (2); the membership of function simulation method based on time-varying system is 0.95, the membership of support vector machine(SVM) prediction method is 0.85, the function simulation method based on time-varying system corresponding output 95 according to the maximum membership method of defuzzification methods is the best prediction model.

4.2   Prediction of Oilfield Production

1. One-month Oil Production
After function simulation method based on time-varying system was selected as the best model, this model can automatically call its related parameters information from the database to predict production of A3 oilfield in December, 1999. Its result is 252,300 tons.

2. Oil Production for Several Months
Following the above several steps, the memberships of function simulation and support vector machine from December 1999 to April 2000 are listed in Table 1.

Table 1. The memberships of function simulation and SVM

Method Time	function simulation	support vector machine (SVM)
1999-12	0.95	0.85
2000-01	0.90	0.70
2000-02	0.94	0.90
2000-03	1.00	0.96
2000-04	1.00	0.98

Then oil productions from December 1999 to April 2000 are predicted by two methods of function simulation and support vector machine, their results are shown in Table 2.

Table 2. Monthly oil production of function simulation and SVM

Method Time	function simulation (ton)	support vector machine (SVM) (ton)
1999-12	252,300	252,000
2000-01	255,800	256,500
2000-02	259,300	259,900
2000-03	258,500	258,900
2000-04	256,800	256,900

5 Conclusion

Based on the study of existing prediction methods of oilfield production, this paper combined the fuzzy mathematical theory with artificial intelligence technology to establish an intelligent prediction model of oilfield production. This model focused on fuzzy reasoning technology, made full use of expert experience, combined the advantages of expert experience with fuzzy technology, and automatically realized intelligent selection of the most appropriate production model for different types and different development stages of the water-drive oilfield by computer. The prediction results are in accordance with the practical situation of the production dynamic changes in oilfield development. The proposed methods in this paper are helpful for oilfield developers to make oilfield development programming and decision. It can be widely applied in the prediction and decision-making of geology, environment, water quality, weather and other areas etc. Moreover, it may also provide study idea with intelligent forecasting of complex dynamic system.

Acknowledgements. The authors are grateful for financial Support from Scientific Reserch Fund of SiChuan Provincial Education Department(11ZA024):Research of Intelligent Predictive Control Theory and Its Application Method, and Scientific Research Fund of Southwest Petroleum University, P. R. China (No. 2010XJZ195): The Research on Intelligent Prediction Method and Its Application in Oilfield Development.

References

1. Zhong, Y.H.: The study of dynamic prediction methods for oilfield development programming in ultrahigh water cut. Southwest Petroleum University Doctoral Dissertation (2008)
2. Narrate, L., Christine, W.C., Ralph, M., et al.: A toolset for construction of hybrid intelligent forecasting systems: application for water demand prediction. Artificial Intelligence in Engineering 13, 21–42 (1999)
3. Yin, Y., Ma, Q.G., Fang, S.Q., et al.: A construction of hybrid intelligent forecasting systems for financial crises. In: 2008 Chinese Control and Decision Conference (2008)
4. Jae, J.A., Suk, J.L., Kyong, J.O., et al.: Intelligent forecasting for financial time series subject to structural changes. Intelligent Data Analysis 13, 151–163 (2009)
5. Wang, H.W.: Intelligent forecasting models-selection system for the portfolio internal structure change. Soft. Computer 11, 1141–1147 (2007)
6. Gai, Z.Y., Cheng, G.J., Wang, Y.: Application of fuzzy expert system to risk predication of drilling. Computer Technology and Development 19, 225–227 (2009)
7. Luo, J., Xie, S.R., Jiang, Q., et al.: Intelligent control engineer and application example. Chemical Industry Press, Beijing (2007)
8. Li, H.: The research of fuzzy expert system and application. Guizhou China University Master Thesis (2006)
9. Han, L.Q., Wang, Z.X., Ye, B.: Intelligent control theory and applications. China Machine Press, Beijing (2007)
10. Cheng, W.S.: The theory of intelligent method and application. Tsinghua University Press, Beijing (2009)

The Research of Modular Software System Reliability Based on HMM

Xiao-mei Zhu, Zhi-gang Guo, and Cheng Yuan

School of Computer Science, Southwest Petroleum University,
School of Economics and Management, Southwest Petroleum University,
School of Sciences, Southwest Petroleum University, Chengdu 610500, China
Zhuxiaomei_75@126.com

Abstract. With the rising demand of software function, software system is becoming larger and more complex. The issue of "Insecticon Syndrome" to software becomes more and more serious, and the research on software reliability has been a focus during the development of software. In this paper, an evaluation model of modular software system reliability based on hidden markov model is proposed. It provides a simple and effective evaluation method for software developers and testers, and it has a wide application value.

Keywords: Hidden markov model (HMM), software, reliability.

1 Introduction

With the rising demand for software function, software system is becoming larger and more complex. In order to improve the software quality, to make it measurable and easy to control, modular development is a kind of effective method. However, the modular software system is usually huge in size, and the issue of "Insecticon Syndrome" is becoming serious increasingly, so reliability analysis to this kind of software system is also a necessary component in the process of software development. Software reliability is point to the probability of the software doesn't cause the system failure in the provision of conditions and time. It is an important index for measuring software quality level, and the evaluation to it has been a research focus. The research method commonly used basically has two kinds. One is the software reliability analysis based on the theory of probability and mathematical statistics, such as using poisson distribution, binomial distribution, markov process, bayesian model and so on. Second is the software reliability analysis based on time series, which commonly use the method as ARMA model, neural network, gray model and so on.

Modular software system is made of multiple modules. During the process of such software running in long-term, the state transitions of software call modules have relatively strong markov characteristics and the breakdown frequency of each module has certain probability statistics characteristics. So this paper studies the software system reliability problems by using a hidden markov model. As a

B.-Y. Cao and X.-J. Xie (Eds.): Fuzzy Engineering and Operations Research, AISC 147, pp. 485–492.
springerlink.com   © Springer-Verlag Berlin Heidelberg 2012

statistical model, HMM is simple and has better statistical properties and explain ability than neural network, support vector machine etc.. It has also succeeded in modeling for complex issues such as speech recognition (Lawrence R, 1989), biological sequence analysis (R Durbin et al., 1998), and natural language processing etc. Therefore, the research in this paper could provide a very good idea about software reliability study for the software engineering employees.

2   Hidden Markov Model

The basic theory of Hidden Markov Model (HMM) is created by Baum et al. in late 1960s to early 1970s.HMM is developed on the basis of Markov chain. However, practical problems are more complex than what Markov chain model describes, and the incident which we observed not one-to-one mapping state but intertwined through a group of probability distribution. Such model is called HMM. It is a double random process, including a description of the transfer among states and another description of the statistics corresponding relationship between state and observed symbols. There isn't a one-to-one relationship between observed symbols and state, therefore, we can only perceive state's nature and characteristics through observed symbols. According to observed symbols whether continuous, HMM can be divided into discrete HMM and continuous HMM. This paper mainly involves the discrete HMM, so mainly introduces discrete HMM.

2.1   Discrete HMM Model

Discrete HMM model is a five factor group, mainly constituted by the following elements:

a) N : The number of states. The collection of states vector is $S = (s_1, s_2, \cdots, s_N)$, and the state at time t recorded as $z_i, z_i \in S$;

b) M : The number of observations. The collection of observations is recorded as $O = (o_1, o_2, \cdots, o_T)$, and the observation at time t is recorded as o_T ;

c) A: State transition probability matrix shown as below. Among them, a_{s_i, s_j} (short as $a_{i,j}$) shows the probability of state transfer from s_i to s_j ;

$$
A = \begin{bmatrix}
a_{s_1,s_1} & a_{s_1,s_2} & \cdots & a_{s_1,s_N} \\
a_{s_2,s_1} & a_{s_2,s_2} & \cdots & a_{s_2,s_N} \\
\vdots & \vdots & \ddots & \vdots \\
a_{s_N,s_1} & a_{s_N,s_2} & \cdots & a_{s_N,s_N}
\end{bmatrix}.
$$

d) B: Observation probability matrix **is** shown as below. Among them, b_{s_i,e_j} (short as $b_{i,j}$) shows the probability of sign e_j being observed when current state is s_i;

$$\mathbf{B} = \begin{bmatrix} b_{s_1,e_1} & b_{s_1,e_2} & \cdots & b_{s_1,e_M} \\ b_{s_2,e_1} & b_{s_2,e_2} & \cdots & b_{s_2,e_M} \\ \vdots & \vdots & \ddots & \vdots \\ b_{s_N,e_1} & b_{s_N,e_2} & \cdots & b_{s_N,e_M} \end{bmatrix}.$$

e) $P = (p_{s_1}, p_{s_2}, \cdots, p_{s_N})$: initial state probability vector. Among them, p_{s_i} (short as p_i) shows the probability of state s_i is selected at the beginning.

And it should meet the following conditions:

1) Randomness, namely

$$\sum_{j=1}^{N} a_{i,j} = \sum_{j=1}^{M} b_{i,j} = \sum_{i=1}^{N} p_i = 1, a_{i,j}, b_{i,j}, p_i \in [0,1] \qquad (1)$$

2) First order feature, namely

$$P(z_{t+1}|z_t, z_{t-1}, \cdots, z_1, \theta) = P(z_{t+1}|z_t, \theta) \qquad (2)$$

$$P(o_t|z_t, z_{t-1}, \cdots, z_1, \theta) = P(o_t|z_t, \theta) \qquad (3)$$

3) Stability, namely

$$\forall t: a_{i,j} = P(z_{t+1} = s_j|z_t = s_i), b_{i,j} = P(o_t = e_j|z_t = s_i) \qquad (4)$$

4) Independence of observations, namely

$$P(o_1, o_2, \cdots, o_T|z_1, z_2, \cdots, z_T, \theta) = \prod_{t=1}^{T} P(o_t|z_t, \theta) \qquad (5)$$

So a HMM can be expressed as $\theta = \{N, M, p, A, B\}$, or $\theta = \{A, B, p\}$ for short.

There are three fundamental problems (evaluation problem, decoding problem, learning problem) with HMM need to be solved. They are the main content of the theory of HMM. This paper mainly use HMM evaluation problem to research modular software reliability.

2.2   Evaluation Problem of HMM

The evaluation problem is a computing problem about evaluating the probability of observation sequence appearing according to established HMM model. It's expressed in mathematical symbols as: Given designated HMM model $\theta = \{A, B, p\}$ and observation sequence $O = (o_1, o_2, \cdots, o_T)$, how to calculate output probability $P(O|\theta)$ of observation sequence O to HMM model θ effectively?

At present, such problems commonly solved by using forward algorithm and back algorithm. This paper mainly use forward algorithm to solve the problem, so it mainly introduces forward algorithm.

Giving HMM model $\theta = \{A, B, p\}$, defining forward variable $\alpha_t(i)$ as appearance probability of partial sequence $O = (o_1, o_2, \cdots, o_T)$ which be observed at time t and in state s_i by using HMM model, that is

$$\alpha_t(i) = P(o_1, o_2, \cdots, o_t, z_t = s_i | \theta), i = 1, 2, \cdots, N, t = 1, 2, \cdots, T \quad . \quad (6)$$

According to the nature of hidden markov model, the recursive formula of forward algorithm is obtained:

$$\alpha_t(j) = \left\{ \sum_{i=1}^{N} [\alpha_{t-1}(i) a_{i,j}] \right\} b_{j,o_t} \quad . \quad (7)$$

So getting the process of forward algorithm as follows:

Step 1. Initialization:

$$\alpha_1(i) = p_i b_{j,o_1}, i = 1, 2, \cdots, N .$$

Step 2. Recursive calculating:

$$\alpha_t(j) = \left\{ \sum_{i=1}^{N} [\alpha_{t-1}(i) a_{i,j}] \right\} b_{j,o_t}, i = 1, 2, \cdots, N, t = 2, 2, \cdots, T .$$

Step 3. Terminate calculating:

$$P(O|\theta) = \sum_{i=1}^{N} [a_T(i) q_i].$$

3   Reliability Model

Software module is the program which defines a series of external interface for accessing and realizes specific function in the internal. A complex software system is usually composed of many different modules. On a mission, software will usually call one or more modules, and sometimes certain key module will be called repeatedly. So each module's importance and work time are different during the task of software system. And for every module, its reliability is not the same as the others, its failure probability statistical characteristic is also different in a certain time. So, we can establish the modular software system reliability analysis model based on hidden markov model through testing the modular software system, analyzing module transfer regularity and failure data statistics during testing.

Assuming software system is divided into a finite number of independent module and the process of calling module is subservient to markov process, that means the system's running status has the characteristic of "stability ineffectiveness" when the module state is known at any time. At the same time, supposing that during the running time of each independent module, all of the module failure probabilities are subordinate to certain discrete distribution such as binomial distribution, poisson distribution etc., and supposing that every software system has certain fault-tolerant function.

Then setting up the hidden markov model of modular software system: $\theta = \{N, M, p, A, B\}$, among them:

N : the number of modules. We use $S = (s_1, s_2, \cdots, s_N)$ to indicate the collection of modules, and denote the state at time t as $z_i, z_i \in S$,;

M : the total number of module failures observed. We use $O = (o_1, o_2, \cdots, o_T)$ to indicate the collection of observations, and denote the sign observed at time t as O_t;

A : module transition probability matrix in software system. Among them, a_{s_i, s_j} (short as $a_{i,j}$) shows the probability of module transition from s_i to s_j;

B : module failure probability matrix. Among them, b_{s_i, e_j} (short as $b_{i,j}$) shows the probability of sign e_j being observed in state s_i;

P : module initial state probability vector. Among them, p_{s_i} (short as p_i) shows the probability of module s_i being selected at the beginning.

If the sum of module failure in software system is greater than L in a certain time, the software is invalid; otherwise the software has self repair function and can continue. So the modular software system failure probability in a certain time T, namely its reliability, is denoted as follow:

$$P = \sum_O P((o_1, o_2, \cdots, o_T)|\theta) \quad , \qquad (8)$$

$$\text{S.T. } o_1 + o_2 + \cdots + o_T \leq L. \qquad (9)$$

4   Data Simulation

A logistics management system has four functional blocks: user management function, scheduling optimization function, transportation management function and purchasing function. Through analysis, it indicates that the process of this system modules calling accord with markov process. Now this system is tested eight hours every day according to work time. And timing unit for hours, we get the software system module transition matrix after three weeks.

$$P = \begin{bmatrix} 0.31 & 0.39 & 0.2 & 0.1 \\ 0.25 & 0.35 & 0.18 & 0.23 \\ 0.3 & 0.1 & 0.33 & 0.27 \\ 0.17 & 0.32 & 0.26 & 0.15 \end{bmatrix}.$$

At the same time, the failure probability distribution function of each module is analyzed according to the test data.

Among them, the function of the number of purchasing function failure obey-ing poisson distribution $P(X_1 = k) = \dfrac{0.02^k e^{-0.02}}{k!}, k \geq 0$.

The function of user management function failure frequency obeying poisson distribution $P(X_2 = k) = \dfrac{0.025^k e^{-0.025}}{k!}, k \geq 0$.

The function of scheduling optimization function failure frequency obeying poisson distribution $P(X_3 = k) = \dfrac{0.036^k e^{-0.036}}{k!}, k \geq 0$.

The function of transportation management function failure frequency obeying poisson distribution $P(X_4 = k) = \dfrac{0.016^k e^{-0.016}}{k!}, k \geq 0$.

At present this system has 1 time automatic fault-tolerant function, namely L = 1. Then the reliability of this system in half a working day (4 hours) is:

$$P = \sum_O P((o_1, o_2, \cdots, o_4) | \theta) \ , \tag{10}$$

$$\text{S.T. } o_1 + o_2 + \cdots + o_8 \leq 1. \tag{11}$$

There are five module failure observation sequences which meet constraint condi-tion (11), and they are (0,0,0,0); (1,0,0,0);(0,1,0,0);(0,0,1,0);(0,0,0,1). We use forward algorithm to calculate the probability of each sequence and get the soft-ware system reliability probability as 0.8843. Then changing the time of software system fault tolerance to 0, 2, 3 separately, and the calculation results is shown in Table 1.

Table 1. Software system reliability calculations

Fault-tolerant number	Number of module failure observation sequences meeting the constraint condition	Software system reliability	Increase of reliability (%)
0	1	0.8237	--
1	5	0.8843	7.36
2	15	0.8368	0.28
3	35	0.8869	0.001

From Table 1, it is known that if the system has twice fault-tolerant functions, namely L=2, there are 15 module failure observation sequences which meet constraint conditions in the system, and the calculation of software system reliability probability is 0.8368; If the system has 3 times fault-tolerant functions, there are 35 module failure observation sequences which meet constraint conditions in the system, and the calculation of software system reliability probability is 0.8869; If the system has no fault-tolerant function, there is only one module failure observation sequence which meet constraint conditions in the system, that is (0,0,0,0), and the calculation of software system reliability probability is 0.8237. At the same time, it is also known from table 1 that the software system reliability increases most when software fault-tolerant times is one, but the increasing size will gradually decrease as the fault-tolerant times increasing, especially when the fault-tolerant times increased to 3, the effect to the increase of software system reliability is very small. So at this time, for improving the software system reliability, we should enhance the reliability of each application modules respectively, rather than increase investment to improve fault-tolerant ability.

5   Conclusion

In this paper, we established a reliability evaluation model based on hidden markov model for modular software system. The result of data simulation shows that this model is simple and practical, and it has strong explanation ability and popularization value. Through analysis we know that it do **a** little to improving software system reliability by continual strengthening of software system fault-tolerant ability after each software module has been developed. The better method is improving the reliability of each module.

The method is simple, but it is noteworthy that if extend the length of time for investigating software reliability, the number of module failure observation sequence meeting requirements should double and redouble, and the computational costs also will increase subsequently, so it need complicated programming to solve the problem.

References

1. Jia, Z.Y., Kang, R.: Software reliability forecasting ARIMA method. Computer Engineering and Applications 35, 17–21 (2008)
2. Wu, Q., Hou, Z.Z., Yu, J.M.: Software reliability model selection based on kohonen network. Computer Applications 10, 2331–2333 (2005)
3. Ma, S.S., Feng, Z., Zhao, S.W.: SVR-based software reliability prediction model. Computer Engineering and Applications 13, 120–123 (2007)
4. Jin, A., Jiang, J.H., Lou, J.G., Zhang, R.: Based ongrey model of software reliability modeling. Computer Applications 3, 690–694 (2009)
5. Wang, G.L.: J. method based on support vector machine prediction of non-stationary time series. Physics 2, 714–719 (2008)
6. Wang, W.L.: S-SVM with the multilayer feedforward network performance comparison of nonlinear regression. System Simulation 1, 256–259 (2008)

Numerical Simulation Study on Pulsed Gas Injection Using Orthogonal Analysis

Bin Jiang[1], Li-jun Huo[2], and Ping Guo[2]

[1] CNOOC Research Institute, Beijing 100027, China
[2] State Key Laboratory of Oil and Gas Geology and Exploitation,
Southwest Petroleum University, Chengdu 610500, China
541759832@qq.com

Abstract. Oil reservoirs having discovered recently and those having not put into development domestically are mainly low permeability or even ultra-low permeability reservoirs that are difficult to exploit, moreover, the exploitation is becoming increasingly difficult. Carbon dioxide injection has become a promising EOR approach since it is cost-effective and can achieve remarkable results, moreover, carbon dioxide can be used repetitively and has good miscibility with crude oil. Most of the sandstone reservoirs implementing gas injection are ultra-low-permeability reservoirs inland, to prevent the injected gas from breakthrough prematurely, the most realistic method is to implement pulsed gas injection, and it has already been applied in Fang 48 area. F180-129 well group of Fang 48 fault block is selected to simulate study profile adjusting using pulsed gas injection after gas breakthrough, and the orthogonal analysis is adopted to optimize the parameters of pulsed gas injection, such as pulsed cycle and gas injection volume. The parameters of the best scheme obtained are as follows: Gas injection time is 4 months, Stopping time is 1 month and gas injected volume per cycle is 0.02 HCPV. Based on the scheme obtained, the optimum pulsed timing is analyzed and research results show that, pulsed gas injection of F180-129 well group will obtain the best effectiveness when it is implemented after gas breakthrough occurs in 1 well or 2 wells, it will achieve higher oil replace ratio as well as higher recovery factor then.

Keywords: Low-permeability oil reservoir, pulsed gas injection, numerical simulation, orthogonal analysis.

1 Introduction

Oil reservoirs having discovered recently and those having not put into development domestically are mainly low permeability or even ultra-low permeability reservoirs that are difficult to exploit, moreover, the exploitation is becoming increasingly difficult. For conventional gas injection, CO2 has the best miscibility with crude oil and it has the effective and cost-effective merits, furthermore, it can be used repetitively. In addition, the increasing voice of

B.-Y. Cao and X.-J. Xie (Eds.): Fuzzy Engineering and Operations Research, AISC 147, pp. 493–503.
springerlink.com   © Springer-Verlag Berlin Heidelberg 2012

greenhouse gas reduction worldwide has brought opportunity to the development of CO_2 flooding technology, CO_2 flooding EOR has extensive application prospect [1, 2].

Most of the sandstone reservoirs adopting gas injection inland are ultra-low permeability reservoirs, to prevent the injected gas from breakthrough prematurely, the most realistically method is to implement pulsed gas injection and it has already been applied in Fang 48 area [3-7].

In this paper, F180-129 well group of Fang 48 fault block is selected to simulate study profile adjusting using pulsed gas injection after gas breakthrough, and the orthogonal test method is adopted to design the simulation schemes. Orthogonal test method can optimize the pulsed parameters based on relatively fewer schemes, and it can also sort different parameters in terms of its effect on oil recovery [8-10].

2   Basic Information of the Pilot Area

2.1   Basic Information of Fang 48 Fault Block

Fang 48 fault block CO_2 flooding pilot area of Songfangtun Oilfield in the peripheral part of Daqing is located in the southeast of Songfangtun Oilfield and belongs to Sanzhao sag to the east of Daqing placanticline regionally. The pilot area lies in the saddle area between two nose-like structures named Songfagntun and Mofantun, and it is a horst block which is sealed by three faults in the north, east and west directions, and there is few faults in the centre.

In early 2002, gas injection test was implemented in Fang 48 pilot area on a small scale and favorable results were gained, therefore, test in enlarged pilot area was carried out in 2007. Thirty production wells and fifteen injection wells were arranged in the enlarged area and the five-spot-pattern with the sizes of 400×250m, 300×150m and 300×100m were adopted. In November 2007, advanced gas injection was carried out successively in injection wells, and in April 2009, oil wells were put into production in succession, but there was no oil response resulted from gas injection, that is to say, the gas injection has had little effect so far. The reason for the ineffectiveness of gas injection may be that, the well spacing of the test area was designed on the basis of fractured well pattern initially, but during the implementation process, considering that fractures will result in early breakthrough of the injected CO_2, fracturing was not implemented to all the well groups. To establish an effective displacement system and develop Fang 48 low permeability reservoir effectively, fracturing optimization research was first conducted. Then, based on the fracturing optimization scheme, pulsed gas injection was selected to perform profile adjusting thus preventing premature gas breakthrough, therefore, the affected area of the injected gas and the utilization ratio of CO_2 can be enhanced, and the oil recovery of the reservoir can be improved consequently.

2.2   Basic Information of F180-129 Well Group

F180-129 well group of Fang 48 fault block is selected to simulate study profile adjusting using pulsed gas injection after gas breakthrough. F180-129 well group has 1 gas injection well and 5 oil production wells. Due to the large invalid grid area of the original geological model and considering the integrity of the well group as well as the injection-production corresponding relationship, this study has only simulated the 18th layer (corresponding to the first layer of the main layer FI7 of the actual formation) that has good connectivity relationship and has both injector and producers. The oil reserve of the 18th layer of the intercepted part of the geological model is 125076 m^3.

3   Fluid Phase Behavior Fitting

We have selected the PVTi Module of the Eclipse software to fit the experimental data of the formation oil PVT test. The fitting mainly involves Heavy Fractions Characterization of formation fluid, Component Lumping, Saturation Pressure Fitting, Experimental Data Fitting of the experiments including Single Flash, Differential Liberation, Constant Composition Expansion and CO2 Injection Expansion, etc. Finally, PVT parameters reflecting the actual property changes of the formation fluid are obtained.

The original composition of the formation fluid of Fang 48 area is shown in Table1, it can be seen that the fluid is the black oil with high heavy component content.

Table 1. Original composition of the formation fluid of Fang 48 area

Original Composition	Moore Composition, %	Weight Composition, wt%
CO_2	0	0
N_2	0.36	0.05
C_1	13.99	1.18
C_2	1.6	0.25
C_3	0.48	0.11
IC_4	1.6	0.49
NC_4	4.51	1.37
IC_5	3.3	1.25
NC_5	1.27	0.48
C_6	1.66	0.75
C_7^+	71.23	94.07

Relative density of $C7^+$=0.8750, Molecular Weight of $C7^+$=252.1.

4   Schematic Design

In the simulation schemes, F180-129 well group was put into production with the wells fractured in 2009, after gas breakthrough happens in the first well (August 2019) pulsed gas injection is implemented. Orthogonal test table L9 (9) of 3 factors and 3 levels is used to design the simulation schemes and 3 different levels of the 3 factors—HCPV injected per cycle, Gas injection time and Stopping time will be analyzed and the optimal scheme will be obtained, moreover, the impact of the above mentioned factors on recovery will be sorted. In the simulation schemes, oil production wells are produced with the bottom hole pressure set as 1MPa, and a well will be automated shut-in when its GOR reaches 3000 m^3/m^3, the simulation will be ended when all the wells are shut-in.

Scheme parameters of pulsed gas injection and the simulation results are shown in Table 2. In the table, it can be seen that the recovery of Continuous gas injection is 29.66%, and two schemes with higher oil recovery are Scheme 1 and Scheme 6, and their recovery are 30.42% and 30.25, respectively.

Table 2. Design parameters of the orthogonal schemes for pulsed gas injection

Scheme No.	HCPV injected per cycle	Gas injection time, month	Stopping time, month	Cumulative gas injection volume, t	Cumulative oil replace ratio, m^3/t	Final recovery, %
1	0.01	2	1	81688.47	0.49	30.42
2	0.01	3	2	62386.08	0.60	28.22
3	0.01	4	3	65612.74	0.58	26.70
4	0.02	2	2	80841.31	0.48	29.39
5	0.02	3	3	72738.98	0.53	28.83
6	0.02	4	1	82842.42	0.48	30.25
7	0.03	2	3	82423.47	0.46	28.90
8	0.03	3	1	85628.92	0.44	28.31
9	0.03	4	2	81688.49	0.48	29.30
Continuous gas injection	—	—	—	93778.59	0.42	29.66
New Scheme 10	0.02	2	1	90711.47	0.44	30.04

5   Results Analysis

5.1   *Intuitive Analysis*

Intuitive analysis is adopted to analyze the simulation results and Table 3 is the intuitive analysis form. Suppose Ki represents the summation of the 3 test indexes (simulation results) of level i of each factor, then mi=Ki/3. The Range of each factor is R=max(mi)-min(mi) (i=1, 2, 3) and the Range of the factor will reflects the level change of the factor on the rangrability of the final recovery.

Table 3. Intuitive analysis form of the forecasting results

Parameter	HCPV injected per cycle		Gas injection time		Stopping time	
	Level	m-value	Level	m-value	Level	m-value
m1	0.01	28.45	2	29.57	1	29.66
m2	0.02	29.49	3	28.45	2	28.97
m3	0.03	28.84	4	28.75	3	28.14
R		0.39		0.82		1.52

In Table 3 it shows that R(Stopping time)>R(Gas injection time)>R(HCPV injected per cycle), which implies that in the 3 selected factors, the factor that affecting recovery most is Stopping time, then follows Gas injection time and

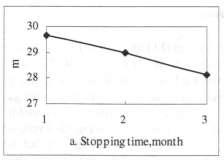

a. Effect of Stopping time on recovery

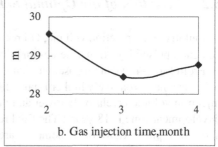

b. Effect of Gas injection time on recovery

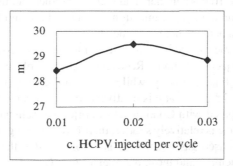

c. Effect of HCPV injected per cycle on recovery.

Fig. 1. Intuitive analysis diagram of recovery

HCPV injected per cycle. For the sake of convenience in intuitive analysis, intuitive analysis diagram with the 3 level of each factor as horizontal coordinate and mi corresponding to each level as vertical coordinate is plotted in Fig.1.

In Fig.1-a, it can be observed that with the increase of Stopping time, there is an overall decline in recovery. Hence, the shorter the Stopping time, the higher the final recovery.

In Fig.1-b, with the increase of Gas injection time, oil recovery decrease first and then increase, and in the figure it shows that when the Gas injection time is 2 months the recovery is the highest.

In Fig.1-c, with the increase of HCPV injected per cycle, the recovery increase first and then decrease, and when injecting 0.02 HCPV per cycle, the recovery is the highest.

According to the analysis results of the effect of those factors on recovery, a new scheme named Scheme 10 is designed with the optimal parameters: Gas injection time is 2 months, Stopping time is 1 month and the gas injected volume per cycle is 0.02 HCPV. Simulation results show that the recovery of this new scheme 10 is 30.04%, the value of which is relatively higher. And thus it indicates that the results obtained from orthogonal optimization method are reliable.

5.2   Selection of the Optimal Scheme

Cumulative gas injection volume, Oil recovery and Oil replace ratio curves of the 9 designed pulsed injection schemes and the new scheme 10 are shown in Fig.2-4.

From Fig.2-4, it can be observed that in the 3 schemes with the highest recovery: Scheme 6 has a relatively higher final recovery, at the same time, its cumulative gas injection volume is relatively lower and oil replace ratio is relatively higher, and its development time is 18 years; The final recovery, cumulative gas injection volume and oil replace ratio of Scheme 1 are approximate with Scheme 6, but its development time is relatively longer, 21 years. Therefore, Scheme 6 is better than Scheme 1; Scheme 10 has a higher final recovery either and its development time is 12 years, but its cumulative gas injection volume is high and the oil replace ratio is very low, so this scheme is not recommended.

According to the analysis above, considering the combined influence of Cumulative gas injection volume, Recovery and Oil replace ratio, Scheme 6 has a relatively higher final recovery while its cumulative gas injection volume is relatively lower, oil replace ratio is relatively higher (it can save about 10 thousand tons of CO_2 compared with Continuous gas injection scheme and Scheme 6), and its development time is relatively shorter, therefore, for Fang 48 fault block, while it is implementing pulsed gas injection it is recommended that it adopts the pulsed parameters of Scheme 6 and those parameters are: Gas injection time is 4 months, Stopping time is 1 month and the gas injected volume per cycle is 0.02 HCPV.

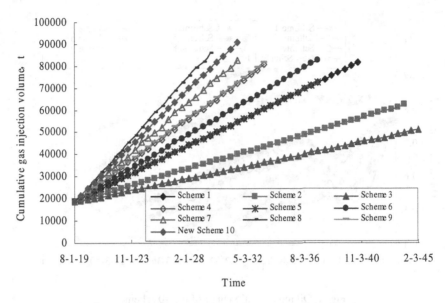

Fig. 2. Cumulative gas injection volume curve of the 10 schemes

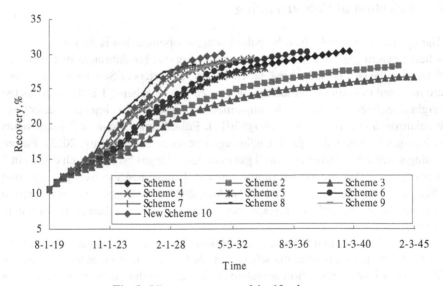

Fig. 3. Oil recovery curve of the 10 schemes

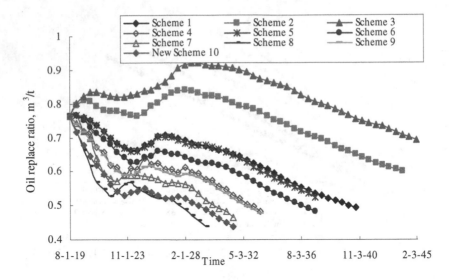

Fig. 4. Oil replace ratio curve of the 10 schemes

6   Selection of Pulsed Timing

The optimal scheme obtained by pulsed scheme optimization is Scheme 6, in this scheme pulsed gas injection is implemented after gas breakthrough in 1 well. To determine the best pulsed timing, based on the parameters of Scheme 6, 4 schemes are designed to optimize pulsed timing. Pulsed timing scheme 1 is the same as the original Scheme 6 and it is implementing pulsed gas injection after gas breakthrough occurs in 1 well (Aug. 2019), Pulsed timing scheme 2 implements pulsed gas injection after gas breakthrough occurs in 2 wells (Aug. 2023), Pulsed timing scheme 3 implements pulsed gas injection after gas breakthrough occurs in 3 wells (Aug. 2024) and Pulsed timing scheme 4 implements pulsed gas injection after gas breakthrough occurs in 4 wells (Aug. 2027). The Cumulative gas injection volume, Recovery and Oil replace ratio curves of the 4 schemes is shown in Fig.5-7.

In Fig.5-7, it sees that Scheme 3 has the lowest recovery among all the schemes. As for Continuous gas injection scheme and Scheme 4, their oil recoveries are not high and their oil replace ratio are relatively lower, therefore these schemes are not recommended. In conclusion, from the point of pulsed timing, F180-129 well group will achieve the best effectiveness while it is implemented after gas breakthrough occurs in 1 well or 2 wells, it will achieve higher oil replace ratio as well as higher recovery factor then.

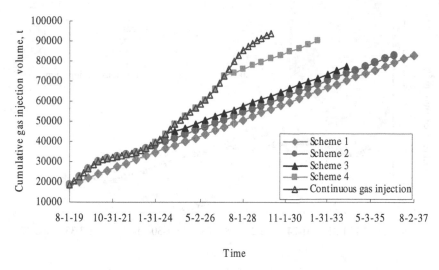

Fig. 5. Cumulative gas injection volume curve of 5 pulsed timing schemes

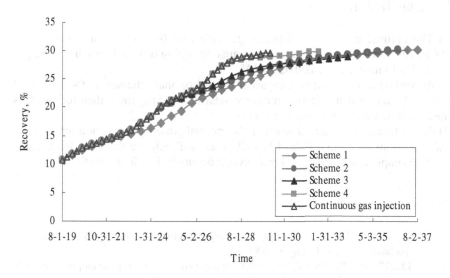

Fig. 6. Oil recovery curve of 5 pulsed timing schemes

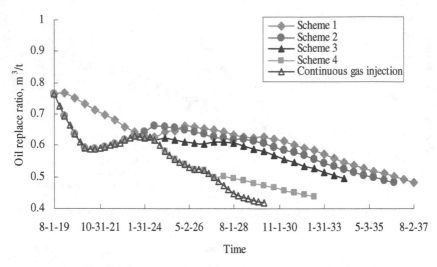

Fig. 7. Oil replace ratio curve of 5 pulsed timing schemes

7 Conclusion

(1) The optimal parameters of pulsed gas injection for the fractured well group F180-129 are: Gas injection time is 4 months, Stopping time is 1 month and the gas injected volume per cycle is 0.02 HCPV.

(2) According to the intuitive analysis of the orthogonal schemes, in the 3 selected factors, the factor that affecting recovery most is Stopping time, then follows Gas injection time and HCPV injected per cycle.

(3) Pulsed timing is optimized based on the optimal pulsed gas injection scheme, and analysis results show that F180-129 well group will achieve the best effectiveness when it is implemented after gas breakthrough occurs in 1 well or 2 wells.

References

1. Li, S.L., Zhang, Z.Q., Ran, X.Q., et al.: Gas injection EOR technology. Sichuan Science and Technology Press, Chengdu (2001)
2. Li, S.L., Zhou, S.X., Du, J.F., et al.: Review and prospects for the development of EOR by gas injection at home and abroad. PGRE 9(2), 1–5 (2002)
3. Guo, P., Li, S.L., Du, Z.M., et al.: Evaluation on IOR by gas injection in low permeability oil reservoir. Journal of Southwest Petroleum Institute 24(5), 46–50 (2004)
4. Li, M.T., Zhang, Y.Z., Yang, Z.H., et al.: Research on CO_2 miscible flooding to enhance oil recovery in low permeability reservoir. Oil Drilling & Production Technology 27(6), 43–46 (2005)
5. Wu, Y.B., Hu, D.D., Chang, Y.W., et al.: Application status of CO_2 flooding in low permeability reservoir to enhance oil recovery. Xinjiang Oil & Gas 6(1), 36–39 (2010)

6. Guo, P., Mo, Z.K., Wang, R.F., et al.: Experimental study on improving recovery of low permeability fracture-type carbonate oil reservoir by pulse hydrocarbon injection. PGRE 11(1), 48–49 (2004)
7. Li, Z.Q., Liu, X.Y., Sun, N., et al.: EOR technology on ultra-heavy oil reservoir and its application with slug steam drive. Fanlt-Block Oil & Gas Field 17(2), 246–249 (2010)
8. Zhang, M.L., Mei, H.Y., Li, M., et al.: Numerical simulation of steam injection in wellblock Wen72-467. Special Oil and Gas Reservoirs 10(4), 46–49 (2003)
9. Cang, H., Yang, Y.L., Chen, J.Q., et al.: Experimental study of optimized steam injection parameters for immiscible displacement. Special Oil and Gas Reservoirs 12(5), 95–98 (2005)
10. Li, B.Z., Li, X.F., Kamy, S., et al.: Optimization of the injection and production schemes during CO_2 flooding for tight reservoir. Journal of Southwest Petroleum University (Science & Technolgy Edition) 32(2), 101–107 (2010)

Research and Application of Parallel Fuzzy Dynamic Time Warping Algorithm in Music Retrieval

Yu-bin Zhong and Yi Xiang

School of Mathematics and Information Sciences, Guangzhou University,
Guangzhou 510006, P.R. China
{zhong_yb,gzhuxiang_yi}@163.com

Abstract. Aiming at the time-consuming matching process and non-completely accurate humming in music retrieval via Query By Humming (QBH), a parallel fuzzy Dynamic Time Warping (DTW) algorithm is proposed to implement approximate matching of melodies. Use relative pitch difference to represent melodies, and use DTW algorithm to realize approximate match between humming pitch difference sequences and target pitch difference sequences. In the process of matching, fuzzy sets and methods are introduced. By constructing membership functions between humming pitch difference and target pitch difference and computing membership grade, we get similarity degree of pitch differences, and then obtain the "Switching Cost" matrix. Finally, we get the matching distance of these two sequences. To accelerate the retrieval speed, parallel algorithms are introduced in the process of matching. The experiment result shows that the introduction of fuzzy methods improves retrieval accuracy and the application of parallel algorithms shortens retrieval time. The accurate rate of humming retrieval based on parallel fuzzy DTW is 80%, which improves a lot compared to traditional matching algorithms.

Keywords: Humming Retrieval, Melody Representation, Dynamic Time Warping Algorithm, Fuzzy Set, Parallel Algorithm.

1 Introduction

As the continual development of computer techniques, demands for new music retrieval methods are put forward in recent years. In traditional music retrieval, such text information as song names, singer names and lyrics (at least one of them) is needed in order to continue retrieving. This retrieval method is not convenient and not natural. Content Based Music Retrieval (CBMR) is a new technology and science which studies how to use such contents of a music file itself as pitches, melodies etc. to realize music retrieval [1]. It has been one of the most important research fields both home and abroad. QBH is a typical CBMR method, which searches music by humming certain part of target music [2]. And key techniques of QBH are melody extracting and approximate match algorithm. In view

B.-Y. Cao and X.-J. Xie (Eds.): Fuzzy Engineering and Operations Research, AISC 147, pp. 505–513.
springerlink.com   © Springer-Verlag Berlin Heidelberg 2012

of the most widely used MP3 files at present, we will extract melodies of target music and humming music to obtain target sequences stored in a database and humming sequences respectively. And then use DTW algorithm to complete match process. In order to reduce error rate while quantifying pitch difference, a reasonable mathematical model -fuzzy set- is introduced to get similarity degree between humming sequences and target sequences. And this retrieval process will be implemented by a parallel algorithm.

2   Melody Extraction and Representation

Melodies consist of a series of musical notes that reflect musical themes. They can well show content characteristics of music. Correctly extracting and representing melodies are key steps in music retrieval, which directly influence the match accuracy rate. Common ways of representing melodies are absolute pitch sequences and relative pitch sequences [2]. Fig.1. shows melody information of a part of one piece of music. The horizontal axis is time, and the vertical axis is absolute pitch. A relative pitch sequence is made up of musical intervals-differences of absolute pitches between the next one and the previous one. In humming retrieval, a relative pitch sequence has more advantages than an absolute pitch sequence. Music files (such as MP3 files) usually consist of vocal part and non-vocal part, while humming music is always unaccompanied. Notice this, a method for splitting vocal part and non-vocal part is first introduced (see reference [3]), and this operation is also called denoising. Then use fast Fourier transform (FFT) to get fundamental frequencies of both de-noised target music and de-noised humming music. By analyzing FFT spectrums after FFT transform, corresponding pitches and then pitch difference sequence are obtained. The detailed step is shown in reference [2] and reference [3].

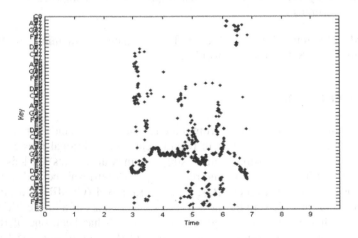

Fig. 1. Melodies of the climax part of a piece of music

Pitch characteristic information for 100 pieces of music are recorded in the target music database, and the pitch difference is represented by semitones, which is denoted $T_i = (t_{i1}, t_{i2}, \cdots, t_{iM_i}), i = 1, 2, \cdots, 100$, where M_i is the length of ith target music pitch difference sequence. Denote humming pitch difference sequence as $H_j = (h_{j1}, h_{j2}, \cdots, h_{jN_j})$, where N_j is the length of jth humming music pitch difference sequence. And here, the pitch difference is represented by cents. Namely, $h_{jk} = 1200 \times \log_2(f_{k+1}/f_k)$, where f_k means the corresponding frequency of kth pitch.

3 Approximate Matching Algorithms of Melodies

In humming music retrieval, users may don't have professional music backgrounds. As a result, there will be a great discrepancy between humming melodies and target melodies. For example, the pitch difference sequence of humming melodies is usually incorrect, which is quite different from sequences of expected melodies. Therefore, it is obvious that finding the right music from a database exactly is very difficult in the fact that the humming melodies are incorrect. Besides, music to be retrieved is often recorded with multi-tones and more voice parts, while humming music is unaccompanied singing. Consequently, the pattern matching algorithm in humming retrieval is a typical approximate matching one [2]. The existing approximate matching algorithms of melodies mainly are: approximate string matching algorithms, such as N-Gram algorithm [5], DP algorithm [6]; statistical model based algorithms, such as Markov model; the dynamic time warping retrieval algorithm [8]; the linear alignment matching algorithm [9] ; the fuzzy DP matching algorithm[4] and so on. These algorithms have their own advantages and shortcomings. After carefully comparing algorithms mentioned above and further studying reference [4], we introduce fuzzy methods and propose a parallel fuzzy DTW approximate matching algorithm.

3.1 The DTW Algorithm

Suppose the target melody sequence in the database and the humming melody sequence are T_i and H_j respectively. Their distance $D = (T_i, H_j)$ can be calculated so as to compare the similarity degree. The smaller the distance is, the higher the similarity degree is. To compute the above distance, $d = (T_i(n), H_j(m))$ need to be calculated first, where $T_i(n)$ and $H_j(m)$ are the nth and the mth elements of T_i and H_j respectively, and $d = (T_i(n), H_j(m))$ is the switching cost. In traditional DTW, Euclidean distance is generally selected as the switching cost. Because of the nonconformity in sequence lengths of target music and humming music in most cases, it is reasonable to compute distance D using dynamic programming method[10]. Subscripts $m = 1 \sim N_j$ of each element in humming sequence H_j are used as the horizontal axis in a 2-dimension system of rectangular coordinates, while the vertical axis denotes subscripts $n = 1 \sim M_i$ of each element in target sequence T_i. Hence, it forms a gridding which is shown in Fig. 2.

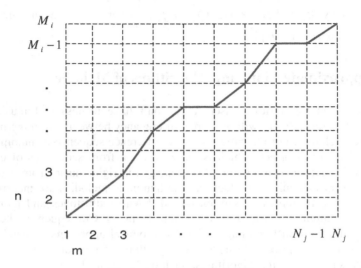

Fig. 2. A search path of DTW

To describe this path (see Fig. 2), we can suppose coordinates of ordered grid points in the path are (m_1, n_1) , (m_2, n_2) , ..., (m_{N_j}, n_{M_i}) , where $(m_1, n_1) = (1,1)$, $(m_{N_j}, n_{M_i}) = (N_j, M_i)$. And define a function $n_i = \sigma(m_i)$, where $m_i = i, i = 1, 2, \cdots, N_j$, $\sigma(1) = 1, \sigma(N_j) = M_i$. To make sure that the path doesn't incline excessively, the slope range is restrained between 0.5 and 2. Therefore, the path solving problem is converted to seek an optimal path function $n_i = \hat{\sigma}(m_i)$ that makes the cumulative distance minimum under the above restrained conditions. The formula for computing cumulative distance $D[(m_k, n_k)]$ at point (m_k, n_k) is shown as below.

$$D[(m_k, n_k)] = d[(T_i(n_k), H_j(m_k))] + D[(m_{k-1}, n_{k-1})], \tag{1}$$

where (m_{k-1}, n_{k-1}) is the previous point of (m_k, n_k) ,and $m_{k-1} = m_k - 1$. n_{k-1} is determined by the following formula.

$$D[(m_{k-1}, n_{k-1})] = \min\{D[(m_{k-1}, n_k)], D[(m_{k-1}, n_k - 1)], D[(m_{k-1}, n_k - 2)]\} \tag{2}$$

Thus, start from $(m_1, n_1) = (1,1)$ to search for (m_2, n_2) , and then for $(m_3, n_3), \ldots \ldots$ An optimal path will be found when reach to point (m_{N_j}, n_{M_i}) .

Implementations of DTW algorithm could be: assigning two $N_j \times M_i$ matrixes storing cumulative distance matrix D and switching cost matrix d respectively. Execute calculations according to the algorithm described above, then $D[(N_j, M_i)]$ is the corresponding matching distance of the optimal matching path.

4   Applications of Fuzzy Methods in Music Retrieval

In most CBMR systems, pitch difference sequences are generally represented by direct quantifying methods. Various pitch difference sequences can be obtained according to different quantifying methods. However, these pitch difference sequences could not be absolutely correct pitch differences. That is to say, direct quantifying methods will introduce quantifying errors, which results in undesired retrieval effects. Fuzzy mathematics is theories and methods that study and deal with fuzziness. Introduce fuzzy sets as a reasonable mathematics model and use fuzzy methods to get similarity degrees between humming sequences and target sequences so as to reduce quantifying errors, then the retrieval accuracy will be improved greatly. Compared to traditional methods, fuzzy methods don't quantify melody data directly but get similarity degrees between humming pitch differences and target pitch differences by measuring their fuzzy memberships, which are switching costs when calculating matching distances. The computation of memberships is realized by defining membership functions.

The fuzzy set is defined as below: the set of humming pitch differences is the variable universe U , and the set of target pitch differences in the database is defined as K . A_k is a fuzzy set –pitch differences on variable universe U can be denoted semitone difference $k \in K$. And A_k is a real-valued function on U . For $\Delta p \in U$ in humming pitch difference sequence, $\mu_{A_k}(\Delta p)$ is the membership grade of Δp to fuzzy set A_k , and μ_{A_k} denotes the membership function of A_k . $A_k(\Delta p)$ is short for $\mu_{A_k}(\Delta p)$.

4.1   Constructions of Membership Functions

There are many ways of constructing membership functions. Reference [4] uses statistical ways to define fuzzy relationships and to construct membership functions. Analyze all pitch differences in database by statistics to get their distribution laws, which are shown in Fig. 3. We can see that the distribution characteristic of pitch differences is approximately symmetric. Absolute values of pitch differences that are equal or greater than 20 just account for 7.8067%, while more than 92.1933% are in $[-20, 20]$. So we can define an integer set as $K = [-20, 20]$, and deem those less than -20 as -20 and those greater than 20 as 20.

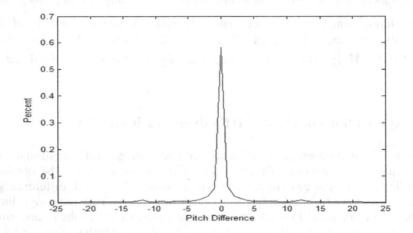

Fig. 3. Distribution laws of pitch differences

(1) The First Construction Method
The study shows that a pitch difference sequence is an approximate normal distribution. Let its density function be $P_k(x)$, and in reference [4] a definition for the membership function is shown as below.

$$A_k(\Delta p) = \frac{P_k(\Delta p)}{\sum\limits_{i \in K} P_i(\Delta p)}, \tag{3}$$

where $P_k(x)$ represents a normal distribution whose mean value is k. Then the similarity degree between humming pitch difference sequence $H_j(m)$ and target pitch difference sequence $T_i(n)$ is defined as

$$S[T_i(m), H_j(m)] = 1 - A_k[H_j(m)], T_i(m) = k. \tag{4}$$

Therefore, the switching cost in DTW algorithm is

$$d = [T_i(n), H_j(m)] = S[T_i(m), H_j(m)]. \tag{5}$$

Thus after computing matching distances of each humming pitch difference sequence H_j and each target pitch difference sequence T_i in database, we can get a matrix and the best matching serial numbers of target music by finding corresponding column subscripts of the minimum distance in each row. Experiments indicate that this membership function construction way is time –consuming and the matching effect is also not so well. Therefore, a new way of construing membership functions is put forward in the following [11].

(2) The Second Construction Method

The pitch difference range of target melody sequences is $[-20, 20]$, and it is can be deemed that $\{-20, -19, \cdots, -2, -1, 0, 1, 2, \cdots, 20\}$ these 41 integer divide pitch differences into 41 grades, which can be denoted as $L = [l_{-20}, l_{-19}, \cdots, l_0, l_1, \cdots, l_{19}, l_{20}]$. The pitch difference that absolutely belongs to grade l_k is $k \in K = [-20, 20]$. The membership function for each grade is $v_{l_k}, l_k \in L$, then

$$v_{l_1}(h) = \begin{cases} 1 & when \quad h \le K(1) \\ [h - K(2)]/[K(1) - K(2)] & when \quad h \in [K(1), K(2)], \\ 0 & else \end{cases} \quad (6)$$

$$v_{l_2}(h) = \begin{cases} [h - K(1)]/[K(2) - K(1)] & when \quad h \in [K(1), K(2)] \\ [h - K(3)]/[K(2) - K(3)] & when \quad h \in [K(2), K(3)], \\ 0 & else \end{cases} \quad (7)$$

$$v_{l_{41}}(h) = \begin{cases} 1 & when \quad h \ge K(41) \\ [h - K(40)]/[K(41) - K(40)] & when \quad h \in [K(40), K(41)], \\ 0 & else \end{cases} \quad (8)$$

where $K(i) = i$, and their graphics are shown in Fig. 4.

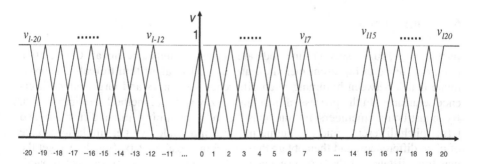

Fig. 4. Graphics of membership functions

4.2   Applications of Parallel Fuzzy DTW Algorithm in Music Retrieval

Although using fuzzy methods to construct the switching cost matrix in DTW algorithm can improve retrieval accuracy to some extent, it increases time costs. Generally, the switching cost matrix is a matrix with higher dimensions, and if the calculation of d is very complex then it will cost a very long computing time. In DTW algorithm, it is needed to compute matching distances between the humming pitch difference sequence and each target pitch difference sequence in database, and then find the minimum distance. Obviously, if the number of target sequences in a database is more, then the longer the computing time will be. Because the computation for each matching distance is independent, it will be effective to use parallel algorithm. That is to say, distributing tasks of computing matching distances over different processors, and the calculation results are gathered after each processor finishes its calculation task, followed by finding the minimal matching distance. Thus retrieval process is realized. Experiments indicate that the introduction of parallel algorithms can improve retrieval efficiency sharply.

5   Experiment Results and Analyses

In this paper, 100 pieces of music both at home and abroad are selected so as to cut out their climax parts and extract their pitch differences to constitute target sequences in database, which contains 37135 pitch differences. 10 persons (half is male and half is female) hum their favorite 10 pieces of music once, which have their counterparts in the database. Thus, there are 100 humming segments. Use the above parallel fuzzy DTW algorithm to conduct music retrieval experiment: the retrieval way that uses the first membership function construction method has just 65% accuracy rate, while the accuracy rate of the second method is 80%. The retrieval time of the second retrieval way is less than that of the former by 10 seconds.

6   Conclusion

In this paper, fuzzy sets and methods are applied in music retrieval. Use relative pitch difference to represent melodies, and use DTW algorithm to realize approximate match between humming pitch difference sequences and target pitch difference sequences. In the process of matching, fuzzy sets and methods are introduced. By constructing membership functions between humming pitch difference and target pitch difference and computing membership grade, we get similarity degree of pitch differences, and then obtain the "switching cost" matrix. Finally, we get the matching distance of these two sequences. To accelerate the retrieval speed, parallel algorithms are introduced in the process of matching. The experiment results show that the introduction of fuzzy methods improves retrieval accuracy and the application of parallel algorithms reduces retrieval time. Two ways of constructing membership functions are proposed in this paper and a comparison of experiment results between them is also made. The results show that: (1) the retrieval accuracy rate of the second construing method is higher than that of the first one by 15%; (2) the parallel fuzzy DTW algorithm proposed in this paper is appropriate for being used in humming-based music retrieval systems, and for retrievals in large-scale music database, it is also very effective.

Acknowledgements. Thanks to the support by keynote Teaching Research Project of Guangzhou University.

References

1. Foote: Content-Based Retrieval of Music and Audio. Multimedia Storage and Archiving systems II. In: Proc. of SPIE, vol. 3229, pp.138-147 (1997)
2. Huang, L.: Research and Implementation of the Content-Based MP3 Music Retrieval. Master degree paper of Xiamen University (2008)
3. Xie, X.-Q.: The singing splitting based on T-F analysis. Master degree paper of Shandong University (2009)
4. Ma, Z.-X., Fu, S.-F., Zhou, L.-H.: A new method using fuzzy approximate melody matching for QBH based music retrieval. Journal of Xidian University 33(1), 85–88 (2006)
5. Uitdenbogerd, A.: Zobel:Melodic Matching Techniques for Large Databases. In: Processing of ACM Multimedia 1999, pp. 57–66 (1999)
6. Lien, J., et al.: Building a Platform for Performance Study of Various Music Information Retrieval Approaches. In: International Symposium on Music Information Retrieval (2003)
7. Rand, W., Birmingham, W.: Statistical Analysis in Music Information Retrieval. In: Proceedings of the 2nd Annual Internationals Symposium on Music Information Retrieval (ISMIR), Bloomington, Indiana, USA, pp. 25–26 (2001)
8. Xu, W.-H., Gao, M.-Y., Zhang, Z.-X.: Intuition Humming Input Music Retrieval Search Engine. In: Proceedings of the Fifth Conference on Artificial Intelligence and Applications, Taiwan, P.R. China, pp. 734–740 (2000)
9. Li, Y., Wu, Y.-D., Liu, B.-L.: A New Method for Approximate Melody Matching and Its Application in Query by Humming System. Journal of Computer Research and Development 40(11), 1554–1560 (2003)
10. He, Q., He, Y.: MATLAB Extending Programming, pp. 340–345. Tsinghua University Press, Beijing (2002)
11. Zhong, Y.-B.: The FHSE model of software system for synthetic evaluating enterprising. Journal of Guangzhou University (Natural Science Edition), 212–225 (2005)

Part IV
OR and Fuzziology

A New Ranking Approach to Fuzzy Posynomial Geometric Programming with Trapezoidal Fuzzy Number

Zeinab Kheiri, Faezeh Zahmatkesh, and Bing-yuan Cao[*]

School of Mathematics and Information Sciences, Guangzhou University,
Key Laboratory of Mathematics and Interdisciplinary Sciences,
Higher Education Institutes, Guangzhou University,
Guangzhou 510006, China
{Zeinab_kheiri,Faeze_zahmatkesh}@yahoo.com,
Caobingy@163.com

Abstract. This paper presents a method for solving posynomial geometric programming with fuzzy coefficients. By utilizing comparison of fuzzy numbers with a new approaching method, the programming with fuzzy coefficients is reduced to the programming with constant coefficient. Then we can solve the programming with fuzzy coefficients using a method to posynomial geometric programming. Finally, one comparative example is used to illustrate the advantage of the new method.

Keywords: Fuzzy posynomial geometric programming, Yager's method, A new approach method for ranking of trapezoidal fuzzy numbers.

1 Introduction

Geometric programming (GP) problems have wide range of application in production planning, location, distribution, risk managements, chemical process designs and other engineer design situations and so on. Since late 1960's, GP has been known. Duffin, Peterson and Zener in 1967 discussed basic theories on GP with engineering application in their books. Another famous book on GP and its application appeared in 1976 (Beightler and phillips). In 1987, Cao first introduced fuzzy geometric programming. There is a good book dealing with fuzzy geometric programming by Cao (2002) which was the most recent book until now. GP is an effective method to solve a particular type of a non-linear programming problem.

Ranking fuzzy numbers plays an important role in linguistic decision making and some other fuzzy application systems such as management, operations research and etc. Many methods have been proposed to deal with ranking fuzzy numbers. Recently a new approach for ranking of trapezoidal fuzzy numbers is proposed by Abbasbandy and Hajjari. Now,using this method to fuzzy geometric programming problems is proposed in this paper .

---

[*] Corresponding author.

B.-Y. Cao and X.-J. Xie (Eds.): Fuzzy Engineering and Operations Research, AISC 147, pp. 517–523.
springerlink.com   © Springer-Verlag Berlin Heidelberg 2012

This paper is organized as follows: Section 2 contains the basic definition with notifications used in the remaining parts of the paper. A review of the Yager's Method and a New Approach Method for ranking of trapezoidal fuzzy numbers is given in Section 3 and 4. In Section 5 we show how to use a new approaching method for solving fuzzy geometric programming problems and explain it by an illustrative example in Section 6. The paper ends with conclusions in Section 7.

2 Preliminaries

Definition 2.1 [3] A fuzzy number is a fuzzy set like $u: R \rightarrow I = [0, 1]$ satisfying:

1. u is upper semi-continuous,
2. $u(x) = 0$ outside some interval $[a, b]$,
3. There are real numbers a; b such that $a \leq b \leq c \leq d$ and
a. $u(x)$ is monotonically increasing on $[a, b]$;
b. $u(x)$ is monotonically decreasing on $[c, d]$;
c. u(x)=1; $b \leq x \leq c$.

The membership function u can be expressed as

$$u(x) = \begin{cases} u_L(x), & a \leq x \leq b, \\ 1, & b \leq x \leq c, \\ u_R(x), & c \leq x \leq d, \\ 0, & otherwise, \end{cases}$$

where $u_L : [a, b] \rightarrow [0,1]$ and $u_R : [c, d] \rightarrow [0,1]$ are left and right membership functions of fuzzy number u. An equivalent parametric form is also given in [3] as follows.

Definition 2.2 [3] *A fuzzy number u in parametric form is a pair* $(\underline{u}, \overline{u})$ *of functions* $\underline{u}(r), \overline{u}(r), 0 \leq r \leq 1$ *which satisfy the following requirements:*

1. $\underline{u}(r)$ is a bounded monotonically increasing left continuous function;
2. $\overline{u}(r)$ is a bounded monotonically decreasing left continuous function;
3. $\underline{u}(r), \overline{u}(r), 0 \leq r \leq 1$.

The trapezoidal fuzzy number $u = (x_0, y_0, \delta, \beta)$ with two defuzzifier x_0, y_0 and left fuzziness $\delta > 0$ and right $\beta > 0$ is a fuzzy set where the membership function serves as

$$u(x) = \begin{cases} \dfrac{1}{\delta}(x - x_0 + \delta), & x_0 - \delta \leq x \leq x_0 \\ 1, & x \in [x_0, y_0] \\ \dfrac{1}{\beta}(y_0 - x + \beta), & y_0 \leq x \leq y_0 + \beta \\ 0, & otherwise \end{cases}$$

and its parametric form is

$$\underline{u}(r) = x_0 - \delta + \delta r, \quad \overline{u}(r) = y_0 + \beta - \beta r.$$

Definition 2.3 *Let $\tilde{A} \in \mathcal{F}(X)$, $\forall \alpha \in [0, 1]$, written as*

$$(\tilde{A})_\alpha = A_\alpha = \{x \in X, \tilde{A}(x) \geq \alpha\}.$$

A_α *is said to be an α-cut set of a fuzzy set $\tilde{A}$. The lower and upper bounds of any α-cut set A_α are represented by*

$$A_\alpha{}^L = \inf\{x \in X, \tilde{A}(x) \geq \alpha\}$$

and

$$A_\alpha{}^u = \sup\{x \in X, \tilde{A}(x) \geq \alpha\}$$

and we suppose that both are finite.

3 Yager's Method for Ranking of Trapezoidal Fuzzy Numbers

A special version of the linear ranking function was first proposed by Yager (1981, [5]) below:

$$R(\tilde{a}) = \frac{1}{2} \int_0^1 \left(\frac{A_\alpha{}^L + A_\alpha{}^u}{2}\right) d\alpha.$$

For trapezoidal fuzzy numbers $\tilde{a} = (x_0, y_0, \delta, \beta)$, we have:

$$R(\tilde{a}) = \frac{2x_0 + 2y_0 - \delta + \beta}{4}.$$

Therefore for any two trapezoidal fuzzy numbers $\tilde{a}$ and $\tilde{b}$, we define the ranking of $\tilde{a}$ and $\tilde{b}$ as:

1. $R(\tilde{a}) \geq R(\tilde{b})$ if and only if $\tilde{a} \geq \tilde{b}$;
2. $R(\tilde{a}) > R(\tilde{b})$ if and only if $\tilde{a} > \tilde{b}$;
3. $R(\tilde{a}) = R(\tilde{b})$ if and only if $\tilde{a} = \tilde{b}$.

Remark 3.1 For two arbitrary trapezoidal fuzzy numbers $\tilde{a}$ and $\tilde{b}$, we have

$$R(k\tilde{a} + \tilde{b}) = kR(\tilde{a}) + R(\tilde{b}).$$

4 New Approach for Ranking of Trapezoidal Fuzzy Numbers [1]

For an arbitrary trapezoidal fuzzy number $u = (x_0, y_0, \delta, \beta)$ with parametric form $u = (\underline{u}(r), \overline{u}(r))$, we define the magnitude of the trapezoidal fuzzy number u as

$$Mag(u) = \frac{1}{2}\left(\int_0^1 (\underline{u}(r) + \overline{u}(r) + x_0 + y_0)r dr\right) = \frac{1}{12}(6x_0 + 6y_0 + \beta - \delta). \quad (1)$$

Therefore, for any two trapezoidal fuzzy numbers u and $v \in$ E, we define the $Mag(.)$ on E as a ranking of u and v:

1. $Mag(u) > Mag(v)$ if and only if $u > v$;
2. $Mag(u) < Mag(v)$ if and only if $u < v$;
3. $Mag(u) = Mag(v)$ if and only if $u = v$.

Remark 4.1 For two arbitrary trapezoidal fuzzy numbers u and v, we have

$$Mag(u + v) = Mag(u) + Mag(v) \ .$$

5   A New Approach to FPGP by Ranking of Trapezoidal Fuzzy Numbers

In this section , we define the fuzzy posynomial geometric programming (FPGP) problem and propose a method for this problem.

Definition 5.1   [2]   *Let $F(R)$ be the set of all trapezoidal fuzzy numbers. The model*

$$\begin{aligned} \widetilde{Min} \quad & \tilde{g}_0(x) \\ s.t \quad & \tilde{g}_i(x) \leq \tilde{1}, \qquad 1 \leq i \leq p, \\ & x > 0, \end{aligned} \qquad (2)$$

where $x = (x_1, x_2, ..., x_m)^T$ is an m-dimensional variable vector, $\tilde{g}_i(x) = \sum_{k=1}^{j_i} \tilde{c}_{ik} \prod_{l=1}^{m} x_l^{\gamma_{ikl}}$ $(1 \leq i \leq p)$ is a fuzzy posynomial function of x, $\tilde{c}_{ik} > 0$ and $\tilde{1} \in F(R)$ and γ_{ikl} an arbitrary real number is called an FPGP problem [1].

The following theorem shows that any FPGP can be reduced to a posynomial geometric programming (PGP) (see [4]).

Theorem 5.1 *The FPGP in (2) is equivalent to the following PGP problem:*

$$\begin{aligned} Min \quad & g_0(x) \\ s.t \quad & \frac{1}{b_i} g_i(x) \leq 1, \qquad 1 \leq i \leq p, \\ & x > 0. \end{aligned} \qquad (3)$$

Here $g_i(x) = \sum_{k=1}^{J_i} c_{ik} \prod_{k=1}^{m} x_l^{r_{ikl}}$ $(0 \leq i \leq p)$ is a posynomial, x is an m-dimensional vector, $c_{ik} > 0$, $b_i > 0$, r_{ikl} an arbitrary real number.

Proof.   Let Q_1 and Q_2 be the set of all feasible solutions to (2) and (3), respectively. Then $x \in Q_1$ if and only if:

$$\sum_{k=1}^{j_i} \tilde{c}_{ik} \prod_{l=1}^{m} x_l^{\gamma_{ikl}} \leq \tilde{b}_i \quad \leftrightarrow$$

$$Mag\left(\sum_{k=1}^{j_i} \tilde{c}_{ik} \prod_{l=1}^{m} x_l^{\gamma_{ikl}}\right) \le Mag(\tilde{b}_i) \leftrightarrow$$

$$\frac{1}{2}\int_0^1 \sum_{k=1}^{j_i}\left(\underline{u}(r) + \overline{u}(r) + c^-_{ik} + c^+_{ik}\right)r\,dr \prod_{l=1}^{m} x_l^{\gamma_{ikl}}$$

$$\le \frac{1}{2}\int_0^1 \left(\underline{u}(r) + \overline{u}(r) + b^-_i + b^+_i\right)r\,dr \leftrightarrow$$

$$\sum_{k=1}^{j_i} \frac{\left(6c^-_{ik} + 6c^+_{ik} + \delta^+ c_{ik} - \delta^- c_{ik}\right)}{\left(6b^-_i + 6b^+_i + \delta^+ b_i - \delta^- b_i\right)} \prod_{l=1}^{m} x_l^{\gamma_{ikl}} \le 1. \tag{4}$$

We denote (4) to following formula:

$$\sum_{k=1}^{j_i} N_k \prod_{l=1}^{m} x_l^{\gamma_{ikl}} \le 1, \qquad 1 \le i \le p$$

$$s.t \;\; N_k = \frac{\left(6c^-_{ik} + 6c^+_{ik} + \delta^+ c_{ik} - \delta^- c_{ik}\right)}{\left(6b^-_i + 6b^+_i + \delta^+ b_i - \delta^- b_i\right)},$$

$$x > 0$$

$$\leftrightarrow x \in Q_2.$$

Hence $Q_1 = Q_2$.

Now we suppose $X^0 = (x_{01}, \ldots, x_{0m})$ to be an optimal feasible solution to (2), then for all $x \in Q_1$, we have:

$$\widetilde{g_0}(x) \ge \widetilde{g_0}(x^0) \leftrightarrow$$

$$\sum_{k=1}^{j_i} \tilde{c}_{ik} \prod_{l=1}^{m} x_l^{\gamma_{ikl}} \ge \sum_{k=1}^{j_i} \tilde{c}_{ik} \prod_{l=1}^{m} x_{0l}^{\gamma_{ikl}} \leftrightarrow$$

$$Mag\left(\sum_{k=1}^{j_i} \tilde{c}_{ik} \prod_{l=1}^{m} x_l^{\gamma_{ikl}}\right) \ge Mag\left(\sum_{k=1}^{j_i} \tilde{c}_{ik} \prod_{l=1}^{m} x_{0l}^{\gamma_{ikl}}\right) \leftrightarrow$$

$$\sum_{k=1}^{j_i} c_{ik} \prod_{l=1}^{m} x_l^{\gamma_{ikl}} \ge \sum_{k=1}^{j_i} c_{ik} \prod_{l=1}^{m} x_{0l}^{\gamma_{ikl}}.$$

We conclude that X^0 is an optimal feasible solution to (3).

Definition 5.2 [2] *Let a PGP be*

$$Min \;\; g_0(x)$$

$$s.t \;\; g_i(x) \le 1, \qquad 1 \le i \le p, \tag{5}$$

$$x > 0,$$

where $g_i(x) = \sum_{k=1}^{J_i} c_{ik} \prod_{k=1}^{m} x_l^{r_{ikl}} (0 \le i \le p)$ *is a posynomial, x is an m-dimensional vector, $c_{ik} > 0$, $b_i > 0$, r_{ikl} an arbitrary real number, Then its dual programming is denoted by:*

$$Max\ d(w) = \prod_{i=0}^{p} \prod_{k=1}^{J_i} \left(\frac{\check{c}_{ik}}{w_{ik}}\right)^{w_{ik}} \prod_{i=1}^{p} w_{i0}{}^{w_{i0}} \tag{6}$$

$$\text{s.t} \quad \sum_{k=1}^{J_0} w_{0k} = 1,$$

$$\sum_{i=0}^{p} \sum_{k=1}^{J_i} \gamma_{ikl} w_{ik} = 0, \qquad 1 \le l \le m,$$

$$W \ge 0.$$

Here $W = \left(w_{01}, w_{02}, \dots, w_{0j_0}, \dots, w_{p1}, \dots, w_{pj_p}\right)$, $w_{i0} = \sum_{k=1}^{J_i} w_{ik}$ $i = 0, \dots, p$
and $w_{00} = \sum_{k=1}^{J_0} w_{0k} = 1.$

Theorem 5.2 [2] *Suppose that the constrained PGP (5) is super-consistent and that x^* is a solution to GP. Then the corresponding DP (6) is consistent and has a solution w^* satisfying*

$$g_0(x^*) = d(w^*),$$

and

$$w^*{}_{ik} = \begin{cases} \dfrac{v_{ik}(x^*)}{g_0(x^*)}, & (i = 0, 1 \le k \le J_0), \\ w^*{}_{i0} v_{ik}(x^*), & (1 \le i \le p, 1 \le k \le J_i), \end{cases}$$

where $v_{ik}(x) = c_{ik} \prod_{l=1}^{m} x_l{}^{\gamma_{ikl}},\ w^*{}_{i0} = \sum_{k=1}^{J_i} w^*{}_{ik}.$
Now we can solve the FPGP problem by the following steps:

Step 1: According to Theorem 5.2, we can change the FPGP into PGP by using a new approaching method.
Step2: Using dual method for solving PGP.

6 Numerical Example

In this section, we give one example to show the numerical advantage of the new method in comparison with Yager's method.

Example 6.1 Consider the following FPGP:

$$Min\ \widetilde{g}_0(x) = (2, 4, 1, 2)x_1 x_2$$
$$\text{s.t}\ \widetilde{g}_1(x) = (2, 5, 1, 2)x_1 x_2{}^{-1} x_3 \le (4, 8, 1, 1), \tag{7}$$
$$\widetilde{g}_2(x) = (2, 5, 1, 1)x_1{}^{-2} + (4, 7, 2, 2)x_1{}^{-2} x_3{}^{-2} \le (4, 7, 1, 2),$$
$$x_1, x_2, x_3 > 0.$$

1) **We use the new approach to (7), then**

$$Min\ \frac{37}{12} x_1 x_2$$
$$\text{s.t}\ \frac{43}{72} x_1 x_2{}^{-1} x_3 \le 1,$$
$$\frac{42}{67} x_1{}^{-2} + \frac{66}{67} x_1{}^{-2} x_3{}^{-2} \le 1,$$
$$x_1, x_2, x_3 > 0,$$

$$g_0(x^*) = d(w^*) = 2.894564,$$

$$x_1^* = 1.119701, x_2^* = 1.253566, x_3^* = 0.838273.$$

2) **We use the Yager's method to (7), then**

$$Min \quad \frac{13}{4}x_1x_2$$

$$s.t \quad \frac{15}{24}x_1x_2^{-1}x_3 \leq 1,$$

$$\frac{14}{23}x_1^{-2} + \frac{22}{23}x_1^{-2}x_3^{-2} \leq 1,$$

$$x_1, x_2, x_3 > 0,$$

$$g_0(x^*) = d(w^*) = 3.099851,$$

$$x_1^* = 1.10335, x_2^* = 0.86445, x_3^* = 1.25356.$$

We saw that the obtained objective function by using the new approaching method is smaller than one by using Yager's method.

7 Conclusion

We used the new approaching method to an FPGP problem. In particular, we emphasize that the obtained objective function is better by using the method than by Yager's method. Based on the obtained results in the last section, we conclude that using the approaching method is useful to solve an FPGP problem.

Acknowledgements. Thanks to the support by National Natural Science Foundation of China (No.70771030 and No. 70271047) and Project Science Foundation of Guangzhou University.

References

1. Abbasbandy, S., Hajjari, T.: A new approach for ranking of trapezoidal fuzzy numbers. Computers and Mathematics with Applications 57, 413–419 (2009)
2. Cao, B.-Y.: Fuzzy geometric programming. Kluwer Academic Publishers (2001)
3. Ma, M., Friedman, M., Kandel, A.: A new fuzzy arithmetic. Fuzzy Sets and Systems 108, 83–90 (1999)
4. Maleki, H.R., Tata, M., Mashinchi, M.: Linear programming with fuzzy variables. Fuzzy Sets and Systems 109, 21–23 (2000)
5. Yager, R.R.: A procedure for ordering fuzzy subsets of the unit interval. Information Science 24, 143–161 (1981)
6. Maleki, H.R., Tata, M., Mashinchi, M.: Linear programming with fuzzy variables. Fuzzy Sets and Systems 109, 21–33 (2000)

A Sharp Upper Bound on the Laplacian and Quasi-Laplacian Spectral Radius of a Graph

Fu-Yi Wei and Mu-huo Liu

Department of Applied Mathematics, South China Agricultural University,
Guangzhou, 510642, China
weifuyi@scau.edu.cn

Abstract. Let $G = (V, E)$ be a simple connected graph with n vertices. The maximum eigenvalue of Laplacian matrix(Quasi-Laplacian matrix), denoted by $\lambda(G)(\mu(G))$, is called the Laplacian(Quasi-Laplacian) spectral radius of G. This paper presents a new sharp upper bound of the largest eigenvalue for both Laplacian and Quasi-Laplacian matrix of a graph.

Keywords: Spectral radius,Laplacian matrix,Quasi-Laplacian matrix, Upper bound.

1 Introduction

Let $G = (V, E)$ be a simple connected graph with the vertex set $V = \{v_1, v_2, \cdots, v_n\}$ and let m be the cardinality of the edge set E. Let Δ and δ denote the maximum and minimum degree of vertices of G, respectively. Specially, if $\Delta = \delta$, then G is called regular. In the following, we assume that the vertices are ordered such that $d(v_i) = d_i$ ($1 \leq i \leq n$), and $d_1 \geq d_2 \geq \cdots \geq d_n$. The symbol $N(v)$ is used to denote the neighbor set of the vertex $v \in V$. Let $m(v)$ denote the average of the degree of the vertices adjacent to v, i.e., $m(v) = \sum\limits_{u \in N(v)} d(u)/d(v)$.

Let $A(G)$ be the adjacency matrix of G and $D(G)$ the diagonal matrix of vertex degree. The Laplacian matrix of G is $L(G) = D(G) - A(G)$ and the Quasi-Laplacian matrix of G is $K(G) = D(G) + A(G)$. The maximum eigenvalue of $L(G)$, denoted by $\lambda(G)$, is called the Laplacian spectral radius of G. $\mu(G)$ is used to denote the largest eigenvalue of $K(G)$, and is called the Quasi-Laplacian spectral radius of G.

The eigenvalues of the Laplacian matrix are important in graph theory, because they have relations to numerous graph invariants. Especially, the largest and the second smallest Laplacian eigenvalue of $L(G)$ probably have the most important information contained in the spectrum of a graph (see

B.-Y. Cao and X.-J. Xie (Eds.): Fuzzy Engineering and Operations Research, AISC 147, pp. 525–531.
springerlink.com   © Springer-Verlag Berlin Heidelberg 2012

[1-3]). In many applications, good bounds for the Laplacian spectral radius and Quasi-Laplacian spectral radius of G are needed.

Since the 1990s, many researchers have investigated upper bounds for $\lambda(G)$ with refer to m, n, Δ, δ. Among the known upper bounds for $\lambda(G)$ are the following:

1. Anderson and Morley's bound [4]:

$$\lambda(G) \leq max\{d_i + d_j : v_i v_j \in E\}. \tag{1}$$

2. Li et al's bound [5]

$$\lambda(G) \leq 2 + \sqrt{(d_1 + d_2 - 2)(d_1 + d_3 - 2)}. \tag{2}$$

3. Liu et al's bound [6]:

$$\lambda(G) \leq \frac{\Delta + \delta - 1 + \sqrt{(\Delta + \delta - 1)^2 + 4(4m - 2\delta(n - 1))}}{2}. \tag{3}$$

4. Shi's bound [7]:

$$\lambda(G) \leq \sqrt{2\Delta^2 + 4m - 2\delta(n - 1) + 2\Delta(\delta - 1)}. \tag{4}$$

5. Das's bound [8]:

$$\lambda(G) \leq \frac{2m}{n - 1} + \frac{n - 2}{n - 1}\Delta + (\Delta - \delta)(1 - \frac{\Delta}{n - 1}). \tag{5}$$

In this paper, we present a new sharp upper bound for both $\lambda(G)$ and $\mu(G)$, and give an example to illustrate that this bound is, in some sense, better than the above listed bounds.

2 Main Results

In generally, if $m = n + c - 1$, then G is called a c-cyclic graph. Specially, a 0-cyclic, 1-cyclic and 2-cyclic graph are known as a tree, unicyclic and bicyclic graph, respectively. In the following, $K_{1,n-1}$ and P_n denotes a start or a path of order n, respectively. T_n denotes the class of trees with order n. By the results of [9] and [10], we can conclude that

Lemma 1. *Let G be a connected graph. Then $\lambda(G) \leq \max\{d(v) + m(v) : v \in V\}$, and the equality holds if and only if G is a regular bipartite graph or a bipartite semiregular graph.*

It is well known that

Lemma 2. *[11] Let G be a connected graph. Then $\Delta + 1 \leq \lambda(G) \leq n$, with the left equality holding if and only if $\Delta = n - 1$.*

The following Theorem is our main result.

Theorem 1. *If G is a connected graph with $n \geq 3$ vertices and m edges, then*

$$\lambda(G) \leq \max\{\Delta + \delta - 1 + \frac{2m - \delta(n-1)}{\Delta}, \delta + 1 + \frac{2m - \delta(n-1)}{2}\}. \quad (6)$$

Moreover, if G is a regular bipartite graph or $G \cong K_{1,n-1}$, then the equality can be obtained.

Proof. By Lemma 1, we only need to prove that $\max\{d(v) + m(v) : v \in V\} \leq \max\{\Delta + \delta - 1 + \frac{2m - \delta(n-1)}{\Delta}, \delta + 1 + \frac{2m - \delta(n-1)}{2}\}$.

Suppose $\max\{d(v) + m(v) : v \in V\}$ occurs at the vertex u. Two cases arise $d(u) = 1$, or $2 \leq d(u) \leq \Delta$.

Case 1. $d(u) = 1$. Suppose that $N(u) = w$. Since $m(u) = d(w) \leq \Delta$, thus $d(u) + m(u) \leq 1 + \Delta \leq \Delta + \delta - 1 + \frac{2m - \delta(n-1)}{\Delta}$, the result follows.

Case 2. $2 \leq d(u) \leq \Delta$. Note that $2m - \delta(n-1) \geq d(u) \geq 2$, and $2m = \sum_{v \in N(u)} d(v) + \sum_{v \notin N(u)} d(v) \geq \sum_{v \in N(u)} d(v) + d(u) + \delta(n - d(u) - 1)$, thus

$$m(u) = \sum_{v \in N(u)} d(v)/d(u) \leq \frac{2m - \delta(n-1) + (\delta - 1)d(u)}{d(u)} = \frac{2m - \delta(n-1)}{d(u)} + (\delta - 1).$$

This follows that $d(u) + m(u) \leq d(u) + \frac{2m - \delta(n-1)}{d(u)} + (\delta - 1)$.

Let $f(x) = x + \frac{2m - \delta(n-1)}{x} + (\delta - 1)$, where $x \in [2, \Delta]$. It is easy to see that $f'(x) = 1 - \frac{2m - \delta(n-1)}{x^2}$. Let $a = 2m - \delta(n-1)$, then $\sqrt{a}$ is the unique positive roots of $f'(x) = 0$. We consider the next three Subcases.

Subcase 1. $\sqrt{a} < 2$. When $x \in [2, \Delta]$, since $f'(x) > 0$, then $f(x) \leq f(\Delta)$.

Subcase 2. $2 \leq \sqrt{a} \leq \Delta$. Then, $f'(x) \leq 0$ for $x \in [2, \sqrt{a}]$, and $f'(x) \geq 0$ for $x \in [\sqrt{a}, \Delta]$. Thus, $f(x) \leq \max\{f(2), f(\Delta)\}$ for $x \in [2, \Delta]$.

Subcase 3. $\sqrt{a} > \Delta$. When $x \in [2, \Delta]$, since $f'(x) < 0$, then $f(x) \leq f(2)$.

Recall that $2 \leq d(u) \leq \Delta$, thus

$$d(u) + m(u) \leq \max\{f(2), f(\Delta)\} = \max\{\delta + 1 + \frac{2m - \delta(n-1)}{2}, \Delta + \delta - 1 + \frac{2m - \delta(n-1)}{\Delta}\}.$$

If $G \cong K_{1,n-1}$, then $\delta + 1 + \frac{2m - \delta(n-1)}{2} = 2 + \frac{n-1}{2} \leq n$ from $n \geq 3$ and $\Delta + \delta - 1 + \frac{2m - \delta(n-1)}{\Delta} = n$. Thus, the equality holds. If G is a regular bipartite graph, since G is connected, then we may assume that $\delta \geq 2$, this implies that $\delta + 1 + \frac{2m - \delta(n-1)}{2} \leq \Delta + \delta - 1 + \frac{2m - \delta(n-1)}{\Delta} = 2\Delta$. Thus, the equality also holds by Lemma 1.

By combining the above discussion, the result follows.

It has been proved in [9] that

Lemma 3. *Suppose* $T \in \mathcal{T}_n$. *If* $\Delta(T) \geq \frac{n-1}{2}$, *then* $\Delta(T) + 1 \leq \lambda(T) \leq \Delta(T) + 2$, *the left equality holds if and only if* $\Delta = n - 1$.

In the following, we give a similar result for general connected graphs

Theorem 2. *Let* G *be a connected graph with* n *vertices and* m *edges. If* $\Delta \geq \frac{2m-(n-1)}{2}$ *and* $\delta = 1$, *then*

$$\Delta + 1 \leq \lambda(G) \leq \Delta + 2, \tag{7}$$

and the left equality holds if and only if $\Delta = n - 1$.

Proof. This Theorem follows immediately from Theorem 1 and Lemma 2.

Clearly, Theorem 2 implies Lemma 3. The next result can be also yielded from Theorem 2.

Corollary 1. *[10] Suppose* $G \in \mathcal{U}_n$. *If* $\Delta(G) \geq \frac{n+1}{2}$, *then* $\Delta(G) + 1 \leq \lambda(G) \leq \Delta(G) + 2$, *the left equality holds if and only if* $\Delta = n - 1$.

Let H_1 be a tree obtained from a start $K_{1,n-3}$ by attaching a path P_3 to the center of $K_{1,n-3}$. The next corollary is an example to illustrate the use of Theorem 2 in the ordering of trees according to their Laplacian spectral radius.

Corollary 2. *[14] 1) Suppose* $T \in \mathcal{T}_n$, *then* $\lambda(T) \leq \lambda(K_{1,n-1})$, *equality holds if and only if* $T \cong K_{1,n-1}$. *2) Suppose* $T \in \mathcal{T}_n \setminus \{K_{1,n-1}\}$, *then* $\lambda(T) \leq \lambda(H_1)$, *equality holds if and only if* $T \cong H_1$.

Proof. By Theorem 2, we have $\lambda(K_{1,n-1}) = n$. If $T \in \mathcal{T}_n \setminus \{K_{1,n-1}\}$, then $n \geq 4$ and $\Delta(T) \leq n - 2$. This implies that $2 + \frac{n-1}{2} < n$ and $\Delta + \frac{n-1}{\Delta} \leq n - 2 + \frac{n-1}{n-2} < n$. Thus, we have $\lambda(T) < n$ by inequality (6). This follows that $\lambda(H_1) < \lambda(K_{1,n-1})$.

Suppose $T \in \mathcal{T}_n \setminus \{K_{1,n-1}, H_1\}$, then $n \geq 5$ and $\Delta(T) \leq n - 3$. This implies that $2 + \frac{n-1}{2} \leq n - 1$ and $\Delta + \frac{n-1}{\Delta} \leq n - 3 + \frac{n-1}{n-3} \leq n - 1$. Thus, we have $\lambda(T) \leq n - 1$ by inequality (6). Combining with Theorem 2, we have $\lambda(T) \leq n - 1 < \lambda(H_1)$ by $\Delta(H_1) = n - 2$. This completes the proof.

Let $\mathbb{G}^*(m, n, \frac{2m-(n-1)}{2}, 1)$ be the classes graphs with $\Delta \geq \frac{2m-(n-1)}{2}$, $m \geq n$ and $\delta = 1$. Next we shall show that the upper bound (7) is better than (1)-(5) in the classes graphs $\mathbb{G}^*(m, n, \frac{2m-(n-1)}{2}, 1)$.

Remark 1. Suppose that there exists $u_0 \in V(G)$ such that $d(u_0) = \Delta$. Since $m \geq n$, then there must exist $v_0 \in N(u_0)$ such that $d(v_0) \geq 2$. This implies that $\Delta + 2 \leq max\{d(u) + d(v) : uv \in E\}$, thus the upper bound (7) is better than (1) in the classes graphs $\mathbb{G}^*(m, n, \frac{2m-(n-1)}{2}, 1)$.

Remark 2. Since $m \geq n$, then there must exist at least one cycle, this implies that $\Delta = d_1 \geq d_2 \geq d_3 \geq 2$. Thus, $\Delta + 2 \leq 2 + \sqrt{(d_1 + d_2 - 2)(d_1 + d_3 - 2)}$, thus the upper bound (7) is better than (2) in the classes graphs $\mathbb{G}^*(m, n, \frac{2m - (n-1)}{2}, 1)$.

Remark 3. Note that $\Delta \leq n - 1$, then $\Delta + 2 \leq 2m - (n-1)$ by $m \geq n$. Thus, $(\Delta + 4)^2 \leq \Delta^2 + 8(2m - (n-1))$, this implies that $\Delta + 2 \leq \frac{\Delta + \sqrt{\Delta^2 + 4(4m - 2(n-1))}}{2}$. Therefore, the upper bound (7) is BETTER than (3) in the graph classes $\mathbb{G}^*(m, n, \frac{2m - (n-1)}{2}, 1)$.

Remark 4. Since $m \geq n$, then there must exist at least one cycle. So we have $8 \leq (\Delta - 2)^2 + 2(m - (n-1)) + 2m$, equivalently, $4 + 4\Delta \leq \Delta^2 + 4m - 2(n-1)$. This implies that $\Delta + 2 \leq \sqrt{2\Delta^2 + 4m - 2(n-1)}$. Therefore, the upper bound (7) is better than (4) in the classes graphs $\mathbb{G}^*(m, n, \frac{2m - (n-1)}{2}, 1)$.

Remark 5. We shall show that the upper bound (7) is better than (5) in the classes graphs $\mathbb{G}^*(m, n, \frac{2m - (n-1)}{2}, 1)$. When $\Delta = n - 1$, it is clear that $\lambda(G) = n$ by Lemma 2. Thus, we only need to show that $2\Delta - \frac{\Delta^2}{n-1} + \frac{2m}{n-1} - 1 \geq \Delta + 2$ when $\Delta \leq n - 2$. Equivalently, we only need to prove that $(\Delta - 3)(n - 1) + 2m - \Delta^2 \geq 0$.

Let $f(x) = (x - 3)(n - 1) + 2m - x^2$, where $\frac{n+1}{2} \leq x \leq n - 2$. When $x \geq \frac{n+1}{2}$, since $f'(x) = n - 1 - 2x$, then $f'(x) < 0$. Thus, $f(x) \geq f(n - 2) = 2m + 1 - 2n > 0$ when $\frac{n+1}{2} \leq x \leq n - 1$.

By combining the above discussion, it follows that the upper bound (7) is better than (5) in the classes graphs $\mathbb{G}^*(m, n, \frac{2m - (n-1)}{2}, 1)$.

By Remarks 1-5, we can conclude that the upper bound (6) is better than (1)-(5) in the classes graphs $\mathbb{G}^*(m, n, \frac{2m - (n-1)}{2}, 1)$. But as shown in the next Example, the upper bound (6) is incomparable with (1), (2), (3), (4), (5), respectively, for the general graphs.

Example 1. Values of $\lambda(G)$ and of the various bounds for a graphs of order $n = 9$ illustrated in Fig. 1 are given in Table. 1.

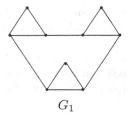

G_1

Fig. 1. A graph G with nine vertices

Table 1. The eigenvalues and upper bound of graph G

λ	(1)	(2)	(3)	(4)	(5)	(6)
G_1 5	6	6	6.4721	6.3246	6.25	6.667

In the following, we will consider the $\mu(G)$ for Quasi-Laplacian matrix of graph.

Lemma 4. *[3] Let G be a connected graph. Then $\mu(G) \leq \max\{d(v) + m(v) : v \in V\}$, and equality holds if and only if G is a regular graph or a bipartite semiregular graph.*

Lemma 5. *[13] Let G be a graph. Then $\lambda(G) \leq \mu(G)$, and the equality holds if and only if G is a bipartite graph.*

By Lemmas 2-5, it can be proved analogously with Theorem 1 that

Theorem 3. *If G is a connected graph with n vertices and m edges, then*

$$\Delta + 1 \leq \mu(G) \leq \max\{\Delta + \delta - 1 + \frac{2m - \delta(n-1)}{\Delta}, \delta + 1 + \frac{2m - \delta(n-1)}{2}\},$$

and the left equality holds if and only if G is a bipartite graph with $\Delta = n - 1$, the right equality can be obtained when $G \cong K_{1,n-1}$ or G is a regular graph.

Corollary 3. *Let G be a connected graph with n vertices and m edges, and $\delta = 1$. If $\Delta \geq \frac{2m+(n-1)}{2}$, then $\Delta + 1 \leq \mu(G) \leq \Delta + 2$, and the left equality holds if and only if G is a bipartite graph with $\Delta = n - 1$.*

3 Conclusion

In this paper, we present a new attainable upper bound for the eigenvalues of the Laplacian matrix and Quasi-Laplacian matrix (See Theorem 1, Theorem 2 and Theorem 3). The result is impossible for the order of the eigenvalues of the Laplacian matrix and Quasi-Laplacian matrix.

Acknowledgement. Thanks to the support by National Natural Science Foundation of China (No. 11071088) and Foundation for Distinguished Young Talents in Higher Education of Guangdong, China (No. LYM10039).

References

1. Alon, N.: Eigenvalues and expanders. Combinatorica 6(2), 83–96 (1986)
2. Chung, F.R.K.: Eigenvalues of graph. In: Proceeding og the International Congress of Mathematicians, Zürich, Switzerland, pp. 1333–1342 (1995)

3. Das, K.C.: The Laplacian Spectrum of a graph. Computers Math. Appl. 48, 715–724 (2004)

4. Anderson, W.N., Morley, T.D.: Eigenvalues of the Paplacian if a graph. Lin. Mult. Algebra 18, 141–145 (1985)

5. Li, J.S., Zhang, X.D.: A new upper bound for eigenvalue of the Laplacian matrix of a graph. Lin. Algebre Appl. 265, 93–100 (1997)

6. Liu, H., Lu, M., Tian, F.: On the Laplacian spectral radius of a graph. Lin. Algebre Appl. 376, 135–141 (2004)

7. Shi, L.S.: Bounds on the (Laplacian) spectral radius of graphs. Lin. Algebre Appl. 422, 755–770 (2007)

8. Das, K.C.: Sharp upper bounds on the spectral radius of the laplacian matrix of graphs. Acta Math. Univ. Comenianae 2, 185–198 (2005)

9. Merris, R.: A note on the Laplacian graph eigenvalues. Lin. Algebra Appl. 285, 33–35 (1998)

10. Pan, Y.L.: Sharp upper bounds for the Laplacian graph eigenvalues. Lin. Algebra Appl. 355, 287–295 (2002)

11. Merris, R.: Laplacian matrices of graphs: a survey. Lin. Algebra Appl., 197–198, 143–176 (1994)

12. Guo, J.M.: On the Laplacian spectral radius of a tree. Lin. Algebra Appl. 368, 379–385 (2003)

13. Merris, R.: Laplacian matrices of graphs: a survey. Lin. Algebra Appl., 197–198, 143–176 (1994)

14. Guo, J.M.: On the Laplacian spectral radius of a tree. Lin. Algebra Appl. 368, 379–385 (2003)

Consensusability Conditions about Output Feedback for Singular Multi-Agent Systems

Chun-Yan Qi

Mathematical and Computational School, Zhanjiang Normal University,
Zhanjiang, Guangdong 524048, P.R. China
qcydwj@sohu.com

Abstract. In this paper, we consider the multi-agent consensus problem for the mobile agents with singular time-invariant systems. Output feedback case for the consensusability of such multi-agent systems is discussed. Necessary and sufficient conditions on consensusability are given based on the structures of the agent dynamics and the (fixed) communication topology described by directed graph.

Keywords: Multi-agent system, consensusability, singular agent system, consensus protocol.

1 Introduction

In recent years, consensus problems for multi-agent systems (MASs) have been applied in many areas, and more and more people are interested in this promising problem [1]. Although many results have been obtained for the agents with linear time-invariant systems under state and output feedback consensus protocols [2,3], the consensus problems of agents with singular time-invariant systems are not considered at present to the best of our knowledge.

The objective of this paper is to study the consensusability of singular time-invariant multi-agent system. Under suitable conditions, the consensusability of the singular system problem is turned into the stability analysis of a closed-loop system. To handle it, a general Riccati equation is employed to study the stability. Under some assumptions, necessary and sufficient conditions on consensusability of singular multi-agent systems are obtained.

2 Preliminaries

First, consider the singular system described in the following form:

$$\Sigma_0 : \begin{cases} E\dot{x}(t) = Lx(t) + Mu(t), \\ y(t) = Hx(t), \end{cases} \tag{1}$$

B.-Y. Cao and X.-J. Xie (Eds.): Fuzzy Engineering and Operations Research, AISC 147, pp. 533–539.
springerlink.com   © Springer-Verlag Berlin Heidelberg 2012

where $x(t), u(t), y(t)$ are the state, control input and output of the system, respectively; E, L, M, H are real constant matrices with appropriate sizes, E may be singular we shall assume that $0 < rank(E) = r \le n$.

Based on [4, 5], we introduce the following definition and lemma:

Definition 1. (a) The pair (E, L) is said to be regular, if $det(sE - L) \ne 0$ for some finite complex number s; If (E, L) is regular, there exist two nonsingular matrices Q, T such that the system (1) is restricted system equivalent to

$$\Sigma_1 : \begin{cases} \dot{x}_1 = L_1 x_1 + M_1 u, \\ y_1 = H_1 x_1, \end{cases} \tag{2}$$

$$\begin{cases} N\dot{x}_2 = x_2 + M_2 u, \\ y_2 = H_2 x_2, \end{cases} \tag{3}$$

where

$$QET = \begin{bmatrix} I_{n_1} & 0 \\ 0 & N \end{bmatrix}, QLT = \begin{bmatrix} L_1 & 0 \\ 0 & I_{n_2} \end{bmatrix}, QM = \begin{bmatrix} M_1 \\ M_2 \end{bmatrix},$$

$$HT = \begin{bmatrix} H_1^T \\ H_2^T \end{bmatrix}^T, Q^{-1}x = \begin{bmatrix} x_1 \\ x_2 \end{bmatrix},$$

where $n_1 + n_2 = n, N \in \mathbb{R}^{n_2 \times n_2}$ is nilpotent, i.e., exists $h \in N^*$ satisfies $N^{h-1} \ne 0, N^h = 0$.

(2), (3) is called slow subsystem and fast subsystem, respectively, and (E, L) is impulse free if and only if $N = 0$.

(b) A regular pair (E, L) is admissible if it is impulse free and stable.

Lemma 1. *The singular system (E, L) is regular, impulse-free, and stable if and only if there exists a matrix P such that*

$$\begin{cases} L^T P + P^T L - P^T M M^T P + H^T C = 0, \\ E^T P = P^T E \ge 0, \end{cases} \tag{4}$$

and furthermore, all the eigenvalues of $(E, L - MM^T P)$ are in the open left half plane and impulse free.

Lemma 2. *System (E, L, M) is stabilizable if there exists a matrix $K \in \mathbb{R}^{r \times n}$ such that all the eigenvalues of $(E, L + MK)$ are in the open left half plane. stabilizable system (E, L, M) is equivalent to the stabilizability of the slow subsystem.*

Lemma 3. *The Laplacian matrix $L_{\mathcal{G}}$ of a digraph $\mathcal{G} = (v, \varepsilon, A)$ has at least one zero eigenvalue and all the nonzero eigenvalues are in the open right half plane. Furthermore, $L_{\mathcal{G}}$ has exactly one zero eigenvalue if and only if $\mathcal{G}$ has a spanning tree.*

3   Main Results

Now we consider a regular and impulse controllable system consisting of agents indexed by $1, \cdots, N$, respectively. The dynamics of the ith agent is described as follows:

$$\begin{cases} E\dot{x}_i(t) = Lx_i(t) + Mu_i(t), \\ y_i(t) = Hx_i(t), \end{cases} \quad (i = 1, \cdots, N), \tag{5}$$

where $x_i(t) \in \mathbb{R}^n$, $u_i(t) \in \mathbb{R}^r$, $y_i(t) \in \mathbb{R}^m$ are the state, control and output of the ith agent, respectively; $E, L \in \mathbb{R}^{n \times n}, M \in \mathbb{R}^{n \times r}, H \in \mathbb{R}^{m \times n}$ are as (1).

Then we consider the output feedback case. Precisely, the consensus protocol of the ith agent is of the following form:

$$u_i(t) = K \sum_{j \in N_i} a_{ij}(y_j(t) - y_i(t)), t \geq 0 (i = 1, \cdots, N), \tag{6}$$

where $K \in \mathbb{R}^{r \times n}$ is a weighted constant matrix. Noticing the property of a_{ij} that $a_{ij} > 0$ if and only if $j \in N_i$, consensus protocol (6) is equivalent to

$$u_i(t) = K \sum_{j=1}^{N} a_{ij}(y_j(t) - y_i(t)), t \geq 0 (i = 1, \cdots, N), \tag{7}$$

Let $u(t) = (u_1^T(t), \cdots, u_N^T(t))^T$. We consider the following admissible control set:

$$U_o = \{u(t) : [0, \infty) \to \mathbb{R}^{rN} | u_i(t) = K \sum_{j=1}^{N} a_{ij}(y_j(t) - y_i(t)), \forall t \geq 0, K \in \mathbb{R}^{r \times m} \\ (i = 1, \cdots, N)\}.$$

Definition 2. For the system (5), if there exists a $u(t) \in U_o$ such that for any initial value $x_i(0), \lim_{t \to \infty} \|x_j(t) - x_i(t)\| = 0 (i, j = 1, \cdots, N)$, then we say that the system (5) is consensusable with respect to U_o.

As follow, we suppose the system (5) is impulse observable and impulse controllable.

Theorem 1. *If the system (5) is consensusable with respect to U_o, then the system (E, L, M, H) is stabilizable and detectable, and if in addition, the eigenvalues of (E, L, M) are not all in the open left half plane, then the topology $\mathcal{G}$ must have a spanning tree.*

Proof: (Necessity): By Definition 2, if the system (5) is consensusable with respect to U_0, then there exist a matrix $K \in \mathbb{R}^{r \times n}$ and consensus protocols

$$u_i(t) = K \sum_{j=1}^{N} a_{ij}(y_j(t) - y_i(t)), t \geq 0 (i = 1, \cdots, N),$$

such that for any $i \neq j$,

$$lim_{t \to \infty} \|x_j(t) - x_i(t)\| = 0 (i, j = 1, \cdots, N).$$

Let $\delta_i(t) \triangleq x_1(t) - x_i(t) (i = 2, \cdots, N)$. Then the above expression is equivalent to $\|\delta_i(t)\| \to 0, t \to \infty (i = 1, \cdots, N)$.

Notice that

$$E\dot{\delta}_i(t) = L\delta_i(t) + MKH[\sum_{j=1}^{N}(a_{ij} - a_{1j})\delta_j(t) - deg_{in}(i)\delta_i(t)](i = 1, \cdots, N).$$

Then

$$E\dot{\delta}(t) \triangleq E \begin{bmatrix} \dot{\delta}_2(t) \\ \vdots \\ \dot{\delta}_N(t) \end{bmatrix} = [I_{N-1} \otimes L - (\Lambda + 1_{N-1} \cdot \alpha^T) \otimes MKH]\delta(t), \quad (8)$$

where 1_{N-1} denotes an $N-1$ dimensional column vector with all components 1,

$$\alpha = (a_{12}, \cdots, a_{1N})^T,$$

$$\Lambda = \begin{bmatrix} deg_{in}(2) & -a_{23} & \cdots & -a_{2N} \\ -a_{32} & deg_{in}(3) & \cdots & -a_{3N} \\ \cdots & & & \cdots \\ -a_{N2} & -a_{N3} & \cdots & deg_{in}(N) \end{bmatrix}.$$

Thus all the eigenvalues of $I_{N-1} \otimes L - (\Lambda + 1_{N-1} \cdot \alpha^T) \otimes MKH$ are in the open left half plane.

Take $\Gamma = \begin{bmatrix} 1 & 0 \\ 1_{N-1} & I_{N-1} \end{bmatrix}$. By the definition of Laplacian, we have

$$\Gamma^{-1} L_{\mathcal{G}} \Gamma = \begin{bmatrix} 0 & -\alpha^T \\ 0 & \Lambda + 1_{N-1} \cdot \alpha^T \end{bmatrix}.$$

Assume that $s_1 = 0, s_2, \cdots, s_N$ are the eigenvalues of Laplacian $L_{\mathcal{G}}$, Then, by (8) the eigenvalues of $L + 1_{N-1} \cdot \alpha^T$ are $s_2, \cdots, s_N$. Thus, there exists an invertible matrix T such that $\Lambda + 1_{N-1} \cdot \alpha^T$ is similar to a Jordan canonical matrix, i.e.,

$$T^{-1}(\Lambda + 1_{N-1} \cdot \alpha^T)T = J = diag(J_1, \cdots, J_s),$$

where $J_k(k = 1, \cdots, s)$, are upper triangular Jordan blocks, whose principal diagonal elements consist of $s_i(i = 2, \cdots, N)$.

Therefore,

$$(T\otimes I_n)^{-1}[I_{N-1}\otimes L-(\Lambda+1_{N-1}\cdot\alpha^T)\otimes MKH](T\otimes I_n) = I_{N-1}\otimes L-J\otimes MKH$$

is an upper triangular block matrix, which together with the properties of Kronecker product implies that the eigenvalues of $(E, I_{N-1}\otimes L-(\Lambda+1_{N-1}\cdot\alpha^T)\otimes MKH)$ are given by the eigenvalues of $(E, L-s_iMKH)(i = 2,\cdots,N)$. Therefore, all the eigenvalues of $(E, L - s_iMKH)(i = 2,\cdots,N)$ are in the open left half plane.

Now, we prove the system (E, L, M, H) is stabilizable and detectable.

By lemma 2, (E, L, M) is stabilizable is equivalent to slow subsystem (L_1, M_1) is stabilizable . In fact, if at least one of $\lambda_i(i = 2,\cdots,N)$, is real, say λ_2, then (E, L, M, H) is stabilizable and detectable since all the eigenvalue of $L_1 - \lambda_2 M_1 K_1 H_1$ are in the open left half plane. If all $\lambda_i(i = 2,\cdots,N)$ are complex numbers, that is, none of their imaginary parts are zeros, then noticing that $\Lambda + 1_{N-1}\cdot\alpha^T$ is a real matrix, the eigenvalues will appear in conjugate pair form. Without loss of generality, we assume that λ_2 and λ_3 are a pair of conjugate roots with $\lambda_2 = e + jd$ and $\lambda_3 = e - jd(j^2 = -1)$. Since

$$\begin{vmatrix} \lambda I_{n_1} - (L_1 - eM_1K_1H_1) & -dM_1K_1H_1 \\ dM_1K_1H_1 & \lambda I_{n_1} - (L_1 - eM_1K_1H_1) \end{vmatrix}$$
$$= |\lambda I_{n_1} - (L_1 - \lambda_2 M_1K_1H_1)| \cdot |\lambda I_{n_1} - (L_1 - \lambda_3 M_1K_1H_1)|,$$

where $\mathbb{C}$ denotes the field of complex numbers, all the eigenvalues of

$$\begin{vmatrix} L_1 - eB_1K_1H_1 & -dM_1K_1H_1 \\ -dM_1K_1H_1 & A_1 - eM_1K_1H_1 \end{vmatrix}$$

are in the open left half plane, since all the eigenvalues of $L_1 - \lambda_2 M_1K_1H_1$ and $L_1 - \lambda_3 M_1K_1H_1$ are in the open left half plane. This together with

$$\begin{bmatrix} L_1 - eM_1K_1H_1 & -dM_1K_1H_1 \\ -dM_1K_1H_1 & -eM_1K_1H_1 \end{bmatrix}$$
$$= \begin{bmatrix} L_1 & 0 \\ 0 & L_1 \end{bmatrix} + \begin{bmatrix} M_1 & 0 \\ 0 & M_1 \end{bmatrix}\begin{bmatrix} -eK_1H_1 & -dK_1H_1 \\ -dK_1H_1 & L_1 - eK_1H_1 \end{bmatrix}$$
$$= \begin{bmatrix} L_1 & 0 \\ 0 & L_1 \end{bmatrix} + \begin{bmatrix} -eM_1K_1 & -dM_1K_1 \\ -dM_1K_1 & -eM_1K_1 \end{bmatrix}\begin{bmatrix} H_1 & 0 \\ 0 & H_1 \end{bmatrix}$$

implies that $(\begin{bmatrix} L_1 & 0 \\ 0 & L_1 \end{bmatrix}, \begin{bmatrix} M_1 & 0 \\ 0 & M_1 \end{bmatrix})$ is stabilizable and $(\begin{bmatrix} L_1 & 0 \\ 0 & L_1 \end{bmatrix}, \begin{bmatrix} H_1 & 0 \\ 0 & H_1 \end{bmatrix})$ is detectable. Thus,

$$2n = Rank\begin{pmatrix} sE - L & 0 & M & 0 \\ 0 & sE - L & 0 & M \end{pmatrix}$$
$$= Rank\begin{pmatrix} sQEP - QLP & 0 & QM & 0 \\ 0 & sQEP - QLP & 0 & QM \end{pmatrix}$$
$$= Rank\begin{pmatrix} \lambda I_{n_1} - L_1 & 0 & 0 & 0 & M_1 & 0 \\ 0 & -I_{n_2} & 0 & 0 & M_2 & 0 \\ 0 & 0 & \lambda I_{n_1} - L_1 & 0 & 0 & M_1 \\ 0 & 0 & 0 & -I_{n_2} & 0 & M_2 \end{pmatrix} = 2(n_2 + n_1), \forall s \in \overline{\mathbb{C}}.$$

Combining this with

$$Rank \begin{pmatrix} \lambda I_{n_1} - L_1 & 0 & M_1 & 0 \\ 0 & \lambda I_{n_1} - L_1 & 0 & M_1 \end{pmatrix} = 2Rank \left(\lambda I_{n_1} - L_1 \ M_1 \right), \forall s \in \overline{\mathbb{C}}$$

gives

$$Rank \left(\lambda I_{n_1} - L_1 \ M_1 \right) = n_1, \forall s \in \overline{\mathbb{C}}$$

Or equivalently, the system (E, L, M) is stabilizable.

Similarly, we can show the detectability of the system (E, L, H). Thus, the system (E, L, M, H) is stabilizable and detectable.

Now, we start to prove the second part of the theorem, that is, if the eigenvalues of (E, L) are not all in the open left half plane, then the topology $\mathcal{G}$ must have a spanning tree. In fact, from Lemma 2, we know that for all $i = 2, \cdots, N, s_i = 0$ or $Res_i > 0$. If the eigenvalues of (E, L) are not all in the open left half plane, then for all $i = 2, \cdots, N, s_i \neq 0$ must be true, since otherwise, there would be an $i \in \{2, \cdots, N\}$ such that $s_i = 0$, which in turn implies that all the eigenvalues of $(E, L) = (E, L - s_i MKH)$ are in the open left half plane. This is a contradiction. Thus, by Lemma 5, $\mathcal{G}$ must have a spanning tree.

From the proof of Theorem 1, it can be seen that by introducing a linear transformation, the consensusability of singular-MASs can be converted into the stability problem of (13), which actually is equivalent to whether there exists a gain matrix $K \in \mathbb{R}^{r \times n}$ such that all the eigenvalues of $(E, L - s_i MKH)(i = 2, \cdots, N)$ are in the open left half plane. The latter is essentially a static output feedback stability problem.

From the above analysis, one can see that the dynamic and communication properties of the agents are key to the consensusability of the singular MASs. Thus, the consensusability research for singular MASs, in essence, comes down to the studies on the dynamic and communication properties of the agents.

4 Conclusion

This paper studied the consensusability of singular multi-agent systems. By using the tools of the algebra graph and singular system theories, the joint impact of the agent dynamic structure and the communication topology on the consensusability was considered and some necessary and sufficient conditions on consensusability of singular multi-agent systems were provided.

Acknowledgements. Thanks to the Supported by the SF of Guangdong Province of China (No.04011425) and the SF of Zhanjiang Normal University of Guangdong Province of China (No.L0802).

References

1. Hong, Y., Hu, J., Gao, L.: Tracking control for multi-agent consensus with an active leader and variable topology. Automatica 42, 1177–1182 (2006)
2. Ma, C., Zhang, J.: Necessary and sufficient conditions for consensusability of linear multi-agent systems. IEEE Trans. on Automatic Control 55, 1263–1268 (2010)
3. Ren, W., Beard, R.: Consensus seeking in multi-agent systems using dynamically changing interaction topologies. IEEE Trans. Automatic Control 50, 665–671 (2005)
4. Dai, L.: Singular Control Systems. Springer, Berlin (1989)
5. Godsil, C., Royle, G.: Algebraic Graph Theory. Springer, New York (2001)

A Note on the Kernelled Quasidifferential in the $n-$Dimensional Space

Chun-ling Song[1] and Zun-quan Xia[2]

[1] Department of Information Sciences and Mathematics,
  Foshan University, Foshan 528000, China
  mary_sohu@sohu.com
[2] Department of Applied Mathematics, Dalian University of Technology,
  Dalian 116024, China
  zqxiazhh@dlut.edu.cn

Abstract. A special class of quasidifferentiable functions on R^n having kernels, g-q.d. functions, are studied in this paper. The corresponding subclasses and augmented class are defined and some closeness is given.

Keywords: Quasidifferentiable function, quasidifferential, kernelled quasidifferential, Demyanov difference.

1 Introduction

A function f on R^n is called quasidifferentiable (for short q.d.), in the sense of Demyanov and Rubinov, if it is directionally differentiable and the directional derivative can be expressed as follows:

$$f'(x;d) = \max_{v \in \underline{\partial} f(x)} < v, d > + \min_{w \in \overline{\partial} f(x)} < w, d >, \forall d \in R^n.$$

The pair of $[\underline{\partial} f(x), \overline{\partial} f(x)]$ is called the quasidifferential of f at x. $\underline{\partial} f(x)$ and $\overline{\partial} f(x)$ are called a subdifferntial and a superdifferential of f at x, respectively.

It is known that the quasidifferential uniqueness is one of the important problems in the study of q.d. functions and optimization in the sense of Demyanov, Polyakova and Rubinov [1-3]. This problem was first time considered in a discussion at IIASA, by Demyanov and Xia in 1984 [4]. There were many reports and publications mentioning or dealing with this subject (see for instance, [5-17], etc).

There are some ways studying this problem, for instance, from the points of view of the quotient space generated by an equivalence class (K-theory), of pairs of sublinear operators (functions, support functions), of the ordering relation of the space of pairs of convex sets and of establishing some criteria by which the representative quasidifferential can be determined automatically. The last point of view is considered in this paper.

B.-Y. Cao and X.-J. Xie (Eds.): Fuzzy Engineering and Operations Research, AISC 147, pp. 541–547.
springerlink.com   © Springer-Verlag Berlin Heidelberg 2012

We denote by $\mathcal{D}f(x)$ the set of all quasidifferentials, $Df(x) = [\underline{\partial}f(x), \overline{\partial}f(x)]$, of a q.d. function f at x. Define $\partial^+ f(x) = (\underline{\partial} + \overline{\partial})f(x) := \underline{\partial}f(x) + \overline{\partial}f(x)$ and $\partial^- f(x) = (\overline{\partial} - \overline{\partial})f(x) := \overline{\partial}f(x) - \overline{\partial}f(x)$. If f is q.d. at x, then $f'(x;d)$ can be written in the form $f'(x;d) = \max_{u \in \underline{\partial}f(x)} u^T d + \min_{v \in \overline{\partial}f(x)} v^T d$, $\forall d \in R^n$ and also in $f'(x;d) = \underline{f}'(x;d) - \overline{f}'(x;d)$, where

$$\underline{f}'(x;d) = \inf_{Df(x) \in \mathcal{D}f(x)} \max_{v \in \partial^+ f(x)} <v,d> = \inf_{Df(x) \in \mathcal{D}f(x)} \delta^*(d \,|\, \partial^+ f(x)),$$

$$\overline{f}'(x;d) = \inf_{Df(x) \in \mathcal{D}f(x)} \max_{w \in \partial^- f(x)} <v,d> = \inf_{Df(x) \in \mathcal{D}f(x)} \delta^*(d \,|\, \partial^- f(x)),$$

(see Xia, 1987, 1993 [14,15]) where $\delta^*(\cdot \,|\, C)$ denotes the support function of a convex set C.

The question is that whether $\underline{f}'(x; \cdot)$ and $\overline{f}'(x; \cdot)$ are sublinear. The answer is not positive, but it is positive for the case of one and two dimensional space. In other words, for the case of one and two dimensional space there exists a quasidifferential, called the kernelled quasidifferential of f at x and denoted by $[\partial^* f(x), \partial_* f(x)]$ such that

$$\underline{f}'(x;d) = \delta^*(d \,|\, \partial_* f(x)), \quad \overline{f}'(x;d) = \delta^*(d \,|\, -\partial^* f(x)), \tag{1}$$

where

$$\partial_* f(x) = \bigcap_{Df(x) \in \mathcal{D}f(x)} \partial^+ f(x), \quad \partial^* f(x) = \bigcap_{Df(x) \in \mathcal{D}f(x)} \partial^- f(x). \tag{2}$$

For the case of one dimensional space the result was given by Gao in 1988 [5]. In two dimensional space, the minimal pairs of quasidifferentials are uniquely determined up to a translation, (see for instance [12, 13]). Later based on the results in [12, 13] it was proved that in two dimensional space the kernelled quasidifferential exists for any q.d. function (see Gao (2001) [6]). For the case of the dimension more than two, generally speaking, there are no kernelled quasidifferential for q.d. functions, but for some special cases, the kernelled quasidifferential exists. In 2005, [7] presented a condition in terms of Demyanov difference, called g-condition, in which the kernelled quasidifferential exists for such a $n-$ dimensional q.d. function ($n \geq 2$). We will present some more properties on this class.

2   A Subclass of Kernelled Quasidifferentiable Functions in High Dimensional Spaces

A set $T \subset R^n$ is called of full measure (with respect to R^n), if $R^n \backslash T$ is a set with measure zero. Let U and $V \subset R^n$ be compact convex and $T \subset R^n$ be a full measure set such that their support functions $\delta^*(\cdot \,|\, U)$ and $\delta^*(\cdot \,|\, V)$ are differentiable on $x \in T$. The Demyanov difference between U and V, denoted by $U \dot{-} V$, is defined by

$$U \dot{-} V = \text{clco}\{\nabla \delta^*(x \mid U) - \nabla \delta^*(x \mid V) \mid x \in T\}, \tag{3}$$

(see [2]).

The following theorem defining a subclass of kernelled q.d. functions is due to Gao [7].

Theorem 1. *Suppose that f is a q.d. function defined on an open set $X \subset R^n$ and there exists a quasidifferential $D_0 f(x) = [\underline{\partial}_0 f(x), \overline{\partial}_0 f(x)]$ of f at $x \in X$ such that*

$$\underline{\partial}_0 f(x) \dot{-} (-\overline{\partial}_0 f(x)) = \underline{\partial}_0 f(x) - (-\overline{\partial}_0 f(x)). \tag{4}$$

Then, the following relations hold

$$\bigcap_{Df(x) \in \mathcal{D}f(x)} \partial^+ f(x) = \partial_0^+ f(x), \qquad \bigcap_{Df(x) \in \mathcal{D}f(x)} \partial^- f(x) = \partial_0^- f(x), \tag{5}$$

and

$$[\partial_0^+ f(x), \partial_0^- f(x)] \in \mathcal{D}f(x), \tag{6}$$

where $\partial^+ f(x) = \underline{\partial} f(x) + \overline{\partial} f(x), \partial^- f(x) = \overline{\partial} f(x) - \overline{\partial} f(x), \partial_0^+ f(x) = \underline{\partial}_0 f(x) + \overline{\partial}_0 f(x), \partial_0^- f(x) = \overline{\partial}_0 f(x) - \overline{\partial}_0 f(x)$.

Definition 1. A subclass of q.d. functions is said to be the G-subclass if it consists of functions satisfying (4), denoted by $\mathcal{G}_X(x)$, and every such a q.d. function is said to be a G-q.d. function.

Proposition 1. *Let f be a q.d. function on R^n, and $x_0 \in R^n$. If $f \in \mathcal{G}_X(x)$, then*

$$\underline{f}'(x; d) = \inf_{Df(x)} \max_{v \in \partial^+ f(x)} <v, d> = \max_{v \in \partial_0^+ f(x)} <v, d>, \tag{7}$$

$$\overline{f}'(x; d) = \inf_{Df(x)} \max_{v \in \partial^- f(x)} <v, d> = \max_{v \in \partial_0^- f(x)} <v, d> . \tag{8}$$

Proof. Since $[\underline{\partial}_0 f(x), \overline{\partial}_0 f(x)] \in \mathcal{D}f(x)$, one has that,

$$\inf_{Df(x)} \max_{v \in \partial^+ f(x)} <v, d> \leq \max_{v \in \partial_0^+ f(x)} <v, d> . \tag{9}$$

On the other hand, for any $[\underline{\partial} f(x), \overline{\partial} f(x)] \in \mathcal{D}f(x)$, the following formulae hold,

$$\begin{aligned}
\partial^+ f(x) &= \underline{\partial} f(x) - (-\overline{\partial} f(x)) \\
&\supset \underline{\partial} f(x) \dot{-} (-\overline{\partial} f(x)) \\
&= \underline{\partial}_0 f(x) \dot{-} (-\overline{\partial}_0 f(x)) \\
&= \underline{\partial}_0 f(x) - (-\overline{\partial}_0 f(x)) \\
&= \partial_0^+ f(x).
\end{aligned} \tag{10}$$

Therefore, one has

$$\max_{v \in \partial^+ f(x)} <v, d> \geq \max_{v \in \partial_0^+ f(x)} <v, d>, \quad \forall [\underline{\partial} f(x), \overline{\partial} f(x)] \in \mathcal{D} f(x). \quad (11)$$

Taking the infimum to (11) over $\mathcal{D} f(x)$, one has

$$\inf_{\mathcal{D} f(x)} \max_{v \in \partial^+ f(x)} <v, d> \geq \max_{v \in \partial_0^+ f(x)} <v, d>. \quad (12)$$

It follows from (9) and (12) that formula (7) holds. For(8), one has

$$\inf_{\mathcal{D} f(x)} \max_{v \in \partial^- f(x)} <v, d> \leq \max_{v \in \partial_0^- f(x)} <v, d>, \quad (13)$$

since $\overline{\partial}_0 f(x) \in \overline{\mathcal{D}} f(x)$. On the other hand, we have

$$\begin{aligned} f'(x; d) &= \max_{v \in \partial^+ f(x)} <v, d> + \min_{v \in \partial^- f(x)} <w, d> \\ &= \max_{v \in \partial_0^+ f(x)} <v, d> + \min_{v \in \partial_0^- f(x)} <w, d>. \end{aligned} \quad (14)$$

Combing with (11), we have

$$\min_{w \in \partial^- f(x)} <w, d> \leq \min_{w \in \partial_0^- f(x)} <w, d>, \quad \forall [\underline{\partial} f(x), \overline{\partial} f(x)] \in \mathcal{D} f(x), \quad (15)$$

i.e.,

$$\max_{w \in \partial^- f(x)} <w, d> \geq \max_{w \in \partial_0^- f(x)} <w, d>, \quad \forall [\underline{\partial} f(x), \overline{\partial} f(x)] \in \mathcal{D} f(x). \quad (16)$$

Taking the infimum to (16) over $\mathcal{D} f(x)$, one has,

$$\inf_{\mathcal{D} f(x)} \max_{v \in \partial^- f(x)} <v, d> \geq \max_{v \in \partial_0^- f(x)} <v, d>. \quad (17)$$

Combing (13) with (17), one has (8). The demonstration is completed.

Now, we introduce two notations $\mathcal{O}_S(\cdot)$ and $\mathcal{CG}_X(\cdot)$ defined on an open set $X \subset R^n$.

$$\mathcal{O}_S(x) = \{f \mid \exists \mathcal{D} f : \mathcal{D} f(x) \in \mathcal{D} f(x) \text{ satistying } (4), \mathcal{D} f(x) \subset (S, S^\perp)\},$$

$$\mathcal{CG}_X(x) = \left\{ f \mid \exists \text{ a finite index set } I, \text{ and } f_i \in \mathcal{G}_X(x), i \in I, \exists \gamma \in \Lambda : f = \gamma((f_i)_{i \in I}) \right\},$$

where S is a subspace of R^n and $S^\perp$ is orthogonally associated to S, $\Lambda = \{\gamma \mid \gamma \text{ is the composite of a finite number of basic operations}\}$, the basic operations include sum, subtraction, multiplication, division, maximal and minimal operations.

The following proposition is copied from Gao [7].

Proposition 2. *Suppose that X_1 and X_2 to be orthogonal complementary subspaces of R^n and $A \subset X_1$, $B \subset X_2$. Then $A \dot{-} B = A - B$.*

It follows from Proposition 2 and the definition of $\mathcal{CG}_X(\cdot)$ that $\mathcal{O}_S(x) \subset \mathcal{G}_X(x) \subset \mathcal{CG}_X(x)$.

Note that the orthogonality of A and B is not necessary for having the equality $A \overset{\cdot}{-} B = A - B$. For example, taking $A = \{(-1,0), (1,0)\}$ and $B = \{(-1,-1), (1,1)\}$ we have $A \overset{\cdot}{-} B = A - B = \mathrm{co}\{(0,1), (-2,-1), (2,1), (0,-1)\}$ but they are not orthogonal.

3 The Property of G-Subclass

If the kernelled quasidifferential exists, some operation rules of kernelled quasidifferential are presented by Xia [15] as follows.

Theorem 2. *Suppose $f_i, i \in I := \{1, \cdots, m\}$, are kernelled quasidifferentiable, $[\partial_* f_i(x), \partial^* f_i(x)]$ are the kernelled quasidifferential of f_i at x and $\lambda \in R^1$. Then one has:*

1) $\partial_* \sum_{i=1}^m f_i(x) = \sum_{i=1}^m \partial_* f_i(x)$ *and* $\partial^* \sum_{i=1}^m f_i(x) = \sum_{i=1}^m \partial^* f_i(x)$;

2) $\partial_*(\lambda f_i)(x) = \lambda \partial_* f_i(x), \quad \partial^*(\lambda f_i)(x) = \begin{cases} \lambda \partial^* f_i(x), & \text{if } \lambda \geq 0, \\ |\lambda| \partial^*(-f_i)(x), & \text{if } \lambda < 0; \end{cases}$

3) $\mathcal{D}_k(f_1 f_2)(x) = |f_1(x)| \mathcal{D}_k(\mathrm{sign} f_1(x) f_2)(x) + |f_2(x)| \mathcal{D}_k(\mathrm{sign} f_2(x) f_1)(x)$, *where* $\mathcal{D}_k f = [\partial_* f, \partial^* f]$;

4) $\mathcal{D}_k(f_1/f_2)(x) = 1/f_2^2(x)[|f_2(x)| \mathcal{D}_k(\mathrm{sign} f_2(x) f_1)(x) + |f_1(x)| \mathcal{D}_k(\mathrm{sign} (-f_1(x))f_2)(x)]$, *where* $\mathcal{D}_k f = [\partial_* f, \partial^* f]$ *and* $f_2(x) \neq 0$;

5) $\partial_*(\max_{i \in I} f_i) = \mathrm{co} \bigcup_{k \in R(x)} [\partial_* f_k(x) + \sum_{i \in R(x) \setminus \{k\}} \partial^* f_i(x)]$ *and* $\partial^*(\max_{i \in I} f_i) = \sum_{k \in R(x)} \partial^* f_k(x)$, *where* $R(x) = \{i \in I \mid f_i(x) = f(x)\}$;

6) $\partial_*(\min_{i \in I} f_i) = -\partial_*(\max\{-f_i(\cdot) \mid i \in I\})(x)$, $\partial^*(\min_{i \in I} f_i) = \partial^*(-\max\{-f_i(\cdot) \mid i \in I\})(x)$.

It follows from the definition of $\mathcal{CG}_X(\cdot)$ and Theorem 2 that the subclass $\mathcal{CG}(\cdot)$ has the kernelled quasidifferential. Specially, one has the following proposition.

Proposition 3. *Let $f = f_1 - f_2$, where $f_1, f_2 : X \to R$ are convex functions. Then, the kernelled quasidifferential exists for f, moreover, one has*

$$\partial_* f(x) = \partial f_1(x) - \partial f_2(x), \tag{18}$$

$$\partial^* f(x) = -\partial f_2(x) + \partial f_2(x), \tag{19}$$

where $\partial f_1(x), \partial f_2(x)$ are the subdifferential of f_1 and f_2 at x in convex analysis, respectively.

Proof. Since f_1 is convex, one has that

$$[\partial f_1(x), \{0\}] \in \mathcal{D} f_1(x).$$

On the other hand, $-f_2$ is concave since f_2 is convex. Moreover,

$$[\{0\}, -\partial f_2(x)] \in \mathcal{D}(-f_2)(x).$$

It is clear from Theorem 1 that for $i = 1, 2$, $f_i \in \mathcal{G}_X(x)$ and

$$\partial_* f_1(x) = \partial f_1(x), \partial^* f_1(x) = \{0\},$$

$$\partial_* f_2(x) = -\partial f_2(x), \partial^* f_2(x) = -\partial f_2(x) + \partial f_2(x).$$

By Theorem 2, one has that f is kernelled quasidifferentiable and the following formula hold:

$$\partial_* f(x) = \partial f_1(x) - \partial f_2(x), \partial^* f(x) = -\partial f_2(x) + \partial f_2(x).$$

Proposition 4. *Suppose that $f_i \in \mathcal{O}_S(x)$, $\lambda \geq 0$, $i = 1, \cdots, m$. Then one has:*

1) $\sum_{i=1}^m f_i \in \mathcal{O}_S(x)$ and $\lambda f \in \mathcal{O}_S(x)$;
2) $-f \in \mathcal{O}_{S^\perp}(x)$.

Proof. Let $[\underline{\partial} f_i(x), \overline{\partial} f_i(x)]$ be the quasidifferential of f at x satisfying $\underline{\partial} f_i(x) \subset S$, $\overline{\partial} f_i(x) \subset S^\perp$. Then by the calculus of quasidifferential one has

$$\underline{\partial}(\sum_{i=1}^m f_i)(x) = \sum_{i=1}^m \underline{\partial} f_i(x) \subset S,$$

$$\overline{\partial}(\sum_{i=1}^m f_i)(x) = \sum_{i=1}^m \overline{\partial} f_i(x) \subset S^\perp.$$

So,

$$\sum_{i=1}^m f_i \in \mathcal{O}_S(x).$$

For $\lambda \geq 0$, one has

$$\underline{\partial}(\lambda f)(x) = \lambda \underline{\partial} f(x) \subset S,$$

$$\overline{\partial}(\lambda f)(x) = \lambda \overline{\partial} f(x) \subset S^\perp.$$

That is to say $\lambda f \in \mathcal{O}_S(x)$.

Similarly, by the quasidifferential calculus, one has

$$\underline{\partial}(-f)(x) = -\overline{\partial} f(x) \subset S^\perp,$$

$$\overline{\partial}(-f)(x) = -\underline{\partial} f(x) \subset S.$$

By the definition of $\mathcal{O}_{S^\perp}$, we have $-f \in \mathcal{O}_{S^\perp}(x)$.

4 Conclusion

The outcome in this paper shows that the class of being kernelled quasidifferential in R^n is enough large.

Acknowledgements. Thanks to the reviewer comments which is important for this paper.

References

1. Demyanov, V.F., Rubinov, A.M.: Quasidifferential calculus, pp. 65–78. Optimization Software. Inc., New York (1986)
2. Demyanov, V.F., Rubinov, A.M.: Constructive Nonsmooth Analysis, pp. 106–186. Verlag Perter Lang, Berlin (1995)
3. Demyanov, V.F., Rubinov, A.M.: Quasidifferentiability and Related Topics, pp. 1–25. Kluwer Academic Publishers, Netherlands (2000)
4. Demyanov, V.F., Xia, Z.-Q.: Minimal quasidifferentials. International Institute for Applied Systems Analysis (IIASA), Laxenburg (1984)
5. Gao, Y.: The Star-Kernel for a Quasidifferentiable Function in One Dimensional Space. J. Math. Research Exposition 8, 152 (1988)
6. Gao, Y., Xia, Z.-Q., Zhang, L.-W.: Kernelled Quasidifferential for a Quasidifferntiable Function in Two-Dimensional Space. Journal of Convex Analysis 8(2), 401–408 (2001)
7. Gao, Y.: Representative of Quasidifferentials and Its Formula for a Quasidifferentiable Function. Set-Valued Analysis 13, 323–336 (2005)
8. Pallaschke, D., Scholtes, S., Urbański, R.: On Minimal Pairs of Compact Convex Sets. Bull. Acad. Polon. Sci. Sér. Sci. Math. 39, 1–5 (1991)
9. Pallaschke, D., Urbański, R.: Some Criteria for the Minimality of Pairs of Compact Convex Sets. Zeitschrift für Operations Research 37, 129–150 (1993)
10. Pallaschke, D., Urbański, R.: Reduction of Quasidifferentials and Minimal Representations. Mathematical Programming 66, 161–180 (1994)
11. Pallaschke, D., Urbański, R.: Minimal pairs of compact convex sets, with applications to quasidifferential calculus. In: Demyanov, V.F., Rubinov, A.M. (eds.) Quasidifferentiability and Related Topics, pp. 173–181. Kluwer Academic Publishers (2000)
12. Crzybowski, J.: On Minimal Pairs of Convex Compact Sets. Archiv der Mathematik 63, 173–181 (1994)
13. Scholtes, S.: Minimal Pairs of Compact Convex Bodies in Two Dimensions. Mathematika 39, 267–273 (1992)
14. Xia, Z.-Q.: The ∗-kernel for quasidifferntiable functions. WP-87-89, IIASA, Laxenburg, Austria (1987)
15. Xia, Z.-Q.: On Quasidifferential Kernels. Demonstratio Mathematica XXVI, 159–182 (1993)
16. Zhang, L.-W., Xia, Z.-Q., Gao, Y., Wang, M.-Z.: Star-Kernels Quasidifferential for a Quasidifferentiable Functions in Two-dimensional Space. Journal of Convex Analysis 9, 139–158 (2002)
17. Zhang, L.W.: Some new results in quasidifferentiable analysis and optimization: Kernel·Equation·Differential. Ph.D. Dissertation, Dalian University of Technology, Dalian, 8–29 (1998)

The Application of GENOCOP in Production Plan

Xiao Yuan and Jin Xiao

Mathematics and Computation Science School,
Zhanjiang Normal University,
Zhanjiang, Guangdong, ZIP 524048, PR china
yuanxiao312@126.com

Abstract. In this paper, in order to solve the expansion problem of electric power companies, we firstly adopt simple linear regression model to fit the next 10-year power demand, secondly build 0-1 Nonlinear integer programming model which is disposed by variables continuization, finally use the GENOCOP algorithm to solve it.

Keywords: Linear Fitting, 0-1 Nonlinear Integer Programming, GENOCOP.

1 Introduction

0-1 mixed integer non-linear programming (MINLP) is a very wide and effective model, which have no unified method to solve with its complexity. There are two kinds of the common solution methods, one is that all mixed variables is transformed into discrete variables, another is that all mixed variables is changed into continuous variables.

Genetic algorithms (GAs)[1] proposed by Holland have attracted considerable attention as global methods for complex function optimization since De Jong considered GAs in a function optimization[2]. In 1992, Michalwicz and Janikow proposed Genetic Algorithm for Numerical Optimization for Constrained Problem (GENOCOP)[3], which not only handles any objective function with any set of linear constraints, but also effectively reduces the search space. Michalewicz also proposed the co-evolutionary genetic algorithm, called GENOCOP III[4], by introducing the concepts of search points and reference points which, respectively, satisfy the linearconstraints and all of the constraints.

Aim at solving an electric power company expansion problem, we build a mixed integer non-linear programming (MINLP), which is disposed by 0-1 variables continuation and solved preferably by the GENOCOP algorithm.

B.-Y. Cao and X.-J. Xie (Eds.): Fuzzy Engineering and Operations Research, AISC 147, pp. 549–555.
springerlink.com
© Springer-Verlag Berlin Heidelberg 2012

2 Problem Posing

Problem posing and data collection

Brite Power is a small investment power company in Finger lake of New York. The Director of the company do some research for whether to expand the company power production in the summer of 2010.

The main requirements are

1. Ensure company have plenty of power before 2020.
2. The time range of the study required at least 10 years.
3. The study is concerned with the electric power total capacity in a year, not every hour or every day of electricity fluctuation.

Brite Power mainly provide the data as follows.

I In early 2010, Brite Power had 60 megawatt(MW) power capacity.
II The company power demand in the past 10 years is listed below:

Table 1. The statistics of the company power demand in 2001 to 2010.

Year	demand(MW)	Year	demand(MW)
2001	36	2006	48
2002	37	2007	53
2003	40	2008	54
2004	42	2009	58
2005	46	2010	61

III The economic conditions of coal-fired plant are operated according to Power Law[5], in other wordsthe new factory constant dollars cost is referred to the zero year dollar, which is abided by the following prediction rules.
Factory cost with capacity for K = [Capacity K/Basic scale factory capacity]0.8 *Basic scale factory cost.
IV Brite Power can build the Factory 5, 10, 15 or 20 (MW) production capacity of the factory. Every year can build a new factory at most.
V Brite Power must also manage factory in addition to increase capacity, and its operating cost bases on the use of capacity. If the demand is D_t MW for the tth years and its total capacity is D_t MW for the tth years, so the operation cost is $(\frac{C_t}{D_t})^{0.5} * D_t * 40$ million dollars according to the constant dollars cost calculation.
VI Suppose that the upper limit of store capacity is 10 MW every year, and the storage fee of each megawatt power is 100 thousand dollars each year. Basically, although the company can buy 5 MW power from the nearby power company each year on the basis of $600000 per megawatt, according to constant dollars cost, the company must meet all the needs of this year.

Need to solve the key problems as follows.

- How to predict the future demand for 10 years?
- What time is increase capacity ? How much should increase ?

3 Model Building

3.1 Problem Analysis

1. Choose the appropriate forecast method to predict the next 10 years demand.
2. What time is to build new factories from 2011 to 2020? What are their capacities?
3. What year need to buy power from the nearby power company to satisfy its needs?

3.2 Problem Analysis

1. To make the system effective and feasible, The company does not consider currency discount problem such as the years inflation rate and the discount rate currencies in the next ten years, and reduce the complexity of the decision-making system.
2. The demand data of company power is reliable and stable.
3. Electric power plants are independent of each other and do not consider their depreciation factors.

3.3 Demand Forecast in Next 10 Years

Through the data of the company power demand from 2001 to 2010 years, we can find they are always increasing. Make a scatter plot chart as follow.

It obviously present a linear growth trend from Figure 1, so we can choose a linear function, $y = ax + b$, to fit the data in Figure 1, and draw a straight line to predict future points, $y = 2.994x + 30.88$, which is obtained by the fitting cftool toolbox of MATLAB, and its graphics is shown in Figure 2.

The coefficient of determination is $R^2 = 0.9908$ which illustrate the fitting effect good. The next ten years demand is predicted as Table 2.

Table 2. The quantitative prediction about the next ten years demand.

Year	Predictor	Year	Predictor
2011	63.814	2016	78.784
2012	66.808	2017	81.778
2013	69.802	2018	84.772
2014	72.796	2019	87.766
2015	75.790	2020	90.760

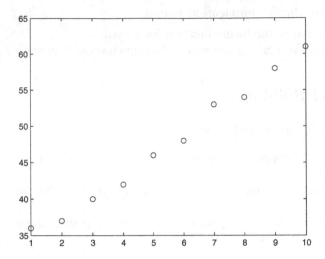

Fig. 1. Demand from 2001 to 2010 year.

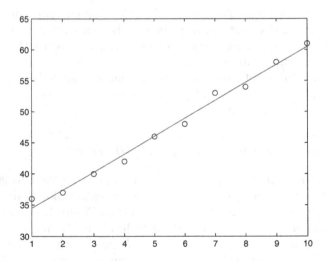

Fig. 2. Linear function fitting graphics.

3.4 Nonlinear Integer Programming Model

Let v_t be storage capacity of the tth year, $t = 1, 2, \cdots, 10$, where $v_0 = 0$.
Let $x_{t1}, x_{t2}, x_{t3}, x_{t4} \in \{0, 1\}$. If a factory is built in the tth year, it equal
to 1, else it equal to 0. Let u_t be the number of power which is purchased
from the nearby power companies in the tth year. Let $C(t)$ be the total cost,
$A(t)$ be the cost for building new factory, $B(t)$ be the operating costs, $D(t)$ be

the cost for purchasing power, $E(t)$ be the inventory cost, $F(t)$ be reducing operation costs for power law in the tth year.

$$C(t) = A(t) + B(t) + D(t) + E(t) + F(t) \qquad (1)$$

where

$$A(t) = (\tfrac{5x_t+10y_t+15z_t+20w_t}{5})^{0.8} * 1800,$$
$$B(t) = (\tfrac{C_t}{D_t})^{0.5} * D_t * 400000,$$
$$D(t) = 600000 * u_t,$$
$$E(t) = 100000 * v_t.$$

Let C be the total charge from 2001 to 2010. Then

$$C = \sum_{t=1}^{10} [A(t) + B(t) + D(t) + E(t)]$$

$$= 1800 \sum_{t=1}^{10} (\tfrac{5x_{t1}+10x_{t2}+15x_{t3}+20x_{t4}}{5})^{0.8} + 400000 \sum_{t=1}^{10} (\tfrac{C_t}{D_t})^{0.5} * D_t$$

$$+600000 \sum_{t=1}^{10} u_t + 100000 \sum_{t=1}^{10} v_t.$$

Finally, 0-1 mixed integer non-linear programming is built as follows:

$$min\ C = 111800 \sum_{t=1}^{10} [(\tfrac{5x_{t1}+10x_{t2}+15x_{t3}+20x_{t4}}{5})^{0.8} + (\tfrac{C_t}{D_t})^{0.5} * D_t + u_t + v_t]$$

$$s.t. \begin{cases} 60 + 5 \sum_{i=1}^{t-1} (x_{i1} + 2x_{i2} + 3x_{i3} + 4x_{i4}) + u_t + v_{t-1} - d_t \geq v_t, \\ \sum_{j=1} 4x_{tj} \leq 1, \\ v_t \leq 10, \\ u_t \leq 5, \\ u_t, v_t \geq 0, \\ x_{t1}, x_{t2}, x_{t3}, x_{t4} = 0, 1. \end{cases}$$

4   Model Solving

Because of no unified and effective method for solving the 0-1 integer non-linear programming, the 0-1 discrete variables method, which can be transformed into some equivalent continuous nonlinear constraint conditions, is adapted to the problem, then GENOCOP algorithm can be used to solve it.

4.1   0-1 Variables Equivalent Continuation

To simplify The complexity of the calculation, we adopt a new method, proposed by Kiwiel K C and Yunkang Sui[6], [7], can transform 0-1 variables into continuous variables. The specific practices as follows:

$$x_{t1}, x_{t2}, x_{t3}, x_{t4} \in \{0,1\}, t = 1, 2, \cdots, 10$$
$$\Updownarrow$$
$$\begin{cases} \alpha_{ij}(x_{ti} - x_{ti}^2) = 0, \\ 0 \leq x_{ti} \leq 1, t = 1, 2, \cdots, 10, j = 1, 2, 3, 4. \end{cases}$$

Then the 0-1 mixed nonlinear integer programming model is transformed into an optimization model with equality constraints, the reduction results as

$$min\ C = 111800 \sum_{t=1}^{10} [(x_{t1} + 2x_{t2} + 3x_{t3} + 4x_{t4})^{0.8} + (\frac{C_t}{D_t})^{0.5} * D_t + u_t + v_t]$$

follows:

$$s.t. \begin{cases} 60 + 5\sum_{i=1}^{t-1}(x_{i1} + 2x_{i2} + 3x_{i3} + 4x_{i4}) + u_t + v_{t-1} - d_t \geq v_t, \\ \sum_{j=1}^{4} x_{tj} \leq 1, \\ v_t \leq 10, \\ u_t \leq 5, \\ u_t, v_t \geq 0, \\ \alpha_{ti}(x_{ti} - x_{ti}^2) = 0, \\ 0 \leq x_{ti} \leq 1, \\ t = 1, 2, \cdots, 10; j = 1, 2, 3, 4. \end{cases}$$

Here α_{tj} are some given positive numbers.

4.2 Solving by the GENOCOP

The problem of the optimization with the linear constraints is used widely, people do a lot of research on it, such as the simplex method, interior point method, iteration method and evolution algorithm, etc. In 1992, Michalwicz and Janikow proposed Genetic Algorithm for Numerical Optimization for Constrained Problem(GENOCOP), which is practically confirmed as a very good method[4] . The main Algorithm is as follows.

Step 1: Population-initialized, Population size N=100, let $t = 0$.

Step 2: Regarding the objective function as fitness function, all of the individual's fitness are calculated, and the optimal value of the individual should be retained.

Step 3: The termination conditions
If $t = 100$, or the optimum values are still invariant after more than 50 times iteration, stop. Else , turn next.

Step 4: Selection
We employ the tournament selection, which selects 60 individuals at random in [1, 100] and then selects the best one among these 60 individuals.

Step 5: Crossing
We employed two crossover operators, they are arithmetic and heuristic respectively. Let and are random numbers in [0, 1]. If $q \leq p_{ca}$, $r \leq p_c$, we employ arithmetic crossover operation, else if $r \leq p_c p_{ca} < q \leq p_{ce}$, we employ heuristic crossover operation. Else, we don't cross.

Step 6: Mutation
After crossing the individual, we mutate individual by using boundary mutation method with a 0.1 probability.

Step 7: Let $t = t + 1$, goto Step 2.

5 Conclusion

The test program is implemented in Matlab 7.0, the finally results are as follows.

The algorithm stop at the 168 times iteration, the optimal total cost is $4580250.338. From $x_{21} = 1, x_{63} = 1$, we can know that the time for establishing the factories with production 10 and 15 megawatt capacity is in 2012 and 2016 respectively, and from $v_1 = 3.814, v_5 = 4.964$, we can know that the time for purchasing 10 and 15 megawatt electric power is in 2011 and 2015, respectively.

The experimental results show that this kind of problem can be firstly dealt with 0-1 variable continuation after built 0-1 mixed nonlinear integer programming, then solved by the GENOCOP algorithm. It is feasible and effective to solve the company capacity expansion problem by the experiment.

References

1. Holland, J.A.: Adaptation in natural and artificial systems, pp. 121–145. Michigan Press, University of Ann Arbor (1975)
2. De Jong, K.A.: Genetic algorithms: A 10 year perspective. In: Proceedings of the First International Conference on Genetic Algorithms, pp. 24–36 (1985)
3. Michalewicz, J.: GENOCOP: A genetic algorithm for numecal optimization problem with linear constrains. Communications of the ACM, 145–173 (1992)
4. Michalewicz, J.: GENOCOP III:A co-evolutionary algorithm for numerical optimization problems with nonlinear constrains. In: Proceedings of the IEEE International Conference on Evolutionary, pp. 674–651 (1996)
5. Sui, Y.-K., Jia, Z.-C.: A Continuous approch to 0-1 linear problem and its solution with genetic algorithm. Mathematics in Practice and Theory, 119–127 (2010)
6. Liu, M.-J.: The model of gas flow in coal seams Basis on power law. Jouranl of Jianzuo Institute of Technology, 46–49 (1994)
7. Kiwiel, K.C., Lindberg, P.O., Nou, A.: Bregman proximal relaximal of large-scale 0-1 problems. Computational Optimization and Applications, 33–44 (2000)

Discrete Time *Geom/G/1* Queue with Second Optional Service and Server Breakdowns

Cai-min Wei[1], Zong-bao Zou[1], and Yi-Yan Qin[2]

[1] Department of Mathematics, Shantou University,
  Shantou 515063, P.R. China
  cmwei@stu.edu.cn,zzbkr123@163.com
[2] College of Business, Guangxi University for Nationalities,
  Nanning 530006, P.R. China
  qinyiyan2002@sina.com.cn

Abstract. This paper studies a discrete time queueing system with server breakdowns in which all the arriving customers demand the first "essential" service while only some of them demand the second "optional" service with probability α. The service time of the first essential service and the second optional service both are independent and arbitrarily random variables. Breakdowns arrive according to a Bernoulli process with rate β when the server is servicing customers and the repair time is an arbitrarily random variable. Using the supplementary variable method, we obtain the probability generating function, the mean of the system size and the waiting time. Finally, it gives some numerical examples to illustrate the effect of the probability α on the mean system size and waiting time.

Keywords: Discrete time queueing system, second optional service, server breakdowns.

1 Introduction

In recent years discrete-time queues have been used extensively to analyze high-speed computer communication networks. Their importance has increased due to the emergence of the broadband integrated service digital networks (B-ISDN), which can provide transfer of video, voice, and data communication through high-speed local area networks (LANs), on-demand video distribution, video telephony communications, etc. The asynchronous transfer mode (ATM) is envisaged as the basic transfer model for implementing B-ISDN. In these systems, all information flow is transmitted in fixed-size units called cells. For more details, see the books [1-4].

Queueing systems with the second optional service have been widely studied. Madan [5] first studied an $M/G/1$ queue with the second optional service

B.-Y. Cao and X.-J. Xie (Eds.): Fuzzy Engineering and Operations Research, AISC 147, pp. 557–568.
springerlink.com   © Springer-Verlag Berlin Heidelberg 2012

in which the first essential service time follows a general distribution, but the second optional service is assumed to be exponentially distributed. Medhi [6] extended Madan's model by considering that the second optional service follows a general distribution. Wang [7] considered Medhi's model with the assumption that the server is subject to breakdowns in which repair time follows an arbitrary distribution. While there are few studies have been in the discrete time queueing systems with the second optional service owing to its complexity. Atencia and Moreno [8] studied the discrete time retial queues with second optional service. Wang and Zhao [9] extended Atencia and Moreno's model by considering starting failures. In recent years, a number of queueing models have been studied, details of which may be seen [10-15].

In practices, the server may well be subject to lengthy and unpredictable breakdowns while serving a customer. At the same time, there are some customers needing the subsidiary service provided by the server. For example, in manufacturing systems, the machine may break down due to machine or job related problems, and some products need deeper processing. Obviously, this manufacturing system constitutes a queueing system with the second optional service and server breakdowns. By using the supplementary variable method and the embedded Markov chain method, we obtain the probability generating functions (P.G.Fs) and the mean values of the system size and the waiting time. Additionally, we also get other important performance indicators, for example, the probabilities that the server is idle, busy, and under repair.

The remainder of this paper is organized as follows. A full description of the model is given in the Section 2. Some important measures performance of the system are obtained in the Section 3. In the Section 4, we give three special models of this model. In the Section 5, we present some numerical results to illustrate the effect of α on the performance of the system. Some conclusions are obtained in the last section.

2   Describing Model

Let us consider a discrete time single server queue that time axis is segmented into a sequence of equal time intervals (called slots). Further, let the time axis be marked by $0, 1, 2, \cdots$. Customers arrive according to a Bernoulli arrival process with rate λ. The first essential service is needed to all arriving customers. As soon as the first service of a customer is complete, he may opt for the second service with probability α or opt to leave the system with probability $1 - \alpha$. The service times of the essential and the second optional services are independent and arbitrarily distributed with probability mass distributions $f_j(j = 1, 2, \cdots)$ and $g_j(j = 1, 2, \cdots)$, and the probability generating functions $G_1(z)$ and $G_2(z)$, respectively. The corresponding mean service times and the second factorial moments are b_1, b_2 and $b_1^{(2)}$, $b_2^{(2)}$, respectively. The server may break down according to a Bernoulli process with rate β when he provides service to the customers, and when the server breaks down, it is sent for repair. When the breakdown occurs at the end of the essential service time, he will provide the essential service with probability $1 - \alpha$ or provide the second optional service with probability α after the server repaired, respectively. In order to avoid this kind of trouble, we assume the service slot which the breakdown happening is invalid. For example, when a breakdown arrives at a customer's j-th remaining service slots, then the remaining service is still j slots after repair. In other words, the service slot the breakdown happening is invalid. Therefore, the means of the actual essential service times and the actual second optional service times are $b_1 + \beta/\bar{\beta}$ and $b_2 + \beta/\bar{\beta}$, respectively.

The customer just being served before server breakdown waits for the server to complete its remaining service. The repair time independent and identically distributed with arbitrary distribution $s_j (j = 1, 2, \cdots)$, the generating function $S(z)$ and the mean repair time s. We assume that after repair the server is as good as new. Immediately after the server is repaired, it starts to serve customers, and the service time is cumulative. We assume various stochastic processes involved in the system are mutually independent.

This model has two deficiencies as following. On the one hand, the breakdowns arrive according to a Bernoulli arrival process with rate β both in the first essential service and the second optional service. On the other hand, the repair times both distributed with arbitrary distribution $s_j (j = 1, 2, \cdots)$.

Let C_n be the server state at time n and defined as follows:

$$C_n = \begin{cases} 0, & \text{if the server is free at time } n, \\ 1, & \text{if the server is busy providing a essential service at time } n, \\ 2, & \text{if the server is busy providing a second optional service at time } n, \\ 3, & \text{if the server is down providing a essential service at time } n, \\ 4, & \text{if the server is down providing a second optional service at time } n. \end{cases}$$

At time n^+, the system can be described by the process $\{L_n = C_n, \xi_n, \eta_n, N_n, n = 0, 1, 2, \cdots, \}$ where N_n denotes the number of customers in the system, if $C_n \in \{1, 2\}$, then ξ_n represents the remaining service time of the customer currently being served and if $C_n \in \{3, 4\}$, ξ_n represents the remaining service time of the customer waiting for the server to complete its remaining service, and η_n corresponds to the remaining repair time. $\{L_n, n = 1, 2, \cdots\}$ is the Markov chain of our queueing system, whose state space is

$$S = \{(0,0); (i,j,k) : i = 1, 2, j \geq 1, k \geq 1; (i,j,l,k) : i = 3, 4, j \geq 1, l \geq 1, k \geq 1\}.$$

We assume that the steady-state distribution exists for Markov chain $\{L_n, n = 1, 2, \ldots\}$ and that it is denoted by

$$\begin{cases} \pi_{0,0} = P(C_n = 0, N_n = 0), \\ \pi_{i,j,k} = P(C_n = i, \xi_n = j, N_n = k); \quad i = 1, 2, \ j \geq 1, \ k \geq 1, \\ \pi_{i,j,l,k} = P(C_n = i, \xi_n = j, \eta_n = l, N_n = k); \quad i = 3, 4, \ j \geq 1, \ l \geq 1, \ k \geq 1. \end{cases}$$

Considering mutually exclusive events they can occur during one slot, we have following Kolmogrov equations.

$$\begin{cases} \pi_{0,0} = \bar{\lambda}\pi_{0,0} + \bar{\lambda}\bar{\alpha}\bar{\beta}\pi_{1,1,1} + \bar{\lambda}\bar{\beta}\pi_{2,1,1}, \\ \pi_{1,j,1} = \lambda f_j \pi_{0,0} + \bar{\lambda}\bar{\beta}\pi_{1,j+1,1} + \bar{\lambda}\bar{\alpha}\bar{\beta}f_j\pi_{1,1,2} + \bar{\alpha}\lambda\bar{\beta}f_j\pi_{1,1,1} + \bar{\lambda}\bar{\beta}f_j\pi_{2,1,2} \\ \qquad + \lambda\bar{\beta}f_j\pi_{2,1,1} + \bar{\lambda}\pi_{3,j,1,1}, \\ \pi_{1,j,k} = \bar{\lambda}\bar{\alpha}\bar{\beta}f_j\pi_{1,1,k} + \bar{\lambda}\bar{\alpha}\bar{\beta}f_j\pi_{1,1,k+1} + \bar{\lambda}\bar{\beta}\pi_{1,j+1,k} + \bar{\lambda}\bar{\beta}\pi_{1,j+1,k-1} \\ \qquad + \lambda\bar{\beta}f_j\pi_{2,1,k} + \lambda\bar{\beta}f_j\pi_{2,1,k+1} + \bar{\lambda}\pi_{3,j,1,k} + \lambda\pi_{3,j,1,k-1}, \\ \pi_{2,j,1} = \alpha\lambda\bar{\beta}g_j\pi_{1,1,1} + \bar{\lambda}\bar{\beta}\pi_{2,j+1,1} + \bar{\lambda}\pi_{4,j,1,1}, \\ \pi_{2,j,k} = \alpha\lambda\bar{\beta}g_j\pi_{1,1,k} + \alpha\lambda\bar{\beta}g_j\pi_{1,1,k-1} + \bar{\lambda}\bar{\beta}\pi_{2,j+1,k} + \lambda\bar{\beta}\pi_{2,j+1,k-1} \\ \qquad + \bar{\lambda}\pi_{4,j,1,k} + \lambda\pi_{4,j,1,k-1}, \\ \pi_{3,j,l,1} = \bar{\lambda}\pi_{3,j,l+1,1} + \bar{\lambda}s_l\pi_{1,j,1}, \\ \pi_{3,j,l,k} = \bar{\lambda}\pi_{3,j,l+1,k} + \lambda\pi_{3,j,l+1,k-1} + \bar{\lambda}\beta s_l\pi_{1,j,k} + \lambda\beta s_l\pi_{1,j,k-1}, \\ \pi_{4,j,l,1} = \bar{\lambda}\pi_{4,j,l+1,1} + \bar{\lambda}s_l\pi_{2,j,1}, \\ \pi_{4,j,l,k} = \bar{\lambda}\pi_{4,j,l+1,k} + \lambda\pi_{4,j,l+1,k-1} + \bar{\lambda}\beta s_l\pi_{2,j,k} + \lambda\beta s_l\pi_{2,j,k-1}, \end{cases} \quad (1)$$

where $k \geq 2$, $j \geq 1$, $\bar{\lambda} = 1 - \lambda$, $\bar{\alpha} = 1 - \alpha$, $\bar{\beta} = 1 - \beta$.

3 Performance Analysis

In the section, we investigate some performance measures of the system in steady state. First of all, we should obtain the steady-state distribution. In order to find the steady-state distribution, let us define the following generating functions:

$$\Pi_1(x,z) = \sum_{j=1}^{\infty}\sum_{k=1}^{\infty}\pi_{1,j,k}x^j z^k, \Pi_2(x,z) = \sum_{j=1}^{\infty}\sum_{k=1}^{\infty}\pi_{2,j,k}x^j z^k,$$

$$\Pi_3(u,x,z) = \sum_{l=1}^{\infty}\sum_{j=1}^{\infty}\sum_{k=1}^{\infty}\pi_{3,j,l,k}u^l x^j z^k,$$

$$\Pi_4(u,x,z) = \sum_{l=1}^{\infty}\sum_{j=1}^{\infty}\sum_{k=1}^{\infty}\pi_{4,j,l,k}u^l x^j z^k,$$

and the auxiliary generating functions

$$\Pi_{1,j}(z) = \sum_{k=1}^{\infty}\pi_{1,j,k}z^k, \quad \Pi_{2,j}(z) = \sum_{k=1}^{\infty}\pi_{2,j,k}z^k, \quad \Pi_{3,j,l}(z) = \sum_{k=1}^{\infty}\pi_{3,j,l,k}z^k,$$

$$\Pi_{4,j,l}(z) = \sum_{k=1}^{\infty}\pi_{4,j,l,k}z^k, \quad \Pi_{3,l}(x,z) = \sum_{j=1}^{\infty}\sum_{k=1}^{\infty}\pi_{3,j,l,k}x^j z^k,$$

$$\Pi_{4,l}(x,z) = \sum_{j=1}^{\infty}\sum_{k=1}^{\infty}\pi_{4,j,l,k}x^j z^k.$$

Multiplying Eqs.(1) by z^k and summing over k, these equations become

$$\Pi_{1,j}(z) = (\bar{\lambda} + \lambda z)\bar{\beta}\Pi_{1,j+1}(z) + \frac{(\bar{\lambda} + \lambda z)\bar{\alpha}\bar{\beta}f_j}{z}\Pi_{1,1}(z) + \frac{(\bar{\lambda} + \lambda z)\bar{\beta}f_j}{z}\Pi_{2,1}(z)$$
$$+ z\lambda f_j\pi_{0,0} + (\bar{\lambda} + \lambda z)\Pi_{3,j,1}(z) - \lambda f_j\pi_{0,0}, \quad (2)$$

$$\Pi_{2,j}(z) = \alpha(\bar{\lambda} + \lambda z)\bar{\beta}g_j\Pi_{1,1}(z) + (\bar{\lambda} + \lambda z)\bar{\beta}\Pi_{2,j+1}(z) + (\bar{\lambda} + \lambda z)\Pi_{4,j,1}(z), \quad (3)$$

$$\Pi_{3,j,l}(z) = (\bar{\lambda} + \lambda z)\Pi_{3,j+1,l}(z) + (\bar{\lambda} + \lambda z)\beta s_l\Pi_{1,j}(z), \quad (4)$$

$$\Pi_{4,j,l}(z) = (\bar{\lambda} + \lambda z)\Pi_{4,j+1,l}(z) + (\bar{\lambda} + \lambda z)\beta s_l\Pi_{2,j}(z). \quad (5)$$

Then, multiplying Eqs.(2)~(5) by x^j and summing over j lead to

$$\frac{x - (\bar{\lambda} + \lambda z)\bar{\beta}}{x} \Pi_1(x, z) = \frac{(\bar{\lambda} + \lambda z)G_1(x)}{z} [\bar{\alpha}\bar{\beta}\Pi_{1,1}(z) + \bar{\beta}\Pi_{2,1}(z)]$$
$$+ \lambda \pi_{0,0} G_1(x)(z - 1) + (\bar{\lambda} + \lambda z)\Pi_{3,1}(x, z) - (\bar{\lambda} + \lambda z)\bar{\beta}\Pi_{1,1}(z), \quad (6)$$

$$\frac{x - (\bar{\lambda} + \lambda z)\bar{\beta}}{x} \Pi_2(x, z) = \alpha(\bar{\lambda} + \lambda z)\bar{\beta}\Pi_{1,1}(z)G_2(x) + (\bar{\lambda} + \lambda z)\Pi_{4,1}(x, z) - (\bar{\lambda} + \lambda z)\Pi_{2,1}(z), \quad (7)$$

$$\Pi_{3,l}(x, z) = (\bar{\lambda} + \lambda z)\Pi_{3,l+1}(x, z) + (\bar{\lambda} + \lambda z)\bar{\beta}s_l \Pi_1(x, z), \quad (8)$$

$$\Pi_{4,l}(x, z) = (\bar{\lambda} + \lambda z)\Pi_{4,l+1}(x, z) + (\bar{\lambda} + \lambda z)\bar{\beta}s_l \Pi_2(x, z). \quad (9)$$

We multiply Eqs.(6) and (7) by u^l and summing over l lead to

$$\frac{u - (\bar{\lambda} + \lambda z)}{u} \Pi_3(u, x, z) = (\bar{\lambda} + \lambda z)\bar{\beta}\Pi_1(x, z)S(u) - (\bar{\lambda} + \lambda z)\Pi_{3,1}(x, z), \quad (10)$$

$$\frac{u - (\bar{\lambda} + \lambda z)}{u} \Pi_4(u, x, z) = (\bar{\lambda} + \lambda z)\bar{\beta}\Pi_2(x, z)S(u) - (\bar{\lambda} + \lambda z)\Pi_{4,1}(x, z). \quad (11)$$

Choosing $u = (\bar{\lambda} + \lambda z)$ in Eqs.(10) and (11), we obtain

$$(\bar{\lambda} + \lambda z)\Pi_{3,1}(x, z) = (\bar{\lambda} + \lambda z)\bar{\beta}\Pi_1(x, z)S(\bar{\lambda} + \lambda z), \quad (12)$$

$$(\bar{\lambda} + \lambda z)\Pi_{4,1}(x, z) = (\bar{\lambda} + \lambda z)\bar{\beta}\Pi_2(x, z)S(\bar{\lambda} + \lambda z). \quad (13)$$

Then, substituting Eqs.(12) and (13) into Eqs.(6) and (7), respectively, we get

$$\frac{x[1 - (\bar{\lambda} + \lambda z)\beta S(\bar{\lambda} + \lambda z)] - (\bar{\lambda} + \lambda z)\bar{\beta}}{x} \Pi_1(x, z)$$
$$= \lambda \pi_{0,0} G_1(x)(z - 1) - (\bar{\lambda} + \lambda z)\bar{\beta}\Pi_{1,1}(z) + \frac{(\bar{\lambda} + \lambda z)G_1(x)}{z} [\bar{\alpha}\bar{\beta}\Pi_{1,1}(z) + \bar{\beta}\Pi_{2,1}(z)], \quad (14)$$

$$\frac{x[1 - (\bar{\lambda} + \lambda z)\beta S(\bar{\lambda} + \lambda z)] - (\bar{\lambda} + \lambda z)\bar{\beta}}{x} \Pi_2(x, z)$$
$$= (\bar{\lambda} + \lambda z)[\alpha\bar{\beta}\Pi_{1,1}(z)G_2(x) - \bar{\beta}\Pi_{2,1}(z)]. \quad (15)$$

Then, choosing

$$x = \frac{(\bar{\lambda} + \lambda z)\bar{\beta}}{1 - (\bar{\lambda} + \lambda z)\beta S(\bar{\lambda} + \lambda z)} = r(z)$$

in Eqs.(14) and (15), we have

$$(\bar{\lambda} + \lambda z)\bar{\beta}\Pi_{1,1}(z) = \lambda \pi_{0,0} G_1[r(z)](z - 1) + \frac{(\bar{\lambda} + \lambda z)G_1[r(z)]}{z} [\bar{\alpha}\bar{\beta}\Pi_{1,1}(z) + \bar{\beta}\Pi_{2,1}(z)], \quad (16)$$

$$(\bar{\lambda} + \lambda z)\bar{\beta}\Pi_{2,1}(z) = (\bar{\lambda} + \lambda z)\alpha\bar{\beta}\Pi_{1,1}(z)G_2[r(z)]. \quad (17)$$

From Eqs.(16) and (17), we obtain the probability generating functions as follows:

$$\Pi_{1,1}(z) = \frac{\lambda \pi_{0,0} G_1[r(z)] z(z-1)}{(\bar{\lambda} + \lambda z)\bar{\beta}\{z - G_1[r(z)][\bar{\alpha} + \alpha G_1[r(z)]]\}}, \tag{18}$$

$$\Pi_{2,1}(z) = \frac{\alpha \lambda \pi_{0,0} G_1[r(z)] G_2[r(z)] z(z-1)}{(\bar{\lambda} + \lambda z)\bar{\beta}\{z - G_1[r(z)][\bar{\alpha} + \alpha G_1[r(z)]]\}}. \tag{19}$$

Then, substituting Eqs.(18) and (19) into (14) and (15), we have

$$\Pi_1(x, z) = \frac{G_1(x) - G_1[r(z)]}{x[1 - (\bar{\lambda} + \lambda z)\beta S(\bar{\lambda} + \lambda z)] - (\bar{\lambda} + \lambda z)\bar{\beta}} \times \frac{\lambda z(1-z) x \pi_{0,0}}{G_1[r(z)][\bar{\alpha} + \alpha G_2[r(z)]] - z}, \tag{20}$$

$$\Pi_2(x, z) = \frac{G_2(x) - G_2[r(z)]}{x[1 - (\bar{\lambda} + \lambda z)\beta S(\bar{\lambda} + \lambda z)] - (\bar{\lambda} + \lambda z)\bar{\beta}} \times \frac{\alpha \lambda z(1-z) G_1[r(z)] x \pi_{0,0}}{G_1[r(z)][\bar{\alpha} + \alpha G_2[r(z)]] - z}. \tag{21}$$

Substituting Eqs.(20) and (21) into (12) and (13), respectively, we get

$$(\bar{\lambda} + \lambda z)\Pi_{3,1}(x, z) = \frac{\{G_1(x) - G_1[r(z)]\}(\bar{\lambda} + \lambda z) S(\bar{\lambda} + \lambda z)}{x[1 - (\bar{\lambda} + \lambda z)\beta S(\bar{\lambda} + \lambda z)] - (\bar{\lambda} + \lambda z)\bar{\beta}} \times \frac{\beta \lambda z(1-z) x \pi_{0,0}}{G_1[r(z)][\bar{\alpha} + \alpha G_2[r(z)]] - z}, \tag{22}$$

$$(\bar{\lambda} + \lambda z)\Pi_{4,1}(x, z) = \frac{\{G_2(x) - G_2[r(z)]\}(\bar{\lambda} + \lambda z) S(\bar{\lambda} + \lambda z)}{x[1 - (\bar{\lambda} + \lambda z)\beta S(\bar{\lambda} + \lambda z)] - (\bar{\lambda} + \lambda z)\bar{\beta}} \times \frac{\beta \alpha \lambda z(1-z) G_1[r(z)] x \pi_{0,0}}{G_1[r(z)][\bar{\alpha} + \alpha G_2[r(z)]] - z}. \tag{23}$$

Substituting Eqs.(22) and (23) into (10) and (11), respectively, and using (20) and (21), we have

$$\Pi_3(u, x, z) = \frac{\{G_1(x) - G_1[r(z)]\}[S(u) - S(\bar{\lambda} + \lambda z)]}{[u - (\bar{\lambda} + \lambda z)]\{x[1 - (\bar{\lambda} + \lambda z)\beta S(\bar{\lambda} + \lambda z)] - (\bar{\lambda} + \lambda z)\bar{\beta}\}} \times \frac{u(\bar{\lambda} + \lambda z)\beta \lambda z(1-z) x \pi_{0,0}}{G_1[r(z)][\bar{\alpha} + \alpha G_2[r(z)]] - z}, \tag{24}$$

$$\Pi_4(u, x, z) = \frac{\{G_2(x) - G_2[r(z)]\}[S(u) - S(\bar{\lambda} + \lambda z)]}{[u - (\bar{\lambda} + \lambda z)]\{x[1 - (\bar{\lambda} + \lambda z)\beta S(\bar{\lambda} + \lambda z)] - (\bar{\lambda} + \lambda z)\bar{\beta}\}} \times \frac{u(\bar{\lambda} + \lambda z)\alpha \beta \lambda z(1-z) G_1[r(z)] x \pi_{0,0}}{G_1[r(z)][\bar{\alpha} + \alpha G_2[r(z)]] - z}. \tag{25}$$

From the expressions $\Pi_1(x, z)$, $\Pi_2(x, z)$, $\Pi_3(u, x, z)$ and $\Pi_4(u, x, z)$, we can obtain theorem 1 as follows:

Theorem 1. In the steady state, we have the probability generating function of the system size is given by

$$\Pi(z) = \frac{G_1[r(z)]\{\bar{\alpha} + \alpha G_2[r(z)]\}(1-z)(1-\rho)}{G_1[r(z)]\{\bar{\alpha} + \alpha G_2[r(z)]\} - z},$$

where

$$r(z) = \frac{(\bar{\lambda} + \lambda z)\bar{\beta}}{1 - (\bar{\lambda} + \lambda z)\beta S(\bar{\lambda} + \lambda z)}, \quad \rho = \lambda b_1[1 + \frac{\beta}{\bar{\beta}}(1 + s)] + \alpha \lambda b_2[1 + \frac{\beta}{\bar{\beta}}(1 + s)].$$

Proof. Let $x = 1, u = 1$ in the (20), (21), (24), and (25), we obtain

$$\Pi_1(1, z) = \frac{1 - G_1[r(z)]}{1 - (\bar{\lambda} + \lambda z)\beta S(\bar{\lambda} + \lambda z) - (\bar{\lambda} + \lambda z)\bar{\beta}} \cdot \frac{\lambda z(1 - z)\pi_{0,0}}{G_1[r(z)][\bar{\alpha} + \alpha G_2[r(z)]] - z},$$

$$\Pi_2(1, z) = \frac{1 - G_2[r(z)]}{1 - (\bar{\lambda} + \lambda z)\beta S(\bar{\lambda} + \lambda z) - (\bar{\lambda} + \lambda z)\bar{\beta}} \cdot \frac{\alpha \lambda z(1 - z)G_1[r(z)]\pi_{0,0}}{G_1[r(z)][\bar{\alpha} + \alpha G_2[r(z)]] - z},$$

$$\Pi_3(1, 1, z) = \frac{\{1 - G_1[r(z)]\}[1 - S(\bar{\lambda} + \lambda z)]}{[1 - (\bar{\lambda} + \lambda z)][1 - (\bar{\lambda} + \lambda z)\beta S(\bar{\lambda} + \lambda z) - (\bar{\lambda} + \lambda z)\bar{\beta}]}$$
$$\times \frac{(\bar{\lambda} + \lambda z)\beta \lambda z(1 - z)\pi_{0,0}}{G_1[r(z)][\bar{\alpha} + \alpha G_2[r(z)]] - z},$$

$$\Pi_4(1, 1, z) = \frac{\{1 - G_2[r(z)]\}[1 - S(\bar{\lambda} + \lambda z)]}{[1 - (\bar{\lambda} + \lambda z)][1 - (\bar{\lambda} + \lambda z)\beta S(\bar{\lambda} + \lambda z) - (\bar{\lambda} + \lambda z)\bar{\beta}]}$$
$$\times \frac{(\bar{\lambda} + \lambda z)\alpha \beta \lambda z(1 - z)G_1[r(z)]\pi_{0,0}}{G_1[r(z)][\bar{\alpha} + \alpha G_2[r(z)]] - z}.$$

Then, summing $\Pi_1(1, z)$ $\Pi_2(1, z)$ $\Pi_3(1, 1, z)$ $\Pi_3(1, 1, z)$ and $\pi_{0,0}$, we have

$$\Pi(z) = \Pi_1(z) + \Pi_2(z) + \Pi_3(1, 1, z) + \Pi_4(1, 1, z) + \pi_{0,0}$$
$$= \frac{G_1[r(z)]\{\bar{\alpha} + \alpha G_2[r(z)]\}(1 - z)\pi_{0,0}}{G_1[r(z)]\{\bar{\alpha} + \alpha G_2[r(z)]\} - z}. \tag{26}$$

Using the normalization condition and L'Hospital rule, we obtain

$$1 = \frac{\pi_{0,0}}{1 - b_1 r'(1) - \alpha b_2 r'(1)}$$

where $r'(1) = \frac{\lambda + \beta \lambda s}{\bar{\beta}}$, and we have

$$\pi_{0,0} = 1 - \lambda b_1[1 + \frac{\beta}{\bar{\beta}}(1 + s)] - \alpha \lambda b_2[1 + \frac{\beta}{\bar{\beta}}(1 + s)].$$

Substituting $\pi_{0,0}$ into (26), we obtain

$$\Pi(z) = \frac{G_1[r(z)]\{\bar{\alpha} + \alpha G_2[r(z)]\}(1 - z)\{1 - \lambda b_1[1 + \frac{\beta}{\bar{\beta}}(1 + s)] - \alpha \lambda b_2[1 + \frac{\beta}{\bar{\beta}}(1 + s)]\}}{G_1[r(z)]\{\bar{\alpha} + \alpha G_2[r(z)]\} - z}.$$

From the expressions $\Pi_1(x, z), \Pi_2(x, z), \Pi_3(u, x, z)$ and $\Pi_4(u, x, z)$ and utilizing the Little's formula $E[W] = \frac{E[L]}{\lambda}$, we can obtain the following theorem.

Theorem 2. If $\rho < 1$, then we have
1) The stationary distribution of the system is given by

$$\pi_{0,0} = 1 - \rho, \quad \Pi_1(1, 1) = \rho_1, \quad \Pi_2(1, 1) = \rho_2, \quad \Pi_3(1, 1, 1) = \rho_3, \quad \Pi_4(1, 1, 1) = \rho_4,$$

where

$$\rho_1 = \lambda(b_1 + \frac{\beta}{\bar{\beta}}), \quad \rho_2 = \alpha\lambda(b_2 + \frac{\beta}{\bar{\beta}}), \quad \rho_3 = \lambda b_1 \frac{\beta}{\bar{\beta}}s, \quad \rho_4 = s\alpha\lambda b_2 \frac{\beta}{\bar{\beta}}.$$

2) The mean size of the system is given by

$$E[L] = \frac{1 - \rho + (b_1 + \alpha b_2)r''(1) + (b_1^{(2)} + 2b_1 b_2 + \alpha b_2^{(2)})[r'(1)]^2}{2(1 - \rho)}.$$

3) The mean time of a customer spends in the system is given by

$$E[W] = \frac{E[L]}{\lambda} = \frac{1 - \rho + (b_1 + \alpha b_2)r''(1) + (b_1^{(2)} + 2b_1 b_2 + \alpha b_2^{(2)})[r'(1)]^2}{2\lambda(1 - \rho)},$$

where

$$r''(1) = \frac{\lambda\beta[\bar{\beta}(2s + s^{(2)}) + 2\lambda(1 + s)(1 + \beta s)]}{\bar{\beta}^2}.$$

If denoted by L_q and W_q the mean number in the system excluding the customer in service and the mean waiting time in the system, respectively, and using the Little's formulas $L_q = E[L] - \rho, W_q = \frac{L_q}{\lambda}$, then we can obtain the following corollary.

Corollary 1. If $\rho < 1$, then the mean number in the system excluding the customer in service and the mean waiting time in the system will be given by, respectively,

$$L_q = \frac{1 - 3\rho + \rho^2 + (b_1 + \alpha b_2)r''(1) + (b_1^{(2)} + 2b_1 b_2 + \alpha b_2^{(2)})[r'(1)]^2}{2(1 - \rho)},$$

$$W_q = \frac{1 - 3\rho + \rho^2 + (b_1 + \alpha b_2)r''(1) + (b_1^{(2)} + 2b_1 b_2 + \alpha b_2^{(2)})[r'(1)]^2}{2\lambda(1 - \rho)}.$$

Corollary 2. If $\rho < 1$, then
1) the probability that the server is idle is

$$\pi_{0,0} = 1 - \lambda b_1[1 + \frac{\beta}{\bar{\beta}}(1 + s)] - \alpha\lambda b_2[1 + \frac{\beta}{\bar{\beta}}(1 + s)];$$

2) the probability that the server is busy is

$$P = (\lambda b_1 + \alpha\lambda b_2)(1 + \frac{\beta}{\bar{\beta}});$$

3) the probability that the server is under repair is

$$R = (\lambda b_1 + \alpha\lambda b_2)\frac{\beta}{\bar{\beta}}.$$

We denote by $A(n)$ the system availability at time n, i.e., the probability the system working at time t. According to the expressions $\Pi_1(x,z), \Pi_2(x,z)$, and $\pi_{0,0}$, it is easy to get the following theorem.

Theorem 3. If $\rho < 1$, then the probability generating function of the $A(n)$ is given by

$$A(x) = \pi_{0,0} + \Pi_1(x,1) + \Pi_2(x,1) = \frac{\bar{\beta}(1-x)(1-\rho) + \lambda x[1+\alpha - G_1(x) - \alpha G_2(x)]}{\bar{\beta}(1-x)}.$$

Corollary 3. The steady state availability of the server is given by

$$A = 1 - \frac{\beta \lambda b_1 s}{\bar{\beta}} - \frac{\alpha \beta \lambda b_2 s}{\bar{\beta}}.$$

4 Special Models

In the section, we present several special cases to study the system parameters on the mean size and the mean waiting times.

Case 1. If the essential service times, the second service times and repair times all are poisson distributions with rate θ_1, θ_2 and σ, respectively, then the mean size and the mean waiting times are given by

$$E[L] = \frac{1 - \rho + (\theta_1 + \alpha\theta_2)r''(1) + [\theta_1^2 + 2\theta_1\theta_2 + \alpha\theta_2^2][r'(1)]^2}{2(1-\rho)},$$

$$E[W] = \frac{1 - \rho + (\theta_1 + \alpha\theta_2)r''(1) + [\theta_1^2 + 2\theta_1\theta_2 + \alpha\theta_2^2][r'(1)]^2}{2\lambda(1-\rho)},$$

where

$$r'(1) = \frac{\lambda + \bar{\beta}\lambda\sigma}{\bar{\beta}}, \quad r''(1) = \frac{\lambda\beta[\bar{\beta}(2\sigma + \sigma^2) + 2\lambda(1+\sigma)(1+\beta\sigma)]}{\bar{\beta}^2},$$

$$\rho = \lambda\theta_1[1 + \frac{\beta}{\bar{\beta}}(1+\sigma)] + \alpha\lambda\theta_2[1 + \frac{\beta}{\bar{\beta}}(1+\sigma)].$$

Case 2. If the essential service times, the second service times and repair times all are constants, with these mean values c_1, c_2 and c_3, respectively, then the mean size and the mean waiting times are given by

$$E[L] = \frac{1 - \rho + (c_1 + \alpha c_2)r''(1) + (c_1^2 - c_1 + 2c_1c_2 + c_2^2 - c_2)[r'(1)]^2}{2(1-\rho)},$$

$$E[W] = \frac{1 - \rho + (c_1 + \alpha c_2)r''(1) + (c_1^2 - c_1 + 2c_1c_2 + c_2^2 - c_2)[r'(1)]^2}{2\lambda(1-\rho)},$$

where

$$r'(1) = \frac{\lambda + \bar{\beta}\lambda c_3}{\bar{\beta}}, \quad r''(1) = \frac{\lambda\beta[\bar{\beta}(c_3 + c_3^2) + 2\lambda(1+c_3)(1+\beta c_3)]}{\bar{\beta}^2},$$

$$\rho = \lambda c_1[1 + \frac{\beta}{\bar{\beta}}(1+c_3)] + \alpha\lambda c_2[1 + \frac{\beta}{\bar{\beta}}(1+c_3)].$$

5 Numerical Results

In this section, based on the special models obtained, we show some numerical examples to illustrate the influence of the probability α on the mean system size and waiting time. We consider the following cases:

Case 1. Choose $\alpha = 0.01, 0.2, 0.3$, $\beta = 0.1, \theta_1 = 0.8, \theta_2 = 0.8$ and $\sigma = 0.2$, vary the values of λ;

Case 2. Choose $\alpha = 0.01, 0.2, 0.3$, $\beta = 0.1, c_1 = 0.8, c_2 = 0.8$ and $c_3 = 0.2$, vary the values of λ.

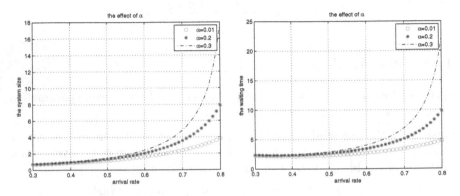

Fig. 1. Case 1 system size   **Fig. 2.** Case 1 waiting time

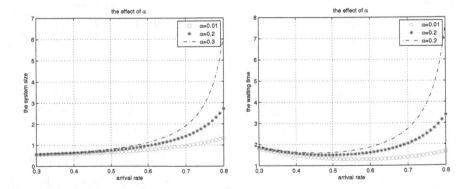

Fig. 3. Case 2 system size   **Fig. 4.** Case 2 waiting time

In the following Fig.1 and Fig.3, we note the $E[L]$ increases to a high level for all the cases; and as λ is fixed, $E[L]$ increases as α increases. From Fig.2 and Fig.4, we also see that $E[W]$ increases to a high level for all the cases, and as λ is fixed, $E[W]$ increases as α increases.

6 Conclusion

In this paper, we have studied discrete time *Geom/G/1* queue with the second optional service and server breakdowns. By using the supplementary variable method and embedded Markov chain method, we have obtained the PGFs and the expected values of the system size and the waiting time. Meanwhile, we get the probabilities that the server is idle, busy, and under repair. We also get the the steady state availability of the server. This model satisfies a lot of manufacturing systems with only production machine. For a manufacturing systems with multiple production machines, we can model a queueing system with multiple servers, and this model is our next works.

Acknowledgement. Thanks to the support by the National Natural Science Foundation of China (No. 71062008).

References

1. Bruneel, H., Kim, B.G.: Discrete-Time Models for Communication Systems Including ATM. Kluwer Academic Publishers, Boston (1993)
2. Woodward, M.E.: Communication and Computer Networks Modelling with Discrete-time Queues. IEEE Computer Society Press, Los Alamitos (1994)
3. Miyazawa, M., Takagi, H. (eds.): Advances in Discrete Queues. Queueing Systems, vol. 18 (1994)
4. Takagi, M.H.: Queueing Analysis: A Foundation of Performance Evaluation. In: Discrete-Time Systems, vol. 3. North-Holland, Amsterdam (1993)
5. Madan, K.C.: An M/G/1 Queue With Second Optional Service. Queueing Systems 34, 37–46 (2000)
6. Medhi, J.A.: A Single Server Poisson Input Queue with A Second Optional Channel. Queueing Systems 42, 239–242 (2002)
7. Wang, J.T.: An M/G/1 Queue with Second Optional Service and Server Breakdowns. Computers and Mathematics with Applications 47, 1713–1723 (2004)
8. Atencia, I., Moreno, P.: A Discrete-time Retrial Queue with 2nd Optional Service. In: Proceeding of Fifth International Workshop on Retrial Queues, pp. 117–121 (2004)
9. Wang, J.T., Zhao, Q.: A Discrete-time Geo/G/1 Retrial Queue with Starting Failures And Second Optional Service. Computers and Mathematics with Applications 53, 115–127 (2007)
10. Wang, J.T., Cao, J., Li, Q.: Reliability analysis of the retrial queue with a server subject to breakdowns and repairs. Queueing Systems 38, 363–380 (2001)
11. Wang, W.L., Xu, G.Q.: The well-posedness of an M/G/1 queue with second optional service and server breakdown. Computers and Mathematics with Applications 57(5), 729–739 (2009)
12. Wu, J.B., Liu, Z.M., Peng, Y.: On the BMAP/G/1 G-queues with second optional service and multiple vacations. Applied Mathematical Modeling 33(12), 4314–4325 (2009)

13. Hassan, A.A., Rabia, S.L., Taboly, F.A.: A discrete time Geo/G/1 retrial queue with general retrial times and balking customers. Journal of the Korean Statistical Society 37(4), 335–348 (2008)
14. Choudhury, G., Deka, K.: An M/G/1 retrial queueing system with two phases of service subject to the server breakown and repair. Performance Evaluation 65(10), 714–724 (2008)
15. Wang, K.H., Yang, D.Y., Pearn, W.L.: Comparison of two randomized policy M/G/1 queues with second optional service, server breakdown and startup. Computers and Mathematics with Applications 234(3), 812–824 (2010)

Fuzzy Risks Analysis Based on Similarity Measures of Generalized Fuzzy Numbers

Lian-sen Zhu and Ruo-ning Xu

School of Mathematics and Information Sciences, Guangzhou University,
Guangzhou 510006, China
zhuliansen@163.com

Abstract. In this paper, we present a novel method for fuzzy risk analysis based on a new similarity measure of generalized fuzzy numbers. Firstly, we present a new method to measure the degree of similarity between generalized fuzzy numbers based on the geometric distance, the center of gravity (COG) distance, the perimeter and the area. We also prove some properties of the proposed similarity measure and use an example to compare with the existing measures. The proposed similarity measure can overcome the drawbacks of the existing methods. Finally we apply the proposed similarity measure to deal with fuzzy risk analysis problems which provides a useful way for dealing with fuzzy risk analysis problems considered the degrees of confidence of decision makers' opinions.

Keywords: Fuzzy risk analysis, generalized fuzzy numbers, similarity measure, linguistic values

1   Introduction

It is obvious that decision makers are normally faced with the lack of precise information to assess the risk of a component made by a manufactory in an uncertain environment. Imprecise evaluations may be attributed to
   1) Unquantifiable information, 2) Incomplete information, 3) No obtainable information, and 4) Partial ignorance [1].
   In order to overcome this problem, the fuzzy set theory pioneered by Zadeh has been extensively used. Fuzzy numbers or fuzzy subsets of the real line R are applied to represent the fuzziness of evaluating values in fuzzy risk analysis problems. Many different approaches have been introduced in this field.
   Schmucker (1984) [11] presented a method for fuzzy risk analysis based on fuzzy number arithmetic operations. Kangari and Riggs (1989) [9] presented a method for constructing risk assessment by linguistic terms. Chen (1996) [7] presented methods for subjective mental workload assessment and fuzzy risk analysis. In recent years, some similarity measures between fuzzy numbers have been proposed. Hsieh and Chen (1999) [15] presented a similarity measure between fuzzy numbers using the "graded mean integration representation distance". Chen and Chen (2003) [5] combined the concepts of the geometric distance and the center of gravity (COG) distance to propose a similarity measure between

B.-Y. Cao and X.-J. Xie (Eds.): Fuzzy Engineering and Operations Research, AISC 147, pp. 569–587.
springerlink.com   © Springer-Verlag Berlin Heidelberg 2012

generalized fuzzy numbers. Wei and Chen (2009) [15] proposed a fuzzy risk analysis method based on a similarity measure with the concept of geometric distance, height and perimeters of generalized fuzzy numbers. Hejazi, Doostparast and Hosseimi (2011) [17] presented a fuzzy risk analysis method based on a similarity measure of generalized fuzzy numbers, which considered the area, perimeter, height and geometric distance of these kinds of fuzzy numbers.

In this paper we present a new method to calculate the degree of similarity measure between generalized fuzzy numbers based on the geometric distance, the center of gravity (COG) distance, the perimeter and the area of generalized fuzzy number. We also prove some properties of the proposed similarity measure. The proposed similarity measure can overcome the drawbacks of the existing methods (see Table1). Based on the proposed similarity measure, we present a new fuzzy risk analysis algorithm operation for dealing with fuzzy risk analysis problems which the values of the evaluating items to be represented by generalized trapezoidal fuzzy numbers. The proposed method provides us a useful way for handling fuzzy risk analysis problems which more realistic results can be obtained.

The rest of this paper is organized as follows: in Section 2, we briefly review the basic concepts of generalized fuzzy numbers and their arithmetic operations, the traditional COG point and introduce the simple center of gravity method (SCGM) to calculate the COG point; in Section 3, we briefly review some existing similarity measures of fuzzy numbers from Lee (1999), Hsieh and Chen (1999), Chen and Chen (2003), Wei and Chen (2009), Hejazi et al (2011); in Section 4, we present a new similarity measure between generalized fuzzy numbers and some properties are proved; comparative examples are presented in Section 5 to illustrate the advantage of the proposed method; in Section 6, we apply the proposed method to deal with fuzzy risk analysis; the conclusions is discussed in Section 7.

2   Preliminaries

In this section, we briefly review the concepts of generalized fuzzy numbers and their arithmetic operations, the traditional COG point and introduce the simple center of gravity method (SCGM) to calculate the COG of fuzzy numbers.

2.1   Generalized Fuzzy Numbers and Their Arithmetic Operations

A generalized fuzzy number $\tilde{A}$ is a fuzzy set of R, and its membership function $\tilde{A}(x)$ satisfies the following criteria:

1) λ-level set of $\tilde{A}$, denoted as $A_\lambda = \{x : \tilde{A}(x) \geq \lambda\}$, is a closed convex interval; and

2)$\exists x \in R$, such that $\tilde{A}(x) = \omega$.where ω denotes the height of the generalized fuzzy number $\tilde{A}$.

The set of all generalized fuzzy numbers is denoted as $\tilde{F}(R)$. From the definition, the λ-level set of fuzzy number $\tilde{A}$ can be represented as $A_\lambda = [a_1(\lambda), a_2(\lambda)]$ for all $\lambda \in (0, \omega]$.

In [3] and [4], Chen represented a generalized trapezoidal fuzzy nuber $\tilde{A}$ as $\tilde{A} = (a,b,c,d;\omega)$, where a, b, c, d are real crisp numbers, ω denotes the height of the generalized trapezoidal fuzzy number $\tilde{A}$, and $0 < \omega \leq 1$.The membership function $\tilde{A}(x)$ satisfies the following conditions:

1) $\tilde{A}(x)$ is a continuous mapping from the universe of discourse X to the closed interval $[0,1]$;

2)$\tilde{A}(x) = 0$, where $-\infty < x \leq a$;   3) $\tilde{A}(x)$ is strictly increasing on $[a,b]$;

4)$\tilde{A}(x) = \omega$, where $b \leq x \leq c$;   5) $\tilde{A}(x)$ is strictly decreasing on $[c,d]$;

6)$\tilde{A}(x) = 0$, where $d \leq x < +\infty$.

If $0 \leq a \leq b \leq c \leq d \leq 1$, then $\tilde{A}$ is called a standardized generalized trapezoidal fuzzy number. If $\omega = 1$, then $\tilde{A}$ becomes a traditional fuzzy number and can be reprsent $\tilde{A} = (a,b,c,d)$. If $b = c$, then $\tilde{A}$ is a generalized triangular fuzzy number. If $a = b$ and $c = d$, then $\tilde{A}$ is called a crisp interval. If $a = b = c = d$, then $\tilde{A}$ is a crisp-value. Logically, the opposite of $\tilde{A} = (a,b,c,d;\omega)$ can be given by $-\tilde{A} = (-d,-c,-b,-a;\omega)$.

In the following, we briefly review some arithmetic operation between fuzzy numbers [2]. Assume that $\tilde{A}_1$ and $\tilde{A}_2$ are generalized trapezoidal fuzzy numbers, where $\tilde{A}_1 = (a_1,b_1,c_1,d_1;\omega_1)$ and $\tilde{A}_2 = (a_2,b_2,c_2,d_2;\omega_2)$.Some arithmetic operations are shown as follows:

1) Generalized fuzzy numbers Addition $\oplus$:

$$\tilde{A}_1 \oplus \tilde{A}_2 = (a_1,b_1,c_1,d_1;\omega_1) \oplus (a_2,b_2,c_2,d_2;\omega_2)$$
$$= (a_1 + a_2, b_1 + b_2, c_1 + c_2, d_1 + d_2; \min(\omega_1,\omega_2)) \tag{1}$$

2) Generalized fuzzy numbers Subtraction $-$:

$$\tilde{A}_1 - \tilde{A}_2 = (a_1,b_1,c_1,d_1;\omega_1) - (a_2,b_2,c_2,d_2;\omega_2) \tag{2}$$
$$= (a_1 - a_2, b_1 - b_2, c_1 - c_2, d_1 - d_2; \min(\omega_1,\omega_2))$$

3) Generalized fuzzy numbers Multiplication $\otimes$:

$$\tilde{A}_1 \otimes \tilde{A}_2 = (a_1,b_1,c_1,d_1;\omega_1) \otimes (a_2,b_2,c_2,d_2;\omega_2) = (a,b,c,d;\min(\omega_1,\omega_2)) \tag{3}$$

$$a = min(a_1 \times a_2, a_1 \times d_2, d_1 \times a_2, d_1 \times d_2),$$
where $b = min(b_1 \times b_2, b_1 \times c_2, c_1 \times b_2, c_1 \times c_2),$
$$c = max(b_1 \times b_2, b_1 \times c_2, c_1 \times b_2, c_1 \times c_2),$$
$$d = max(a_1 \times a_2, a_1 \times d_2, d_1 \times a_2, d_1 \times d_2).$$

It is obvious that if $a_1, b_1, c_1, d_1, a_2, b_2, c_2, d_2$ are all positive-real numbers, then

$$\tilde{A}_1 \otimes \tilde{A}_2 = (a_1 \times a_2, b_1 \times b_2, c_1 \times c_2, d_1 \times d_2; min(\omega_1, \omega_2)) \qquad (3')$$

4) Generalized fuzzy numbers Division $\div$:

The inverse of the fuzzy number $\tilde{A}_2$ is $\frac{1}{\tilde{A}_2} = (\frac{1}{d_2}, \frac{1}{c_2}, \frac{1}{b_2}, \frac{1}{a_2}; \omega_2)$, where a_2, b_2, c_2 and d_2 are all nonzero positive real numbers or nonzero negative real numbers. Especially $a_1, b_1, c_1, d_1, a_2, b_2, c_2$ and d_2 are all nonzero positive real numbers, then the division is:

$$\tilde{A}_1 \div \tilde{A}_2 = (a_1 \div d_2, b_1 \div c_2, c_1 \div b_2, d_1 \div a_2; min(\omega_1, \omega_2)) \qquad (4)$$

2.2   Tradition and Simple Center of Gravity Method

The tradition COG method is very useful to deal with defuzzification problems and fuzzy rank problems.

Yager (1978) presented a method for calculating the center of gravity of a fuzzy number $\tilde{A}$, which formula is shown as follows:

$$x_{\tilde{A}}^* = \frac{\int x \mu_{\tilde{A}(x)} dx}{\int \mu_{\tilde{A}(x)} dx}, \qquad (5)$$

where $\mu_{\tilde{A}(x)}$ is the membership function of $\tilde{A}$ and $\mu_{\tilde{A}(x)} : x \rightarrow [0,1]$.

Chen (1998) presented a method for calculating the centric point of fuzzy numbers.

Let us consider a generalized fuzzy number $\tilde{A} = (a, b, c, d; \omega)$ with the membership function $f_{\tilde{A}}(x)$, shown as follows:

$$f_{\tilde{A}}(x) = \begin{cases} f_{\tilde{A}}^L(x), a \leq x \leq b \\ \omega, \quad b \leq x \leq c \\ f_{\tilde{A}}^R(x), c \leq x \leq d \\ 0, \quad otherwise \end{cases} \qquad (6)$$

where $f_{\tilde{A}}^L(x) : [a,b] \rightarrow [0,1]$ and $f_{\tilde{A}}^R(x) : [a,b] \rightarrow [0,1]$. Because $f_{\tilde{A}}^L(x)$ and $f_{\tilde{A}}^R(x)$ is continuous and monotonous, the inverse function $g_{\tilde{A}}^L(x)$ of $f_{\tilde{A}}^L(x)$ and $g_{\tilde{A}}^R(x)$ of $f_{\tilde{A}}^R(x)$ exists.

Then, the centric point $(x_{\tilde{A}}, y_{\tilde{A}})$ of the fuzzy number $\tilde{A}$ is defined as follows:

$$x_{\tilde{A}} = \frac{\int_{a}^{b}(xf_{\tilde{A}}^{L})dx + \int_{b}^{c}xdx + \int_{c}^{d}(xf_{\tilde{A}}^{R})dx}{\int_{a}^{b}(f_{\tilde{A}}^{L})dx + \int_{b}^{c}dx \int_{c}^{d}(f_{\tilde{A}}^{R})dx}, \tag{7}$$

$$y_{\tilde{A}} = \frac{\int_{0}^{\omega}(yg_{\tilde{A}}^{L})dy + \int_{0}^{\omega}yg_{\tilde{A}}^{R}dy}{\int_{0}^{\omega}(g_{\tilde{A}}^{L})dy + \int_{0}^{\omega}(g_{\tilde{A}}^{R})dy}. \tag{8}$$

Although the tradition COG method can be used to deal with defuzzification problems and fuzzy rank problems, there are some drawbacks, that is, we can see that it cannot directly calculate the COG of a crisp interval or a real number, namely, the denominators will be zero according to (5), (7) and (8).

Chen and Chen (2003) proposed the simple center of gravity method (SCGM) to overcome the drawbacks of tradition COG method which is based on the concept of the medium curve, shown as follows.

Assume that there is a generalized trapezoidal fuzzy number $\tilde{A} = (a,b,c,d;\omega)$, and then we can obtain the COG point of $\tilde{A}$ as follows:

$$y_{\tilde{A}}^{*} = \begin{cases} \dfrac{\omega \times (\dfrac{c-b}{d-a}+2)}{6}, & \text{if } a \neq d \text{ and } 0 < \omega \leq 1 \\ \dfrac{\omega}{2}, & \text{if } a = d \text{ and } 0 < \omega \leq 1 \end{cases} \tag{9}$$

$$x_{\tilde{A}}^{*} = \frac{y_{\tilde{A}}^{*}(b+c)+(a+d)(\omega-y_{\tilde{A}}^{*})}{2\omega}. \tag{10}$$

3   A Review of Existing Similarity Measures between Fuzzy Numbers

In the following, we briefly review existing methods of similarity measures between fuzzy numbers, shown as follows:

Chen (1996) presented a similarity measure between traditional trapezoidal fuzzy numbers $\tilde{A} = (a_1,a_2,a_3,a_4)$ and $\tilde{B}(b_1,b_2,b_3,b_4)$. Based on the geometric distance, where the degree of similarity $S(\tilde{A},\tilde{B})$ between the fuzzy numbers $\tilde{A}$ and $\tilde{B}$ can be calculated as follows:

$$S(\tilde{A},\tilde{B}) = 1 - \frac{\sum_{i=1}^{4}|a_i - b_i|}{4}, \tag{11}$$

where $S(\tilde{A},\tilde{B}) \in [0,1]$. If $\tilde{A}$ and $\tilde{B}$ are triangular fuzzy numbers, where $\tilde{A} = (a_1,a_2,a_3)$ and $\tilde{B}(b_1,b_2,b_3)$, the degree of similarity $S(\tilde{A},\tilde{B})$ between $\tilde{A}$ and $\tilde{B}$ can be calculated as follows:

$$S(\tilde{A},\tilde{B})=1-\frac{\sum_{i=1}^{3}\left|a_i-b_i\right|}{3}. \tag{12}$$

The lager the value of $S(\tilde{A},\tilde{B})$, the more the similarity between the fuzzy numbers $\tilde{A}$ and $\tilde{B}$.

Lee (1999) proposed a similarity measure between traditional trapezoidal fuzzy numbers, where the degree of similarity $S(\tilde{A},\tilde{B})$ between the fuzzy numbers

$\tilde{A}=(a_1,a_2,a_3,a_4)$ and $\tilde{B}(b_1,b_2,b_3,b_4)$ is calculated as follows:

$$S(\tilde{A},\tilde{B})=1-\frac{\left\|\tilde{A}-\tilde{B}\right\|_{lp}}{\left\|U\right\|}\times 4^{-1/p}, \tag{13}$$

where U is the universe of discourse

$$\left\|\tilde{A}-\tilde{B}\right\|_{lp}=(\sum_{i=1}^{4}\left(\left|a_i-b_i\right|\right)^{p})^{1/p}, \text{ and } \left\|U\right\|=\max(U)-\min(U). \tag{14}$$

The lager the value of $S(\tilde{A},\tilde{B})$, the more the similarity between the trapezoidal fuzzy numbers $\tilde{A}$ and $\tilde{B}$.

Hsieh and Chen (1999) presented a similarity measure using the "graded mean integration-representation distance", where the degree of similarity $S(\tilde{A},\tilde{B})$ between the fuzzy numbers $\tilde{A}$ and $\tilde{B}$ is calculated as follows:

$$S(\tilde{A},\tilde{B})=\frac{1}{1+d(\tilde{A},\tilde{B})}, \tag{15}$$

where $d(\tilde{A},\tilde{B})=\left|p(\tilde{A})-p(\tilde{B})\right|$ which $p(\tilde{A})$ and $p(\tilde{B})$ are the mean integration representation of $\tilde{A}$ and $\tilde{B}$, respectively. If $\tilde{A}$ and $\tilde{B}$ are trapezoidal fuzzy numbers, where $\tilde{A}=(a_1,a_2,a_3,a_4)$ and $\tilde{B}(b_1,b_2,b_3,b_4)$, then the mean integration representation $p(\tilde{A})$ and $p(\tilde{B})$ of $\tilde{A}$ and $\tilde{B}$ are defined as follows:

$$p(\tilde{A})=\frac{a_1+2a_2+2a_3+a_4}{6}, \tag{16}$$

$$p(\tilde{B})=\frac{b_1+2b_2+2b_3+b_4}{6}. \tag{17}$$

If $\tilde{A}$ and $\tilde{B}$ are triangular fuzzy numbers, where $\tilde{A}=(a_1,a_2,a_3)$ and $\tilde{B}(b_1,b_2,b_3)$, with the same machination we have:

$$p(\tilde{A})=\frac{a_1+4a_2+a_3}{6}, \tag{18}$$

$$p(\tilde{B})=\frac{b_1+4b_2+b_3}{6}. \tag{19}$$

The lager the value of $S(\tilde{A}, \tilde{B})$, the more the similarity between the fuzzy numbers $\tilde{A}$ and $\tilde{B}$.

Chen and Chen (2003) presented a similarity measure between generalized trapezoidal fuzzy numbers. According to (9) (10), they calculate the COG points $(x^*_{\tilde{A}}, y^*_{\tilde{A}})$ and $(x^*_{\tilde{B}}, y^*_{\tilde{B}})$ of the generalized trapezoidal fuzzy numbers $\tilde{A}$ and $\tilde{B}$ respectively, where $\tilde{A} = (a_1, a_2, a_3, a_4; \omega_{\tilde{A}})$ and $\tilde{B} = (b_1, b_2, b_3, b_4; \omega_{\tilde{B}})$ .then, the degree of similarity $S(\tilde{A}, \tilde{B})$ between the fuzzy numbers $\tilde{A}$ and $\tilde{B}$ is calculated as follows:

$$S(\tilde{A}, \tilde{B}) = (1 - \frac{\sum_{i=1}^{4}|a_i - b_i|}{4}) \times (1 - |x^*_{\tilde{A}} - x^*_{\tilde{B}}|)^{B(S^{\tilde{A}}, S^{\tilde{B}})} \times \frac{\min(y^*_{\tilde{A}}, y^*_{\tilde{B}})}{\max(y^*_{\tilde{A}}, y^*_{\tilde{B}})}, \qquad (20)$$

where $B(S_{\tilde{A}}, S_{\tilde{B}})$ is defined as follows:

$$B(S_{\tilde{A}}, S_{\tilde{B}}) = \begin{cases} 1, if S_{\tilde{A}} + S_{\tilde{B}} > 0 \\ 0, if S_{\tilde{A}} + S_{\tilde{B}} = 0 \end{cases}, \text{ and } S_{\tilde{A}} = a_4 - a_1, S_{\tilde{B}} = b_4 - b_1. \qquad (21)$$

The lager the value of $S(\tilde{A}, \tilde{B})$, the more the similarity between the generalized trapezoidal fuzzy numbers $\tilde{A}$ and $\tilde{B}$.

Wei and Chen (2009) proposed a method for calculating the similarity of two fuzzy numbers $\tilde{A} = (a_1, a_2, a_3, a_4; \omega_{\tilde{A}})$ and $\tilde{B} = (b_1, b_2, b_3, b_4; \omega_{\tilde{B}})$. If $0 \le a_1 \le a_2 \le a_3 \le a_4 \le 1$ and $0 \le b_1 \le b_2 \le b_3 \le b_4 \le 1$, then the degree of $S(\tilde{A}, \tilde{B})$ is calculated as follows:

$$S(\tilde{A}, \tilde{B}) = (1 - \frac{\sum_{i=1}^{4}|a_i - b_i|}{4}) \times \frac{\min(P(\tilde{A}), P(\tilde{B})) + \min(\omega_{\tilde{A}}, \omega_{\tilde{B}})}{\max(P(\tilde{A}), P(\tilde{B})) + \max(\omega_{\tilde{A}}, \omega_{\tilde{B}})}, \qquad (22)$$

where $S(\tilde{A}, \tilde{B}) \in [0,1]$, $P(\tilde{A})$ and $P(\tilde{B})$ are the perimeters of generalized trapezoidal fuzzy numbers of $\tilde{A}$ and $\tilde{B}$, respectively. Which are defined as follows:

$$P(\tilde{A}) = \sqrt{(a_1 - a_2)^2 + \omega_{\tilde{A}}^2} + \sqrt{(a_3 - a_4)^2 + \omega_{\tilde{A}}^2} + (a_3 - a_2) + (a_4 - a_1) \quad , \qquad (23)$$

$$P(\tilde{B}) = \sqrt{(b_1 - b_2)^2 + \omega_{\tilde{B}}^2} + \sqrt{(b_3 - b_4)^2 + \omega_{\tilde{B}}^2} + (b_3 - b_2) + (b_4 - b_1). \qquad (24)$$

The lager the value of $S(\tilde{A}, \tilde{B})$, the more the similarity between the generalized trapezoidal fuzzy numbers $\tilde{A}$ and $\tilde{B}$.

Hejazi et al (2011) presented a method for calculating the similarity of two fuzzy numbers $\tilde{A}$ and $\tilde{B}$ based on the modal delineated by Wei and Chen (2009), which is calculated as follows:

$$S(\tilde{A}, \tilde{B}) = (1 - \frac{\sum_{i=1}^{4}|a_i - b_i|}{4}) \times \frac{\min(P(\tilde{A}), P(\tilde{B}))}{\max(P(\tilde{A}), P(\tilde{B}))} \times \frac{\min(A(\tilde{A}), A(\tilde{B})) + \min(\omega_{\tilde{A}}, \omega_{\tilde{B}})}{\max(A(\tilde{A}), A(\tilde{B})) + \max(\omega_{\tilde{A}}, \omega_{\tilde{B}})} \quad , \qquad (25)$$

where $S(\tilde{A},\tilde{B}) \in [0,1]$, $P(\tilde{A})$, $P(\tilde{B})$ are the same as(23),(24),and $A(\tilde{A})$, $A(\tilde{B})$ are the areas of the two fuzzy numbers, which are calculated as follows:

$$A(\tilde{A}) = \frac{1}{2}\omega_{\tilde{A}}(a_3 - a_2 + a_4 - a_1), \tag{26}$$

$$A(\tilde{B}) = \frac{1}{2}\omega_{\tilde{B}}(b_3 - b_2 + b_4 - b_1). \tag{27}$$

The lager the value of $S(\tilde{A},\tilde{B})$, the more the similarity between the generalized trapezoidal fuzzy numbers $\tilde{A}$ and $\tilde{B}$.

4 New Similarity Measure between Generalized Trapezoidal Fuzzy Numbers

In this section, we present a new method to calculate the degree of similarity between generalized fuzzy numbers and prove five properties of the proposed similarity measure. Assume that there are two generalized trapezoidal fuzzy numbers $\tilde{A} = (a_1, a_2, a_3, a_4; \omega_{\tilde{A}})$ and $\tilde{B} = (b_1, b_2, b_3, b_4; \omega_{\tilde{B}})$ in the universe of discourse X, where $0 \le a_1 \le a_2 \le a_3 \le a_4 < +\infty(-\infty < a_1 \le a_2 \le a_3 \le a_4 \le 0)$ and $0 \le b_1 \le b_2 \le b_3 \le b_4 < +\infty(-\infty < b_1 \le b_2 \le b_3 \le b_4 \le 0)$. The degree of similarity between $\tilde{A}$ and $\tilde{B}$ can be calculated as follows:

$$S(\tilde{A},\tilde{B}) = \begin{cases} \sqrt{[1 - d_g(\tilde{A}',\tilde{B}')] \times [1 - d_{COG}(\tilde{A}',\tilde{B}')]} \times \frac{\min(P(\tilde{A}'),P(\tilde{B}'))}{\max(P(\tilde{A}'),P(\tilde{B}'))} \\ \qquad \times e^{\frac{\min(A(\tilde{A}'),A(\tilde{B}'))}{\max(A(\tilde{A}'),A(\tilde{B}'))}-1}, \max(A(\tilde{A}'),A(\tilde{B}')) \neq 0 \\ \sqrt{[1 - d_g(\tilde{A}',\tilde{B}')] \times [1 - d_{COG}(\tilde{A}',\tilde{B}')]}, \max(A(\tilde{A}'),A(\tilde{B}')) = 0 \end{cases} \tag{28}$$

The method for calculating the degree of similarity $S(\tilde{A},\tilde{B})$ is now presented as follows:

Step 1: standardize each generalized trapezoidal fuzzy numbers $\tilde{A}$, $\tilde{B}$ into standardized generalized trapezoidal fuzzy numbers $\tilde{A}'$ and $\tilde{B}'$, shown as follows:

$$\tilde{A}' = (\frac{a_1}{k}, \frac{a_2}{k}, \frac{a_3}{k}, \frac{a_4}{k}; \omega_{\tilde{A}}) = (a_1', a_2', a_3', a_4'; \omega_{\tilde{A}}), \tag{29}$$

$$\tilde{B}' = (\frac{b_1}{k}, \frac{b_2}{k}, \frac{b_3}{k}, \frac{b_4}{k}; \omega_{\tilde{B}}) = (b_1', b_2', b_3', b_4'; \omega_{\tilde{A}}), \tag{30}$$

$$k = \max(\lceil |a_i| \rceil, \lceil |b_i| \rceil, 1), \tag{31}$$

where $|a_i|$, $|b_i|$ denote the absolute value of a_i, b_i, $\lceil |a_i| \rceil$, $\lceil |b_i| \rceil$ denote taking the upper bound of $|a_i|, |b_i|$ respectively and $1 \le i \le 4$.

Step 2: calculate the geometric distance between two standardized generalized trapezoidal fuzzy numbers $\tilde{A}'$ and $\tilde{B}'$, shown as follows:

$$d_g(\tilde{A}',\tilde{B}') = \frac{\sum_{i=1}^{4}|a_i' - b_i'|}{4}, 1 \leq i \leq 4 \cdot \tag{32}$$

Step 3: based on (9) (10), we can get COG point $(x_{\tilde{A}'}^*, y_{\tilde{A}'}^*)$, $(x_{\tilde{B}'}^*, y_{\tilde{B}'}^*)$ of $\tilde{A}'$, $\tilde{B}'$ respectively. Calculate the distance between them, shown as follows:

$$d_{COG}(\tilde{A}',\tilde{B}') = \sqrt{(x_{\tilde{A}'}^* - x_{\tilde{B}'}^*)^2 + (y_{\tilde{A}'}^* - y_{\tilde{B}'}^*)^2} \cdot \tag{33}$$

Step 4: Calculate the perimeter and area of standardized generalized trapezoidal fuzzy numbers $\tilde{A}'$, $\tilde{B}'$ respectively according to (23) (24) (26) (27).

Step 4 then the degree of similarity between $\tilde{A}$ and $\tilde{B}$ can be calculated as (28):

$$S(\tilde{A},\tilde{B}) = \begin{cases} \sqrt{[1 - d_g(\tilde{A}',\tilde{B}')] \times [1 - d_{COG}(\tilde{A}',\tilde{B}')]} \times \frac{\min(P(\tilde{A}), P(\tilde{B}))}{\max(P(\tilde{A}), P(\tilde{B}))} \\ \qquad \times e^{\frac{\min(A(\tilde{A}'), A(\tilde{B}'))}{\max(A(\tilde{A}'), A(\tilde{B}'))} - 1}, \max(A(\tilde{A}'), A(\tilde{B}')) \neq 0 \\ \sqrt{[1 - d_g(\tilde{A}',\tilde{B}')] \times [1 - d_{COG}(\tilde{A}',\tilde{B}')]}, \max(A(\tilde{A}'), A(\tilde{B}')) = 0 \end{cases}$$

The proposed similarity measure $S(\tilde{A},\tilde{B})$ integrates the concepts of the geometric distance, the center-of-gravity distance, the perimeter and the area. If the standardized generalized trapezoidal fuzzy numbers $\tilde{A}'$ and $\tilde{B}'$ are real numbers (i.e. $\max(A(\tilde{A}'), A(\tilde{B}')) = 0$), then we don't consider the perimeters and the areas. If either $\tilde{A}'$ or $\tilde{B}'$ is a generalized trapezoidal fuzzy number (i.e. $\max(A(\tilde{A}'), A(\tilde{B}')) \neq 0$), then we must consider the perimeters and the areas. The lager the value of $S(\tilde{A},\tilde{B})$, the more the similarity between the generalized trapezoidal fuzzy numbers $\tilde{A}$ and $\tilde{B}$.

Let $\tilde{A} = (a_1, a_2, a_3, a_4; \omega_{\tilde{A}})$ and $\tilde{B} = (b_1, b_2, b_3, b_4; \omega_{\tilde{B}})$ be two standardized generalized trapezoidal fuzzy numbers (otherwise we can translate them into standardized according to Step 1. The proposed similarity measure has the following properties:

Property 4.1. $S(\tilde{A},\tilde{B}) \in [0,1]$.

Proof: Obviously, there is $0 \leq \frac{\sum_{i=1}^{4}|a_i - b_i|}{4} \leq 1$, $0 \leq \frac{\min(P(\tilde{A}), P(\tilde{B}))}{\max(P(\tilde{A}), P(\tilde{B}))} \leq 1$ and $0 \leq \frac{\min(A(\tilde{A}), A(\tilde{B}))}{\max(A(\tilde{A}), A(\tilde{B}))} \leq 1$.

$$0 \le y_{\tilde{A}}^{*} = \frac{\omega \times (\frac{a_3 - a_2}{a_4 - a_1} + 2)}{6} \le \frac{\omega_{\tilde{A}} \times (1+2)}{6} = \frac{\omega_{\tilde{A}}}{2} \le 0.5,$$

$$0 \le x_{\tilde{A}}^{*} = \begin{cases} \dfrac{y_{\tilde{A}}^{*}(a_3 + a_2) + (a_4 + a_1)(\omega_{\tilde{A}} - y_{\tilde{A}}^{*})}{2\omega_{\tilde{A}}}, (a_3 + a_2 - a_4 - a_1) > 0, 0 \le y_{\tilde{A}}^{*} \le \dfrac{\omega_{\tilde{A}}}{2} \\[2mm] \le (a_4 + a_1) \cdot \omega_{\tilde{A}} + (a_3 + a_2 - a_4 - a_1) \cdot \dfrac{\omega_{\tilde{A}}}{2} \\[2mm] = \dfrac{a_1 + a_2 + a_3 + a_4}{4} \le 1 \\[2mm] \dfrac{(a_4 + a_1)\omega_{\tilde{A}}}{2\omega_{\tilde{A}}} = \dfrac{(a_4 + a_1)}{2} \le 1, \quad (a_3 + a_2 - a_4 - a_1) > 0, 0 \le y_{\tilde{A}}^{*} \le \dfrac{\omega_{\tilde{A}}}{2} \end{cases}.$$

That is $0 \le x_{\tilde{A}}^{*} \le 1$. For the same reason, there are $0 \le y_{\tilde{B}}^{*} \le 0.5$ and $0 \le x_{\tilde{B}}^{*} \le 1$.

So, $0 \le d_{COG}(\tilde{A}, \tilde{B}) \le 1$. Then

$$S(\tilde{A}, \tilde{B}) = \begin{cases} \sqrt{\left[1 - d_g(\tilde{A}', \tilde{B}')\right] \times \left[1 - d_{COG}(\tilde{A}', \tilde{B}')\right]} \times \dfrac{\min(P(\tilde{A}'), P(\tilde{B}'))}{\max(P(\tilde{A}'), P(\tilde{B}'))} \\ \qquad \times e^{\frac{\min(A(\tilde{A}'), A(\tilde{B}'))}{\max(A(\tilde{A}'), A(\tilde{B}'))} - 1} \\ \qquad \le \sqrt{(1-0) \times (1-0)} \cdot 1 \cdot e^{1-1} = 1, \ \max(A(\tilde{A}'), A(\tilde{B}')) \ne 0 \\[2mm] \sqrt{\left[1 - d_g(\tilde{A}', \tilde{B}')\right] \times \left[1 - d_{COG}(\tilde{A}', \tilde{B}')\right]} \le \sqrt{(1-0) \times (1-0)} = 1, \max(A(\tilde{A}'), A(\tilde{B}')) = 0 \end{cases},$$

$$S(\tilde{A}, \tilde{B}) = \begin{cases} \sqrt{\left[1 - d_g(\tilde{A}', \tilde{B}')\right] \times \left[1 - d_{COG}(\tilde{A}', \tilde{B}')\right]} \times \dfrac{\min(P(\tilde{A}'), P(\tilde{B}'))}{\max(P(\tilde{A}'), P(\tilde{B}'))} \times e^{\frac{\min(A(\tilde{A}'), A(\tilde{B}'))}{\max(A(\tilde{A}'), A(\tilde{B}'))} - 1} \\ \qquad \ge \sqrt{(1-1) \times (1-1)} \cdot 1 \cdot e^{1-1} = 0, \ \max(A(\tilde{A}'), A(\tilde{B}')) \ne 0 \\[2mm] \sqrt{\left[1 - d_g(\tilde{A}', \tilde{B}')\right] \times \left[1 - d_{COG}(\tilde{A}', \tilde{B}')\right]} \le \sqrt{(1-1) \times (1-0)} = 0, \max(A(\tilde{A}'), A(\tilde{B}')) = 0 \end{cases}.$$

Therefore there is $S(\tilde{A}, \tilde{B}) \in [0,1]$.

Property 4.2. $S(\tilde{A}, \tilde{B}) = 1$ if and only if $\tilde{A} = \tilde{B}$.

Proof: 1) If $\tilde{A}$ and $\tilde{B}$ are identical, then $\tilde{A} = \tilde{B}$, that is, $a_i = b_i, 1 \le i \le 4$, $x_{\tilde{A}}^{*} = x_{\tilde{B}}^{*}$, $y_{\tilde{A}}^{*} = y_{\tilde{B}}^{*}$ and $\omega_{\tilde{A}} = \omega_{\tilde{B}}$. Thus, $\sum\limits_{i=1}^{4} |a_i - b_i| / 4 = 0$, $d_{COG}(\tilde{A}', \tilde{B}') = 0$, $\min(P(\tilde{A}), P(\tilde{B})) = \max(P(\tilde{A}), P(\tilde{B}))$ and $\min(A(\tilde{A}), A(\tilde{B})) = \max(A(\tilde{A}), A(\tilde{B}))$. The degree of similarity between $\tilde{A}$ and $\tilde{B}$ is calculated as follows:

$$S(\tilde{A},\tilde{B}) = \begin{cases} \sqrt{\left[1-d_g(\tilde{A}',\tilde{B}')\right]\times\left[1-d_{COG}(\tilde{A}',\tilde{B}')\right]}\times\dfrac{\min(P(\tilde{A}'),P(\tilde{B}'))}{\max(P(\tilde{A}'),P(\tilde{B}'))} \\ \qquad \times e^{\frac{\min(A(\tilde{A}'),A(\tilde{B}'))}{\max(A(\tilde{A}'),A(\tilde{B}'))}-1} \\ \quad = \sqrt{(1-0)\times(1-0)}\cdot 1\cdot e^{1-1} = 1,\ \max(A(\tilde{A}'),A(\tilde{B}')) \neq 0 \\ \sqrt{\left[1-d_g(\tilde{A}',\tilde{B}')\right]\times\left[1-d_{COG}(\tilde{A}',\tilde{B}')\right]} = \sqrt{(1-0)\times(1-0)} = 1,\ \max(A(\tilde{A}'),A(\tilde{B}')) = 0 \end{cases}.$$

2) If $S(\tilde{A},\tilde{B}) = 1$, then

$$S(\tilde{A},\tilde{B}) = \begin{cases} \sqrt{\left[1-d_g(\tilde{A}',\tilde{B}')\right]\times\left[1-d_{COG}(\tilde{A}',\tilde{B}')\right]}\times\dfrac{\min(P(\tilde{A}'),P(\tilde{B}'))}{\max(P(\tilde{A}'),P(\tilde{B}'))}\times e^{\frac{\min(A(\tilde{A}'),A(\tilde{B}'))}{\max(A(\tilde{A}'),A(\tilde{B}'))}-1} = 1, \\ \qquad\qquad\qquad \max(A(\tilde{A}'),A(\tilde{B}')) \neq 0 \\ \sqrt{\left[1-d_g(\tilde{A}',\tilde{B}')\right]\times\left[1-d_{COG}(\tilde{A}',\tilde{B}')\right]} = 1,\ \max(A(\tilde{A}'),A(\tilde{B}')) = 0. \end{cases}$$

It implies that $a_i = b_i, 1 \leq i \leq 4$, $x_{\tilde{A}}^* = x_{\tilde{B}}^*$ $y_{\tilde{A}}^* = y_{\tilde{B}}^*$ and $\omega_{\tilde{A}} = \omega_{\tilde{B}}$. Thus, $\sum_{i=1}^{4}|a_i - b_i|\big/4 = 0$, $d_{COG}(\tilde{A}',\tilde{B}') = 0$, $\min(P(\tilde{A}),P(\tilde{B})) = \max(P(\tilde{A}),P(\tilde{B}))$ and $\min(A(\tilde{A}),A(\tilde{B})) = \max(A(\tilde{A}),A(\tilde{B}))$. Therefore $\tilde{A} = \tilde{B}$, that is, $\tilde{A}$ and $\tilde{B}$ are identical.

Property 4.3. $S(\tilde{A},\tilde{B}) = S(\tilde{B},\tilde{A})$.

Proof: It is obviously that the proposition is true.

Property 4.4. $S(-\tilde{A},-\tilde{B}) = S(\tilde{A},\tilde{B})$.

Proof: $-\tilde{A} = (-a_4,-a_3,-a_2,-a_1;\omega_{\tilde{A}}), -\tilde{B} = (-b_4,-b_3,-b_2,-b_1;\omega_{\tilde{B}})$ .Then we get

$$d_g(-\tilde{A},-\tilde{B}) = 1 - \frac{\sum_{i=1}^{4}|-a_i-(-b_i)|}{4} = 1 - \frac{\sum_{i=1}^{4}|-a_i+b_i|}{4} = 1 - \frac{\sum_{i=1}^{4}|a_i-b_i|}{4} = d_g(\tilde{A},\tilde{B})$$
,

$$y_{-\tilde{A}}^* = \begin{cases} \dfrac{\omega_{\tilde{A}}\times\left[\dfrac{-a_3-(-a_2)}{-a_4-(-a_1)}+2\right]}{6}, & \text{if } -a_4 \neq -a_1 \text{ and } 0 \leq \omega_{\tilde{A}} \leq 1 \\ \dfrac{\omega_{\tilde{A}}}{2}, & \text{if } -a_4 = -a_1 \text{ and } 0 \leq \omega_{\tilde{A}} \leq 1 \end{cases}$$

$$= \begin{cases} \dfrac{\omega_{\tilde{A}}\times\left[\dfrac{a_3-a_2}{a_4-a_1}+2\right]}{6}, & if a_4 \neq a_1 \text{ and } 0 \leq \omega_{\tilde{A}} \leq 1 \\ \dfrac{\omega_{\tilde{A}}}{2}, & if a_4 = a_1 \text{ and } 0 \leq \omega_{\tilde{A}} \leq 1 \end{cases}$$
,

$$= y_{\tilde{A}}^* ,$$

$$x^*_{-\tilde{A}} = \frac{y^*_{-\tilde{A}} \cdot [-a_3 + (-a_2)] + [-a_4 + (-a_1)] \cdot (\omega_{\tilde{A}} - y^*_{-\tilde{A}})}{2\omega_{\tilde{A}}}$$

$$= \frac{-[y^*_{\tilde{A}} \cdot (a_3 - a_2) + (a_4 - a_1) \cdot (\omega_{\tilde{A}} - y^*_{\tilde{A}})]}{2\omega_{\tilde{A}}}$$

$$= -x^*_{\tilde{A}}.$$

For the same reason, we have $x^*_{-\tilde{B}} = -x^*_{\tilde{B}}$, $y^*_{-\tilde{B}} = y^*_{\tilde{B}}$.

$$d_{COG}(-\tilde{A}, -\tilde{B}) = \sqrt{(x^*_{-\tilde{A}} - x^*_{-\tilde{B}})^2 + (y^*_{-\tilde{A}} - y^*_{-\tilde{B}})^2}$$

Then

$$= \sqrt{(x^*_{\tilde{A}} - x^*_{\tilde{B}})^2 + (y^*_{\tilde{A}} - y^*_{\tilde{B}})^2}$$

$$= d_{COG}(\tilde{A}, \tilde{B}).$$

Obviously, there is $P(-\tilde{A}) = P(\tilde{A}), P(-\tilde{B})) = P(\tilde{B})$, $A(-\tilde{A}) = A(\tilde{A})$ and $A(-\tilde{B})) = A(\tilde{B})$. Therefore, there is $S(-\tilde{A}, -\tilde{B}) = S(\tilde{A}, \tilde{B})$.

Property 4.5. If $\tilde{A}$ and $\tilde{B}$ are two generalized trapezoidal fuzzy numbers with the same geometric shape, the same height and the same offset d, where $d = |a_i - b_i|$, then $S(\tilde{A}, \tilde{B}) = 1 - d_g(\tilde{A}, \tilde{B})$.

Proof. Since $d = |a_i - b_i|$, due to (23) (24) (26) (27) (32) (33), we have:

$$\min(P(\tilde{A}), P(\tilde{B})) = \max(P(\tilde{A}), P(\tilde{B})), \min(A(\tilde{A}), A(\tilde{B})) = \max(A(\tilde{A}), A(\tilde{B})),$$

$$y^*_{\tilde{A}} = y^*_{\tilde{B}}, x^*_{\tilde{A}} = x^*_{\tilde{B}} + d.$$

so, the degree of similarity measure between $\tilde{A}$ and $\tilde{B}$ is as follows:

$$S(\tilde{A}, \tilde{B}) = \begin{cases} \sqrt{[1 - d_g(\tilde{A}, \tilde{B})] \times [1 - d_{COG}(\tilde{A}, \tilde{B})]} \times \dfrac{\min(P(\tilde{A}), P(\tilde{B}))}{\max(P(\tilde{A}), P(\tilde{B}))} \\ \quad \times e^{\frac{\min(A(\tilde{A}), A(\tilde{B}))}{\max(A(\tilde{A}), A(\tilde{B}))} - 1}, \max(A(\tilde{A}), A(\tilde{B})) \neq 0 \\ \sqrt{[1 - d_g(\tilde{A}, \tilde{B})] \times [1 - d_{COG}(\tilde{A}, \tilde{B})]}, \max(A(\tilde{A}), A(\tilde{B})) = 0 \end{cases}$$

$$= 1 - d.$$

Therefore, there is $S(\tilde{A}, \tilde{B}) = 1 - d_g(\tilde{A}, \tilde{B})$.

5   A Comparison of Similarity Measures

In this section, we make a comparison between the proposed similarity measure methods with other six methods using 18 sets of fuzzy numbers which are shown in Fig1 [5][15] [17] and then we will analyze the quality of the results of each method.

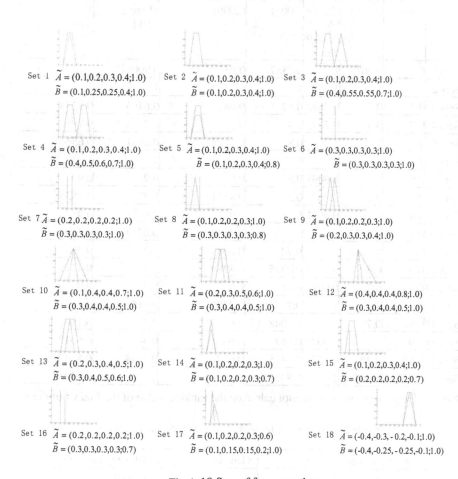

Set 1 $\tilde{A} = (0.1,0.2,0.3,0.4;1.0)$
$\tilde{B} = (0.1,0.25,0.25,0.4;1.0)$

Set 2 $\tilde{A} = (0.1,0.2,0.3,0.4;1.0)$
$\tilde{B} = (0.1,0.2,0.3,0.4;1.0)$

Set 3 $\tilde{A} = (0.1,0.2,0.3,0.4;1.0)$
$\tilde{B} = (0.4,0.55,0.55,0.7;1.0)$

Set 4 $\tilde{A} = (0.1,0.2,0.3,0.4;1.0)$
$\tilde{B} = (0.4,0.5,0.6,0.7;1.0)$

Set 5 $\tilde{A} = (0.1,0.2,0.3,0.4;1.0)$
$\tilde{B} = (0.1,0.2,0.3,0.4;0.8)$

Set 6 $\tilde{A} = (0.3,0.3,0.3,0.3;1.0)$
$\tilde{B} = (0.3,0.3,0.3,0.3;1.0)$

Set 7 $\tilde{A} = (0.2,0.2,0.2,0.2;1.0)$
$\tilde{B} = (0.3,0.3,0.3,0.3;1.0)$

Set 8 $\tilde{A} = (0.1,0.2,0.2,0.3;1.0)$
$\tilde{B} = (0.3,0.3,0.3,0.3;0.8)$

Set 9 $\tilde{A} = (0.1,0.2,0.2,0.3;1.0)$
$\tilde{B} = (0.2,0.3,0.3,0.4;1.0)$

Set 10 $\tilde{A} = (0.1,0.4,0.4,0.7;1.0)$
$\tilde{B} = (0.3,0.4,0.4,0.5;1.0)$

Set 11 $\tilde{A} = (0.2,0.3,0.5,0.6;1.0)$
$\tilde{B} = (0.3,0.4,0.4,0.5;1.0)$

Set 12 $\tilde{A} = (0.4,0.4,0.4,0.8;1.0)$
$\tilde{B} = (0.3,0.4,0.4,0.5;1.0)$

Set 13 $\tilde{A} = (0.2,0.3,0.4,0.5;1.0)$
$\tilde{B} = (0.3,0.4,0.5,0.6;1.0)$

Set 14 $\tilde{A} = (0.1,0.2,0.2,0.3;1.0)$
$\tilde{B} = (0.1,0.2,0.2,0.3;0.7)$

Set 15 $\tilde{A} = (0.1,0.2,0.3,0.4;1.0)$
$\tilde{B} = (0.2,0.2,0.2,0.2;0.7)$

Set 16 $\tilde{A} = (0.2,0.2,0.2,0.2;1.0)$
$\tilde{B} = (0.3,0.3,0.3,0.3;0.7)$

Set 17 $\tilde{A} = (0.1,0.2,0.2,0.3;0.6)$
$\tilde{B} = (0.1,0.15,0.15,0.2;1.0)$

Set 18 $\tilde{A} = (-0.4,-0.3,-0.2,-0.1;1.0)$
$\tilde{B} = (-0.4,-0.25,-0.25,-0.1;1.0)$

Fig. 1. 18 Sets of fuzzy numbers

Table 1. A comparison of the calculation results of the proposed similarity measure with existing methods

Sets	Methods							
	Chen's Method (1996)	Lee's Method (1999)	Hsieh and Chen's Method1999)	Chen&Chen's Method (2001)	Wei and Chen's Method (2009)	Hejazial's Method (2011)	.et	The proposed method
Sets 1	0.975	0.9617	1	0.8357	0.95	0.900411		0.72017
Sets 2	1	1	1	1	1	1		1
Sets 3	0.7	0.5	0.7692	0.42	0.682	0.6465		0.5234
Sets 4	0.7	0.5	0.7692	0.49	0.7	0.7		0.7
Sets 5	1	1	1	0.8	0.8428	0.6681		0.6566
Sets 6	1	*	1	1	1	1		1
Sets 7	0.9	0	0.9091	0.81	0.9	0.9		0.9
Sets 8	0.9	0.5	0.9091	0.54	0.8411	0.37		0.2835
Sets 9	0.9	0.6667	0.9091	0.81	0.9	0.9		0.9
Sets 10	0.9	0.8333	1	0.9	0.7833	0.6957		0.4004
Sets 11	0.9	0.75	1	0.72	0.8003	0.7165		0.3949
Sets 12	0.9	0.8	0.9375	0.8325	0.8289	0.7361		0.4779
Sets 13	0.9	0.75	0.9091	0.81	0.9	0.9		0.9
Sets 14	1	1	1	0.7	0.7209	0.619		0.5133
Sets 15	0.95	0.75	1	0.9048	0.6215	0.443		0.1962
Sets 16	0.9	0	0.9091	0.567	0.63	0.441		0.6012
Sets 17	0.95	0.75	0.9524	0.5425	0.61747	0.3924		0.514776
Sets 18	0.975	0.9617	1	0.8357	*	*		0.72017

Note: * Denotes the method cannot calculate the ranking value of the fuzzy numbers.

Table 2. A 9-member linguist term Set (Schmucker, 1984).

Linguistic terms	Generalized trapezoidal fuzzy numbers
Absolutely-low	(0, 0, 0, 0; 1.0)
Very-low	(0, 0, 0.02, 0.07; 1.0)
Low	(0.04, 0.1, 0.18, 0.23; 1.0)
Fairly-low	(0.17, 0.22, 0.36, 0.42; 1.0)
Medium	(0.32, 0.41, 0.58, 0.65; 1.0)
Fairly-high	(0.58, 0.63, 0.80, 0.86; 1.0)
High	(0.72, 0.78, 0.92, 0.97; 1.0)
Very-high	(0.93, 0.98, 1.0, 1.0; 1.0)
Absolutely-high	(1.0, 1.0, 1.0, 1.0; 1.0)

Table 3. Linguistic values of the probability of failure, the degree of confidence of the three sub-components evaluated by three decision-makers and the evaluating items w_i of the three subcomponents (Chen and Chen, 2003).

Subcomponent A_i	Decision-makers			Linguistic value of the w_i severity of loss
A_1	E_1	E_2	E_3	W_1
	$\tilde{R}_{11}$=low $\omega_{11}=1.0$	$\tilde{R}_{12}$=medium $\omega_{12}=0.8$	$\tilde{R}_{13}$=fairly-low $\omega_{13}=0.9$	low
A_2	$\tilde{R}_{21}$= fairly-high $\omega_{11}=0.6$	$\tilde{R}_{22}$= fairly-low $\omega_{22}=0.7$	$\tilde{R}_{23}$= medium $\omega_{23}=0.8$	fairly-high
A_3	$\tilde{R}_{31}$= medium $\omega_{31}=0.7$	$\tilde{R}_{32}$= high $\omega_{32}=0.9$	$\tilde{R}_{33}$= high $\omega_{33}=0.8$	Very-low

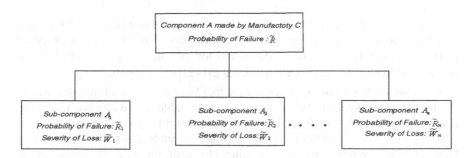

Fig. 2. The structure of fuzzy risk analysis

Table 4. The degree of similarity between $\tilde{R}'$ and each linguistic term in the 9-member linguistic term set

Linguistic term X_i	Degree of similarity $S(\tilde{R}', X_i)$
Absolutely-low	0.2287
Very-low	0.3082
Low	0.4881
Fairly-low	0.6407
Medium	0.6736
Fairly-high	0.4511
High	0.3915
Very-high	0.1379
Absolutely-high	0.0975

In the following and we can see the drawbacks of the existing similarity measures from Table1.

From Set 1 of Fig.3, we can see that $\tilde{A}$ and $\tilde{B}$ are different generalized fuzzy numbers. However, if we use Hsieh and Chen's method, the degree of similarity measure will be 1which is incorrect obviously.

From Set 5 of Fig.3, we can see that $\tilde{A}$ and $\tilde{B}$ are different generalized fuzzy numbers. However, by using Lee's, Hsieh and Chen's and Chen's method, they get the same degree of similarity (i.e., $S(\tilde{A},\tilde{B})=1$) which is incorrect.

From Set 6 of Fig.3, we can see that Lee's method can not calculate the degree of similarity between the two fuzzy numbers due to the fact that the denominator $\|u\|$ will become zero from (14).

From Set 8 of Fig.3, it can be seen that the result of Wei and Chen's method is 0.8411 which does not coincide with the intuition of human being.

From Set 5 and Set 8 of Fig.3, we can see that Set 5 is more similar than Set 8 by the intuition of human being. However, from Table 1, we can see that the results of Wei and Chen's show that Set 8 is more similar than Set 5 which is incorrect.

From Set 7, Set 9 and Set 13 of Fig.3, we can see that $\tilde{A}$ and $\tilde{B}$ have the same shape and the same offset $d = 0.1$ in the X-axis, respectively. By using the proposed method, we can find that the proposed method has the good property that the degree of similarity between $\tilde{A}$ and $\tilde{B}$ is equal to $1-\|d\|=1-0.1=0.9$.

From Set 14, Set 15 of Fig.3, we can see that Set 14 is more similar than Set 15 by the intuition of human being. However, from table 1, we can see that if we apply Chen and Chen's method, then we can see that it get an incorrect result.

Comparing Set 15 and Set 17 of Fig.3, it can be seen that Set 17 is more similar than Set 15 by the intuition of human being. However, the result of the method by Hejazi .et al shows that Set 15 is more similar than Set 15 which is incorrect.

From Set 10, 11, 14, 15 of Fig.3, using Hsieh and Chen's method, it shoes that the degree of similarity measure of $\tilde{A}$ and $\tilde{B}$ is one which is incorrect.

From Set 18 of Fig.3, we can see that $-\tilde{A} = (-0.4, -0.3, -0.2, -0.1;1)$,

$-\tilde{B} = (-0.4, -0.25, -0.25, -0.1;1)$ which is the inverse of Set1. By applying the proposed method, we have $S(-\tilde{A},-\tilde{B}) = S(\tilde{A},\tilde{B})$ which satisfy the intuition of human being. So the proposed method has the good property.

6   Fuzzy Risk Analysis Based on the Proposed Similarity Measure

In this section, we apply the proposed similarity measure of generalized trapezoidal fuzzy numbers to deal with fuzzy risk analysis problems. Let us consider the structure of risk analysis shown in Fig.2 [9].

According to Schmucker [11], each subcomponent A_i is evaluated by two evaluating items, i.e., "probability of failure" and "severity of loss", where the linguistic term R_i denotes the probability of failure of the sub-component A_i and the linguistic term W_i denotes the severity of loss of the sub-component A_i, respectively. Zhang [14] used trapezoidal fuzzy numbers to represent linguistic terms. Table2 illustrate the linguistic terms and their corresponding trapezoidal fuzzy numbers.

In the following, Assume that we want to evaluate the probability of failure of the sub-component A_i. The algorithm is now presented as follows:

Step 1: using the fuzzy weighted mean method and the generalized fuzzy number arithmetic operations to integrate the evaluating items $\tilde{R}_i$ and $\tilde{W}_i$ of each sub-component A_i made by manufactory C_j, where $1 \le i \le n$, shown as follows:

$$\tilde{R} = \frac{\sum_{i=1}^{n} \tilde{W}_i \otimes \tilde{R}_i}{\sum_{i=1}^{n} \tilde{W}_i} = (r_1, r_2, r_3, r_4; \omega_1)$$

(34)

If $r_j > 1$, then standardize according to (29) (30) (31), where $1 \le j \le 4$.

Step 2: Use the proposed similarity measure [i.e., (28)] to evaluate the degree of similarity between the fuzzy number $\tilde{R}$ and each linguistic term shown in Table 2. Translate the fuzzy number $\tilde{R}$ into a linguistic term, the probability of failure of the component A is equal to the linguistic term with the largest degree of similarity with respect to $\tilde{R}$.

In the following, we use the proposed method to deal with fuzzy risk analysis problem considered the degree of confidence of the decision maker's opinions.

Example 6.1. Consider the structure of fuzzy risk analysis as shown in Fig. 2, where the component A consists of three sub-components A_1, A_2, A_3 and we want to evaluate the probability of failure $\tilde{R}$ of the component A.

Assume that there are three decision-makers E_1, E_2, E_3 to evaluate the probability of failure of three sub-component that A_1, A_2, A_3 as shown in Table 3, where the value ω_{ij} denotes the degree of confidence with the probability of each evaluated which decision-maker E_j evaluates the probability of failure $\tilde{R}_{ij}$ of sub-component A_i. For convenience, we denote $\tilde{R}_{ij} = (a_{ij}, b_{ij}, c_{ij}, d_{ij}; \omega_{ij})$, where $1 \le i \le 3, 1 \le j \le 3$. Then, the average probability of failure $\tilde{R}_i$ of sub-component A_i is calculated as follows:

$$\tilde{R}_i = \frac{\sum_{j=1}^{3} \tilde{R}_{ij}}{3} = (\frac{\sum_{j=1}^{3} a_{ij}}{3}, \frac{\sum_{j=1}^{3} b_{ij}}{3}, \frac{\sum_{j=1}^{3} c_{ij}}{3}, \frac{\sum_{j=1}^{3} d_{ij}}{3}; \frac{\sum_{j=1}^{3} \omega_{ij}}{3})$$

(35)

$$= (a_i, b_i, c_i, d_i; \omega_i)$$

where $1 \le i \le 3$.

According to Table3 and (35), we can get:

$\tilde{R}_1 = (0.1767, 0.2433, 0.3733, 0.4333; 0.9)$, $\tilde{R}_3 = (0.5867, 0.6567, 0.8067, 0.8633; 0.8)$.

Based on the average probability $\tilde{R}_i$ of each sub-component A_i, where $1 \le i \le 3$, we can use the proposed fuzzy risk analysis algorithm to deal with the fuzzy risk analysis problem, shown as follows:

Step1. Based on (1) (3') (4) (34), Table2 and3, the probability of failure $\tilde{R}$ of component A is:

$$\tilde{R} = \left[\frac{\tilde{R}_1 \otimes low \oplus \tilde{R}_2 \otimes fairly - high \oplus \tilde{R}_3 \otimes very - low}{low \oplus faith - high \oplus very - low} \right]$$

$$= (0.1845, 0.2889, 0.74971, 1.1494; 0.7)$$

Based on (29) (30) (31), we can transform $\tilde{R}$ into standardized generalized trapezoidal fuzzy numbers $\tilde{R}'$, where

$$k = (\lfloor 0.1845 \rfloor, \lfloor 0.2889 \rfloor, \lfloor 0.74017 \rfloor, \lfloor 1.1494 \rfloor, 1; 0.7) = 2$$

$$\tilde{R}' = (0.09225, 0.1445, 0.37485, 0.5747; 0.7)$$

Step2. According to (33), the degree of similarity between the standardized generalized trapezoidal fuzzy numbers $\tilde{R}'$ and the linguistic term "absolute-low" is evaluated as follows: $S(\tilde{R}', Absolutely - low) = 0.2287$.

In the same way, we can obtain the other degrees of similarity, respectively, as shown in Table4. Obviously, we can see $S(\tilde{R}', Medium) = 0.6736$ is the largest value. Therefore the generalized trapezoidal fuzzy numbers $\tilde{R}$ is translated into the linguistic term "Medium". It means that the probability of failure of the component A is medium. The calculation is reasonable by the intuition of the human being.

7 Conclusion

In this paper, we have presented a new similarity measure between generalized trapezoidal fuzzy numbers. We have proven five properties of the proposed similarity measure. We use 18 sets of generalized fuzzy numbers to compare the calculation results of the proposed method with the existing similarity measures. The proposed method can overcome the drawbacks of the existing methods. We also apply the proposed similarity measure to present a new fuzzy risk analysis algorithm for dealing with fuzzy risk analysis problems. We can see that the proposed fuzzy risk analysis method is flexible and intelligent due to the fact that it considers the degrees of confidence of decision makers' opinions.

Acknowledgements. Thanks to Project Science Foundation of Guangzhou University.

References

1. Chen, S.-J., Hwang, C.-L.: Fuzzy multiple attribute decision making. Springer, New York (1992)
2. Chen, S.H.: Operations on fuzzy numbers with function principal. Tamkang Journal of Management Sciences 6(1), 13–25 (1985)
3. Chen, S.H.: Ranking generalized fuzzy number with graded mean integration. In: Proceedings of the Eighth International Fuzzy Systems Association World Congress, Taipei, Taiwan, vol. 2, pp. 899–902 (1999)
4. Chen, S.J., Chen, S.M.: A new method to measure the similarity between fuzzy numbers. In: Proceedings of the 10th IEEE International Conference on Fuzzy Systems, Proceedings of the Eighth International Fuzzy Systems, Integration, Melbourne, Australia (2001)
5. Chen, S.J., Chen, S.M.: Fuzzy risk analysis based on similarity measures of generalized fuzzy numbers. IEEE Transactions on Fuzzy Systems 11(1), 45–56 (2003)
6. Chen, S.J., Chen, S.M.: Fuzzy risk analysis based on the ranking of generalized trapezoidal fuzzy numbers. Applied Intelligence 26(1), 1–11 (2007)
7. Chen, S.M.: New methods for subjective mental workload assessment and fuzzy risk analysis. Cybernetics and Systems 27(5), 449–472 (1996)
8. Hsieh, C.H., Chen, S.H.: Similarity of generalized fuzzy numbers with graded mean integration representation. In: Proceedings of the 8th International Fuzzy Systems Association World Congress, Taipei, Taiwan, Republic of China, vol. 2, pp. 551–555 (1999)
9. Kangari, R., Riggs, L.S.: Construction risk assessment bylinguistics. IEEE Transactions on Engineering Management 36(2), 126–131 (1989)
10. Lee, H.S.: An optimal aggregation method for fuzzy opinions ofgroup decision. In: Proceedings of the 1999 IEEE International Conference on Systems, Man, and Cybernetics, vol. 3, pp. 314–319 (1999)
11. Schmucker, K.J.: Fuzzy sets, natural language computations, and risk analysis. Computer Science Press, MD (1984)
12. Tang, T.C., Chi, L.C.: Predicting multilateral trade credit risks:comparisons of Logit and Fuzzy Logic models using ROC curve analysis. Expert Systems with Applications 28(3), 547–556 (2005)
13. Wei, S.H., Chen, S.M.: A new similarity measure betweengeneralized fuzzy numbers. In: Proceedings of the Joint 3rd International Conference on Soft Computing and Intelligent Systems and 7th International Symposium on Advanced Intelligent Systems, Tokyo, Japan, pp. 315–320 (2006)
14. Zhang, W.R.: Knowledge representation using linguistic fuzzy relations. Ph.D. Dissertation, University of South Carolina, USA (1986)
15. Wei, S.H., Chen, S.M.: A new approach for fuzzy risk analysis based on similarity measures of generalized fuzzy numbers. Expert Systems with Applications 36, 589–598 (2009)
16. Lee, H.S.: An optional aggregation method for fuzzy opinions of group decision. In: Proceedings of the 1999 IEEE International Conference on Systems, Man and Cybernetics, vol. 3, pp. 314–319 (1999)
17. Hejazi, S.R., Doostparast, A., Hosseini, S.M.: An improved fuzzy risk analysis based on a new similarity measures of generalized fuzzy numbers. Expert Systems with Applications (2011)

Part V
Guess and Review

Three Guess of Fuzzy Geometric Programming

Bing-yuan Cao

School of Mathematics and Information Science,
Guangzhou University and
Key Laboratory of Mathematics and Interdisciplinary
Sciences of Guangdong Higher Education Institutes,
Guangzhou University, Guangdong.
ZIP 510006, P.R. China
caobingy@163.com

Abstract. Based on his experience in researching fuzzy geometric programming for many years, the author sums up the following three guesses. I. Fuzzy reverse geometric programming optimal (satisfactory) solution has properties of fuzzy geometric programming; II. Fuzzy relation geometric programming holds; III. Fuzzy geometric programming classification exists. They do not seem to hold, but the author guesses they do hold, and some studies have been found. In order to draw on collective wisdom, he raised these problems for world colleagues to explore, expecting to find efficiently quick algorithms and a complete proof.

Keywords: Fuzzy geometric programming, fuzzy relation, fuzzy classification, fuzzy non-convex, guess.

1 Introduction

Fuzzy geometric programming, since 1987 proposed in conference in IFSA'87, has been on its researching way through 24 years, while the fuzzy relation geometric programming, determined by the author to his PhD dissertation topic (doctoral thesis), was published in May 2005 on IEEE on Fuzzy Systems International Conference. Fuzzy geometric programming classification method, since 2000, has been a problem for the author strugglingly to think over. The author in this article, from three aspects, develops a question for each, constituting three guesses. These three problems should not be underestimated, because a lot of research has been done for them, but nothing has been found for them. To solve them, it may take a lot of time and effort, and even a lifetime, but it does not necessarily make a difference. Now the author advances the three following guesses, hoping to resolve them or providing some good ideas.

B.-Y. Cao and X.-J. Xie (Eds.): Fuzzy Engineering and Operations Research, AISC 147, pp. 591–594.
springerlink.com   © Springer-Verlag Berlin Heidelberg 2012

2 Propose of the Problem

Definition 1. *We consider the following to be reverse posynomail geometric programming:*

$$(P) \qquad \min \ g_0(x)$$
$$g_i(x) \leq 1(1 \leq i \leq p'),$$
$$g_i(x) \geq 1(p' + 1 \leq i \leq p),$$
$$x > 0.$$

Here $x = (x_1, x_2, \cdots, x_m)^T$ *of* m-*dimensional variable vector,* $g_i(x) = \sum_{i=1}^{J_i} v_{ik}(x)(0 \leq i \leq p)$ *are positive polynomial, and*

$$v_{ik}(x) = \begin{cases} c_{ik} \prod_{l=1}^{m} x_l^{\gamma_{ikl}}, & (1 \leq k \leq J_i; 0 \leq i \leq p'), \\ c_{ik} \prod_{l=1}^{m} x_l^{-\gamma_{ikl}}, & (1 \leq k \leq J_i; p' + 1 \leq i \leq p), \end{cases}$$

where its coefficients $c_{ik} \geq 0$, *index* γ_{ikl} *are real numbers.* $-\gamma_{ikl}$ *is index, corresponding to the reverse inequality* $g_i(x) \gtrsim 1$ *of each* x_l. *(P) is fuzzified, and there is exists*

$$\widetilde{\min} \ g_0(x)$$
$$s.t. \ g_i(x) \lesssim 1(1 \leq i \leq p'), \qquad (1)$$
$$g_i(x) \gtrsim 1(p' + 1 \leq i \leq p),$$
$$x > 0,$$

where $x = (x_1, x_2, \cdots, x_m)^T$ *remains* m-*dimensional variable vector,* "$\lesssim$" *and* "$\gtrsim$", *respectively,* "*roughly* $\leq$" *and* "*roughly* $\geq$"; *and* $g_i(x) = \sum_{i=1}^{J_i} v_{ik}(x)$ *$(0 \leq i \leq p)$ is polynomial. We call it the reverse fuzzy geometric programming.*

Definition 2. *We call*

$$\min \bigvee_{i=1}^{m} (c_i \wedge x_i^{\gamma_m})$$
$$s.t. \ x \circ A = b, \qquad (2)$$
$$0 \leq x_i \leq 1(1 \leq i \leq m)$$

a $(\vee, \wedge)$ *(max-min) fuzzy relation GP, where* $A = (a_{ij})(0 \leq a_{ij} \leq 1, 1 \leq i \leq m, 1 \leqslant l \leqslant n)$ *is an* $(m \times n)$-*dimensional fuzzy matrix,* $x = (x_1, x_2, \cdots, x_m)$ *an* m-*dimensional variable vector,* $c_i \geq 0)$ *and* $b = (b_1, b_2, \cdots, b_n)(0 \leq b_j \leq 1)$ *an* n-*dimensional constant vector,* γ_i *an arbitrary real number, and composition operator is* "$\circ$" $(\vee, \wedge)$, *i.e.,*

$$\bigvee_{i=1}^{m} (x_i \wedge a_{ij}) = b_j(1 \leq j \leq n).$$

For the programming problem mentioned above, we propose the three following guesses,

3 Guess

I. Optimal (satisfactory) Solution Problem in Fuzzy Reverse Geometric Programming

In fuzzy posynomial geometric programming case, we have proved that it has many other good properties of a fuzzy local optimal (satisfactory) solution, i.e., the global optimal (satisfactory) solution, by which some fast algorithms are found. But the fuzzy reverse geometric programming, including fuzzy reverse posynomial geometric programming, is not necessarily the case of the above properties, because they are not necessarily fuzzy convex. In author's opinion, fuzzy reverse geometric programming would be likely to hold good properties of fuzzy geometric programming through a series of transformation.

Guess 1

Under a certain transformation, the fuzzy local optimal (satisfactory) solution of fuzzy reverse geometric programming (1) is its global fuzzy optimal (satisfactory) solution.

II. Fuzzy Relation Geometric Programming Problem

When the relation operator is $M(\cdot, +)$,

$$\sum_{k=1}^{J} v_k^{w_k} \leq \sum_{k=1}^{J} w_k \cdot v_k \tag{3}$$

is ordinary arithmetic geometric inequality, fuzzy geometric programming is extended to fuzzy relations geometric programming. If you can prove that the operator $M(\wedge, \vee)$ and $M(\cdot, \vee)$ is substituted for (3), (3) holds. So we can extend (1) to be fuzzy relation geometric programming similar to the definition of fuzzy geometric programming.

Guess 2

By modifying an axiomatic system, in which it proves that the above inequality for improvement holds. Thus fuzzy geometric programming (1) will be able to expand into a fuzzy relation geometric programming.

III. Fuzzy Geometric Programming Classification Method

The following programming, Linear, nonlinear and fuzzy linear, can be used for classification, for which there are many results. But as for fuzzy geometric programming for classification, there is no work in this area. It is known that, by geometric shape of the fuzzy two-dimensional or three-dimensional geometric programming, geometric programming is actually a fuzzy boundary fluctuations under the curve or (super) surface constraints, such that objective function reaches an optimal case.

Guess 3

Call fuzzy geometric programming classification method a classification of fuzzy geometric programming. So we assert that fuzzy geometric programming will be applied to a pattern classification.

4 Conclusion

For Guess 1, we have some proven ideas; Guess 2 and Guess 3 has also been some solution ideas. We would like enthusiastic scholars to join in the discussion.

Acknowledgements. Supported by National Natural Science Foundation of China (No. 70771030 and No. 70271047) and Project Science Foundation of Guangzhou University.

References

1. Peterson, E.L.: The origins of geometric programming. Annals of Operations Research 105, 15–19 (2001)
2. Cao, B.Y.: Solution and theory of question for a kind of fuzzy positive geometric program. In: Proceedings of the 2nd IFSA Conference, Tokyo, Japan, vol. 1, pp. 205–208 (1987)
3. Cao, B.Y.: Fuzzy geometric programming. Kluwer Academic Publishers, Dordrecht (2001)
4. Advances in fuzzy geometric programming. In: Fuzzy Information and Engineering. AISC, vol. 40, pp. 497–502. Springer (2007)
5. Cao, B.Y.: Fuzzy geometric programming (I). Int. J. of Fuzzy Sets and Systems 53(2), 135–154 (1993)
6. Yang, J.H., Cao, B.Y.: Geometric programming with fuzzy relation equation constraints. In: Proceedings of the IEEE International Conference on Fuzzy Systems, pp. 557–560 (2005)
7. Cao, B.Y.: Antinomy in posynomial geometric programming. Advances in Systems Science and Applications 4(1), 7–12 (2004)
8. Kou, G., Peng, Y., Chen, Z.X., Shi, Y.: Multiple criteria mathematical programming for multi-class classification and application in network intrusion detection. Information Sciences 179, 371–381 (2009)
9. Cao, B.Y.: Optimal Models and Methods with Fuzzy Quantity. Springer Science Business (2010)

Factor Neural Network Theory and Its Applications

Zeng-liang Liu[*]

Foreign Affairs Office of National Defense University,Beijing, 100091, China
chinaphd@tom.com
Beijing Insititute of Technology, 100081, China
lgchinaphd@tom.com

Abstract. The Factor Neural Network Theory (FNN) and its applications are introduced in this paper. The Factor Neural Network theory to model intelligent computation was founded by Liu Zengliang, etc [2,7]. Based on this theory, intelligence problems such as the knowledge representation, fuzzy reasoning and associated learning could be quantitatively described and simulated. And the unified description of the logical and visual thinking could be implemented. This theory made significant progress in exhibit the intelligence science theorem, the academic view and the research approach of FNN. It motivates the development of intelligence science and other related fields. It also helps the development of the nation's economy. The project obtained prominent recognition and acknowledgement from L.A. Zadeh, Koczy [2], Li Guojie[3]and other domain experts.

Keywords: Factor space approach for knowledge representation, Factor Neural Networks (FNN), intellectualized computing.

1 Research Background

This theory belongs to the domain of intelligence science. We did research in this project for the following two reasons [3]:

First, scientific research detected that fuzzy theory and neural network theory respectively have their own superiority, but they also have deficiency. New intelligent theory approach was urgently needed to make up the deficiency.

Since 1980s, the fuzzy logic and neural network had become two significant branches in the intelligence research. They have their own advantage, but they also have their deficiency. The fuzzy logic held fuzziness characteristic of the human brain thought, and it has advantage in the high-level knowledge description. It may simulate human's comprehensive reasoning to deal with complicated

---

[*] Liu Zengliang Vice-president of Fuzzy Information and Engineering Branch of ORSC, president of Intelligence system Engineering Branch of CAAI, Professor of National Defense University, Professor of Beijing University of Science and Technology, Professor of Tsinghua University, china. Special field: Fuzzy Information and Engineering.

B.-Y. Cao and X.-J. Xie (Eds.): Fuzzy Engineering and Operations Research, AISC 147, pp. 595–604.
springerlink.com   © Springer-Verlag Berlin Heidelberg 2012

problems which computers may difficult to solve. And it expanded the computer applications to human science, social sciences, complex system and so on. But the fuzzy logic also has weakness. In membership function and regulation, it's difficult for auto-produce and self-learning. The neural network can simulate the human neural network, attempts to exploit skill which is computer automated study to prompt forward a stride, it has displayed the new foreground and new mentality in the aspect of pattern recognition, gathered analysis and personally study, but there also exists deficiency in logical thinking and the high-level knowledge description. As a new direction for combination of the two research domain, many famous researchers such as B.Kosko, Y.Yamamoto, M.M.Gupta, T.Inoue, T.Teh from America, Japan, Canada and other countries, all successively started this aspect research work.

Second, the researchers have a new assumption based on the idea of "the factor space"[1].

Whether it could find a new way to display the advantage of the fuzzy theory and the neural network, and to make up their deficiency? Whether it could establish a "Cartesian coordinate system" described by the meta-knowledge just like the "Cartesian coordinate system" in mathematics and mechanics? What is the mathematical theory of the knowledge representation and knowledge acquisition? Whether it could form a neural network system which can understand and process high-level knowledge? Can fuzzy rule and membership function created by training? Can the logical reasoning be parallel processed? To solve these questions, we need some original and innovation thought.

With the idea of "original innovation", we established a new approach to describe the knowledge unit which is the "Cartesian coordinates", created a new approach to utilize the high level knowledge qualitative inference, the synthetic judgment and the intelligence processing —this new frame is factor neural network. This subject is one of the key technical approaches for intelligence science. This subject was listed in the significant project of State Natural Science Foundation-- fuzzy information processing and the machine intelligence in 1988 and 2009 (subject number 6988007, 90818025,) and in the National 863 Program in 1991 (subject number 863-306-602-120). Later, this subject successively obtained support from the State Natural Science Foundation project in 1993, 2001 (subject number 60173057) and 2004 (subject number 60572162) respectively.

2   The Architecture of Factor Neural Network Theory

For many years, with the establishment of the theoretical system about Factor Neural Network and intelligent computation and its application, the research successively obtains many significant international leading-level theories and technical approaches innovation achievement. From research aspect, 365 papers and 10 monographs were published. The main academic thought and the opinion have been cited widely by the academe and obtained confirmation in the significant background application. Meanwhile,18 achievements had successively obtained The State Natural Science Award and Military Award for Science and Technology Progress in P.R.CHINA (first Class 3 times, second Class 15 times), one theory monograph wins the Chinese Books Prize, and one scientific research achievement "fuzzy control computer system "wins the gold prize in the national 863 high tech achievement exhibitions in china. All of them have brought significant influence for impel the knowledge development and accommodate the need of the national security.

Before evaluating USRC, it's necessary to analyze science research capability elements of universities [7].

2.1 Overview of Research Motivation

Whether can we find a new way to display the advantage of the fuzzy theory and the neural network and make up their deficiency? Whether can establish a "Cartesian coordinate system" described by knowledge cell just like the "Cartesian coordinate system" in mathematics and mechanics? What is the mathematic theory of the knowledge representation and knowledge acquisition? Whether can form a neural network system which can understand and process advanced knowledge? Can fuzzy rule and subordinate functional creation by training? Can the logical inference carry on the parallel processing? To solve these questions, we need original and innovation thought.

Taking the application which involves the national security background and based on the whole research of the fuzzy theory and neural network, the research systematically analyzes the advantage and disadvantage of the fuzzy theory and neural network theory. Meanwhile, basing on the "factor space" thought proposed by Professor Peizhuang Wang, the research achieved a kind of Factor Neural Network theory and its application through the reconciled research of the fuzzy information processing and the association study mechanism. It includes:

1) From the fundamental question of the intelligent description --- knowledge cell description, to establish a kind of "Cartesian coordinate system" approach described by high-level knowledge cell --- factor method to knowledge representation, and study its mathematical theory;
2) Form an intelligent processing approach to energetically utilizing high-level knowledge qualitative inference, synthetic judgment and associated study mechanism--- which the new frame we called as the Factor Neural Network, and study its mathematic theory, make it to certain extent reflect the logical reasoning of the human brain and the mental image characteristic, develop the advantage of the fuzzy theory and the neural network method, make up its deficiency;
3) Form a kind of uncertain knowledge acquisition approach, which we named as theory of knowledge fuzzy-stochastic project acquisition, and study its mathematic character;
4) Combine the theory innovation achievement and the application background; form a series of application theoretical method;

In the research of the Factor Neural Network and intelligent computation, we should specially pay attention to the following features:

(1) Strict mathematic reasoning;
(2) Exhibit the advantage of fuzzy logic and neural network;
(3) Conduct application and implementation.

2.2   Research Achievements

After many years of research, based on the application background, the research group established the new system of the factor neural network theory, proposed the knowledge factor representation theoretical method, the factor neural network theory method and the method of knowledge fuzzy-stochastic project acquisition, solved several difficult problems in mathematics such as "the fuzzy logic expressed by neural network and the logic explanation of the neural network running conclusion, knowledge cell description, the concept connotation and the extension unification description, the fuzzy set universe of discourse choice and the subordinate function creation", intensified and developed the research in fuzzy theory and neural network, published many high-quality research and application achievements.

2.2.1   Establish the Knowledge Factor Representation Theory[1,2,4]

Whether it can establish a "Cartesian coordinate system" described by knowledge cell just like the "Cartesian coordinate system" in mathematics and mechanics? To solve the quantification cell description problem of the high-level knowledge in mathematics theory, L.A.Zadeh may use the fuzzy set theory to describe the extension of the general conception, but it doesn't solve the important problem such as the choice and the transform of the extension representation theory in the concept mathematic description. And in mathematic research it is always difficult to represent the intention. The knowledge factor representation method we proposed on the basis of the factor space can solve the above problem and unify the description of the concept extension and intension. Our major contribution as follows:

∗ Proposed analytical method of the intelligent system's three essential factors on the basis of the research to the basic concepts of the intelligence, the intelligent body, and the intelligent system.

∗ Based on the concept of the factor space and the factor space cane proposed by Professor Peizhuang Wang, we formed a theory frame to descript the high-level knowledge quantification cell. We named factor f as the element to describe the object set O, and consider that f has its state space in its domain U. X (f) may constitute the factor space $\{X (f) \mid f \in F\}$ in the view of factor set F. The factor may operate, the factor space can "change dimension", the factor space as the knowledge described mathematic model has proven its mathematic nature.

∗ Propose the knowledge factor representation atom pattern M (o) (O,V,X (f)) and the relational pattern R (R, Mi (o) $\mid$ i=1,2, …,n). And the fact, the concept, the inference, the judgment, the knowledge dynamic process all can be quantify described in the atomic pattern and the relational pattern.

＊Study the mathematic theory abort the knowledge factor representation. With mathematical instrument such as the group, link, and category, the knowledge representation theory can realize the identical factor collection, the different factor collection, and the factor cane net.

The research indicated that the knowledge factor representation method is an object-oriented knowledge representation method in both generality and specialty. Its generality shows that this analysis representation method is adaptable for any system, its specialty performance shows that individual question should be analyzed independently, each specific object may correspond a specific factor set or space. This kind of knowledge representation method not only has the strict mathematic theory to be used as the basis, but also has the explicit description object as the background. It can make the concept connotation and extension obtain the unified reflection under the frame. At the same time, the knowledge factor representation method may also reflect the object level structure and describe the dynamic and static relations between the different objects. Thus, this kind of knowledge representation method is a promising knowledge representation method in the expert system and intelligent engineering research.

2.2.2   Establishment of the Factor Neural Network Theory[2,3,7]

(1) Propose a kind of object-oriented factor neuron FN and the factor neural network FNN model, proposed the factor neural network theory frame to carry on the formalized description for the basic concept such as FN and FNN, studied its mathematic description theory. The factor neuron is a kind of information processing unit. With the support of the knowledge factor analysis and factor representation theory, it can combine the knowledge memory and the knowledge application. In biology neuron can be the basic unit of the intelligent information processor; the factor neuron is defined as the reflection thing and its relational basic unit. The factor neural network constituted by the factor neuron is a kind of general neural network system, it can reflect the mechanism which human recognize things and its thought characteristic.

(2) As the fundamental research of the factor neural network, this subject investigated into the collection value and fuzzy value factor neural network system which are basically not yet touched in present neural network research field. The research indicated that, these two neural networks have their important application background and theory value. The collection value factor neural network proposed the new way and processing method in uncertainty (stochastic, fuzzy, incomplete and so on) information processing aspect. The sector value contains much more information than the simple point value, therefore to some extent it can unify knowledge processing method based on the probability, general fuzzy and evidence theory. But fuzzy linguistic value factor neural network has a richer connotation, may process the complex information system directly which two fuzzy sets can describe, and has the stronger information processing function compared with the sector value factor neural network. Thus, these two kind of neural network researches made the nerve network have more obvious approaching in the aspect of reflecting objective world and human senior thinking activity. Also from the

network knowledge processing angle, defined and studied such as the analysis neural network problems. These are a kind of beneficial attempt without doubt to the high level development of the neural network.

(3) Study the factor neural network automaton nature. Study the factor neural network and the automaton relation, unified two kind of operations with different running mechanisms essence on the function equivalence concept, and through 2-FNN, u-FNN realized its equivalent transformation. The significance of this research shows that we further understand the factor neural network is not only a kind of operation network but also a development network with memory inference function. The factor neural network had succession characteristic like the automaton, it found a possible method for the logical thinking explanation of the image thinking characteristic, provided the theory basis for the factor neural network digital circuit realization, and found a method for the factor neural network hardware design. The comprehension of the automaton neural network drew out the automaton factor neural network analysis, found a possible way for the images thinking solution of the logical question, and also can play a positive role in the automaton theory development.

(4) Propose the entropy theory of the factor neural network. Under the information theory produced a kind of description theory to the dynamic behavior of the Factor Neural Network such as memory and learning. Meanwhile, through the entropy analysis to the 2-FNN and u--FNN network, as well as the related research between the network energy function and entropy function, basically affirmed the feasibility of developing the factor neural network entropy theory so as to provide the reasoning for the stability analysis of the factor neural network and the model structure.

(5) Study the high-level knowledge memory and dynamic information processing nature of the factor neural network.

∗ The mechanism, characteristic, construction method and running characteristic of the analysis and simulation factor neural network.

∗ The inference mechanism in analysis factor neuron, as well as the non- definite inference method in analysis factor neural network.

∗ Simulation factor neural network nature, fuzzy association memory and fuzzy association mapping.

∗ Basic problem such as the stability of the factor neural network.

(6) Using the study function of the factor neural network, explored the knowledge acquisition neural network new way. Also did thorough research on the rule conduction of the fuzzy control system and the subordinate function creation, and confirmed the feasibility of this method through the practice.

(7) Study the factor neural network realization strategy, produced the electronic realization method to each kind of factor neuron defined in the research, also produced the electronic realization strategy and the computer software realization method to the two values, successive value, sector value, linguistic value logic and its inference network.

(8) Propose a theory combining fuzzy logic with neural network. This research discussed the logical function of different level neural network, as well as the factor neural network realization of different level logic relations. Produce the

corresponding relations and the equivalence realization between 2-FNN and two values logic, u-FNN and the successive value logic, i-SFNN and the collection value logic, 1-KFNN and between the linguistic value logic. At the same time also studied the logical function of the analysis and simulation factor neural network. Especially indicated the neural network and the fuzzy logic inference can be interlinked through realizing strategic research of the successive value inference network, sector value inference network, linguistic value dynamic inference network as well as fuzzy association, memory realization, and so on. The neural network can realize the fuzzy logic inference, and the fuzzy logic inference can translate into the neural network realization. Thus, to some extent, this kind of research linked the relation between the logical thinking and the image thinking, laid the foundation for developing the new generation of expert system with study association function, the intelligent machine and the controller.

3 Theory Application and Implementation

Facing to the important application background intellectualized application is a big characteristic of this project. The practice confirmed that relying on the significant background "construction with application" theory innovation carries out the path validity, and provides a series of valid arithmetic model for the engineering. Capture a batch of key technique method problems. Through the intellectualized application and the high technique application, got the extensive verification in the practice, confirmed important function of the creative theory method to the national defense and the economic construction. Some applications of this research are as follows.

(1) Proposed a demand foresee and decision--making new method which can fit macroscopically complex problem and be of fuzzy, uncertain characteristic, and provided the quantification research technique for the medium and long-term strategic direction automated system and the complex problem decision-making.

(2) Proposed a new self-adapted fuzzy gathering track vector recognition method with multi-factors fuzzy information characteristic, and provided the essential technical method for directing the essential factor digitization, graph processing intellectualization, aviation accident "fuzzy forecast" and the airplane breakdown fuzzy pattern distinguishes project.

(3) Explored an implementation mechanism of the intelligent system self-study system and the factor neutral network realizing new way, discussed one new method of the fuzzy rule withdraws and the subordinate functional creation, and lain the foundation for the complex, flexible system intelligence control realization such as airplane refueling in the air, spacecraft airborne docking.

(4) Proposed an essential technology method for realizing the high level knowledge fuzzy reasoning with high speed real-time processing, explored the complex inference system mechanism such as exploration successive value inference network, the sector value inference network and the linguistic value dynamic inference network, provided the effective algorithm of the fuzzy reasoning with realization for developing the high performance platform like "the fuzzy association processor".

With this theory, we captured key technology of a series of intelligent application projects such as "TsingHua satellite and small satellite posture controlling and track rectifying deviation controlling", some research achievements and technology such as "intelligent graphic work station and electronic map storehouse "and" strategy commands automatic development" get the domestic leading position.

This theory achievement has generality in applications. In uncertain amount of information processing and flexible controlling field of military and civil complicated system, it has the extensive prospect of popularization and application. It can apply to extensively in intelligence systems such as the fuzzy controller with fuzzy information processing and associative learning mechanism, fuzzy recognizer, intellectual device and expert system of new generation, etc, can be used in experienced and controlling automated system, such as burning of stove kiln controlling system, chemical reaction controlling system, the non-linear time variant controlling system, for instance, the aircraft posture controlling, etc.

The achievements have already been implemented in the key research project "the intelligence graph work station and the electronic map databases". According to the command of effective troops demand, making use of the fuzzy reasoning judgment intelligence method captured the road intersection of different map; the contour breaks quantified judgment, etc. several intelligence graph processing key technique problems, belonging to domestic origination, being the leading position in the technique. The achievement acquires the first-class award of science & technology advancement   in china.

In key topic "strategic conductor automation development research", for the first time using fuzzy following shadow concluding and decision-making method of the combination of the fixed nature and fixed quantity, argued scientifically the medium and long-term strategic conductor automation development   strategy.

In the project "the electronic key" product development, the project with full of difficulties fills up the domestic blank greatly. The system adoption fuzzy identifying and reasoning theories method captured the key techniques such as the key power and frequency noise influence. The project's achievement acquired "85 important science and technology achievement gold prizes of whole army", its function comes to all requests of the soldier object, and eliminated sending telegram with hand key, copy receiving in the whole army successfully, having the important   performance.

In the project" The aviation maintaining safety management assistance decision support system", Implementation the achievement plays a good function to" fuzzy forecasting" of airplane accident and " fuzzy mode identification" of the airplane's broken-down piece veins, that project gets the second-class award of science & technology advancement in china.

For 863 projects of nation "The fuzzy control calculator system". Carried out 20,000,000 times/second high-speed fuzzy reasoning, the achievement gets the gold prizes of 863 high-tech achievement exhibition of national science committee.

For the national postdoctoral science fund subject, researched successfully "fuzzy information processing and fuzzy computer", and carried on the handstand control experiment successfully. The experts think: study model fuzzy reasoning controller according to the FNN, is a kind of new intelligence controller not depending on being controlled accurate mathematics model of object and having the study function.

In the aviation science and technology project" study model fuzzy reasoning control research according to the FNN ", this project have passed the department class experts' authentication, and achieved advanced level internationally.

The achievements have already been used for" the very low code rate PSTN net visual telephone" pressing processing and coordination arithmetic, for "certain missile control equipment", for "satellite and tiny moonlets" carriage control and the orbit departure control, for the national natural science fund project" network offend and defend theories model research according to the FNN "; for the national natural science fund project" the fixed amount research of the information safe risk valuation according to FNN ".

Over all, this project stands in the frontier of science with full of difficulties in the field of the intelligence science. The related research achievements have already emerged important implementation in economy and social efficiency.

4 The Value in Science

First: Through the innovated research method of the depth fuzzy theory, we successfully solved many difficult problems in the developing of fuzzy theory. And it has significant influence for promoting the fuzzy theory. Majorly shows:

(1) Utilizing the knowledge factor representation method which this research released, solved mathematics difficult problem well in the representation of fuzzy concept connotation and the fuzzy regular variable extraction and so on, and established the corresponding mathematic theory.
(2) Utilizing the factor neural network method which this research proposed, solved association method between the fuzzy logic and the neural network, and established theorization in order to parallel network realization which the fuzzy released.
Second: Developed the new field of research about the neural network, promoted the development of neural network. Major shows:
(1) Proposed the new model of factor neural network, established parallel processing model method about the high-level knowledge of neural network, developed the ability of neural network processing high-level knowledge. It has laid the foundation of theory on neural network project application.
(2) Developed the field of neural network research, this research has proposed the models of neural network, like the set value type, fuzzy linguistic value type, simulated type and so on, it has developed the field of the neural network research greatly.
(3) Promoted cross of discipline between neural network and fuzzy theory, proposed association method between the fuzzy logic and the nerve network. It has laid the foundation of theory for the logic interpretation of neural network conclusion and the neural network logic science of establishment.

5   Conclusion

In this paper, we introduced the Factor Neural Network Theory and its applications. The factor neural network theory had become a new branch subject which provided a new characteristic methodology and theory architecture for the knowledge representation, acquisition, processing and other fundamental science research. It also provided a new approach to the application.

Acknowledgments. National Natural Science Foundation (90818025, 60773127, 60572162).

References

1. Wang, P.Z.: Factor Spaces Approach to Knowledge Representation. Fuzzy Sets and Systems 36(1), 113–124 (1990)
2. Liu, Z.L., Liu, Y.C.: Factor neural network theory and its realization tactics. Beijing Normal University Press (1992)
3. Liu, Z.L., Liu, Y.C.: the theory research combining fuzzy logic with neural network. BUAA Press (1996)
4. Wang, P.Z.: Fuzzy Set and the Falling Shadow of Random Sets. Beijing Normal University Press, Beijing (1985)
5. Liu, Z.L., Liu, Y.C.: Principle of the fuzzy expert system. The BUAA Press (1994)
6. Li, H.-X., Phillip Chen, C.L., Lee, E.S.: Factor Spaces Theory and Its Applications to Fuzzy Information Processing: Two Kinds of Factor Space Canes. Computers and Mathematics with Applications 40, 835–843 (2000)
7. Vukadinovic, K., Teodorovic, D.: A neural network approach to the vessel dispatching problem. European Journal of operational Research 102, 473–487 (1997)

Author Index